Information Systems Engineering and Management

Volume 14

The book series "Information Systems Engineering and Management" (ISEM) publishes innovative and original works in the various areas of planning, development, implementation, and management of information systems and technologies by enterprises, citizens, and society for the improvement of the socio-economic environment.

The series is multidisciplinary, focusing on technological, organizational, and social domains of information systems engineering and management. Manuscripts published in this book series focus on relevant problems and research in the planning, analysis, design, implementation, exploration, and management of all types of information systems and technologies. The series contains monographs, lecture notes, edited volumes, pedagogical and technical books as well as proceedings volumes.

Some topics/keywords to be considered in the ISEM book series are, but not limited to: Information Systems Planning; Information Systems Development; Exploration of Information Systems; Management of Information Systems; Blockchain Technology; Cloud Computing; Artificial Intelligence (AI) and Machine Learning; Big Data Analytics; Multimedia Systems; Computer Networks, Mobility and Pervasive Systems; IT Security, Ethics and Privacy; Cybersecurity; Digital Platforms and Services; Requirements Engineering; Software Engineering; Process and Knowledge Engineering; Security and Privacy Engineering, Autonomous Robotics; Human-Computer Interaction; Marketing and Information; Tourism and Information; Finance and Value; Decisions and Risk; Innovation and Projects; Strategy and People.

Indexed by Google Scholar. All books published in the series are submitted for consideration in the Web of Science.

For book or proceedings proposals please contact Alvaro Rocha (amrrocha@gmail.com).

Pethuru Raj · Alvaro Rocha · Simar Preet Singh ·
Pushan Kumar Dutta · B. Sundaravadivazhagan
Editors

Building Embodied AI Systems: The Agents, the Architecture Principles, Challenges, and Application Domains

Springer

Editors
Pethuru Raj
Reliance Jio Platforms
Bengaluru, Karnataka, India

Alvaro Rocha
University of Lisbon
Lisbon, Portugal

Simar Preet Singh
Bennett University
Greater Noida, Uttar Pradesh, India

Pushan Kumar Dutta
Department of Electronics
and Communication Engineering
Amity University
Kolkata, West Bengal, India

B. Sundaravadivazhagan
Department of Information and Technology
University of Technology and Applied
Sciences
AlMussanah, Oman

ISSN 3004-958X ISSN 3004-9598 (electronic)
Information Systems Engineering and Management
ISBN 978-3-031-68255-1 ISBN 978-3-031-68256-8 (eBook)
https://doi.org/10.1007/978-3-031-68256-8

This Springer imprint is published by the registered company Springer Nature Switzerland AG
The registered company address is: Gewerbestrasse 11, 6330 Cham, Switzerland

Contents

Building Embodied AI Systems: The Agents, the Architectural Principles, Challenges and Application Domains

Anubhav Bewerwal, Amit Kumar, and Manoj Kumar

Abstract The topic of embodied AI is experiencing tremendous growth due to recent advancements in computer vision, machine learning, and artificial intelligence. Intel Labs and Facebook AI Research (FAIR) have been leading new initiatives in the field of embodied AI. The definition of "embodied" is "giving an idea a tangible or visible form." "Embodied AI" is just "AI for virtual robots," to put it simply. More precisely, Embodied AI is the area of study that deals with AI challenges for virtual robots that can move, see, talk, and communicate with other virtual robots. These simulated robots are subsequently applied to real-world robots as solutions. AI subfields have been largely divided due to a variety of constraints. Embodied AI, on the other hand, unites a number of interdisciplinary disciplines, including robotics, computer vision, reinforcement learning, navigation, physics-based simulations and Natural Language Processing (NLP). Computer Vision techniques have played a major role in the evolution of Embodied AI as a research area, even if it needs to succeed in numerous AI subfields.

Keywords Kernel point convolution · Residual neural networks · Contrastive language-image pretraining · MASK R-CNN · Proximal policy optimization · DDPPO

A. Bewerwal (✉)
Department of Computer Science and Engineering, Graphic Era Hill University, Bhimtal Campus, Nainital, Uttarakhand, India
e-mail: anubhavbewerwal@gehu.ac.in

A. Kumar
Department of Computer Science and Engineering, Ajay Kumar Garg Engineering College, Ghaziabad, UP, India
e-mail: kumaramit2@akgec.ac.in

M. Kumar
School of Computer Science, FEIS, University of Wollongong in Dubai, Dubai Knowledge Park, Dubai, UAE

MEU Research Unit, Middle East University, Amman 11831, Jordan

M. Kumar
e-mail: manojkumar@uowdubai.ac.ae

P. Raj et al. (eds.), *Building Embodied AI Systems: The Agents, the Architecture Principles, Challenges, and Application Domains*, Information Systems Engineering and Management 14, https://doi.org/10.1007/978-3-031-68256-8_1

1 The Connection Between Natural Language Processing and Computer Vision

Natural language processing tasks are seen as more diverse from a technical standpoint compared to those in computer vision. This diversification includes the examination of abstract semantics, morphology, and segmentation abilities, as well as the finding of variable syntax. Complex tasks in natural language processing include direct machine translation, dialogue interface learning, digital information extraction, and rapid key summarization. Three major, interconnected processes are involved in the coupling of computer vision with natural language processing: recognition, reconstruction, reorganization.

Recognition: In this procedure, items in the image are given digital labels. For 2D objects, examples of identification include handwriting or facial recognition; for 3D assignments, problems like moving object recognition are addressed, which aids in autonomous robotic handling.

Reconstruction: It is the process of generating a 3D scene using inputs from specific visual images by combining digital shading, numerous angles of view, and sensory depth information. A 3D digital model is the end product, which is subsequently utilized for additional processing.

Reorganization: This procedure is the division of raw pixels into data groups that symbolize the blueprint for a preset setup. Tasks of higher level encompass semantic segmentation, that may have some overlap with identification processes. On the other hand, lower-level tasks involve detection of corners, edges and contours.

2 Algorithms Used in Semantic Segmentation in 3D Point Cloud

Point cloud data is comprised of an extensive collection of positional coordinates, each representing a specific point in physical space. Due to its ability to retain spatial depth information, which is valuable for an agent, it serves as an excellent data format for Embodied AI. Enabling an agent to train a semantic vision system on point clouds can enhance the agent's understanding of a scene.

2.1 Convolutional Neural Network (CNN)

A convolution is a matrix operation in which a new matrix is created by convolving a kernel over an existing matrix. Because it takes into account each pixel's neighbors, this process is useful for applying machine learning to photos. It is well-known for producing filters that recognize various image edges and turn them into a representation that a thick neural network (Multi-Layer Perception Neural Network) can employ.

2.2 *Kernel Point Convolution (KPC)*

It seemed sense that convolutional neural networks, which have been successfully used for perceptual purposes since the 2010s, would also perform well for point clouds. However, the implementation of a convolution in a sparse dataset of higher dimensional order was not clear-cut. The kernel point convolution (KPC) [1], which convolves around a portion of the point cloud and uses about 7 points as the kernel, proved to be successful. Other methodologies encompass multi-layer perceptron (MLP) networks, which directly employ an MLP on the data points, projection convolution networks, which initially transform point data into a voxel or image format, and graph convolution neural networks, which consider the interconnections among the data points [2].

3 Visual Navigation in Embodied AI

Computer vision is commonly employed in the field of Embodied AI with architectures such as Residual Neural Networks, Contrastive Language-Image Pretraining (CLIP), or MASKRCNN.

3.1 *Residual Neural Networks (ResNet)*

It is now possible to construct convolutional neural networks (CNNs) with a large number of convolutional layers, even reaching thousands. These networks outperform shallower networks due to the implementation of the Residual Network (ResNet) design [3], which effectively addresses the problem of 'vanishing gradient'. Subsequent successful architectural designs in the field of convolutional neural networks (CNNs) have incorporated extra layers in order to effectively reduce the error rate. While this approach is efficient for lower layer counts, as the number of layers increases, a common problem in deep learning called the Vanishing/Exploding gradient becomes apparent. The gradient either approaches zero or gets quite large. Consequently, the error rates for both training and test increase as the number of layers increases.

The plot above demonstrates that a 56-layer CNN exhibits a greater error rate on both training and testing datasets compared to a 20-layer CNN architecture. The

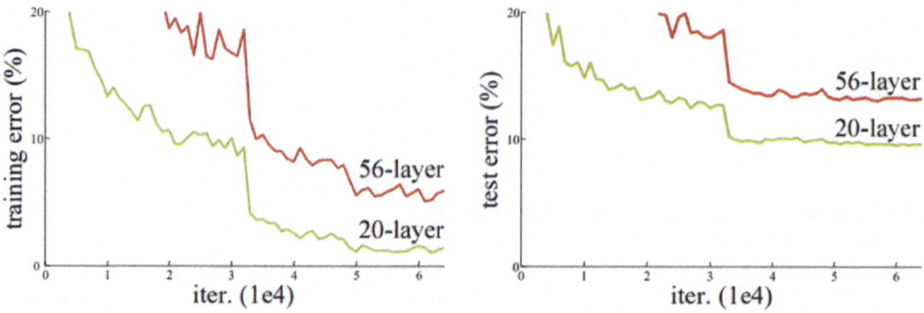

Fig. 1 On CIFAR-10, training error (left) and test error (right) with 20-layer and 56-layer "plain" networks, respectively. Test error and training error are increased in the deeper network [3]

authors of ResNet architecture were able to conclude that the vanishing/exploding gradient is the cause of the mistake rate after conducting additional analysis.

The design proposes the concept of Residual Blocks to specifically tackle the issue of vanishing/exploding gradient. In this network, we utilize a technique known as skip connections. The skip connection directly connects the activations of one layer to the activations of following layers, bypassing some intermediate layers. Consequently, a residual block is formed. The residual blocks are arranged in a stack to form ResNets. The concept underpinning this network is to allow the network to adapt to the residual mapping instead of requiring the layers to learn the underlying mapping. So, let the network fit, rather than using, H(x) as the initial mapping,

$$F(x) := H(x) - x \text{ which gives } H(x) := F(x) + x.$$

Adding this kind of skip connection has the benefit of allowing regularization to bypass any layers that degrade architecture performance. As a result, training a very deep neural network is achieved without the vanishing/exploding gradient issues.

Network design: This network adds a shortcut connection to a 34-layer simple network design that was modelled by the VGG-19. The architecture is subsequently transformed into a residual network by these shortcut links.

Fig. 2 Residual learning: a building block [3]

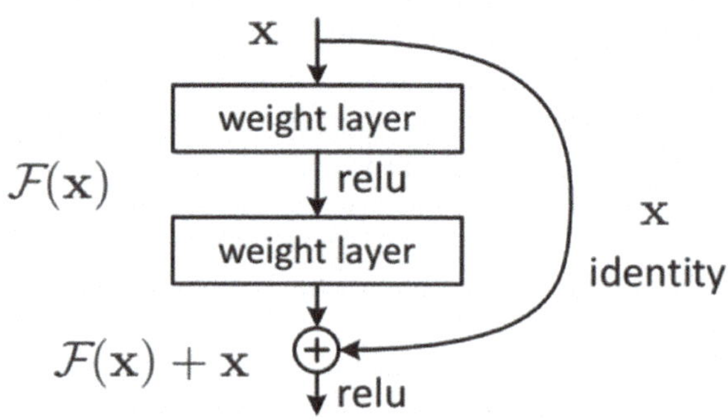

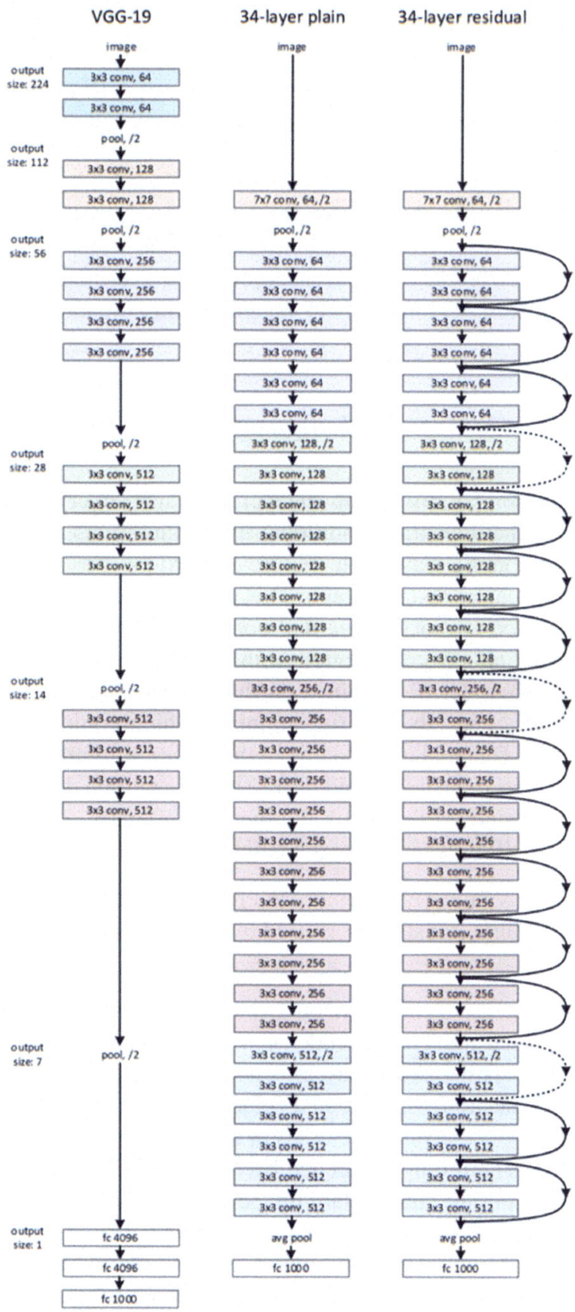

Fig. 3 Illustrative network structures for the ImageNet dataset. Left: the VGG-19 model [4] (19.6 billion FLOPs) as a reference. Middle: a plain network with 34 parameter layers (3.6 billion FLOPs). Right: a residual network with 34 parameter layers (3.6 billion FLOPs). The dotted shortcuts increase dimensions [3]

A 152-layer ResNet was employed on the ImageNet dataset, which has less parameters but is eight times deeper than VGG19. Using the ImageNet test set, an ensemble of these ResNets produced an error of just 3.7%, winning the ILSVRC 2015 competition. Because of its extremely deep representation, it also produces a 28% relative improvement on the COCO object detection dataset. The outcome also demonstrated that shortcut connections would be able to address the issue brought on by adding layers, as the error rate on the ImageNet Validation Set similarly drops as the layers are increased from 18 to 34 layers, in contrast to the simple network. With the lowest top-5 error rate of 3.57%, according to the results on the ImageNet Test Set, ResNet's architecture won the 2015 ImageNet classification competition.

3.2 *Contrastive Language-Image Pretraining (CLIP)*

The comparatively tiny handcrafted and supervised datasets (ImageNet, COCO, etc.) that constrain traditional image classifiers. CLIP [5] recommends employing more publicly available picture and text pairs as training datasets (e.g., images with informative text captions). This method might employ roughly 400 million training instances from the internet, compared to ~100k training samples of MS-COCO. When an image is paired with a label that describes the related image in plain language, CLIP is trained on these pairs. CLIP is shown a set of (supposed as N) images and labels during training, which could result in N^2 potential image-text pairs. The goal is to maximize the cosine similarity of the N pairings, each of which consists of an image and its true label (a written description). The goal is to minimize the cosine similarity for all other (N^2-N) pairs.

CLIP uses two essential building components to accomplish this:

1. A Vision Transformer (ViT)-based visual image encoder
2. A text transformer is used by a text encoder.

In the first stage, referred known as 'contrastive pre-training', the visual and text encodings are simultaneously taught to predict the proper matches.

The trained text encoder serves as a zero-shot classifier during the testing phase. The possible classes are the class names (or descriptions) of the target test dataset. The question, 'A photo of a {object}', is used with the class names, and {object} can be any of the various classes in the target dataset (e.g. plane, automobile, dog, bird, etc.). The prompt is needed because labels used for pre-training CLIP were picture descriptions, but standard image datasets like MS-COCO contain single word labels as class names (e.g., cat, dog). The link between these two is provided by a prompt. Furthermore, a prompt offers more information than a single label. As an aside, the authors point out that the absence of label context in many existing datasets causes ambiguity (take datasets with the term 'boxer' as an example).

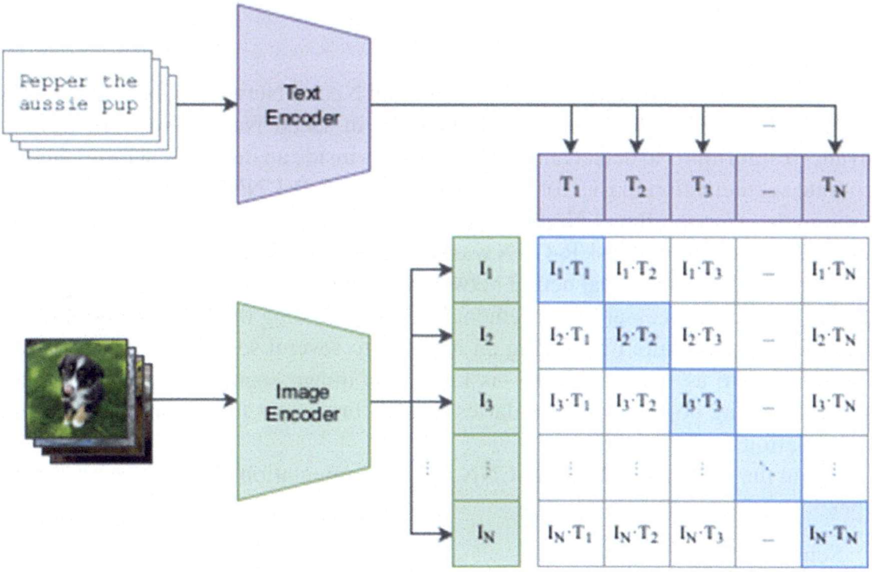

Fig. 4 Contrastive pre-training jointly optimises image and text pairs [5]

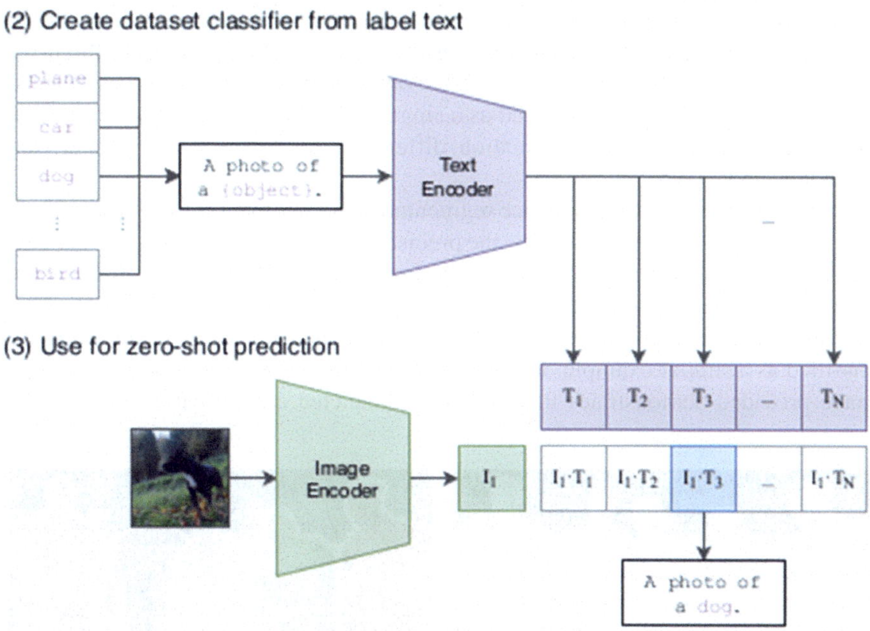

Fig. 5 Test phase single shot classification [5]

3.3 Mask R-CNN

Mask R-CNN [6] is a cutting-edge Convolutional Neural Network (CNN) that excels in the task of photo segmentation. This form of Deep Neural Network utilizes advanced algorithms to detect and identify objects inside an image, generating highly accurate segmentation masks for each instance. Mask R-CNN, or Mask RCNN, is a cutting-edge Convolutional Neural Network (CNN) that excels at both picture and instance segmentation. Mask R-CNN was built upon the foundation of Faster R-CNN, a region-based convolutional neural network. Comprehending the concept of picture segmentation is a necessary requirement for understanding the functioning of Mask R-CNN. The procedure of dividing an image into several segments—set of pixels, also referred to as image objects—is known as image segmentation in computer vision. Locating objects and boundaries (lines, curves, etc.) is accomplished using segmentation.

Within the category of Mask R-CNN image segmentation, there are two primary types:

1. Semantic Segmentation
2. Instance Segmentation

Semantic Segmentation: Semantic Division With semantic segmentation, every pixel is categorized into a predetermined set without distinguishing between different instances of an object. Put differently, semantic segmentation pertains to the pixel-by-pixel recognition and grouping of related objects into a single class. Every object in the picture below was categorized as a single individual. Semantic segmentation, also known as background segmentation, differentiates the objects of an image from its background.

Instance Segmentation: Instance segmentation, sometimes referred to as instance recognition, encompasses two tasks: the precise detection of every object in an image and the accurate segmentation of each individual instance. Therefore, it is the merging of object classification, object location, and object detection. To put it in other words, this type of segmentation takes an additional step in distinguishing each object that is classified as a similar example. In the context of Instance Segmentation, the sample image provided demonstrates that all objects depicted are humans. However, this

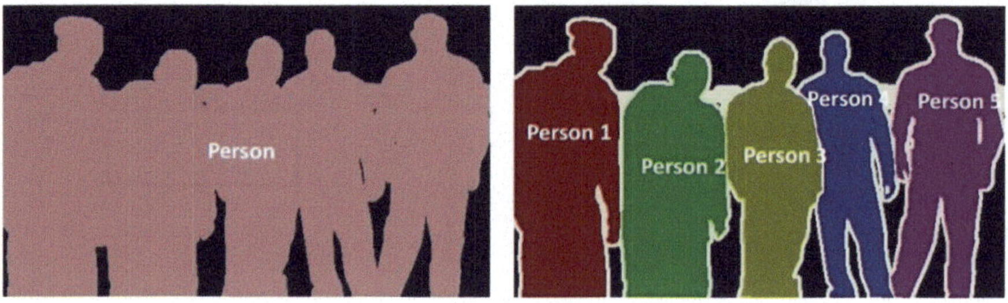

Fig. 6 Differences between semantic segmentation and instance segmentation [6]

segmentation technique treats each person as a separate and distinct entity. Semantic segmentation, also known as foreground segmentation, focuses on extracting the subjects of a picture rather than the background.

Working of Mask R-CNN: Mask R-CNN was built using Faster R-CNN as its foundation. Mask R-CNN introduces an additional branch that generates the object mask, while Faster R-CNN just generates two outputs for each candidate item: a class label and a bounding-box offset. Due to the discrepancy between the additional mask output and the class and box outputs, it is essential to fetch excerpt of a greater precise spatial arrangement of an object. A supplementary wing for foreseeing object masks (Region of Interest) is incorporated into the preexisting wing for recognizing bounding boxes in Mask R-CNN, which is an enhanced version of Faster R-CNN.

Advantages of Mask R-CNN:

- Simplicity: Training Mask R-CNN is an easy task.
- Performance: On every task, Mask R-CNN performs better than any single-model entry now in use.
- Efficiency: The technique adds very little overhead to Faster R-CNN and is incredibly efficient.
- Flexibility: Mask R-CNN can be easily used to a variety of tasks. For instance, the same system may be used to estimate human pose using Mask R-CNN.

The essential component of Mask R-CNN is the precise orientation of pixels, that is absent in Fast/Faster R-CNN. Mask R-CNN employs an identical two-step procedure, commencing with the initial RPN stage. In the second stage, Mask R-CNN produces a binary mask for each region of interest (ROI) while also predicting the class and box offset. Mask predictions serve as the foundation for categorization in most contemporary systems. Moreover, because to the flexibility of the Faster R-CNN framework in accommodating various customizable architectural designs, the creation and training of Mask R-CNN is straightforward. Furthermore, the mask

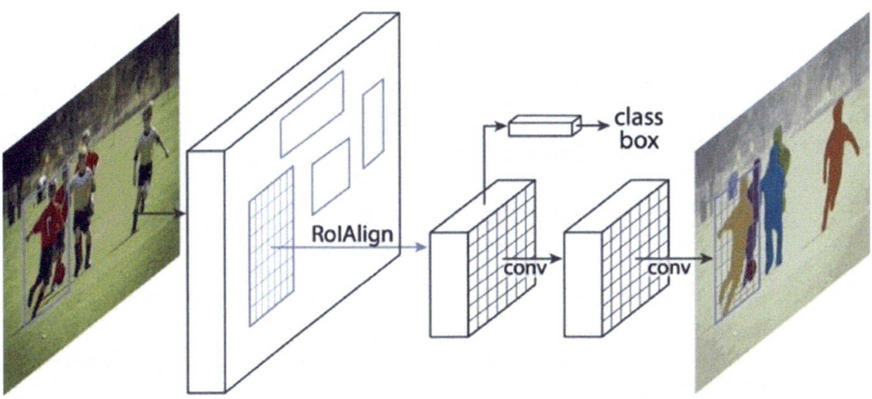

Fig. 7 Mask R-CNN—The mask R-CNN structure for instance segmentation [6]

branch's little computing burden enables the implementation of a fast system and rapid experimentation.

4 Reinforcement Learning

Reinforcement Learning (RL) is a crucial field in machine learning that focuses on training intelligent agents to optimize their anticipated future rewards within a certain environment. The agent's interactions with the environment are documented and utilized to generate the agent's training data. A notable differentiation between conventional machine learning and the current model predictions lies in the fact that the actions resulting from the predictions have the potential to impact the dataset of the agent. Reinforcement learning (RL) is a powerful learning technique that involves the agent sampling the environment to get training data. Thanks to this type of learning, engineers can effectively optimize agents across various problem domains. These areas encompass robotics, controls, navigation, AI alignment, and various other fields. Successful application of reinforcement learning results in the creation of an agent that explores all possible options and exploits the most beneficial ones.

4.1 Proximal Policy Optimization (PPO)

With supervised learning, the cost function can be quickly put into practice, apply gradient descent to it, and be very sure that only minor adjustments are needed to the hyperparameters to achieve outstanding results. The path to success in reinforcement learning is less clear-cut since the algorithms contain numerous moving components that are challenging to debug and require significant adjusting to produce desirable outcomes. In order to minimize the cost function while guaranteeing that the departure from the prior policy is minimal, PPO [7] attempts to compute an update at each step, striking a balance between sample complexity, ease of tweaking, and ease of implementation.

PPO uses the following unique function:

$$\boldsymbol{LCLIP(\theta) = E^t\left[min(rt(\theta))A^t, clip(rt(\theta), 1 - \varepsilon, 1 + \varepsilon)A^t)\right]}$$

- θ is the policy parameter
- E^t denotes the empirical expectation over timesteps
- r_t is the ratio of the probability under the new and old policies, respectively
- A^t *is the estimated advantage at time* t
- ε *is a hyperparameter, usually 0.1 or 0.2.*

By eliminating the KL penalty and the requirement for adaptive updates, this aim streamlines the algorithm and implements a Trust Region update method that is consistent with stochastic gradient descent. Even though it is simpler to develop, this approach has demonstrated the best performance in testing on continuous control tasks.

Algorithm 1 PPO-Clip

1: Input: initial policy parameters θ_0, initial value function parameters ϕ_0
2: **for** $k = 0, 1, 2, \ldots$ **do**
3: Collect set of trajectories $\mathcal{D}_k = \{\tau_i\}$ by running policy $\pi_k = \pi(\theta_k)$ in the environment.
4: Compute rewards-to-go $\hat{R}_t$.
5: Compute advantage estimates, $\hat{A}_t$ (using any method of advantage estimation) based on the current value function V_{ϕ_k}.
6: Update the policy by maximizing the PPO-Clip objective:

$$\theta_{k+1} = \arg\max_{\theta} \frac{1}{|\mathcal{D}_k|T} \sum_{\tau \in \mathcal{D}_k} \sum_{t=0}^{T} \min\left(\frac{\pi_\theta(a_t|s_t)}{\pi_{\theta_k}(a_t|s_t)} A^{\pi_{\theta_k}}(s_t, a_t),\ \ g(\epsilon, A^{\pi_{\theta_k}}(s_t, a_t)) \right),$$

typically via stochastic gradient ascent with Adam.
7: Fit value function by regression on mean-squared error:

$$\phi_{k+1} = \arg\min_{\phi} \frac{1}{|\mathcal{D}_k|T} \sum_{\tau \in \mathcal{D}_k} \sum_{t=0}^{T} \left(V_\phi(s_t) - \hat{R}_t \right)^2,$$

typically via some gradient descent algorithm.
8: **end for**

The pseudo code of the Proximal Policy Optimization (PPO) algorithm, as stated in the PPO documentation is as follows [8]

5 Deep Reinforcement Learning

Deep Reinforcement Learning (DRL) is a reinforcement learning technique that approximates the value or policy function using deep neural networks. A type of deep reinforcement learning is called Decentralized Distributed Proximal Policy Optimization (DDPPO).

5.1 Decentralized Distributed Proximal Policy Optimization (DDPPO)

The technique known as Decentralized Distributed Proximal Policy Optimization (DD-PPO) [9] is utilized for the decentralized reinforcement learning within artificial environments with abundant resources. Because DD-PPO is synchronous (no computation is ever 'stale'), decentralized (no centralized server is needed), and distributed (uses numerous machines), it is conceptually straightforward and simple to implement. For reinforcement learning, proximal policy optimization, or PPO, is a policy gradient technique. The goal was to create an algorithm that used only first-order optimization and had the data efficiency and dependable performance of TRPO.

Let $r_t(\theta)$ denote the probability ratio $r_t(\theta) = \frac{\pi_\theta(a_t|s_t)}{\pi_{\theta_{old}}(a_t|s_t)}$ so $r(\theta_{old}) = 1$.

TRPO maximizes a 'surrogate' objective: $L^v(\theta) = \hat{\mathbb{E}}_t[\frac{\pi_\theta(a_t|s_t)}{\pi_{\theta_{old}}(a_t|s_t)})\hat{A}_t] = \hat{\mathbb{E}}_t[r_t(\theta)\hat{A}_t]$

DD-PPO is a general abstraction that involves the next steps: for any step k, a worker n possesses a copy of the parameters, θ_n^k, computes the gradient, $\delta\theta_n^k$, and updates θ using this gradient. DD-PPO is a general abstraction that involves the next steps: at step k, a worker n possesses a copy of the parameters, θ_n^k, computes the gradient, $\delta\theta_n^k$, and updates θ using this gradient:

$$\begin{aligned}\theta_n^{k+1} &= \text{ParamUpdate}(\theta_n^k, \text{AllReduce}(\delta\theta_1^k, \ldots, \delta\theta_N^k)) \\ &= \text{ParamUpdate}(\theta_n^k, \frac{1}{N}\sum_{i=1}^{N}\delta\theta_i^k)\end{aligned}$$

where, ParamUpdate refers to any first-order optimization technique, such as gradient descent. AllReduce gives an operation for reducing, such as calculating the mean, over all instances of a variable and distributes the outcome to all workers.

6 A Look at the Prominent Vision Systems Used in the Past

6.1 AlexNet Architechture

The Convolutional Neural Network Architecture that emerged victorious in the 2012 LSVRC competition was named AlexNet [10].

The AlexNet contains 8 layers with weights;

5 convolutional layers, and

3 fully connected layers.

The ReLu activation function is applied to each layer, except for the final layer, which uses a softmax function to produce a probability distribution over the 1000 class labels. Dropout is applied in the initial two fully linked layers. Max-pooling is applied following the first, second, and fifth convolutional layers, as depicted in the image above. The kernels of the second, fourth, and fifth convolutional layers are only connected to the kernel maps from the previous layer that are stored on the same GPU. Each kernel map in the second layer is linked to the kernels of the third convolutional layer. Each neuron in the layer above is linked to every other neuron in the fully interconnected levels.

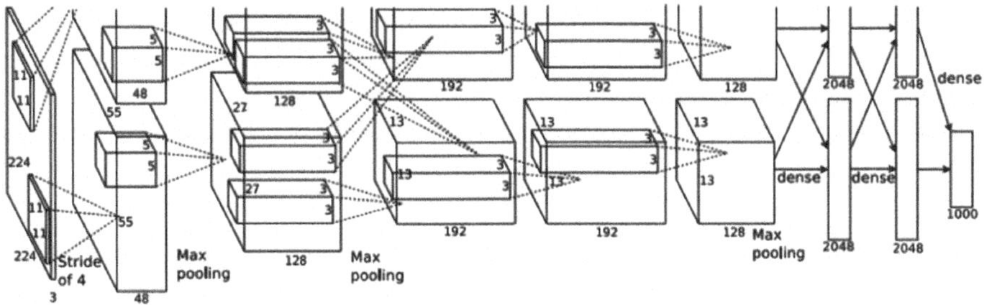

Fig. 8 An architecture diagram of the CNN that clearly illustrates the roles that each GPU is responsible for. Layer components are rendered by one GPU at the top of the figure and by another GPU at the bottom. The GPUs only exchange information at specific layers. The network's input is 150,528-dimensional, and 253,440–186,624–64,896–64,896–43,264–4096–4096–1000 gives the number of neurons in the network's remaining layers [10]

Dataset

An image collection called ImageNet has more than 15 million high-resolution, annotated photos divided into 22,000 categories. The goal of this competition is for researchers to obtain the lowest top-1 and top-5 error rates using a subset of ImageNet's photos. A 256 × 256 RGB image serves as AlexNet's input. This implies that every image in the test set and the training set must have a dimension of 256 × 256. In other words, everything must be transformed to 256 by 256 before the network can be trained with it.

ReLU

Utilizing ReLU (Rectified Linear Unit) Nonlinearity is a key component of the AlexNet. The conventional method of training a neural network model was to employ sigmoid or tanh activation functions.

Fig. 9 Training error versus epoch number for the neural network model [10]

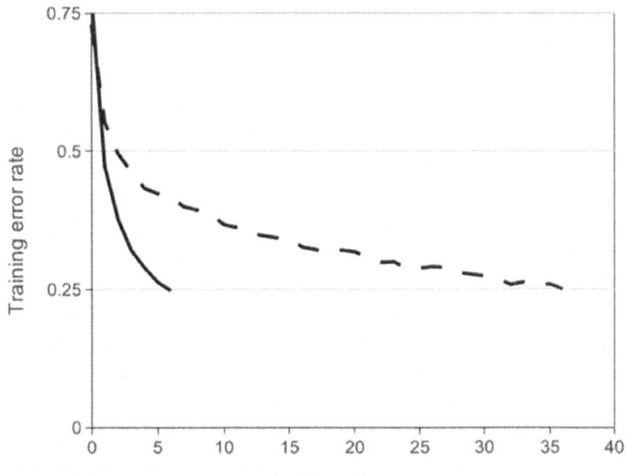

AlexNet demonstrated that deep CNNs could be trained far more quickly utilizing ReLU nonlinearity than saturating activation functions like sigmoid or tanh. The CIFAR-10 dataset is used for testing.

Here's why it trains more quickly while using ReLUs. The function ReLU is provided by

$$f(x) = max(0, x)$$

The two functions' graphs:

1. Tanh
2. ReLU

Fig. 10 Graph of the function: TanH

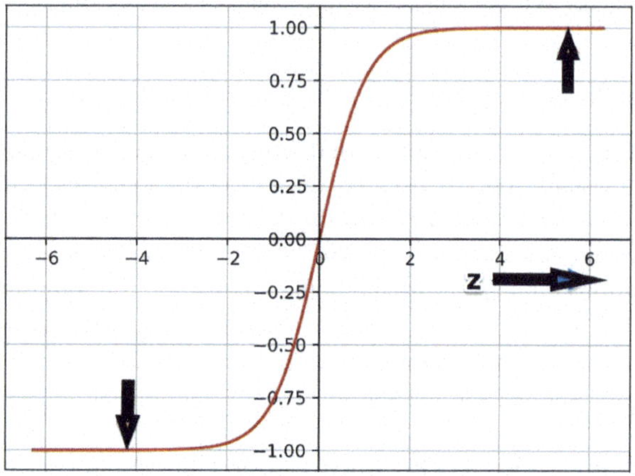

Graph of the function: TanH

Fig. 11 Graph of the function: ReLU

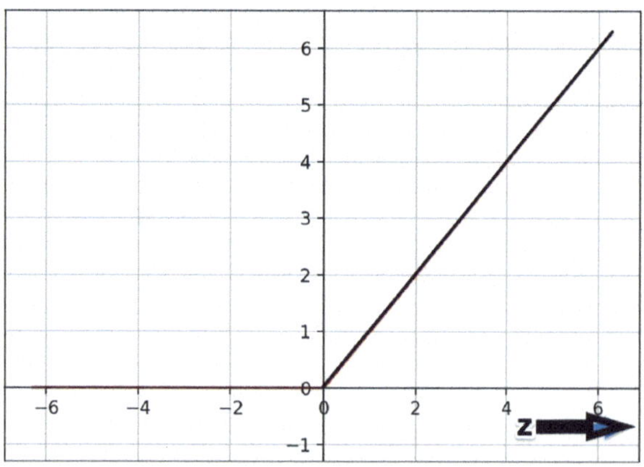

Graph of the function: ReLU

The tanh function reaches its maximum or minimum values with extremely high or low values of z, respectively. The function's slope tends to approach zero with a high degree of proximity in some regions. This could potentially result in a delay in the progress of gradient descent. For higher positive values of z, the gradient of the ReLU function is significantly different from zero. This enables a more rapid convergence of the optimization process. The slope remains constant at 0 for negative values of z, although the majority of neurons in a neural network ultimately have positive values. Similarly, the ReLU function is superior to the sigmoid function for the same reason.

The issue of overfitting. With 60 million parameters, AlexNet had a significant overfitting problem.

Two techniques to lessen overfitting are:

1. Data Augmentation
2. Dropout

1. **Data Augmentation**

The authors augmented the training set by 2048 instances by the use of horizontal reflections and visual translations. In addition, the RGB channel intensities were modified through the application of Principle Component Analysis (PCA) on the RGB pixel values. This resulted in a reduction of the top-1 error rate by almost 1%.

2. **Dropout**

A dropout was the second method AlexNet employed to prevent overfitting. It entails setting each hidden neuron's output to zero with a probability of 0.5. In this sense, the neurons that are "dropped out" do not take part in backpropagation or the forward pass. Thus, the neural network samples a different design each time an input is supplied. By employing this technique, neurons are deactivated with a predetermined likelihood. This implies that throughout each iteration, the neurons that are "deactivated" do not have any impact on either the forward pass or the backpropagation process.

6.2 Deep Convolutional Neural Networks Using ImageNet Classification

The AlexNet architecture revolutionized machine learning research by significantly outperforming its competitors in image identification accuracy on the ImageNet dataset [11]. The authors utilized two fundamental principles: the indispensability of GPU-accelerated training and the need for strong semantic priors in picture classification. At the time, convolutional neural networks may not have been the optimal selection for identifying photographs of objects other than characters, such as vehicles, boats, pets, and so on. Furthermore, investigations into the utilization of Nvidia GPU CUDA programming for neural networks were still in their nascent phases, commencing solely in 2007. The authors made a notable breakthrough in the field of artificial intelligence by harnessing the confluence of multiple technologies.

Due to its ability to replicate the firing pattern of real neurons in the human brain, the sigmoid was widely favored for design during that period. The deviation from the sigmoid activation function is a continuation of a pattern in which progress in machine learning is achieved by addressing practical problems, such as image classification, rather than relying on biological inspiration. Given that Embodied AI encounters several real-world challenges akin to those faced by a growing kid, it is logical. The importance of cropping the input training data to permute it was emphasized. Currently, cropping remains an essential process in data augmentation as it enhances the model's ability to generalize. The inherent nature of the random cropping solution suggests that the model's data input technique is of utmost importance.

6.3 Image Recognition Using Deep Residual Learning (ResNet)

The deep learning convolutional neural network was frequently enhanced in depth to optimize performance, employing models such as Visual Geometry Group (VGG) [4]. This continued until the addition of depth ceased to enhance the performance. Furthermore, it was noted that the decline in performance caused by the ongoing expansion of the CNN was not due to overfitting, as the training error actually rose instead of decreasing with the rise in depth. The significance of these lower weights was insignificant; so, their gradient was small and would vanish. It was peculiar that a larger network should have the capability to encompass the identity function together with the smaller network, at the very minimum. The increase in performance will reach a maximum point as the model becomes deeper, and it is crucial that the performance does not decline any further.

The authors note that this poses a challenge for the network to effectively learn the identity function due to the weights' initial activation being in close proximity to 0. Furthermore, the underlying layers have a diminished role in the model's logic, as the neural network is primarily responsive to the input. In view of these findings, the authors propose the utilization of the Residual Neural Network, which has a general architecture resembling that of a convolutional neural network. Additionally, it incorporates skip connections that facilitate the learning of neuronal distinctions. More precisely, rather than the network determining the mapping from x to y, it would determine the position of $\tilde{x}$ such that $x + \tilde{x} = y$. Consequently, a residual block is created throughout the network to depict this newly formed mapping.

6.4 *PointNet: 3D Classification and Segmentation Using Deep Learning on Point Sets*

The first model to investigate applying neural networks to point cloud data was the PointNet [2] model. Previous models would transform point cloud data into various formats, including meshes and voxel representations, before consuming the data. Point clouds are a more conventional representation of the data, according to the authors, who claimed that information was lost during translation. According to them, the point cloud data format better captures the characteristics of the data, matching the methods used by the Depth and LiDAR sensors to gather it.

6.5 *Kernel Point Convolution—KPC*

A collection of data points in space, often generated by a Lidar sensor, is called a point cloud. The XYZ coordinate is one of the set's required features, along with a few optional features like the point's color (RGB), intensity, etc. Among other applications, point clouds are helpful for 3D modeling of objects, aerial sensor scanning of broad areas, and scene interpretation for autonomous vehicles. The majority of state-of-the-art (SOTA) models for image segmentation and classification utilise convolutional neural networks, or CNNs. But we can't use CNNs with point clouds because of their special structure (we'll talk about the distinctions in the next section). Thus, the search for deep learning methods to process point clouds gave rise to a whole field of study.

KPNN [1] presents a new NN layer that works directly with point cloud data. It demonstrates how to use this layer into a DNN to perform segmentation and classification tasks and produce SOTA outcomes. By developing a convolution operation that operates directly on the point cloud (without projecting the point cloud to a lower dimension), the KPConv layer attempts to address the difficulties encountered. Similar to CNNs, KPConv operates on an intuitive level. Each pixel in a CNN is represented by a collection of values called "channels," which are then processed using k filters. The dot product of the channels of the adjacent pixels, as well as the pixel itself, as determined by the filters, is the new representation of the pixel. In a KPConv layer, every point is multiplied by k kernel points (like filters) and has f_i characteristics (like channels). The product of all the kernel values and the neighbor's characteristics yields the new representation of the point.

Description

Each point in the point cloud has an activated KPConv layer. Upon receiving a point (including its location and features) and its neighboring points (with the location and features of each neighbor), the system generates additional features for the given point. The neighborhood of a point consists of all the points located inside a sphere of a given radius centered on that point. Similar to a convolution layer filter, the

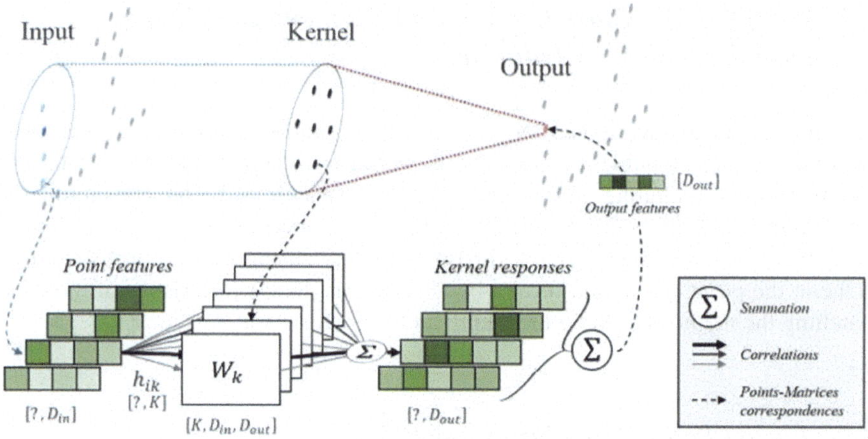

Fig. 12 An example of the correlations between each kernel point and its neighboring point [1]

positions of the neighboring elements are aligned with the given point (the XYZ values of each neighbor are aligned relative to the supplied point location).

A KPConv layer is defined by many kernel points, which are analogous to filters in a CNN. The kernel points are uniformly distributed in a spherical pattern around the given point. Each point strives to stay within the sphere centered on the provided point, while also maximizing the distance between itself and other points. Similar to filters, every kernel point possesses a collection of weights that are utilized to affect the neighboring points. The weights of each kernel point are assigned to its neighboring points based on their proximity to the kernel point. Consequently, a correlation function (explained later) determines the extent of influence of each kernel point. The correlation function determines whether a neighboring point contributes to the calculation of the correlation and whether a kernel point influences a given point.

The new characteristics of a particular point are computed as the product of all features of its neighbors, multiplied by their correlation and learnt weights. Each point's features, including RGB and intensity, are assigned to the point's attributes at the input layer. Similar to how RGB channels are entered into a CNN. Following the initial KPConv layer operation on the point cloud, each point possesses additional features (think of them as CNN channels) that are utilized to compute the subsequent KPConv layer.

Rigid Kernel

The regular points around every point in the point cloud are called kernel points. Every kernel point in the point cloud has a weight matrix that is applied to the nearest neighbors of each point. The locations of the kernel points are found by solving an optimization problem. The kernel points must be as widely apart as feasible while still being as near to the center as feasible (which is the location of the provided

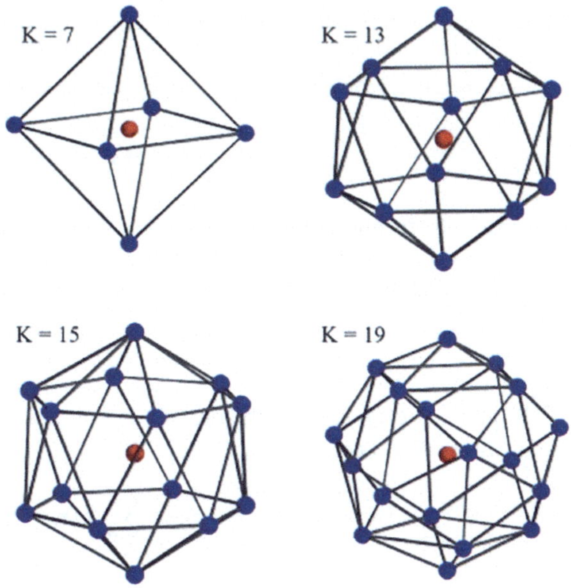

Fig. 13 Rigid kernel point positions [1]

point). It is required that one of the points be in the middle. The orderly arrangement of kernel points in a sphere is associated with the rigid kernel. Examples of kernel point placements with various K values are as follows.

By employing several kernel points, it is possible to differentiate information from different directions of the neighboring points that have been analyzed. The systematic arrangement of the kernel points produces a uniform resolution in all directions. Each kernel point is associated with a distinct set of weights, represented by a matrix of dimensions [f_{in}, f_{out}]. Here, f_{in} refers to the number of features in the current layer, while f_{out} represents the number of features in the subsequent layer. In a manner akin to CNN, every kernel possesses a collection of weights that correspond to the dimensions of the input channels and the output channels.

Kernel Function

The neighbor position x_i is taken by the kernel function, which centers it on the supplied point x. The following formula determines neighbor x_i's centered position:

$$y_i = x_i - x$$

The kernel function $g(Y_i)$ adds a varied ratio (correlation) to the kernel weights based on the neighbor location. The weighted sum of all the kernel weights with the kernel correlations yields the kernel value for a certain neighbor:

$$g(y_i) = \Sigma h\left(y_i, \widetilde{x_k}\right) W_k \tag{1}$$

where x_k is the location of the kernel point, W_k is the weight matrix of kernel k, and $h(y_i, x_k)$ is a correlation function. Next, the neighbor features f_i are multiplied by the kernel value that has been determined for every neighbor y_i:

$$g(x_i - x) f_i$$

The new representation of the provided point x is calculated as the sum of all the kernel values of x's neighbors:

$$(F * g)(x) = \sum_{x_i \in Nx} g(x_i - x) f_i \tag{2}$$

Correlations

The correlation between each neighbor, x_i, and a kernel point, x_k, is computed by the correlation function. The contribution of each kernel to the overall kernel value of each neighbor is measured using the correlations. The locations of each point (XYZ) are used to determine the correlation.

$$h\left(y_i, \widetilde{x_k}\right) = max\left(0, 1 - \frac{\|y_i - \widetilde{x_k}\|}{\sigma}\right) \tag{3}$$

where the influence distance of the kernel points, denoted by the hyper-parameter σ, is specified. This hyper-parameter is determined by the point cloud's density. Higher values are produced by this correlation function when the points are near to one another and its value falls between [0, 1]. They stated these two essential conditions for the correlation function.

Mathematical Implementation of Assembling all

Every point has a maximum of n_{nei} neighbors for a batch size of n points. The quantity of kernel points is k. F_i is a representation of each point's characteristics. Each neighborhood's characteristics are represented by neighb_x (dim(neighb_x) = [n, n_{nei},f_i]). Each kernel point's weights are represented by kernel_weights. dim(kernel_ weights) = [$f_{o\gamma}$, k, f_{in}].

Neighborhood centers are given by dim(neighbors) = [n, n_{nei},3].

Use the formula diffs = neighbors-kernels to get the differences between each neighborhood and kernel point. dim(diffs) = [n, n_{nei}, k, 3], dim(kernels) = [k, 3]. After applying the L1 norm to the differences, compute the correlation by adding the differences along the 4ブ^{h} (XYZ axis). r = min (1 − Σ||diffs||/σ, 0). dim(correlations) = [n, k, n_a_b_a_i].

The following formula is used to determine each point's updated representation:

$$\begin{aligned} \boldsymbol{dim(new\_representation)} &= [\boldsymbol{n}, \boldsymbol{f}_{\boldsymbol{out}}]; \boldsymbol{new\_representation} \\ &= \boldsymbol{\Sigma}(\boldsymbol{correlationsneighb\_x})^{\boldsymbol{T}} \boldsymbol{kernel\_weights}. \end{aligned}$$

Deformable Kernel

KPConv implements a deformable version that is similar to deformable CNN [12]. For every kernel point, the deformable version of KPConv produces a shift, which is added to each kernel point's location to establish its new location. The function Δ(x) produces k shifts in relation to the supplied point. Three values make up each shift (one value for each axis); for example, Δ(x) produces 3k values, which are then reshaped to (3, k). The area of the specified point x is used to calculate the shifts.

The changes that are needed to be made to use the deformable version are in the main KPConv function and the kernel function:

$$(\boldsymbol{F} * \boldsymbol{g})(\boldsymbol{x}) = \sum_{x_i \in Nx} \boldsymbol{g}_{\boldsymbol{deform}}(\boldsymbol{x} - \boldsymbol{x}_i, \boldsymbol{\Delta}(\boldsymbol{x}))\boldsymbol{f}_i \quad (4)$$

$$\boldsymbol{g}_{\boldsymbol{deform}}\left(\boldsymbol{y}_i, \boldsymbol{\Delta}(\boldsymbol{x})\right) = \sum_{k<K} \boldsymbol{h}\left(\boldsymbol{y}_i, \widetilde{\boldsymbol{x}}_k + \boldsymbol{\Delta}_k(\boldsymbol{x})\right)\boldsymbol{w}_k \quad (5)$$

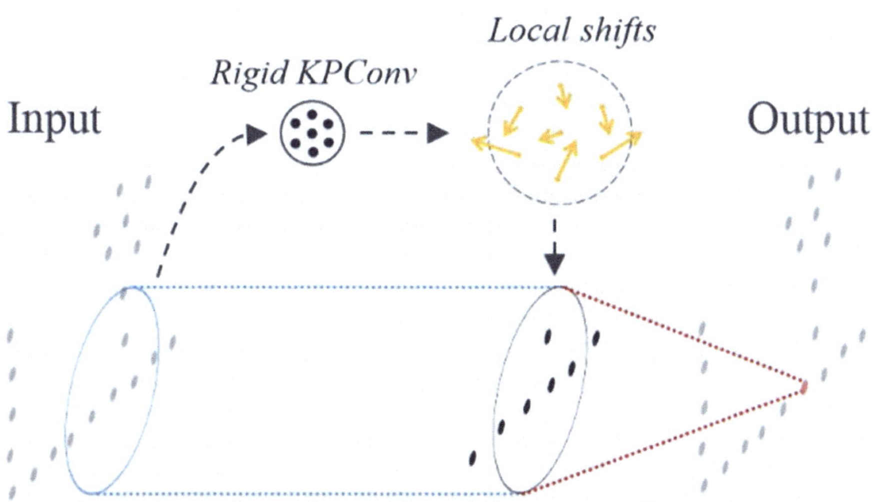

Fig. 14 The deformable KPConv method is demonstrated with 2D point data [1]

A rigid KPConv layer with the same input parameters as the deformable parameters calculates the offset function $\triangle(x)$, and its output dimension is 3K. The new locations of each kernel point are determined by reshaping the output of $\triangle(x)$ to $(K, 3)$. The new location of each kernel point is computed as the product of the calculated shifts and the original location (as in the rigid version).

How KPConv handles point cloud's challenges?

Sparse: In contrast to KNN neighborhoods, radius neighborhoods offer resilience to changing density. The KPConv layer uses a per-dataset radius to sample neighborhoods.

Unstructured: By forming a sphere around each point in the point cloud, unstructured-KPConv raises the significance of the relationships between the neighbors and the kernel points while decreasing the value of the interior structure of the neighborhood.

Unordered: The points' order is irrelevant to the unordered-KPConv layer. To create a new embedding, it considers the qualities of each point as well as those of its surroundings.

7 Embodied AI for Vision

Embodied AI highlights the idea that an agent's sensory interactions with its surroundings should be the source of artificial intelligence. Typically, the field investigates how Embodied AI is used in isolated settings, such Habitat AI. This will enhance the overall performance of Embodied AI models. Nonetheless, the utilization of Embodied AI is infrequently explored in domains like computer vision, which often feature more mature machine learning research. Embodied AI in computer vision is based on the notion that the model may act to enhance its visual comprehension of objects it is unsure of and can utilize historical observations to guide present observations. Most computer vision systems use the same training paradigm as 'Alexnet.' The majority of noteworthy developments, like those with PointNet and KPC, have resulted from altering the model architecture. The 'static dataset paradigm' refers to the idea that all of these models operate from a static viewpoint, where the data is constant between experiments.

With an embodied perspective, the agent can reflect on past observations and take action to see things and locations it is unsure of from a better angle. Because it is doing different actions, there is a noticeable difference in the sequence of observations between batches. But before all of that, the objective of this chapter will be to simply enable an Embodied AI to navigate the environment effectively, which is a huge development issue. Developing navigation techniques that may be applied to the training of an embodied visual system is the aim of the experiments. One of the fundamental skills an agent must possess is the ability to navigate around a point correctly.

References

1. Thomas, H.: KPConv: flexible and deformable convolution for point clouds. arXiv.org, 18 April 2019 (2019). http://arxiv.org/abs/1904.08889
2. Qi, C.R.: PointNet: deep learning on point sets for 3D classification and segmentation. arXiv.org, 02 December 2016 (2016). http://arxiv.org/abs/1612.00593
3. He, K.: Deep residual learning for image recognition. arXiv.org, 10 December 2015 (2015). https://arxiv.org/abs/1512.03385
4. Simonyan, K.: Very deep convolutional networks for large-scale image recognition. arXiv.org, 04 September 2014. https://arxiv.org/abs/1409.1556
5. Radford, A.: Learning transferable visual models from natural language supervision. arXiv.org, 26 February 2021 (2021). https://arxiv.org/abs/2103.00020
6. He, K.: Mask R-CNN. arXiv.org, 20 March 2017. https://arxiv.org/abs/1703.06870
7. Schulman, J.: Proximal policy optimization algorithms. arXiv.org, 20 july 2017. https://arxiv.org/abs/1707.06347
8. Proximal policy optimization—spinning up documentation. https://spinningup.openai.com/en/latest/algorithms/ppo.html
9. Wijmans, E.: DD-PPO: learning near-perfect PointGoal navigators from 2.5 billion frames. arXiv.org, 01 November 2019. https://arxiv.org/abs/1911.00357
10. Krizhevsky, A., Sutskever, I., Hinton, G.E.: ImageNet classification with deep convolutional neural networks. Commun. ACM (2017). https://doi.org/10.1145/3065386
11. Deng, J., Dong, W.: ImageNet: a large-scale hierarchical image database. In: IEEE Conference Publication|IEEE Xplore, 01 June 2009. https://ieeexplore.ieee.org/document/5206848
12. Mishra, D.: Deformable convolutions demystified - towards data science. Medium, 16 December 2021. https://towardsdatascience.com/deformable-convolutions-demystified-2a77498699e8

Anubhav Bewerwal The author is a seasoned writer in the field of artificial intelligence (AI). With a background in engineering, he brings a wealth of knowledge and experience to his writing. His chapter in the book—Building Embodied AI Systems: The Agents, the Architectural Principles, Challenges and Application Domains, titled Computer Vision (CV) and Natural Language Processing (NLP) Algorithms, delves into the introduction to CV and NLP, offering insightful analysis and practical applications. His research areas are Machine learning and Recommender Systems. Based in Uttarakhand, the author continues to push the boundaries of AI research and advocate for its responsible use, ensuring that technology serves humanity's best interests

Amit Kumar is a prominent researcher in the areas of Machine learning and Biometric Securities. He received the M.Tech. degree in Rajiv Gandhi Proudyogiki Vishwavidyalaya, India, in 2015. He has published various research papers focused on computer vision and pattern

Manoj Kumar completed his Ph.D. from The Northcap University and M.Sc. (Information Security and Digital Forensics) from Technological University Dublin, Ireland in 2013. Dr. Kumar is having more than 13.5+ years of research, teaching, and corporate experience. Currently, Dr. Kumar holds the position of Associate Professor at the University of Wollongong in Dubai. In addition, he serves as the Research Cluster Head for Network and Cyber Security. Dr. Kumar is listed among the top 2% of computer scientists worldwide.

Demystifying Embodied AI

Sunil Sable and Mitesh Ikar

Abstract Our chapter seeks to explore a revolutionary paradigm in artificial intelligence (AI). That goes beyond the usual models by stressing the integration of sensory perception, motor abilities, and contextual awareness. Our chapter starts with an introduction to Embodied AI unraveling its transformative potential. We establish Foundations of Embodied AI that delve into the core principles of sensory-motor integration as well as situatedness and which accentuate the inseparable relationship between intelligent systems and specific environmental contexts. Then come methodologies in Embodied AI that reveal a plethora of approaches ranging from sensory-motor loops to bio-inspired models. Learning in Embodied AI foregrounds continuous, reciprocal interactions between agents and their surroundings. Developed perception systems include advances made in computer vision, haptic feedback, attention mechanisms allowing machines to be actively engaged with their environments. The narrative opens into decision making and control processes within Embodied AI incorporating reinforcement learning, evolutionary algorithms and active perception. It also identifies challenges and open problems that become a basis for critical discussions about ethical considerations in Embodied AI. Finally, exploration concludes by shedding light on diverse applications and future directions, envisioning a landscape where machines not only process information but actively engage, adapt, and learn from their embodied experiences, challenging preconceptions and demystifying the profound implications of Embodied AI.

Keywords Embodied Artificial Intelligence · Sensorimotor integration · Context-aware systems · Adaptive decision-making · Bio-inspired robotics

S. Sable (✉) · M. Ikar (✉)
Department of Chemical Engineering, Vishwakarma Institute of Technology, Pune, Maharashtra 411037, India
e-mail: sunil.sable@vit.edu

M. Ikar
e-mail: mitesh.ikar21@vit.edu

P. Raj et al. (eds.), *Building Embodied AI Systems: The Agents, the Architecture Principles, Challenges, and Application Domains*, Information Systems Engineering and Management 14, https://doi.org/10.1007/978-3-031-68256-8_2

1 Introduction to Embodied AI

Embodied Artificial Intelligence (Embodied AI) marks a significant departure from traditional artificial intelligence models that focus solely on computational power and abstract reasoning. This approach seeks to give machines a physical presence, enabling them to interact with the world much like humans. Embodied AI goes beyond conventional algorithms and code, integrating sensory perception and motor capabilities to create intelligent entities capable of navigating and manipulating their environment. A crucial aspect of Embodied AI is the integration of sensory systems, allowing machines to perceive and interpret their surroundings. Brooks et al.'s research underscores the importance of sensory-motor integration, asserting that cognition arises from the dynamic interplay between a system and its environment. This integration allows machines to gather information from the physical world, leading to a more nuanced understanding of context and facilitating adaptive decision-making [1]. In addition to sensory integration, Embodied AI places significant emphasis on motor capabilities, enabling machines to physically interact with their environment. Pfeifer and Bongard argue that the embodiment of intelligence necessitates the ability to act upon the world, establishing a profound connection between perception, action, and cognition. This embodiment, they suggest, fosters intelligence through the continuous loop of sensing, processing, and acting [2].

Embodied AI is not limited to robotics; it extends to virtual agents and simulations. Steels' research explores the development of embodied agents in linguistic contexts, demonstrating how the physical grounding of language in sensorimotor experiences can lead to language capabilities in artificial entities. This highlights the broad applicability of embodied principles across various domains within AI [3]. Fundamental aspect of Embodied AI is the concept of "situatedness," where intelligence is inherently tied to the specific context in which an agent operates. Clark's research argues that intelligent behavior is not solely a result of internal cognitive processes but is profoundly influenced by the external environment. Consequently, by embedding intelligence within the physical and contextual framework, Embodied AI aims to create more adaptive and context-aware systems [4].

Embodied AI represents a revolutionary shift in the AI paradigm, emphasizing the integration of sensory perception and motor capabilities to create intelligent entities capable of effectively navigating and interacting with their environment. This approach challenges the traditional mind–body dichotomy in AI and opens new avenues for developing intelligent systems intimately connected to the physical world.

2 Foundations of Embodied AI

This holistic approach aims to emulate the interactive and adaptive nature of human intelligence, establishing a profound connection between computational processes and real-world engagement. The underpinnings of Embodied AI involve synthesizing cognitive mechanisms with physical agency, departing significantly from traditional AI models. The forefront of Embodied AI highlights the principle of sensory-motor integration, where the seamless connection between perception and action is crucial for intelligent behavior. Metta et al.'s influential work underscores this integration, suggesting that cognition emerges from the dynamic interplay between a system and its environment. This framework enables machines not only to process information but also actively engage with and interpret their surroundings, fostering a nuanced understanding of context and adaptive decision-making [5]. Motor capabilities, as emphasized by Ijspeert et al.'s research, play a pivotal role in shaping Embodied AI's foundational principles. According to this perspective, the embodiment of intelligence requires machines to physically interact with their environment. This interaction establishes a continuous loop where sensory input informs cognitive processes, leading to purposeful actions. Granting machines the ability to act upon their environment allows Embodied AI to transcend the limitations of traditional computational models, enabling a more dynamic and interactive form of artificial intelligence [6].

Table 1 Explains Foundations of Embodied AI Embodied. AI's applicability extends beyond physical robots to virtual agents and linguistic domains, as demonstrated by Cangelosi and Schlesinger's work. Their research explores embodied language learning, showcasing how language capabilities can emerge through the integration of sensorimotor experiences. This expansion of Embodied AI principles into diverse domains highlights its versatility and potential to revolutionize various aspects of artificial intelligence, extending beyond the confines of physical robotics [7]. The foundational concept of "situatedness" is a key tenet of Embodied AI, asserting that intelligence is inherently tied to the specific context in which an agent operates. Beer's research delves into the significance of situated intelligence, arguing that intelligent Behavior is not solely a product of internal cognitive processes but is profoundly influenced by the external environment. By considering the situated nature of intelligence, Embodied AI strives to create systems that are contextually aware and adaptable, bridging the gap between abstract reasoning and real-world application [8]

Recent advancements in biologically inspired Embodied AI models introduce elements from the natural world into artificial systems. Haikonen's work, for instance, explores the incorporation of brain-inspired architectures to enhance the cognitive capabilities of machines. This interdisciplinary approach signifies a deeper integration of biological insights into AI, pushing the boundaries of Embodied AI beyond mere emulation of human intelligence to the creation of more organic and adaptive artificial cognition [9].

Table 1 Foundations of Embodied AI

Concept	Meaning	Example
Sensory-motor integration	The seamless connection between perception and action, crucial for intelligent behavior	A robot uses its vision sensors to navigate through a cluttered environment, avoiding obstacles
Motor capabilities	The ability of machines to physically interact with their environment	A robotic arm uses its grippers to pick up and manipulate objects
Contextual awareness	The understanding of the specific environment in which an agent operates	A self-driving car uses its sensors to identify other vehicles, pedestrians, and traffic signs, making informed decisions about how to navigate
Biological inspiration	The incorporation of insights from the natural world into artificial systems	A brain-inspired AI model uses neural networks to process information and learn from experience

3 Methodologies in Embodied AI

Embodied Artificial Intelligence (Embodied AI) leverages unique methodologies to bridge the gap between artificial systems and the dynamic nature of the real world. This innovative field encompasses diverse approaches, from the implementation of sensory-motor loops to the development of adaptive systems that integrate seamlessly with their environment. Examining the methodologies in Embodied AI sheds light on the diverse strategies employed to imbue machines with the capacity for interactive and contextually aware behavior.

A fundamental methodology in Embodied AI involves the integration of sensory-motor loops, as highlighted by the work of Pfeifer and Scheier. They emphasize the importance of sensorimotor coordination, asserting that cognition emerges from the reciprocal relationship between a system and its environment. By incorporating sensory inputs and motor outputs, machines gain the ability to perceive their surroundings and act upon them, setting the stage for a more embodied form of intelligence. This methodology represents a departure from traditional AI approaches that focus solely on abstract reasoning [10]. A notable approach to realizing Embodied AI methodologies is through the development of robotic platforms. The work of Hoffman et al. introduces the concept of morphological computation, where the physical design of a robot contributes significantly to its cognitive abilities. By structuring robots with adaptable bodies, these researchers demonstrate how the physical form can enhance the computational capabilities of the system. This methodology underscores the importance of the robot's embodiment in shaping its interactions with the environment and, consequently, its intelligent behavior [11].

Table 2 explain Methodologies in Embodied AI. The integration of neural networks into Embodied AI methodologies has also gained prominence. In the study by Oudeyer and Kaplan, the authors explore the role of intrinsically motivated learning in the development of sensorimotor coordination. They argue that an agent can autonomously discover its sensorimotor contingencies, leading to the emergence of adaptive Behavior. This methodology aligns with the idea that intelligence arises from an agent's interaction with its environment, emphasizing the importance of self-directed learning in achieving embodied cognition [12]. Furthermore, a bio-inspired methodology has been influential in Embodied AI, drawing inspiration from natural systems to design intelligent machines. The research of Floreano and Keller explores the application of evolutionary robotics, where artificial agents evolve through a simulated natural selection process. This methodology allows machines to adapt to their environment over time, mirroring the evolutionary principles observed in biological systems. By incorporating evolutionary strategies, Embodied AI methodologies aim to create systems that can continuously adapt and optimize their performance in response to changing conditions [13].

The concept of enactive cognition, proposed by Varela et al., introduces a methodology rooted in the idea that cognition is an embodied and embedded process. Enactive systems actively create and maintain their own cognitive structures through their interactions with the environment. This methodology challenges traditional cognitive science by emphasizing the reciprocity between an agent and its surroundings, highlighting the inseparability of cognition from the embodied experience [14]. Moreover, recent advances in machine learning, particularly reinforcement learning, have been applied to Embodied AI methodologies. In their work, Mnih et al. demonstrate the training of agents to interact with complex environments using deep reinforcement learning. This methodology allows machines to learn optimal actions through trial and error, adapting their Behavior based on feedback from the environment. The combination of deep learning and reinforcement learning represents a powerful methodology in creating intelligent agents capable of navigating and interacting with their surroundings [15].

4 Learning in Embodied AI

Learning in Embodied AI is not confined to traditional computational processes; instead, it involves a continuous, reciprocal interaction between agents and their surroundings. This approach seeks to imbue machines with the ability to learn and adapt in real-world contexts, challenging conventional notions of learning in artificial systems. A fundamental aspect of learning in Embodied AI involves the incorporation of sensory-motor loops, as proposed by Pfeifer and Scheier. This methodology underscores the importance of learning through active interaction with the environment. In this approach, agents acquire knowledge not only through processing abstract information but also by experiencing and responding to the physical world. The integration of sensory inputs and motor outputs enables agents to adapt their Behavior based on real-time interactions, resulting in a form of learning deeply rooted in embodied experiences [16]. The concept of reinforcement learning has gained prominence in

Table 2 Methodologies in Embodied AI

Methodology	Meaning	Example
Sensory-motor integration	The incorporation of sensory inputs and motor outputs to enable machines to perceive their surroundings and act upon them	A robot uses its cameras and microphones to gather information about its environment, and then uses its actuators to move around and manipulate objects
Morphological computation	The idea that the physical design of a robot contributes significantly to its cognitive abilities	A robot with a flexible body is better able to adapt to its environment and perform complex tasks
Intrinsically motivated learning	The idea that an agent can autonomously discover its sensorimotor contingencies, leading to the emergence of adaptive behavior's	A robot is programmed to explore its environment and learn from its own experiences
Bio-inspired evolutionary robotics	The application of evolutionary principles to design intelligent machines	Artificial agents are created and then allowed to evolve through a simulated natural selection process. The fittest agents are selected to reproduce, and their offspring inherit their traits
Enactive cognition	The idea that cognition is an embodied and embedded process, where systems actively create and maintain their own cognitive structures through their interactions with the environment	A robot is programmed to learn about its environment by interacting with it. For example, it may learn to avoid obstacles by bumping into them
Deep reinforcement learning	A machine learning technique that allows agents to learn optimal actions through trial and error, adapting their behavior based on feedback from the environment	A robot is trained to play a video game using deep reinforcement learning. The robot learns to play the game by trial and error, and is rewarded for winning

Embodied AI, providing a robust framework for agents to learn through trial and error. In their influential work, Sutton and Barto define reinforcement learning as a process where agents learn to make decisions by receiving feedback in the form of rewards or punishments. This methodology is particularly well-suited for Embodied AI, allowing agents to adapt their Behavior based on the consequences of their actions in a given environment. Reinforcement learning thus enables machines to acquire complex skills and Behaviors through interactive experiences [17].

Table 3 explains Learning in Embodied AI. Evolutionary algorithms, as discussed by Nolfi and Floreano, provide another avenue for learning in Embodied AI. Evolutionary algorithms emulate the principles of natural selection, allowing artificial agents to undergo a process of selection and variation to optimize their performance. This methodology enables machines to evolve and adapt over time, leading to the emergence of intelligent Behavior's that are finely tuned to the specific demands of their environments. Evolutionary learning in Embodied AI reflects a dynamic and adaptive approach to acquiring cognitive capacities [18]. In the context of embodied language learning, Cangelosi and Parisi propose a methodology where agents learn linguistic capabilities through sensorimotor experiences. This approach aligns with the idea that language comprehension and production are inherently tied to the physical engagement of agents with their surroundings. By grounding language in sensory-motor experiences, agents in Embodied AI can acquire linguistic skills in a manner that is more aligned with human cognitive development [19].

The exploration of intrinsic motivation in learning is another compelling facet of Embodied AI methodologies. Oudeyer and Kaplan delve into the idea that agents can autonomously explore their environments based on intrinsic motivators, fostering the discovery of sensorimotor contingencies. This methodology emphasizes the importance of self-directed learning, allowing agents to autonomously seek out information and develop adaptive Behavior's without explicit external guidance. Intrinsic motivation adds a layer of autonomy to learning in Embodied AI, enabling agents to explore and learn in a manner that aligns with their inherent motivations [12]. Biologically inspired models contribute significantly to learning in Embodied AI,

Table 3 Learning in Embodied AI

Methodology	Meaning	Example
Sensory-motor loop integration	Agents learn by actively interacting with the environment and receiving feedback through their senses	A robot learns to walk by trial and error, using its vision and proprioceptive sensors to correct its movements
Reinforcement learning	Agents learn to make decisions by receiving rewards or punishments for their actions	A robot learns to play a game by trial and error, receiving rewards for winning and punishments for losing
Evolutionary algorithms	Agents evolve and adapt over time through a process of selection and variation	A population of robots is created and allowed to compete with each other. The fittest robots are selected to reproduce, and their offspring inherit their traits
Embodied language learning	Agents learn linguistic capabilities through sensorimotor experiences	A robot learns to understand and generate language by grounding it in its physical interactions with the world
Intrinsic motivation	Agents autonomously explore their environments based on intrinsic motivators, such as curiosity or creativity	A robot is programmed to explore its environment and learn from its own experiences
Hebbian learning	Synaptic connections between neurons are strengthened when they are simultaneously activated	A neural network in a robot learns to associate sensory inputs with motor outputs by repeatedly experiencing them together

with the concept of Hebbian learning playing a prominent role. In Hebbian learning, synaptic connections are strengthened between neurons that are simultaneously activated, as proposed by Hebb. This methodology has been applied to neural networks in Embodied AI, allowing artificial agents to learn and reinforce connections based on the correlation of sensory inputs and motor outputs. Hebbian learning mimics the neural plasticity observed in biological systems, enabling machines to adapt their neural architecture based on experiences [20].

5 Perception Systems

Embodied Artificial Intelligence (Embodied AI) places a robust emphasis on perception systems, acknowledging their crucial role in enabling machines to interpret and interact with their environment. In contrast to traditional artificial intelligence models heavily reliant on abstract reasoning, perception systems in Embodied AI are crafted to capture sensory inputs, fostering a deeper integration between agents and the external world. This exploration of perception systems delves into the methodologies and advancements that empower machines to perceive, interpret, and act upon their surroundings.

A foundational aspect of perception systems in Embodied AI lies in the integration of sensory-motor loops, as proposed by Pfeifer and Scheier. This methodology underscores that intelligence emerges not only from internal processing but crucially from the dynamic interaction between an agent and its environment. By incorporating sensory inputs and linking them to motor outputs, perception systems enable machines to acquire information from their surroundings, setting the stage for contextually rich and adaptive Behavior [21]. Advancements in computer vision play a pivotal role in enhancing perception systems within Embodied AI. LeCun et al. discuss the evolution of convolutional neural networks (CNNs), highly effective in image and pattern recognition tasks. These networks, inspired by the structure of the human visual system, allow machines to process visual information hierarchically, capturing complex features and patterns. The integration of CNNs into perception systems empowers machines to discern and interpret visual stimuli with increasing accuracy [22].

Table 4 explains Perception Systems. Beyond visual perception, the incorporation of auditory perception systems is a significant aspect of Embodied AI. In their work, Blum and Moore underscore the importance of auditory processing for intelligent Behavior. Auditory perception systems enable machines to recognize and interpret sound patterns, contributing to a more comprehensive understanding of the environment. This multimodal approach, integrating both visual and auditory perception, allows agents to navigate a broader spectrum of real-world scenarios [23]. The concept of active perception, as introduced by Dana, adds a dynamic dimension to perception systems in Embodied AI. Active perception emphasizes that agents actively explore their environment to gather relevant information, rather than being passive recipients of sensory inputs. This methodology aligns with the embodied nature of intelligence, where agents dynamically engage with their surroundings to acquire information actively. Active perception systems facilitate more efficient and purposeful exploration of the environment, leading to more informed decision-making [24].

Research in haptic perception further extends the scope of Embodied AI's perception systems. Kuchenbecker et al. discuss advancements in haptic feedback systems that enable machines to sense and interpret the sense of touch. Integrating haptic perception allows agents to interact with the physical world, discerning properties like texture, hardness, or temperature. By expanding perception to include the sense of touch, Embodied AI systems gain a more comprehensive understanding of their surroundings, enabling a richer and more nuanced interaction with the environment [25]. The integration of attention mechanisms in perception systems is another

Table 4 Perception systems

Methodology	Meaning	Example
Sensory-motor loop integration	Agents perceive the world by acting on it and receiving feedback through their sensors	A robot uses its vision sensors to navigate a maze, adjusting its movements based on the feedback it receives
Advancements in computer vision	CNNs are used to extract features from images and videos, allowing machines to recognize objects and patterns	A robot uses a CNN to identify different types of objects in a scene, such as people, furniture, and vehicles
Auditory perception systems	Agents use their hearing to perceive the world, identifying and interpreting sound patterns	A robot uses its microphones to identify the direction of a sound and the type of object producing it
Active perception	Agents actively explore their environment to gather relevant information, rather than passively receiving sensory inputs	A robot moves its head and eyes to focus on specific objects in a scene, gathering more information about them
Haptic perception	Agents use their sense of touch to perceive the world, discerning properties like texture, hardness, or temperature	A robot uses its grippers to feel the size and shape of an object, determining how to manipulate it
Attention mechanisms	Agents focus on relevant information in their environment, while ignoring distractions	A robot uses an attention mechanism to focus on the person it is speaking to, ignoring background noise and other visual stimuli
Bio-inspired perception models	Agents use perception models inspired by the human brain, such as hierarchical predictive coding	A robot uses a hierarchical predictive coding model to predict what it will see next, allowing it to respond more quickly to changes in its environment

noteworthy advancement. Tsotsos et al. delve into the role of attention in visual perception, emphasizing the selective focus on relevant information. Attention mechanisms enable Embodied AI systems to allocate computational resources to specific regions of interest, mimicking the human ability to focus on salient details. This targeted approach enhances the efficiency of perception systems, enabling machines to process relevant information more effectively [26]. Additionally, the application of bio-inspired perception models contributes to the development of more sophisticated Embodied AI systems. In their study, Noesselt al. explore the concept of hierarchical predictive coding, inspired by the hierarchical structure of the human brain. This approach involves the generation of predictions at different levels of abstraction, allowing perception systems to reconcile expectations with actual sensory inputs. The incorporation of predictive coding enhances the anticipatory capabilities of Embodied AI, enabling agents to proactively respond to changes in their environment [27].

Perception systems are at the core of Embodied AI, shaping the way machines interact with and interpret the world. The integration of sensory-motor loops, advancements in computer vision, multimodal approaches, active perception, haptic feedback, attention mechanisms, and bio-inspired models collectively contribute to the development of more sophisticated and contextually aware perception systems. As Embodied AI continues to advance, the integration of these diverse methodologies promises to create machines that not only perceive their surroundings but actively engage with and adapt to the dynamic sensorimotor landscape.

6 Decision-Making and Control

Embodied Artificial Intelligence (Embodied AI) extends beyond conventional models by emphasizing decision-making and control processes that are intimately tied to an agent's interaction with the environment. In this paradigm, decision-making is not solely a computational task but a dynamic process shaped by real-time sensory inputs, motor outputs, and the agent's physical embodiment. This exploration delves into the methodologies and advancements in decision-making and control within Embodied AI, shedding light on how agents navigate the interactive landscape.

A foundational element of decision-making in Embodied AI involves the integration of sensory-motor loops, as proposed by Pfeifer and Scheier. This approach underscores the idea that decision-making emerges from the ongoing interplay between an agent and its surroundings. By incorporating sensory inputs and linking them to motor outputs, agents can make decisions that are contextually informed and dynamically responsive to the environment. The integration of sensory-motor loops aligns decision-making with the embodied nature of intelligence [16]. Reinforcement learning, as discussed by Kaelbling et al., is a pivotal methodology in decision-making and control within Embodied AI. Reinforcement learning enables agents to learn optimal actions through trial and error, receiving feedback in the form of rewards or punishments. This methodology allows agents to adapt their decision-making strategies based on the consequences of their actions in a given environment, leading to the development of complex and adaptive Behaviors [28].

Table 5 explains Decision Making and Control. Evolutionary algorithms also play a significant role in decision-making within Embodied AI. Nolfi and Floreano discuss the application of evolutionary algorithms to optimize decision-making strategies in artificial agents. Through a process of selection and variation inspired by natural selection, agents evolve and adapt over time, leading to the emergence of decision-making Behaviors that are finely tuned to the demands of their environments. Evolutionary algorithms provide a means for agents to develop effective decision-making capabilities through an iterative and adaptive process [29]. The concept of active perception, introduced by Ballard, extends into decision-making processes within Embodied AI. Active perception emphasizes that agents actively explore their environments to gather relevant information, contributing to more informed decision-making. This approach aligns decision-making with the notion that agents dynamically engage with their surroundings to acquire information actively. Active decision-making allows agents to purposefully explore and adapt their strategies based on real-time sensory inputs [24].

Moreover, bio-inspired models, such as predictive coding, contribute to decision-making in Embodied AI. Friston introduces the concept of predictive coding, where agents generate predictions about future sensory inputs based on their internal models. Decision-making involves reconciling these predictions with actual sensory inputs, enabling agents to anticipate and respond proactively to changes in their environment. Predictive coding enhances decision-making by incorporating an anticipatory aspect, allowing agents to prepare for upcoming events [30]. The integration of attention mechanisms is another critical aspect of decision-making in Embodied AI. It allows agents to selectively focus on relevant information in their environment. Tsotsos et al. discuss the role of attention in visual perception, and this concept extends into decision-making processes. Attention mechanisms enable agents to allocate computational resources to specific regions of interest, facilitating more efficient and effective decision-making by focusing on salient details [31].

The concept of affordances, as introduced by Gibson, provides a framework for decision-making that is deeply rooted in the agent's perception of its environment. Affordances are the action possibilities that the environment offers to an agent. Decision-making, in this context, involves perceiving and exploiting these affordances to achieve goals. The integration of affordances into decision-making processes aligns with the embodied nature of intelligence, where decisions are informed by the immediate context and action possibilities in the environment [32]. Decision-making and control in Embodied AI represent a departure from traditional computational models by emphasizing the dynamic interaction between agents and their environments. The integration of sensory-motor loops, reinforcement learning, evolutionary algorithms, active perception, bio-inspired models, attention mechanisms, and affordances collectively shape decision-making processes. As Embodied AI continues to advance, these diverse methodologies promise to create agents that

Table 5 Decision making and control

Methodology	Description	Example
Sensory-motor loop integration	Agents make decisions based on their current sensory inputs and the desired motor outputs	A robot decides to turn left after seeing a car coming towards it
Reinforcement learning	Agents learn optimal actions through trial and error, receiving feedback in the form of rewards or punishments	A robot learns to navigate a maze by trial and error, receiving a reward for reaching the goal and a punishment for hitting a wall
Evolutionary algorithms	Agents evolve and adapt over time to develop decision-making strategies that are finely tuned to their environment	A population of robots is created and allowed to compete with each other. The robots that make the best decisions are selected to reproduce, and their offspring inherit their traits
Active perception	Agents actively explore their environment to gather relevant information for decision-making	A robot moves its head and eyes to focus on a specific object in order to learn more about it
Bio-inspired models	Agents use decision-making models inspired by the human brain, such as predictive coding	A robot uses a predictive coding model to anticipate the next sensory inputs and make decisions based on those predictions
Attention mechanisms	Agents selectively focus on relevant information in their environment for decision-making	A robot uses an attention mechanism to focus on the speaker's face while listening to a conversation, ignoring background noise and other visual stimuli
Affordances	Agents make decisions based on the action possibilities that the environment offers	A robot decides to use its grippers to pick up a cup after seeing it on a table

not only make decisions based on computational processes but actively engage, adapt, and navigate the interactive landscape.

7 Challenges and Open Problems

Embodied Artificial Intelligence (Embodied AI) has made significant strides, yet it faces a spectrum of challenges and open problems that shape the trajectory of its development. From addressing issues in sensorimotor integration to enhancing adaptability and scalability, the field confronts complex hurdles. This exploration sheds light on the critical challenges and open problems that researchers grapple with as they navigate the future frontier of Embodied AI.

7.1 Sensorimotor Integration and Real-World Complexity

One of the primary challenges in Embodied AI is achieving robust sensorimotor integration in diverse and dynamic environments. The work of Metta et al. highlights the complexity of translating sensory inputs into meaningful motor actions, especially in real-world scenarios. Ensuring that agents can seamlessly navigate and interact with their surroundings remains a challenge, requiring advancements in perception systems that can handle the intricacies of the physical world [5].

7.2 Adaptability and Generalization

Embodied AI systems often face difficulties in adapting to novel situations and generalizing learned behaviors. The study by Parisotto and Salakhutdinov underscores the challenge of achieving true adaptability, where agents can apply their knowledge across a range of contexts. The open problem lies in developing methods that enable Embodied AI to generalize learning experiences and adapt rapidly to new and unseen environments [33].

7.3 Scalability and Computational Complexity

As Embodied AI systems become more sophisticated, scalability becomes a significant challenge. The work of Floreano et al. discusses the computational demands associated with the evolution of complex behaviors in artificial agents. The open problem revolves around devising scalable algorithms that can handle the increasing computational complexity of large-scale Embodied AI systems, allowing for the development of more intricate and capable agents [34].

7.4 Long-Term Autonomy and Continual Learning

Ensuring long-term autonomy in Embodied AI systems is a pressing challenge. Traditional machine learning models often struggle with retaining knowledge over extended periods. The research by Parisotto and Salakhutdinov emphasizes the need for continual learning mechanisms that enable agents to accumulate knowledge and experiences over time, evolving their capabilities in an ever-changing environment [33].

7.5 *Multimodal Integration*

While progress has been made in integrating different modalities such as vision, audition, and touch, achieving seamless multimodal integration remains a challenge. The study by Cangelosi and Schlesinger delves into the complexities of embodied language learning, emphasizing the need for effective integration of diverse sensory inputs. The challenge lies in developing models that can exploit the synergies between different modalities, enhancing the richness of perception and interaction [7].

Table 6 explains Challenges and Open Problems in Embodied AI.

7.6 *Ethical Considerations and Bias*

As Embodied AI systems increasingly interact with humans, ethical considerations become paramount. The work of Luxton DD et al. highlights the potential for biases to emerge in AI systems, reflecting the biases present in training data. Ensuring fairness, transparency, and accountability in Embodied AI decision-making processes is an open problem, requiring the development of ethical frameworks to guide the deployment of intelligent agents in real-world scenarios [35].

7.7 *Interdisciplinary Collaboration*

Embodied AI draws inspiration from various disciplines, including robotics, neuroscience, and cognitive science. However, achieving effective interdisciplinary collaboration remains a challenge. The research by Ijspeert et al. emphasizes the need for a cohesive approach that integrates insights from different fields. The open problem involves fostering collaboration to develop a unified understanding of intelligence that spans both biological and artificial systems [6].

7.8 *Benchmarking and Evaluation Metrics*

As the field evolves, the lack of standardized benchmarking and evaluation metrics poses a challenge. The study by Beattie et al. highlights the importance of robust evaluation methodologies to assess the performance of Embodied AI systems consistently. Developing standardized benchmarks and metrics is an open problem critical for objectively comparing and advancing the capabilities of different approaches [36].

Table 6 Challenges and open problems

Challenge	Description	Possible solutions
Sensorimotor integration and real-world complexity	Agents cannot reliably translate sensory inputs into meaningful motor actions in complex real-world environments	Develop more robust and reliable sensorimotor integration systems that can handle the complexities of the real world
Adaptability and generalization	Agents cannot easily adapt to novel situations and generalize learned behaviors	Develop methods that enable Embodied AI to generalize learning experiences and adapt rapidly to new and unseen environments
Scalability and computational complexity	Large-scale Embodied AI systems can be computationally expensive to train and deploy	Develop scalable algorithms and hardware that can handle the increasing computational demands of Embodied AI
Long-term autonomy and continual learning	Agents cannot retain knowledge and adapt over extended periods of time	Develop continual learning mechanisms that enable Embodied AI to accumulate knowledge and experiences over time, evolving their capabilities in an ever-changing environment
Multimodal integration	Agents cannot effectively integrate different modalities such as vision, audition, and touch	Develop models that can exploit the synergies between different modalities, enhancing the richness of perception and interaction
Ethical considerations and bias	Embodied AI systems can be biased and unfair, reflecting the biases present in training data and the design of the system	Develop ethical frameworks and guidelines to guide the development and deployment of Embodied AI systems, ensuring fairness, transparency, and accountability
Interdisciplinary collaboration	Embodied AI draws inspiration from various disciplines, but achieving effective interdisciplinary collaboration remains a challenge	Foster collaboration between researchers from different fields to develop a cohesive understanding of intelligence that spans both biological and artificial systems
Benchmarking and evaluation metrics	There is a lack of standardized benchmarking and evaluation metrics for Embodied AI systems	Develop standardized benchmarks and metrics to assess the performance of Embodied AI systems consistently

(continued)

Table 6 (continued)

Challenge	Description	Possible solutions
Robustness in real-world environments	Embodied AI systems must operate in diverse and unpredictable real-world environments	Conduct research on adversarial robustness, safety, and reliability in the deployment of Embodied AI agents
Human-AI interaction and user acceptance	It is important to understand and enhance human-AI interaction as Embodied AI systems increasingly engage with humans	Design embodied agents that can interact seamlessly with humans, considering factors such as user experience, trust, and acceptance

7.9 Robustness in Real-World Environments

Embodied AI systems must operate in diverse and unpredictable real-world environments. Ensuring the robustness of these systems in the face of uncertainties, adversarial conditions, and dynamic changes is a significant challenge. The work of Amodei et al. discusses the need for research on adversarial robustness, safety, and reliability in the deployment of Embodied AI agents [37].

7.9.1 Human-AI Interaction and User Acceptance

As Embodied AI systems increasingly engage with humans, understanding and enhancing human-AI interaction is crucial. The research by Steels discusses the development of embodied agents in linguistic contexts. The challenge lies in designing systems that not only perform intelligently but also interact seamlessly with humans, considering factors such as user experience, trust, and acceptance [38].

Embodied AI stands at the intersection of numerous challenges and open problems, ranging from sensorimotor integration to ethical considerations. Addressing these hurdles requires a concerted effort from researchers across multiple disciplines. As the field progresses, the resolution of these challenges will pave the way for more sophisticated, adaptable, and ethically sound Embodied AI systems that can seamlessly navigate and interact with the complex world they inhabit.

8 Other Challenges in Embodied AI

Tables 7, 8, 9 and 10 explains various other challenges in Embodied AI.

Table 7 Sensory realism- limitations and challenges

Limitation/challenge	Meaning	Example	Impact
Resolution	Sensor's ability to distinguish fine details	Cameras, LiDAR, touch sensors	Hinders tasks like facial recognition, object detection, tactile feedback
Accuracy	Reliability of sensor measurements	GPS systems in urban canyons, dense foliage	Immersion and task performance affected by inaccuracies
Robustness to environmental conditions	Sensor's ability to function across diverse environments	Extreme temperatures, humidity, low light	Limits applicability and reliability under certain conditions
Difficulty integrating multiple sensor modalities	Combining data from vision, audio, touch for a cohesive understanding	Data formats, temporal synchronization, calibration	Inconsistent and unrealistic experiences due to integration challenges
Processing power	Ability to handle large amounts of data generated by high-resolution sensors	Real-time processing demands for AR/ autonomous vehicles	Limits sensor capabilities due to insufficient processing power
Power consumption	Amount of power consumed by sensors	Mobile and battery-powered devices	Limits prolonged use due to battery drain

Table 8 Real-world navigation- challenges and considerations

Challenge	Meaning	Example
Handling complex, unstructured environments	Traditional navigation methods struggle in complex environments. Robots or autonomous systems must rely on a combination of sensor data to create detailed and accurate maps	Urban canyons or indoor spaces with limited satellite visibility
Responding to dynamic changes and events in real-time	Navigation systems must be capable of adapting to sudden obstacles, changes in the environment, or unexpected events	Autonomous vehicles need to respond instantly to the movement of other vehicles, pedestrians, and unexpected road conditions
Exploration challenges and uneven/unsafe terrain navigation	Robots designed for exploration, search and rescue, or disaster response must navigate through areas with uneven terrain, obstacles, or potential safety hazards	A Mars rover traversing a cratered Martian terrain with loose rocks and crevasses

Table 9 Complexity versus scalability in autonomous systems

Area	Challenge	Consequences
Software complexity	– Algorithmic complexity – Large scale machine learning models – Debugging difficulty	– Modular and maintainable software design – Efficient code optimization – Robust testing and debugging tools
Hardware complexity	– Sensor and actuator coordination – Power consumption optimization – Compatibility across diverse components	– Standardized communication protocols – Modular hardware design – Efficient power management techniques
Cost-effective scaling	– Balancing functionality with costs – Limiting expensive components – Maintaining performance	– Innovative sensor and processor solutions – Resource-efficient algorithms – Low-cost manufacturing techniques
Accessibility	– User-friendly interfaces – Robustness for diverse users – Reduced reliance on expertise	– Simplified user interactions – Adaptability to user needs – Intuitive controls and feedback
Customized versus generalizable designs	– Tailored performance might lack flexibility – Difficulty adapting to new situations	– Modular design with adaptable algorithms – Hardware compatible with diverse scenarios – Balance specialization and generability

9 Applications of Embodied AI

Embodied Artificial Intelligence (Embodied AI) has emerged as a transformative paradigm with diverse applications across various domains. By imbuing artificial agents with the capacity for sensorimotor interaction and environmental engagement, Embodied AI transcends traditional computational approaches. This exploration delves into the applications of Embodied AI, showcasing its versatility and impact across different fields.

9.1 Robotics and Autonomous Systems

Embodied AI finds extensive application in the field of robotics, where agents interact with the physical world in real-time. The research by Metta et al. highlights the development of the iCub humanoid robot, emphasizing its open platform for research in embodied cognition. Robots equipped with Embodied AI capabilities can navigate environments, manipulate objects, and respond dynamically to changing scenarios.

Table 10 Hardware design challenges in autonomous systems

Area	Challenge	Consequences
Size and weight	– Limited portability and maneuverability – Reduced efficiency	– Miniaturization techniques – Lightweight materials – Efficient component layout
Degrees of Freedom (DOF)	– Difficulty handling complex tasks – Compromised maneuverability	– Optimal balance between DOF and space/weight – Efficient actuator and joint design – Software-assisted motion control
Power	– Limited operating time – Dependence on external power sources	– Energy-efficient components – Low-power algorithms – Alternative energy sources
Production costs	– Limited market reach – Reduced profitability	– Optimized manufacturing processes – Cost-effective materials – Streamlined assembly and logistics
Specialty materials	– Availability and cost constraints – Potential environmental impact	– Exploring alternative materials – Engineering substitutes – Supplier diversification
Quality control	– System failures and malfunctions – Safety hazards	– Rigorous testing protocols – Standardized manufacturing – Continuous process monitoring

These applications range from manufacturing and logistics to exploration and rescue missions [5].

9.2 Healthcare and Rehabilitation

In healthcare, Embodied AI contributes to the development of assistive technologies and rehabilitation tools. The work of Molteni F et al. discusses the use of robotic exoskeletons in neurorehabilitation, demonstrating how embodied systems can enhance physical therapy for patients with motor impairments. Embodied AI allows these devices to adapt to the individual needs of patients, providing personalized and interactive rehabilitation experiences [39].

9.3 Virtual and Augmented Reality

Embodied AI plays a crucial role in enhancing virtual and augmented reality experiences. The study by Yanhong Liu et al. explores the integration of embodied models for visual perception in virtual environments. By incorporating realistic sensorimotor interactions, virtual reality systems become more immersive, allowing users to engage with virtual spaces in a natural and intuitive manner. This application extends to training simulations, gaming, and virtual prototyping [40].

9.4 Education and Learning Environments

Embodied AI contributes to the development of intelligent tutoring systems and interactive learning environments. The research by Kramer LL et al. discusses the ethical considerations in developing embodied conversational agents for coaching and psychotherapy. In educational settings, embodied agents can provide personalized feedback, adapt to individual learning styles, and create interactive learning experiences that go beyond traditional instructional methods [41].

9.5 Human–Robot Collaboration

Collaboration between humans and robots is a burgeoning application area for Embodied AI. The study by Tenorth et al. explores the integration of robots into human environments, emphasizing the importance of natural communication and interaction. Embodied AI enables robots to understand human gestures, respond to verbal cues, and work alongside humans in shared spaces. This collaborative approach is valuable in industries such as manufacturing, where humans and robots cooperate on complex tasks [42].

Table 11 explains Applications of Embodied AI.

9.6 Language Acquisition and Natural Language Processing

Embodied AI contributes to language acquisition models, enabling artificial agents to understand and generate language in context. The work of Cangelosi and Schlesinger explores the development of embodied agents in linguistic contexts. Applications include natural language processing systems that can comprehend and generate human-like language, enhancing communication between humans and machines in various domains, including customer service and virtual assistants [7].

Table 11 Applications of Embodied AI

Domain	Application	Example
Robotics and autonomous systems	Navigation, manipulation, exploration	Self-driving cars, robotic arms, search and rescue robots
Healthcare and rehabilitation	Assistive technologies, rehabilitation tools	Robotic exoskeletons, prosthetics, virtual reality therapy
Virtual and augmented reality	Immersive experiences, training simulations	Virtual reality headsets, augmented reality glasses, haptic feedback devices
Education and learning environments	Intelligent tutoring systems, interactive learning environments	Personalized feedback, adaptive learning, virtual classrooms
Human–robot collaboration	Collaborative tasks, shared spaces	Manufacturing robots, service robots, exoskeletons
Language acquisition and natural language processing	Language comprehension, generation, translation	Natural language processing systems, virtual assistants, chatbots
Environmental monitoring and exploration	Data collection, task completion in hazardous environments	Planetary rovers, underwater robots, drones
Gaming and entertainment	Immersive gaming experiences, virtual worlds	Intelligent non-player characters (NPCs), procedurally generated content, interactive environments
Social interaction and companion robots	Social conversations, emotional recognition, companionship	Eldercare robots, educational robots, museum robots
Neuroscience and cognitive science	Insights into human cognition, new avenues for research	Embodied models of animal locomotion, brain-computer interfaces, cognitive prosthetics

9.7 Environmental Monitoring and Exploration

In environmental science, Embodied AI aids in monitoring and exploring remote or hazardous environments. The research by Ijspeert et al. discusses the simulation and robotics studies of salamander locomotion. Embodied agents, whether robotic or virtual, can navigate challenging terrains, collect data, and perform tasks in environments unsuitable for direct human intervention. This application is valuable in tasks such as planetary exploration, disaster response, and ecological monitoring [6].

9.8 Gaming and Entertainment

Embodied AI enhances gaming experiences by creating more immersive and responsive virtual environments. The work of Beattie et al. introduces DeepMind Lab, a platform for the development of intelligent agents in 3D environments. Embodied agents in gaming environments can exhibit adaptive Behavior's, learn from player interactions, and contribute to the creation of realistic and engaging gaming experiences [36].

9.9 Social Interaction and Companion Robots

Embodied AI contributes to the development of social robots designed to interact with humans in social contexts. The study by Steels discusses evolving grounded communication for robots. Socially intelligent robots can engage in conversations, recognize emotions, and provide companionship. These applications extend to areas such as eldercare, where companion robots can enhance the well-being of individuals through social interaction [43].

9.9.1 Neuroscience and Cognitive Science

Embodied AI serves as a valuable tool for researchers in neuroscience and cognitive science. The work of Pfeifer and Scheier emphasizes the understanding of intelligence through embodied systems. Embodied models provide insights into the interplay between the body, the brain, and the environment, contributing to a deeper understanding of human cognition and inspiring new avenues for research [44].

The applications of Embodied AI span a broad spectrum, from robotics and healthcare to education, entertainment, and beyond. By integrating sensorimotor capabilities and environmental awareness, Embodied AI enhances the interaction between artificial agents and the real world, paving the way for innovative solutions in various domains.

10 Future Directions and Trends

Embodied Artificial Intelligence (Embodied AI) has witnessed significant progress, yet its evolution continues to unveil new frontiers and possibilities. As researchers push the boundaries of intelligence grounded in sensorimotor experiences, several future directions and emerging trends are shaping the trajectory of Embodied AI. This exploration delves into the potential avenues that promise to define the future of Embodied AI, with a focus on advancements, challenges, and the integration of novel technologies.

10.1 Integration of Advanced Sensor Technologies

The future of Embodied AI is closely tied to the integration of advanced sensor technologies. LiDAR, radar, and advanced vision systems are anticipated to play a pivotal role in enhancing the perception capabilities of embodied agents. The work of A. Vakil et al. discusses the potential of multimodal sensor fusion to provide a more comprehensive understanding of the environment. This trend aligns with the quest for more sophisticated sensorimotor integration, enabling agents to navigate complex real-world scenarios with increased precision [45].

10.2 Enhanced Motor Capabilities and Dexterous Manipulation

Advancements in robotic systems and prosthetics are likely to result in agents with enhanced motor capabilities. Research by Lee J et al. highlights the potential of soft robotics for achieving dexterous manipulation in embodied agents. Future directions in Embodied AI involve the development of robots and autonomous systems with fine-grained motor control, enabling them to perform intricate tasks and interact seamlessly with objects in their environment [46].

10.3 Continual Learning and Lifelong Adaptation

Continual learning mechanisms are poised to become a core aspect of Embodied AI systems, facilitating lifelong adaptation to changing environments. The study by Parisotto and Salakhutdinov emphasizes the importance of neural map structures for deep reinforcement learning. Future trends involve the exploration of algorithms that enable agents to accumulate knowledge over time, adapting their Behavior's based on a continual stream of experiences. This shift toward lifelong learning aligns with the goal of creating more adaptable and resilient intelligent systems [47].

10.4 Explainability and Interpretability

As Embodied AI systems become more sophisticated, the need for explainability and interpretability grows. Ethical considerations and user trust demand that intelligent agents provide transparent explanations for their decisions and actions. Research by Lipton discusses the importance of interpretability in machine learning models. Future trends in Embodied AI involve the development of models that not only exhibit intelligent Behavior's but also offer clear and understandable rationales for their actions, fostering human-AI collaboration [48].

10.5 Biohybrid Systems and Neuromorphic Computing

The convergence of Embodied AI with biohybrid systems and neuromorphic computing holds exciting possibilities. The study by Will Riherd explores neuromorphic engineering, which mimics the architecture and functioning of the human brain. Future directions include the integration of neuromorphic hardware in Embodied AI systems, enabling more energy-efficient and brain-inspired processing. Biohybrid systems that combine biological and artificial components may open new avenues for achieving cognitive capabilities beyond traditional computing [49].

Table 12 explains Future Directions and Trends in Embodied AI.

10.6 Ethical and Societal Implications

As Embodied AI becomes more integrated into daily life, addressing ethical and societal implications is crucial. Research by Provoost S et al. emphasizes the ethical considerations in developing embodied conversational agents. Future trends involve interdisciplinary collaboration to establish ethical frameworks, guidelines, and regulations for the deployment of Embodied AI in various domains. Addressing issues of bias, fairness, and privacy will be paramount to ensure responsible and equitable use of intelligent agents [50].

10.7 Human-AI Synergy and Collaboration

The future of Embodied AI envisions seamless synergy between humans and intelligent agents. The research by Steels discusses evolving grounded communication for robots. Future trends involve designing systems that not only understand human gestures and language but actively collaborate with humans in shared tasks.

Table 12 Future directions and trends

Trend	Description	Potential impact
Integration of advanced sensor technologies	LiDAR, radar, and advanced vision systems will enhance the perception capabilities of embodied agents	More precise navigation, manipulation, and interaction with the environment
Enhanced motor capabilities and dexterous manipulation	Advancements in robotics and prosthetics will lead to robots and autonomous systems with fine-grained motor control	Ability to perform intricate tasks and interact seamlessly with objects
Continual learning and lifelong adaptation	Embodied AI systems will be equipped with continual learning mechanisms to adapt to changing environments over time	More adaptable and resilient intelligent systems
Explain-ability and interpretability	Embodied AI models will offer clear and understandable rationales for their decisions and actions	Increased transparency and trust in human-AI collaboration
Biohybrid systems and neuromorphic computing	The integration of neuromorphic hardware and biohybrid systems in Embodied AI will enable more energy-efficient and brain-inspired processing	New avenues for achieving cognitive capabilities beyond traditional computing
Ethical and societal implications	Ethical frameworks, guidelines, and regulations will be established to ensure the responsible and equitable use of embodied AI systems	Address issues of bias, fairness, and privacy
Human-AI synergy and collaboration	Embodied AI systems will be designed to actively collaborate with humans in shared tasks	Symbiotic relationship where humans and embodied agents leverage each other's strengths
Edge computing and decentralized intelligence	Decentralized Embodied AI systems that can operate efficiently on edge devices will be developed	Reduced latency, improved scalability, and wider applicability in edge AI domains
Interdisciplinary research collaborations	Collaboration between experts from diverse fields will drive innovation and breakthroughs in Embodied AI	Holistic understanding of embodied intelligence
Open-source development and accessibility	Accessible tools, platforms, and resources will be made available to promote the advancement of Embodied AI	Collaborative ecosystem and accelerated innovation

This collaborative approach extends beyond mere assistance to create a symbiotic relationship where humans and embodied agents leverage each other's strengths [51].

10.8 Edge Computing and Decentralized Intelligence

Advancements in edge computing and decentralized intelligence are expected to influence the future landscape of Embodied AI. The study by Satyanarayanan discusses the paradigm of edge computing for enabling real-time processing on devices. Future directions involve the development of decentralized Embodied AI systems that can operate efficiently on edge devices, reducing latency and promoting scalability. This trend aligns with the growing emphasis on edge AI for applications in robotics, healthcare, and other domains [52].

10.9 Interdisciplinary Research Collaborations

The complexity of Embodied AI necessitates increased interdisciplinary collaboration. The work of Ijspeert et al. explores the simulation and robotics studies of salamander locomotion. Future trends involve fostering collaboration between experts in robotics, neuroscience, computer science, and other fields. Interdisciplinary research initiatives will contribute to a holistic understanding of embodied intelligence, driving innovation and breakthroughs in Embodied AI [6].

10.9.1 Open-Source Development and Accessibility

Promoting open-source development and accessibility will be crucial for the widespread adoption and democratization of Embodied AI. The research by Metta et al. emphasizes the open platform for research in embodied cognition. Future trends involve the creation of accessible tools, platforms, and resources that enable researchers, developers, and enthusiasts to contribute to the advancement of Embodied AI. This openness fosters a collaborative ecosystem and accelerates innovation in the field [5].

The future of Embodied AI is marked by a convergence of technological advancements, ethical considerations, and interdisciplinary collaboration. As researchers navigate these future directions and emerging trends, the potential for Embodied AI to revolutionize industries, improve human–machine interaction, and contribute to a deeper understanding of intelligence remains promising.

11 Ethical Considerations in Embodied AI

Embodied Artificial Intelligence (Embodied AI) brings forth a new frontier in technology, intertwining machines with the physical world through sensorimotor interactions. As this field advances, ethical considerations become paramount in ensuring responsible and humane development and deployment of intelligent agents. This exploration delves into the ethical dimensions of Embodied AI, addressing key concerns, potential risks, and frameworks for responsible innovation.

11.1 Transparency and Explainability

One of the foremost ethical considerations in Embodied AI is the need for transparency and explainability. As agents make decisions based on complex interactions in the physical world, understanding the rationale behind those decisions becomes crucial. Lipton emphasizes the importance of interpretability in machine learning models. Ethical frameworks must prioritize the development of systems that can elucidate their decision-making processes, fostering trust and accountability [48].

11.2 Fairness and Bias Mitigation

Embodied AI systems, like any machine learning models, are susceptible to biases present in training data. Provoost S et al. discuss the ethical considerations in the development of embodied conversational agents. It is imperative to address biases to ensure fairness and equity, especially when these systems interact with humans. Ethical guidelines should promote continuous monitoring, identification, and mitigation of biases throughout the lifecycle of Embodied AI applications [50].

11.3 Privacy and Data Security

Embodied AI often involves the collection and processing of sensory data from the environment. This raises concerns about privacy and data security. As highlighted by Floridi, ethical frameworks for Embodied AI should prioritize robust data protection measures, ensuring that sensitive information is handled responsibly. Consent mechanisms, anonymization, and secure storage practices should be integral components of ethical guidelines [53].

11.4 Human-AI Interaction and Autonomy

The ethical implications of human-AI interaction are multifaceted. Embodied AI systems are designed to collaborate with humans in various contexts, ranging from healthcare to manufacturing. Steels discusses evolving grounded communication for robots, emphasizing the need for ethical considerations in the development of socially intelligent agents. Ensuring that these interactions respect human autonomy, preferences, and cultural norms is a vital aspect of ethical Embodied AI design [43].

11.5 Accountability and Liability

Determining accountability and liability for actions taken by Embodied AI systems poses a significant ethical challenge. As agents interact in real-world environments, the consequences of their actions can be profound. Ethical frameworks, as proposed by Jobin et al., should establish clear guidelines for assigning responsibility when errors or adverse outcomes occur. This includes defining the roles of developers, manufacturers, and users in ensuring responsible AI deployment [54].

Table 13 explains Ethical Considerations in Embodied AI.

11.6 Long-Term Autonomy and Adaptability

The ethical dimension of long-term autonomy in Embodied AI systems involves considerations of continual learning and adaptability. Parisotto and Salakhutdinov emphasize the importance of developing neural map structures for deep reinforcement learning. Ethical guidelines should address how agents accumulate knowledge over time, ensuring that continual learning aligns with ethical principles and societal values [47].

11.7 Societal Impact and Inclusivity

The broader societal impact of Embodied AI necessitates ethical frameworks that account for inclusivity and accessibility. Johnson et al. underscore the importance of considering diverse user populations. Ethical considerations should involve efforts to minimize societal inequalities, ensure access to AI benefits, and prevent the exacerbation of existing disparities through thoughtful design and deployment [50].

Table 13 Ethical considerations in Embodied AI

Ethical consideration	Description	Potential risks	Mitigation strategies
Transparency and explainability	Agents should be able to explain their decision-making processes to foster trust and accountability	Lack of trust, misunderstandings, and biases	Develop interpretable models, provide mechanisms for users to understand and feedback on decisions, and establish ethical guidelines for model development and deployment
Fairness and bias mitigation	Agents should be free from biases to ensure fairness and equity	Biased decisions and discrimination	Monitor, identify, and mitigate biases throughout the lifecycle of Embodied AI applications. Develop ethical guidelines for data collection, model training, and evaluation
Privacy and data security	Robust data protection measures should be in place to protect sensitive information	Privacy breaches, data leaks, and misuse of data	Implement consent mechanisms, anonymize data, and secure storage practices. Develop ethical guidelines for data handling and privacy protection
Human-AI interaction and autonomy	Human-AI interactions should respect human autonomy, preferences, and cultural norms	Loss of autonomy, manipulation, and cultural insensitivity	Develop ethical guidelines for human-AI interaction, including the right to consent, the right to know, and the right to withdraw. Design systems that are aligned with human values and cultural norms
Accountability and liability	Clear guidelines for assigning responsibility when errors or adverse outcomes occur are needed	Lack of accountability and legal liability	Develop ethical guidelines for accountability and liability in Embodied AI systems. Define the roles of developers, manufacturers, and users in ensuring responsible AI deployment
Long-term autonomy and adaptability	Continual learning and adaptability should be aligned with ethical principles and societal values	Unintended consequences, loss of control, and misuse of power	Develop ethical guidelines for the development and deployment of long-term autonomous Embodied AI systems. Ensure that continual learning mechanisms are aligned with human values and societal goals

(continued)

Table 13 (continued)

Ethical consideration	Description	Potential risks	Mitigation strategies
Societal impact and inclusivity	Embodied AI systems should be designed and deployed in a way that minimizes societal inequalities and ensures access to AI benefits for all	Exacerbation of existing inequalities, social unrest, and job losses	Develop ethical guidelines for the societal impact of Embodied AI systems. Consider diverse user populations in the design and deployment of these systems
Environmental sustainability	Environmentally responsible AI technologies should be developed to promote sustainability in the deployment of Embodied AI systems	High energy consumption, environmental pollution, and resource depletion	Develop ethical guidelines for the environmental impact of Embodied AI systems. Encourage the development of energy-efficient AI technologies
Cross-disciplinary collaboration	Interdisciplinary collaboration is needed to address complex ethical challenges in Embodied AI	Siloed thinking, lack of coordination, and conflicting perspectives	Facilitate collaboration among researchers, policymakers, ethicists, and other stakeholders. Develop ethical frameworks that incorporate diverse perspectives and disciplines
Public engagement and education	Public awareness and understanding of Embodied AI technologies are crucial for fostering trust and informed decision-making	Public fear, distrust, and misinformation	Implement public engagement and education programs to raise awareness and understanding of Embodied AI technologies. Develop ethical guidelines for public communication and engagement

11.8 Environmental Sustainability

As Embodied AI systems become more prevalent, their environmental impact becomes a relevant ethical concern. The energy consumption associated with training and operating complex models raises questions about sustainability. Roy discusses neuromorphic computing as a potential solution for energy-efficient processing. Ethical guidelines should encourage the development of environmentally responsible AI technologies, promoting sustainability in the deployment of Embodied AI systems [55].

11.9 Cross-Disciplinary Collaboration

The ethical challenges in Embodied AI require cross-disciplinary collaboration among researchers, policymakers, ethicists, and other stakeholders. Ijspeert et al. emphasize the importance of interdisciplinary research in the development of robotic systems. Ethical frameworks should facilitate collaboration to address complex challenges, ensuring that diverse perspectives contribute to the formulation of ethical guidelines that stand up to scrutiny from various disciplines [6].

Table 14 explains in detailed Safety and Ethics in Autonomous Systems.

11.9.1 Public Engagement and Education

Ethical considerations extend beyond the development community to the general public. Public awareness and understanding of Embodied AI technologies are crucial for fostering trust and informed decision-making. Ethical frameworks should include provisions for public engagement, education, and open dialogue about the ethical implications of Embodied AI, as advocated by Floridi and Taddeo [56].

The ethical considerations in Embodied AI are integral to shaping the responsible and sustainable development of intelligent agents. Addressing transparency, fairness, privacy, accountability, and societal impact requires a collaborative effort involving researchers, developers, policymakers, and the public. Ethical frameworks should evolve dynamically, staying abreast of technological advancements and societal values to ensure that Embodied AI aligns with humanistic principles.

12 Conclusion

Embodied Artificial Intelligence (Embodied AI) signifies a profound departure from conventional AI frameworks, introducing a transformative paradigm that seamlessly integrates sensory perception, motor capabilities, and contextual awareness. The focal point on sensory-motor integration underscores the dynamic interplay between agents and their surroundings, fostering adaptive decision-making and a nuanced comprehension of context. The wide-ranging applicability of Embodied AI, spanning from robotics to virtual agents and linguistic contexts, underscores its versatility across diverse domains. The foundational principle of "situatedness" underscores the intrinsic connection between intelligence and specific environmental contexts, nurturing the evolution of systems that are not merely adaptive but also inherently context-aware. Various methodologies within Embodied AI, encompassing advancements in computer vision, haptic feedback, attention mechanisms, and bio-inspired models, contribute to crafting sophisticated perception systems. These systems empower agents to actively engage, adapt, and navigate their surroundings,

Table 14 Safety and ethics in autonomous systems- challenges and considerations

Area	Challenge	Potential consequences	Mitigation strategies
Fail-safe designs	Addressing edge cases and unforeseen circumstances	Uncontrolled malfunctions, unintended consequences	– Develop robust fault detection and response mechanisms – Define rigorous safety thresholds – Prioritize risk assessment and mitigation
Validation testing	Covering all possible real-world scenarios	Insufficiently tested systems, increased risk of accidents	– Implement diverse testing environments and protocols – Prioritize real-world testing – Maintain a balance between validation and timely deployment
Security vulnerabilities	Cyber-attacks, unauthorized access, data breaches	Compromised systems, privacy violations, safety hazards	– Implement strong cybersecurity measures and encryption standards – Regularly update software and patch vulnerabilities – Strike a balance between connectivity and security
Ethical data collection	Balancing data needs with ethical considerations	Unfair data practices, misuse of personal information, privacy violations	– Establish guidelines for responsible data collection: transparency, purpose limitation, data minimization – Develop data anonymization techniques – Prioritize ethical data collection frameworks
Privacy and consent	Protecting individual privacy in public spaces	Intrusion on privacy rights, erosion of trust	– Implement privacy-preserving technologies – Anonymize collected data – Obtain clear and informed user consent – Balance data-driven benefits with respect for privacy

effectively closing the divide between computational models and real-world interaction. The evolution of decision-making in Embodied AI, incorporating reinforcement learning, evolutionary algorithms, and active perception, signifies a dynamic departure toward systems that actively navigate and adapt within their environments. This departure from traditional computational models elevates the agility and adaptability of intelligent agents. Embodied AI marks a groundbreaking paradigm shift in the AI landscape, pledging the development of intelligent systems not only proficient in processing information but actively immersed in engagement, adaptation, and learning from their embodied experiences. As research progresses in this field, the potential for creating increasingly context-aware, adaptive, and interactive AI systems expands, ushering in novel possibilities for the seamless integration of artificial intelligence into both our physical and digital realms.

Acknowledgements Authors would like to thank Prof. (Dr.) Sunil Sable, Department of Chemical Engineering, Vishwakarma Institute of Technology, Pune.

References

1. Brooks, R.A., Breazeal, C., Marjanović, M., Scassellati, B., Williamson, M.M.: The cog project: building a humanoid robot. In: Nehaniv, C.L. (eds.) Computation for Metaphors, Analogy, and Agents. CMAA 1998, vol. 1562. Lecture Notes in Computer Science. Springer, Berlin (1999). https://doi.org/10.1007/3-540-48834-0_5
2. Pfeifer, R., Bongard, J.: How the Body Shapes the Way We Think: A New View of Intelligence. MIT Press (2006)
3. Steels, L.: Evolving grounded communication for robots. Trends Cogn. Sci. **7**(7), 308–312. https://www.cell.com/trends/cognitive-sciences/fulltext/S1364-6613(03)00129-3 (2003)
4. Clark, A.: Being There: Putting Brain, Body, and World Together Again. MIT Press (1996)
5. Metta, G., Natale, L., Nori, F., Sandini, G., Vernon, D., Fadiga, L., von Hofsten, C., Rosander, K., Lopes, M., Santos-Victor, J., Bernardino, A., Montesano, L.: The iCub humanoid robot: an open-systems platform for research in cognitive development. Neural Netw. **23**(8–9), 1125–1134 (2010). https://doi.org/10.1016/j.neunet.2010.08.010. Epub 2010 Sep 22 PMID: 20864311
6. Ijspeert, A.J., Crespi, A., Cabelguen, J.M.: Simulation and robotics studies of salamander locomotion: applying neurobiological principles to the control of locomotion in robots. Neuroinformatics **3**(3), 171–195 (2005). https://doi.org/10.1385/NI:3:3:171. PMID: 16077158
7. Cangelosi, A., Schlesinger, M.: Developmental Robotics: From Babies to Robots. MIT Press. https://direct.mit.edu/books/book/3435/Developmental-RoboticsFrom-Babies-to-Robots (2015)
8. Beer, R.D.: The dynamics of active categorical perception in an evolved model agent. Adapt. Behav. **11**(4), 209–243 (2003). https://doi.org/10.1177/1059712303114001
9. Haikonen, P.O.: The Cognitive Approach to Consciousness. Lawrence Erlbaum Associates. https://psycnet.apa.org/record/2003-02663-000 (2003)
10. Pfeifer, R., Scheier, C.: Understanding Intelligence. MIT Press. https://direct.mit.edu/books/book/2319/Understanding-Intelligence (1999a)
11. Hoffman, G.: Embodied cognition for autonomous interactive robots. Top. Cogn. Sci. **4**(4), 759–772 (2012). https://doi.org/10.1111/j.1756-8765.2012.01218.x. Epub 2012 Aug 14 PMID: 22893571

12. Oudeyer, P.Y., Kaplan, F.: What is Intrinsic Motivation? A Typology of Computational Approaches. Front. Neurorobot. **1**, 6. https://doi.org/10.3389/neuro.12.006.2007 (2007). (PMID: 18958277; PMCID: PMC2533589)
13. Floreano, D., Keller, L.: Evolution of adaptive Behavior in robots by means of Darwinian selection. PLoS Biol. **8**(1), e1000292 (2010). https://doi.org/10.1371/journal.pbio.1000292
14. Varela, F.J., Thompson, E., Rosch, E.: The Embodied Mind: Cognitive Science and Human Experience. MIT Press. https://psycnet.apa.org/record/1991-98659-000 (1991)
15. Mnih, V., et al.: Human-level control through deep reinforcement learning. Nature **518**, 529–533 (2015)
16. Pfeifer, R., Scheier, C.: Understanding Intelligence. MIT Press. https://direct.mit.edu/books/book/2319/Understanding-Intelligence (1999b)
17. Sutton, R.S., Barto, A.G.: Reinforcement Learning: An Introduction. MIT Press. https://psycnet.apa.org/record/2019-19679-000 (2018)
18. Nolfi, S., Floreano, D.: Evolutionary Robotics: The Biology, Intelligence, and Technology of Self-Organizing Machines. MIT Press (2000)
19. Cangelosi, A., Parisi, D.: The processing of verbs and nouns in neural networks: Insights from synthetic brain imaging. Brain Lang **89**(2), 401–408 (2004). ISSN 0093-934X, https://doi.org/10.1016/S0093-934X(03)00353-5
20. Morris, R.G., Hebb, D.O.: The organization of behavior, Wiley: New York; 1949. Brain Res. Bull. **50**(5–6), 437 (1999). https://doi.org/10.1016/s0361-9230(99)00182-3. PMID: 10643472
21. Pfeifer, R., Scheier, C.: Understanding Intelligence. MIT Press. https://direct.mit.edu/books/book/2319/Understanding-Intelligence (1999c)
22. LeCun, Y., Bengio, Y., Hinton, G.: Deep learning. Nature **521**, 436–444 (2015). https://doi.org/10.1038/nature14539
23. Blum, T., Moore, R.: Auditory Scene Analysis: The Perceptual Organization of Sound. MIT Press. https://direct.mit.edu/books/book/3887/Auditory-Scene-AnalysisThe-Perceptual-Organization (2019)
24. Ballard, D.H.: Animate vision. Artif. Intell. **48**(1), 57–86 (1991). ISSN 0004-3702. https://doi.org/10.1016/0004-3702(91)90080-4
25. Kuchenbecker, K.J., Fiene, J., Niemeyer, G.: Improving contact realism through event-based haptic feedback. IEEE Trans. Vis. Comput. Graph. **12**(2), 219–230 (2006). https://doi.org/10.1109/TVCG.2006.32
26. Tsotsos, J.K., Culhane, S.M., Wai, W.Y.K., Lai, Y., Davis, N., Nuflo, F.: Modeling visual attention via selective tuning. Artif. Intell. 78(1–2), 507–545 (1995). ISSN 0004-3702. https://doi.org/10.1016/0004-3702(95)00025-9
27. Noesselt, T., Shah, N.J., Jäncke, L.: Top-down and bottom-up modulation of language related areas–an fMRI study. BMC Neurosci. **4**, 13 (2003). https://doi.org/10.1186/1471-2202-4-13. PMID:12828789;PMCID:PMC166149
28. Kaelbling, L.P., Littman, M.L., Moore, A.W.: Reinforcement learning: a survey. J. Artif. Intell. Res. **4**, 237–285 (1996)
29. Nolfi, S., Floreano, D.: Evolutionary Robotics: The Biology, Intelligence, and Technology of Self-Organizing Machines. MIT Press. https://direct.mit.edu/books/book/4241/Evolutionary-RoboticsThe-Biology-Intelligence-and (2000)
30. Friston, K.: A theory of cortical responses. Phil. Trans. R. Soc. B360815–836 (2005). https://doi.org/10.1098/rstb.2005.1622
31. Tsotsos, J.K., Culhane, S.M., Wai, W.Y.K., Lai, Y., Davis, N., Nuflo, F.: Modelling visual attention via selective tuning. Artif. Intell. 78(1–2), 507–545 (1995). ISSN 0004-3702. https://doi.org/10.1016/0004-3702(95)00025-9
32. Gibson, J.J.: The Ecological Approach to Visual Perception. Houghton Mifflin (2014)
33. Parisotto, E., Salakhutdinov, R.: (2017). Neural Map: Structured Memory for Deep Reinforcement Learning. https://doi.org/10.48550/arXiv.1702.08360 (2017)
34. Floreano, D., Dürr, P., Mattiussi, C.: Neuroevolution: from architectures to learning. Evol. Intel. **1**, 47–62 (2008). https://doi.org/10.1007/s12065-007-0002-4

35. Luxton, D.D.: Ethical implications of conversational agents in global public health. Bull. World Health Organ. **98**(4), 285–287 (2020). https://doi.org/10.2471/BLT.19.237636. Epub 2020 Jan 27. PMID: 32284654; PMCID: PMC7133471
36. Beattie, C., Leibo, J.Z., Teplyashin, D., Ward, T., Wainwright, M., Küttler, H., Lefrancq, A., Green, S., Valdés, V., Sadik, A., Schrittwieser, J., Anderson, K., York, S., Cant, M., Cain, A., Bolton, A., Gaffney, S., King, H., Hassabis, D., Legg, S., Petersen, S.:. DeepMind Lab. https://doi.org/10.48550/arXiv.1612.03801 (2016)
37. Amodei, D., Olah, C., Steinhardt, J., Christiano, P., Schulman, J., Mané, D.: Concrete Problems in AI safety. arXiv:1606.06565 (2016)
38. Steels, L.: Evolving grounded communication for robots. Trends Cogn. Sci. 7(7), 308–312 (2003). ISSN 1364-6613. https://doi.org/10.1016/S1364-6613(03)00129-3
39. Molteni, F., Gasperini, G., Gaffuri, M., Colombo, M., Giovanzana, C., Lorenzon, C., Farina, N., Cannaviello, G., Scarano, S., Proserpio, D., Liberali, D., Guanziroli E.: Wearable robotic exoskeleton for overground gait training in sub-acute and chronic hemiparetic stroke patients: preliminary results. Eur. J. Phys. Rehabil. Med. **53**(5), 676–684 (2017). https://doi.org/10.23736/S1973-9087.17.04591-9. Epub 2017 Jan 24. PMID: 28118698
40. Liu, Y., Song, S., Yang, L., Bian, G., Yu, H.: A novel dynamic gesture understanding algorithm fusing convolutional neural networks with hand-crafted features. J. Vis. Commun. Image Represent. **83**, 103454 (2022). ISSN 1047-3203. https://doi.org/10.1016/j.jvcir.2022.103454
41. Kramer, L.L., Ter Stal, S., Mulder, B.C., de Vet, E., van Velsen, L.: Developing embodied conversational agents for coaching people in a healthy lifestyle: scoping review. J. Med. Internet Res. **22**(2), e14058 (2020). https://doi.org/10.2196/14058.PMID:32022693;PMCID:PMC7055763
42. Tenorth, M., Beetz, M.: KnowRob: a knowledge processing infrastructure for cognition-enabled robots. Int. J. Robot. Res. **32**(5), 566–590 (2013). https://doi.org/10.1177/0278364913481635
43. Steels, L.: Evolving grounded communication for robots. Trends Cogn. Sci. **7**(7), 308–312 (2003). https://doi.org/10.1016/s1364-6613(03)00129-3. PMID: 12860189
44. Pfeifer, R., Scheier, C.: Understanding Intelligence. MIT Press. https://direct.mit.edu/books/book/2319/Understanding-Intelligence (2001)
45. Vakil, A., Liu, J., Zulch, P., Blasch, E., Ewing, R., Li, J.: A survey of multimodal sensor fusion for passive RF and EO information integration. IEEE Aerosp. Electron. Syst. Mag. **36**(7), 44–61 (2021). https://doi.org/10.1109/MAES.2020.3006410
46. Lee, J., Kim, S., Lee, J., Yang, D., Park, B.C., Ryu, S., Park, I.: A stretchable strain sensor based on a metal nanoparticle thin film for human motion detection. Nanoscale **6**(20), 11932–11939 (2014). https://doi.org/10.1039/c4nr03295k. Epub 2014 Sep 1 PMID: 25175360
47. Parisotto, E., Salakhutdinov, R.: Neural Map: Structured Memory for Deep Reinforcement Learning. arXiv:1702.08360 (2017)
48. Lipton, Z.C.: The Mythos of Model Interpretability. arXiv:1606.03490 (2016)
49. Riherd, W.: Neuromorphic computing. Elements (2022)
50. Provoost, S., Lau, H.M., Ruwaard, J., Riper, H.: Embodied conversational agents in clinical psychology: a scoping review. J. Med. Internet Res. **19**(5), e151 (2017). https://doi.org/10.2196/jmir.6553.PMID:28487267;PMCID:PMC5442350
51. Steels, L.: Evolving grounded communication for robots. Trends Cogn. Sci. **7**(7), 308–312 (2003)
52. Satyanarayanan, M.: The emergence of edge computing. Computer **50**(1), 30–39 (2017). https://doi.org/10.1109/MC.2017.9
53. Floridi, L.: Soft ethics, the governance of the digital and the general data protection regulation. Philos. Trans. R. Soc. A: Math. Phys. Eng. Sci **376**(2133), 20180081 (2018)
54. Jobin, A., Ienca, M., Vayena, E.: The global landscape of AI ethics guidelines. Nat. Mach. Intell. **1**, 389–399 (2019). https://doi.org/10.1038/s42256-019-0088-2
55. Roy, K., Jaiswal, A., Panda, P.: Towards spike-based machine intelligence with neuromorphic computing. Nature **575**, 607–617 (2019). https://doi.org/10.1038/s41586-019-1677-2
56. Floridi, L., Taddeo, M.: What is data ethics? Philos. Trans. R. Soc. A: Math. Phys. Eng. Sci. **374**(2083), 20160360 (2016)

Dr. Sunil Sable is working as an Assistant Professor in Chemical Engineering Department, at Vishwakarma Institute of Technology, Pune, Maharashtra, India. Prior to this, he has worked as a "Research Associate" on CSIR Government of India sponsored research project at Department of Chemical Engineering, Thapar University, Patiala, Punjab, India. His research areas are polymer blends and composites, biodegradation and eco-toxicity studies of polymeric materials, ultrasound assisted processes, solid-liquid extraction, solid-waste management, waste water treatment, Human Computer interaction, and Internet of Things (IoT). Dr. Sunil has around 11 years of teaching and 3 years of research experience. Till date, he has published 14 research papers in Peer-reviewed SCI Journals and 8 research papers in international conferences in India and abroad. Also, he has published/file 2 Indian Patents and 4 Book Chapters. He has attended various National/International Conferences/Faculty Development Programs/Short-Term Training Program/Webinars. Also, he has conducted various Faculty Development Programs (FDP) and Short-Term Training Programs (STTP). He received his doctoral degree in Chemical Engineering from Thapar Institute of Engineering and Technology, Patiala, Punjab, India. Also, he earned his graduation and postgraduation degree from Savitribai Phule Pune University, Pune, Maharashtra, India.

Mitesh Ikar is expertise in Human Computer Interaction domain with work focusing in Computer Vision (OpenCV), Speech Recognition, Email Automation with all 3 works presented at international conferences and published in IEEE Xplore with indexing in Scopus. Also, he has published/ file 1 Indian Patent. He has won 3rd prize (Rs 50,000) in UST Xperiment 6.0 competition by competing with 350+ students and employees' team. His works are related to language models like CTRL, DialoGPT, Embodied Artificial Intelligence, etc. presented in International Conference which have been publish in Springer Nature.

Some recent works from IEEE Xplore includes:

(1) Hand Gesture-based Virtual Mouse using Open CV:Worked on creating a hand gesture-based virtual mouse using OpenCV. This project aimed to enhance human-computer interaction (HCI) by allowing users to control the mouse cursor through hand gestures. [https://doi.org/10.1109/IDCIoT56793.2023.10053488]

(2) Streamlining Email Workflow with Voice Recognition Technology and Website-Email Autometa Solutions:A project that aimed to empower users with voice recognition technology, making email workflows more efficient through website, 'Email Autometa Solutions'. The focus was likely on improving productivity and communication. [https://doi.org/10.1109/ICCCNT56998.2023.10307334]

(3) Advancing Multimodal Fusion in Human-Computer Interaction:Explored integrating various modalities (such as eye tracking, lips detection, speech recognition, and voice synthesis) for intelligent cursor control and auditory feedback in HCI. [https://doi.org/10.1109/ICCCNT56998.2023.10306457]

Navigating the Nexus: Unravelling Challenges, Ethics, and Applications of Embodied AI in Drone Technology Through the Lens of Computer Vision

G. Arun Sampaul Thomas, S. Muthukaruppasamy, S. Sathish Kumar, Beulah J. Karthikeyan, and K. Saravanan

Abstract Significant progress has been made in the subject of embodied artificial intelligence (AI) over the past several years, with a special emphasis placed on the integration of embodied AI with computer vision technologies. It explores the complex landscape of embodied artificial intelligence, concentrating on the difficulties and ethical concerns that occur during its creation, the categorization of the industry in which it operates, and the various application scenarios that it can be used for. It focuses on the integration of embodied AI with computer vision and artificial intelligence in drone technology. Embodied artificial intelligence is an area of artificial intelligence that is distinguished by the presence of systems that not only process information but also interact with their surroundings in a manner that mimics human cognition and behaviour. As the level of sophistication of these systems continues to rise, several issues are emerging, one of which is the ethical implications that are involved with their deployment. This article sheds light on the significance of developing appropriate norms to regulate the development and implementation of embodied AI by addressing the ethical aspects connected to privacy, security, and the potential misuse of embodied AI. This article looks at the industrial classification of embodied artificial intelligence to provide insights into the various industries that are utilising this technology. Embodied artificial intelligence has a wide variety of

G. Arun Sampaul Thomas (✉) · S. Sathish Kumar · B. J. Karthikeyan
Department of Artificial Intelligence and Machine Learning, J.B. Institute of Engineering and Technology, Hyderabad, Telangana, India
e-mail: arunsam.infotech@gmail.com

S. Sathish Kumar
e-mail: sathish.ai_ml@jbiet.edu.in

B. J. Karthikeyan
e-mail: beulah.ai_ml@jbiet.edu.in

S. Muthukaruppasamy
Department of EEE, Velammal Institute of Technology, Tamil Nadu, India

K. Saravanan
Department of CSE, College of Engineering, Anna University, Guindy, Chennai, India

P. Raj et al. (eds.), *Building Embodied AI Systems: The Agents, the Architecture Principles, Challenges, and Application Domains*, Information Systems Engineering and Management 14, https://doi.org/10.1007/978-3-031-68256-8_3

applications, ranging from manufacturing and healthcare to education and entertainment. These solicitations demonstrate the potential of embodied AI to revolutionise a variety of different companies and industries. In the context of drone technology, a particular emphasis is placed on the integration of embodied artificial intelligence with computer vision. The goal of this investigation is to investigate the ways in which these technologies might work together to produce intelligent and autonomous agents that can execute complicated tasks in settings that are constantly changing. In addition to this, it investigates applications of computer vision and embodied artificial intelligence in the field of drone technology. Examples include monitoring the environment, responding to natural disasters, and conducting surveillance and reconnaissance services. Furthermore, it explores the revolutionary influence of these applications, with an emphasis on the potential for increased efficiency, safety, and accuracy in a variety of different fields. The article also explores the role that embodied artificial intelligence plays in addressing issues that are unique to drone technology. These challenges include navigation in unpredictable situations, obstacle recognition and avoidance, and decision-making in real time. The incorporation of embodied artificial intelligence (AI) and computer vision into drones presents a possible option for solving these problems, paving the way for drone systems that are more technically advanced and adaptable. In its primary function, it offers a detailed review of the difficulties and ethical problems that are inherent in the process of developing and deploying embodied artificial intelligence. In it, the categorization of embodied AI within the industry is investigated, with a focus on the many uses of this technology. Additionally, the integration of embodied AI and computer vision in drone technology is examined in greater detail. This article helps to provide a more in-depth understanding of the growing landscape of embodied artificial intelligence and the potential influence that it could have on many sectors and technologies by covering these elements.

Keywords Embodied artificial intelligence · Computer vision ethical considerations in AI · Industry categorization · Human-like cognition and behaviour · Privacy and security · Drone technology

1 Introduction

In the beginning of robot technology, machines were unbendingly modified to complete explicit undertakings, without any type of knowledge or flexibility. Despite, as Artificial Intelligence based Machine and deep Learning innovations advanced, the mix of these strategies with robots achieved huge progress. Artificial intelligence empowers robots to see and decipher their environmental factors through different sensors, including cameras, LIDAR, and ultrasonic sensors, among others. These sensors give constant information, which Artificial Intelligence calculations can investigate to settle on informed choices.

ML, then again, engages robots with the capacity to gain from information and encounters. Through ML calculations like managed learning, solo learning, and support learning, robots can secure information, work on their presentation, and adjust to evolving conditions. As robots accumulate more information and experience, they can refine their activities, prompting improved independence and effectiveness [1].

The coordination of Artificial Intelligence and Machine Learning methods with robots has opened a great many applications across different businesses. How about we investigate probably the most prominent use cases:

(a) **Assembling and Industry**: Autonomous robots play a crucial role in the assembly line for tasks such as welding, collecting, painting, and quality monitoring. These robots can work quickly, precisely, and consistently, which boosts production efficiency and cuts down on mistakes.
(b) **Healthcare**: Robots have found their way into healthcare settings, assisting in surgeries, carrying out repetitive tasks, and aiding in patient care. With Artificial Intelligence and Machine Learning, these robots can break down clinical information, make findings, and even foresee well-being patterns, adding to more viable and customized medical services [2, 3].
(c) **Agriculture**: Independent robots are reforming agribusiness by performing undertakings like planting, reaping, and checking crops. They can navigate through fields, detect pests, and optimize resource utilization, thereby enhancing crop yields and sustainable farming practices.
(d) **Logistics and Warehousing**: AI-powered robots are transforming the logistics industry by automating tasks such as sorting, packing, and warehouse management. These robots can work collaboratively with human workers, streamlining operations and reducing labour costs.
(e) **Autonomous Vehicles**: Due to the incorporation of machine learning algorithms and artificial intelligence, these vehicles possess the capability to autonomously navigate, detect their environment, and execute decisions in real-time. This establishes a foundation for a forthcoming era characterised by improved transportation efficacy and a decline in auto accidents.

2 Challenges and Ethical Considerations

Robotics, once bound to the domains of sci-fi, has arisen as a ground-breaking innovation that is reforming different businesses and reshaping our regular day-to-day existences. Robots are at this point not simple mechanical creatures customized to follow fixed ways; they have developed into modern machines with the capacity to think, learn, and adjust. The mix of Artificial Intelligence and Machine Learning strategies with actual robots has prepared them for their independence in performing assignments. This article investigates the interesting universe of robot technology, digging into the progressions that have prompted robots to become wise substances fit for independent errand execution. Humans and robots are beginning to work together more intimately as artificial intelligence (AI) moves from the back offices of information technology (IT) to the public. For both organisations and societies, artificial intelligence will bring about enormously transformative benefits, and it will completely reimagine the way in which we perform our jobs. Every industry is experiencing a proliferation of competitive shifts with the introduction of new technologies such as robotics, simulations for decision-making, facial recognition, language translation, and drones [4–6].

2.1 Motivation for Embodied Agents

Deep learning and machine learning have been responsible for many of the advancements in artificial intelligence that have occurred over the past ten years (for example, semantic segmentation, object detection, and image captioning). Both machine learning and deep learning have proven successful because of the growing amounts of data (for example, YouTube, Flickr, and Facebook) and the growing amounts of computer power (for example, central processing units, graphics processing units, and teraflops). On the other hand, this kind of "Internet data" (which includes photographs, videos, and text that have been curated from the internet) does not originate from a first-person human perspective in the real world. As a result of the fact that data is scrambled, randomised, arriving from satellites, coming from selfie shots, and coming from Twitter feeds, none of these things are how a human understanding of the world works. Machine learning techniques, on the other hand, are attempting to feed this data into natural language processing, CV, and navigation difficulties. Although "Internet data" and "Internet AI" have been responsible for a significant amount of advancement in these areas, the data and approaches that have been utilised are not the most appropriate. The approaches that are used in machine learning do not correspond to the ways in which people learn. By observing, moving, interacting, and communicating with other people, humans acquire knowledge. The only way for humans to learn is through sequential experiences, not through events that are mixed and random. Embodied artificial intelligence is based on the idea that embodied agents, sometimes known as virtual robots, should be able to learn in the same way that people do. Therefore, it is vital to have the insights of professionals in the fields of cognitive science and psychology. Considering this, it is necessary for virtual robots to acquire knowledge by observing, moving, communicating, and interacting with the environment, just like people do [7].

Although "embodied AI" employs an approach that is distinct from that of "Internet AI," it is nevertheless able to reap the benefits of many of the achievements that Internet AI has achieved. Currently, computer vision and natural language processing are effective for certain tasks. The advancements that have been made in CV and NLP have significantly increased the possibility of success in embodied AI. In addition, there are now a great number of realistic 3D settings that can be used as simulated environments for the purpose of training embodied and artificial intelligence. SUNCG, Matterport3D, iGibson, Replica, Habitat, and DART are some of the settings that fall under this category. The settings that have been utilised in earlier research simulators are not quite as realistic as these scenarios, which are significantly more realistic. These datasets are already widely available to the public, which significantly boosts the likelihood of success in the field of embodied artificial intelligence.

3 Embodied AI Industry Categorisation

Personal companions and children: firms such as Jibo and Mabu, both developed by Catalia Labs, offer companionship and care to children and hospital patients. A significant amount of money and innovation are converging in this sector because of the requirements for Natural Language Processing (NLP), the requirements for empathy sensing, and the domain expertise that is required for human contact. Growing artificial general intelligence in these service contexts is an important strategy to control the catastrophic automated general intelligence takeover. This is since a robot never completely know if their purpose of, for example, making the patient feel peaceful, is 100% satisfied permanently [8].

Robots having significant physical capabilities are referred to as operations or movers. The supply chain or operations are the primary focus whether it be in robots or in the software and tools that enable them. There is a large concentration of the well-known Roomba, which is a consumer robot that is used for cleaning homes. There are several "old school" labs that are working in this area, such as Charles River Analytics and Boston Dynamics, which provide their associated datasets, institutional knowledge, and theories that are already in existence. Environmental Sensing Systems: These businesses are primarily concerned with engaging with the natural environment. They also entail sensing and feedback between intelligent systems and the real world. To environmental sensing and adaptation, Honda is now focusing on developing technologies in this area. There is also Grid Bots, which is a company that creates robots for use in industrial and military settings underwater. A linked and intelligent system for agriculture was created by Mavrx using high-resolution crop photography. This makes me wonder whether the aims of sensing, balancing, complicated feedback, and flexible responsiveness to extraneous events are "fertile ground" for artificial intelligence theory [9].

AI firms that are focused on the legal system or applications for document management and analysis are referred to as legal AI. At first glance, it appears that this does not really push the notions of artificial intelligence and is more about the practical application of machine learning (ML). It is quite clear that there is a great deal of money at this location, and in environments where there is both money and data, artificial intelligence will flourish [10].

This cluster contains the game software pioneer Velan Studios, which offers software for the integration of virtual reality and robotics. As part of this cluster, virtual reality is also included. In addition to that, it encompasses several companies that specialise in augmented reality, virtual reality, and mixed reality and use powerful machine learning to offer their solutions, most frequently for gaming. Empathetic avatars are currently being investigated by gaming laboratories such as Sony PlayStation and others. Will any of these businesses ever be able to collect sufficient data to develop artificial general intelligence? Even if it isn't true, the unusual combination of sensing, sympathetic systems, and feedback is fertile ground for the notion of artificial intelligence [11].

(a) **Artificial Intelligence Software Developers**: This group of businesses is at the heart of the field of artificial intelligence and robotics. They create solutions for artificial intelligence, speech, and industrial processes. There are several companies, such as Pony AI, Curious AI, Osaro Inc., and Twenty Billion Neurons GmbH, which happened to be my personal favourite name. As soon as I get the opportunity, I would like to go more into the promises and insights that are specific to each performer in this category.
(b) **Process Automation and Consulting**: This group of enterprises focuses on robotics and either develops products in-house or consults on the creation of goods for industrial, financial, and manufacturing process automation employing artificial intelligence, but most of the time low-intelligence robots. It is possible that AI theorists will need to look to this group to gain a better understanding of the practical feedback mechanisms for AI development as more complex robots are built [12].

There are a growing number of businesses, investors, and funds that have recently been established with the purpose of specialising in robots and artificial intelligence. In this instance, Robik is investigating the possibility of using intelligent robots for last-mile delivery, and Toyota is working on developing and investing in autonomous vehicles. It is possible that the theories that these corporations use to deal with ambiguity could be useful input for artificial general intelligence theory.

This cluster is all about interactive discussion between semi-intelligent artificial intelligence and robot-embodied artificial intelligence, as indicated by the name of the cluster, which is Natural Language and Chat Technology. Theorised artificial intelligence in this space might be the most data-enabled for advanced sentience can be studied a lot about singularity (for example, the easily consumable content of Avogadro Corp). On the other hand, it is the use of this advanced technology with embodiment that has the potential to inform a new set of theories for the development of artificial intelligence.

There is a cluster of aerial drone companies that are used for security, rescue, and surveillance. These companies are relatively modest, but they are expanding. Innovation in this sector is primarily concentrated in San Francisco. Neural Robotics, Iris Automation, and Aeroxo are some of the companies that are involved. About the things that we should presume robots are able to perceive and respond to, the computer vision that is required for this subject would be an interesting addition [13].

This is a fast heat map of each cluster, which displays the timing, funding, and number of business sums, as well as some summary facts about the network that are displayed in the Fig. 1.

Clusters	Num. Companies	Founding Year Median	Inv. Rcvd. Count Total	Inv. Rcvd. Amt.		Inv. CAGR (2014 - 2017)
				Total	Median	
personal companion / child	33	2014	49	$195.6M	$5.5M	51.1%
operator / movers	32	2009	50	$493.8M	$2.7M	12.3%
environmental sensor systems	31	2009	24	$127.4M	$4.0M	-20.7%
document / analysis	30	2013	42	$127.2M	$2.1M	136.4%
virtual reality / work	30	2014	31	$3.1B	$2.5M	-1.4%
AI software developers	22	2015	23	$193.1M	$2.5M	12%
process automation / consulting	20	2010	13	$93.1M	$4.6M	55.9%
Chinese / manufacturing	19	2011	17	$322.3M	$9.5M	N/A
autonomous vehicles	19	2013	14	$186.0M	$9.1M	-41.4%
NLP / Chat	19	2012	22	$119.7M	$13.8M	51.5%

Fig. 1 Embodied agent industry stream cluster versus timelines statistics

4 Application and Use Cases of Embodied AI

Embodied AI can power personal robotics, virtual assistants, and autonomous cars. With NLP, computer vision, and robotics, embodied AI can solve problems more easily. Real-world Embodied AI is our goal. Our ideal physical agent can interact with humans in spoken language and conduct real-world activities. Cases include search and rescue. A firefighter asks, "Is there smoke in any room?" First, the robot must comprehend the query. To grasp "room," language acquisition and understanding must be based on the environment. If the agent is puzzled or wants help, it must ask the firefighter. Agents ask, "What subset of rooms are you interested in?" Firefighters could say, "The rooms in the burning house." Or this agent has used enough 'common sense' to recognise that the burning building's rooms need to be inspected for smoke. Once the agent knows its goal, it must navigate each chamber. After arriving in each room, it must use its egocentric perception to determine if there is smoke. It returns to the firefighter and says, "Yes, in one room."

AI solves many real-world issues. Technology allows the AI market, but the implementation of use cases will define its business. The most important AI application case for manufacturing, healthcare, retail, financial services, and others is predictive analytics. In predictive analytics, data, machine learning, statistical methods, and historical data are used to forecast outcomes. A delivery truck may break down at any time. The Fig. 2 infographic shows the top three AI use cases by industry. This firefighting scenario demonstrates various duties that fall under the umbrella of embodied AI [14].

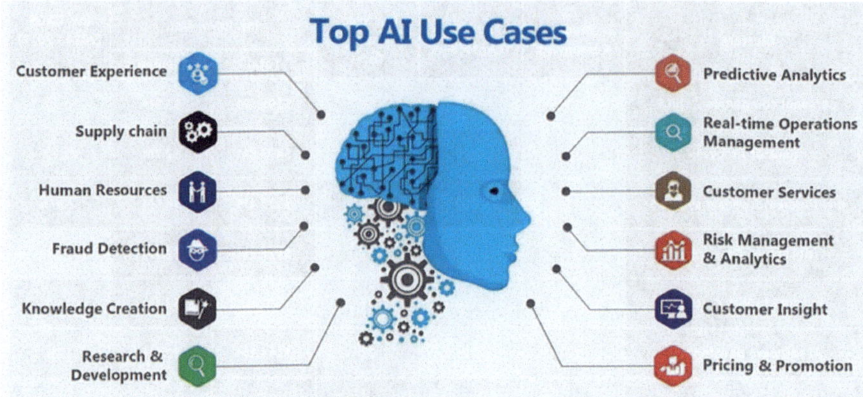

Fig. 2 Top AI use cases in current scenario

4.1 Virtual World Simulator

Placing embodied beings directly in the physical environment is a straightforward approach to training them. This is helpful, but real-world robot training is tedious, risky (robots can break), time-consuming, and resource-intensive (robots and environments both demand resources and time). It's also hard to replicate, especially for unusual edge cases. As an alternative, embodied agents could be taught to perform tasks in lifelike simulations before being deployed in the actual world. Training on a simulator is risk-free and inexpensive, and it can run at speeds orders of magnitude faster than real-time when run in parallel over a cluster. After an approach has been refined and validated in a virtual environment, it can be brought to life on actual operational systems.

4.2 Information Resources and Models

Internet AI has made great strides thanks to datasets. When it comes to embodied AI, simulators will take over from databases. Environments are scanned in three dimensions to create datasets. These files depict various 3D scenarios, such as those found in a home, laboratory, room, or even the outdoors. Unfortunately, agents cannot "walk" or otherwise interact with these 3D scans. The embodied agent can physically traverse the environment and engage in interactions with it using simulators. The embodied agents inhabit and engage with the datasets that are brought into a simulator. Because of the simulator, the agent can perceive its surroundings, navigate them, and engage with them. With the help of the simulator, the agent communicates with both other agents and humans.

Simulators bring environments to life. Nevertheless, various kinds of representations of the environment exist.

(a) ***Not organised***: Everything can be preserved and even compressed in an environment. Habitat is one such organisation. It can manage jobs with a large time horizon. Metric representations of the environment are common in this paradigm, which is the current standard.
(b) ***Topology***: A network of important nodes can depict an ecosystem. In a topological environment, each node represents a physical location, and the system can retrieve other nodes from the graph using observations. Only the locations linked to the nodes are stored in the graph; no metrics whatsoever are kept there. Strategies for navigation, including "landmark navigation" instead of metric representations, are possible in topological contexts.
(c) ***Mental Map/Spatial Memory***: Based on its own subjective perceptions, a mapper can create a detailed map of its surrounding area. The spatial memory records the physical layout of a given area. Using the agent's first-person observations across time, the cognitive map constructs a top-down, metric- and semantic-egocentric belief about the world. The agent's belief about its environment is updated at each step. Because of this, the agent can refine its environment model as it explores.

4.3 Perception in Action

Agents can be generated in any part of the environment, and it's possible that they won't be able to "see" the pixels that contain the answer to their visual goal right away. Therefore, to succeed, the agent needs to act, namely manipulating the pixels it will observe. The agent's visual input must be translated into the appropriate action according to its understanding of the question, its perception of the world, and the underlying physical limits. What the agent notices in its surroundings is a direct result of what it does in its environment. The data distribution is being controlled by the agent. The pixels that the agent sees are within its control. This contrasts with curated online static datasets, where factors such as viewpoint variations of items, etc. are less tightly controlled. Staying consistently resilient in the face of visual variation is a challenge that active perception faces.

4.4 Minimal Gains

In contrast to supervised learning methods like object detection or picture recognition, these agents do get rewards instantly for every action. Reward density is low for agents in most environments (reinforcement learning). An RL algorithm's goal is to help an agent get the most out of its surroundings in terms of rewards. There are settings where the agent is constantly fed rewards. In some cases, the agent is only rewarded when they reach the end of the task (such as "walk to the car"), which results in few rewards. It can be tougher to learn the desired behaviour when rewards are few. Additionally, it has the potential to make exploration harder [15].

In the realm of embodied AI, several goals can be achieved. Listed below are a few of the current assignments.

(a) **Visual odometry**: Visual odometry is simply a way to specify that a visual sensor, such as a camera, should be used for odometry to ascertain the amount of distance that has been travelled. The odometrical measure of a distance is the distance travelled with respect to a fixed point. Visual odometry, then, works on the premise that the starting point is known. For decades, researchers in computer vision and robotics have been interested in visual odometry (VO) since it is a crucial method for position estimation. It is a popular addition to many robots' navigation systems, such as Global Positioning Systems (GPS), Inertial Navigation Systems (INS), wheel odometry, and others. Extensive research and development of reliable VO systems has taken place in the past three decades.

(b) **Worldwide Deployment**: Estimating an autonomous agent's position from a map of its surroundings and observations made by the agent is known as localization. To carry out activities farther down the pipeline, like planning, exploring, and navigating, autonomous agents need the capability to localise under uncertainty. One of the most basic issues in robotics is localization. Numerous practical uses of localization have been identified, including self-driving cars, industrial robots, and delivery drones. Comparatively, VO operates under the assumption that the starting position is known, whereas the global localization problem operates under the assumption that it is unknown. There has been a lot of study on global localization, yet the issue remains unsolved.

(c) **Directions via Eyesight**. Whether they're operating in a simulated world or the real world, navigation in three dimensions is a crucial feature for robots. All animals, including humans, are capable of deftly navigating complex, ever-changing surroundings in search of various objectives. Even in unfamiliar settings, animals can move about with ease and purpose, creating mental models of their surroundings as they go. To understand AI, it is necessary to understand these internal representations. See our overview of embodied navigation for additional details on visual navigation. [16]

(d) **Linguistic Pursuits**. More and more, our environment is filled with AI systems that can make judgements and carry them out automatically. Because of this, there is an immediate need for universal methods of interacting with AI, the most attractive of which is the use of human language for this purpose. Scalability requires agents to have a grounded and embodied grasp of language, which allows them to connect words to reality and actions. On the other hand, researching grounded language is one of the most difficult problems in AI. [17]

(e) **Visual Navigation with Instructions**. One of the long-standing aspirations of robotics and AI is the possibility of verbally instructing a robot to perform a task with a fair probability of success. Even though we've come a long way, there are still several big technical hurdles to jump before robots can do everyday jobs.

Innovative methods for bridging the gap between visual perception and motor control in unstructured, novel settings will be essential. This task is also known as Vision-and-Language Navigation (VLN) in its navigational variant.

At last, answering questions with one's body. In embodied quality assurance, a randomly placed 3D agent is given a query (such as "What colour is the car?") and then created. For the agent to reply to a question like "Orange!" it needs to intelligently investigate its surroundings, collect data using first-person (egocentric) vision, and then deliver an answer. This complex task calls for a wide range of artificial intelligence abilities, including active perception (the agent needs to move around to perceive the car, so it controls which pixels it perceives), language understanding, goal-driven navigation, common-sense reasoning, and the ability to ground language into actions [18].

4.5 Internet-Based Embodied Agent Powered by AI

When contrasted with Internet AI, embodied AI presents unique difficulties. Internet AI is trained using picture datasets (like ImageNet) that contain static images. In contrast to these well-framed, high-quality static images, films created by embodied AI rely on egocentric perception, unsteady, and poorly produced. When comparing embodied AI to internet AI, the egocentric perception's inherent features cause significant alterations. Embodied AI aims to enable action by an embodied agent (e.g., a robot) in an environment, while Internet AI focuses on pattern detection in images, videos, and text using datasets often selected from the internet. The end goal is to apply all the improvements in machine learning and computer vision to embodied AI, as stated in Fig. 3, and make it work for internet AI.

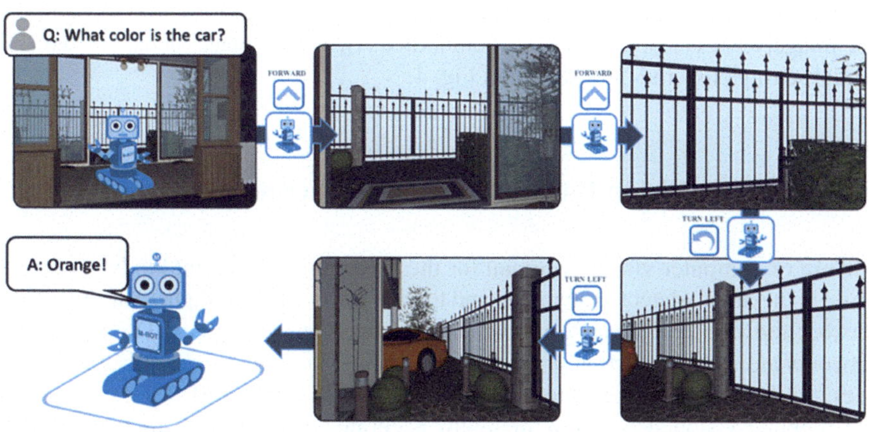

Fig. 3 Internet based embodied AI agent to identify a gate

(a) **Agent and Ecosystem**: An agent is nothing more than a concept with the ability to operate in a certain setting. Because it operates through first-person views, sensors, and interactions, we call this agent an egocentric agent, virtual robot, simulated agent, or virtual agent. Agents are like players in a game. An environment, which is merely an abstraction, nevertheless represents a world with numerous locations, rooms, and objects. Interaction, navigation, and language comprehension are only a few of the numerous things that can be achieved in an environment. Reinforcement Learning can be learnt from more about agents and environments [19].
(b) **Self-Centred Viewing**: Unlike allocentric AI, embodied AI is self-centred. One way to describe egocentric perception is from a first-person perspective. Objects are encoded in relation to the agent in egocentric perception. Items (such as the front door or the room's centre) are encoded via allocentric perception in relation to other items. While an allocentric view could be helpful for AIs that have a complete picture of their surroundings, embodied agents can only remember what they've seen through their own eyes. A map is not available to an embodied agent unless the agent itself creates it as it moves through the world. It is only fitting that embodied AI mirror the emphasis on the self in the embodiment thesis.

4.6 Computer Vision AI Based Embodied Agent

Experts in computer vision describe embodied AI as computer programmes that operate like humans in three-dimensional spaces, making decisions based on subjective perceptions that evolve in response to the agent's activities. With the help of embodied AI, virtual robots and egocentric assistants are trained in a lifelike 3D simulator before being brought into the real world to put their talents to use. This paves the way for a paradigm shift from "Internet AI," which relies on static datasets like ImageNet, COCO, and VQA, to "embodied AI," which places agents in realistic virtual settings like the ones shown in Fig. 4.

5 Computer Vision Based AI in Drone Technology Use Case

The use of computer vision is crucial for the detection of different kinds of objects while in the air. To detect, categorise, and track objects while in flight, a drone neural network and high-performance onboard image processing are utilised. Drones' neural networks aid in the detection of all sorts of objects, including cars, hills, buildings, trees, things on or near the water's surface, and varied terrain [20–22].

Computer vision also aids in the highly accurate detection of living beings, such as humans, whales, terrestrial creatures, and other marine mammals. The propulsion and navigation systems, global positioning system (GPS), sensors, cameras,

Fig. 4 Computer vision based embodied AI agent for realistic environment

programmable controllers, and other equipment necessary for automatic flights are all integral parts of a self-flying drone's construction. The use of a drone's camera and sensors allows for the collection of data, which is subsequently processed to derive actionable insights. Automatically extracting, analysing, and comprehending relevant information from one or more images processed by computer vision technology is what this process is called [23].

5.1 *Improving Drone Navigation and Object Detection with Deep Machine Learning*

The use of computer vision algorithms supported by ML and DL is revolutionising the drone business. Drones can take pictures of all sorts of things, and this helps algorithms learn from those pictures. Annotations are applied to the items so that drones can recognise them using computer vision. The drone can identify its surroundings, determine its course of action, and avoid hazards thanks to the extensive labelling of various objects [24, 25].

Computer vision has many uses in drone technology, including object tracking, self-navigation, and systems for detecting and avoiding obstacles. When a drone is using computer vision to navigate itself, it will keep tabs on nearby objects and identify potential hazards so that it can avoid collisions. A drone that follows objects in the air collects data in real time, runs it through an onboard intelligence system in real time, and then uses that data to make a judgement that doesn't require human intervention. However, self-navigation drones can figure out the best method to get where they're going with the help of AI-enabled computer vision innovations; all they need are pre-defined GPS coordinates for their starting and ending spots. Similarly, avoiding collisions is an issue that GPS navigation alone cannot resolve. As a result, unmanned aerial vehicles (UAVs) collide with fixed and moving objects in their surroundings, including trees, buildings, high-rise poles, and other UAVs. In this case, a large number of data sets are required to teach the drone to recognise and avoid a broad range of objects and obstacles, both stationary and in motion, while it is flying at high speeds. Assuming you are working with a reliable data annotation service that can supply you with accurately annotated data, training an AI model for autonomous flight is well within your reach.

(i) **Urban Planning and Smart City Drones**

Modern conveniences, such as interconnected dwellings and machinery that can be operated from the palm of the user's hand, are the building blocks of smart cities. With the use of advanced security surveillance using AI-based cameras for instant face recognition and tracing undesired things, most of the infrastructure utilities are either fully or partially automated. Drones also play a role in urban management by providing aerial views of city houses and landscapes, which architects and civil engineers can use to create the most practical floor plans. In addition, it is essential for traffic monitoring and for defining the city route so that people travel easily and without problems [26, 27].

(ii) **Aerial Work and Modern Farming Using Drones**

Drones powered by artificial intelligence are revolutionising crop and plant health monitoring in the agricultural sector. Farmers are utilising drones to precisely time the application of fertiliser by pinpointing the optimal planting window. Drones have many uses in agriculture, including crop monitoring, soil assessment, fertility testing, and crop protection. Drones are also being used to plant trees by releasing seedpods into ready-made soil. This helps farmers save time and money, which in turn increases agricultural productivity.

(iii) **Aerial Work Platform for the Building and Real Estate Sectors**

Construction businesses are expected to invest over $11.2 billion in drones over the next five years, which will fuel the growth of the drone industry across several sectors and provide manufacturers with an opportunity to produce more drones. Artificial intelligence (AI) drones used by construction firms can quickly scan and map terrain, a task that would take humans days to accomplish by hand. And soon, drones will be able to provide precise location and information for the creation of self-guided equipment, giving construction trucks a bird's-eye perspective of the projects. In a similar vein, real estate agencies are utilising drones to capture aerial footage of houses and businesses, as well as maps and information about the area, so that potential buyers can get a feel for the area and see what it's like to live there before they commit to a purchase [12].

(iv) **Ai-Powered Drone for the Defence and Military Industry**

Drones, which are unmanned aerial vehicles used for battle or bombardment, are gaining popularity in the defence and military industries. The use of drones for border patrols, security surveillance, storm tracking, and safety checks is already commonplace, though. More often than not, they end up in the food supply as well. A reliable navigation and object identification system based on computer vision is essential for the efficient development of drones. At this time, only a small number of drones can reliably detect chemicals or gather intelligence.

(v) **Tracking People and Recognising their Faces**

From a security standpoint, drones can also be deployed to monitor people in public places like parks or society. Its camera can recognise faces, so it can spot suspicious individuals or follow their emotions as they fly. Another technology is required to train an artificial intelligence model to identify individuals based on their facial characteristics in order to develop drones for face recognition. Once again, these autonomous flying devices can find humans in a sea of people thanks to tagged photos used to build an AI model for drones.

(vi) **Drones for Defence and Monitoring**

Drones are adding to the list of ways they improve people's lives by monitoring and protecting those in dangerous regions or even in their own homes. The use of drones as surveillance cameras allows for the early detection and tracking of suspicious activity, including but not limited to acts of violence, theft, and protests.

6 Conclusion

Recent developments in artificial intelligence have been a driving force behind the initial advances of embodied AI. It will be a big step forward in the process of enabling robots to effectively learn how to interact with the real world when embodied artificial intelligence surpasses the achievements made by internet-based artificial intelligence. This will give robots the ability to learn how to interact with the real world. Applications of drones are becoming increasingly significant in a variety of industries, and drones powered by artificial intelligence have the potential to deliver game-changing results by enabling users to connect from remote locations and offer real-time data for prompt action. In the future, however, drones will bring benefits to a variety of important businesses and contribute to the resolution of humanity's most pressing issues. The development of more effective training techniques for autonomous flying capabilities and advancements in computer vision technology will make this possible. It is necessary to have training data for a drone to identify various things. In addition, these training data consist of annotated photographs of a variety of items that a drone can identify with the use of computer vision and travel about while avoiding obstacles in its route. To ensure that drones can readily detect a wide variety of things and can be utilised for a variety of reasons that are beneficial to people, there are several companies that provide image annotation services. An incredible level of precision in the annotation of the data is a goal that these organisations try to achieve.

References

1. Embodied Intelligence Unravelling the nexus of AI and robotics (2023). https://medium.com/@mindful_maven/embodied-intelligence-unraveling-the-nexus-of-ai-and-robotics-for-autonomous-task-mastery-a4bbbf7ee6c
2. Khan, M.S., et al.: A review of embodied AI in healthcare: applications, challenges, and future directions. IEEE Trans. Syst. Man. Cybern.: Syst. **52**(10), 2339–2353 (2022)
3. Arun Sampaul Thomas, G., Harold Robinson, Y.: Real-Time Health System (RTHS) centred Internet of Things (IoT) in Healthcare Industry: Benefits, Use Cases and Advancements in 2020. Springer's Multimedia Technologies on the Internet of Things Environment (Scopus Indexed) (2020). ISSN: 978-981-15-7965-3
4. Chang, W., et al.: Embodied AI for obstacle recognition and avoidance in drone navigation (2023). arXiv:2301.03245
5. Das, A., et al.: Ethical considerations for embodied AI in the context of privacy and security (2023). arXiv:2302.01234
6. Ma, X., et al.: Ethical considerations for embodied artificial intelligence. Artif. Intell. **298**, 103451 (2021)
7. Overview of embodied AI (2021). https://medium.com/machinevision/overview-of-embodied-artificial-intelligence-b7f19d18022
8. A look at the current state of embodied AI companies (2018). https://medium.com/@bethaniemaples/a-quick-look-at-the-current-state-of-embodied-ai-companies-abd4850bf143
9. Ren, Z., et al.: Embodied AI for industrial automation: A review of recent advancements and future directions (2023). arXiv:2304.06789
10. Arun Sampaul Thomas, G., Harold Robinson, Y.: IoT, Big Data, Blockchain and Machine Learning Besides Its Transmutation with Modern Technological Applications. Springer Book Chapter-Internet of Things and Big Data Applications-Part of the Intelligent Systems Reference Library book series (ISRL), vol. 180, pp. 47–63 (2020). ISSN: 978-3-030-39118-8
11. Sun, Y., et al.: Embodied AI for entertainment and gaming: a survey of current trends and future challenges. ACM Comput. Surv. (CSUR) **56**(1), 1–35 (2023)
12. Wang, Y., et al.: Embodied AI for manufacturing: a review. IEEE Trans. Ind. Inf. **17**(8), 5245–5258 (2021)
13. Li, T., et al.: Applying embodied AI and computer vision for drone-based surveillance and reconnaissance (2023). arXiv:2303.04567
14. AI – Top use cases and technologies used today (2018). https://medium.com/@Brian.johnson_62680/artificial-intelligence-ai-top-use-cases-and-technologies-used-today-3c22e1a63e78
15. Howard, A., et al.: Visually-guided robotic manipulation with deep reinforcement learning (2021). arXiv:2102.09326
16. Chen, T., Li, J.: Embodied artificial intelligence: a review of progress, challenges, and applications (2021). arXiv:2104.07228
17. Wang, B., et al.: A survey on embodied AI for collaborative learning in education. J. Educ. Behav. Stat. **47**(4), 341–373 (2022)
18. Arun Sampaul Thomas, G., et al.: Classification and analysis of pistachio species through neural embedding-based feature extraction and small-scale machine learning techniques. Int. J. Image Graph. Scopus, March 2023 (2023), Online ready, ISSN (print): 0219-4678|ISSN (online): 1793-6756. https://doi.org/10.1142/S0219467824500323
19. Liu, Y., et al.: Safe and efficient embodied navigation in dynamic environments with deep reinforcement learning (2021). arXiv:2103.13501

20. How AI-based drone works – AI drone use cases (2020). https://medium.com/vsinghbisen/how-ai-based-drone-works-artificial-intelligence-drone-use-cases-7f3d44b8abe3
21. Gupta, A., et al.: Embodied AI with computer vision for drone-based environmental monitoring (2022). arXiv:2203.06783
22. Arun SamPaul Thomas, G.: SPF-ECP: self-organized, predictive, and fast adaptive – enhanced clustering protocol for DTMN routing. Wirel. Pers. Commun. (2016). Springer Publications, ISSN: 0929-6212, 1572-834X, IF - 0.7
23. Hwang, S., et al.: Integrating embodied AI and computer vision for improved drone object recognition and tracking. J. Intell. Rob. Syst. **104**(1), 1–18 (2022)
24. Jiang, B., et al.: Embodied AI for real-time decision-making in drone-assisted natural disaster response (2022). arXiv:2202.04311
25. Arun Sampaul Thomas, G.: Intelligent prediction approach for diabetic retinopathy using deep learning based convolutional neural networks algorithm by means of retina photographs. Comput. Mater. & Continua J. Tech Science Press **66**(02) (2020), ISSN: 1546-2218, SCI Index with IF 4.89, Scopus Indexed with IF 3.8
26. Arun Sampaul Thomas, G., Muthukaruppasamy, S.: Green IoT use case approaches for blockchain technology taking industry 5.0 to the next level. In: Green Blockchain Technology for Sustainable Smart Cities, pp. 119–141. Elsevier (2023), ISSN: 978-0-323-95407-5
27. Arun Sampaul Thomas, G., Muthukaruppasamy, S., Saravanan, K., Muleta, N.: Revolutionizing smart grids with big data analytics: a case study on integrating renewable energy and predicting faults. In: Kumar Sharma, D., Sharma, R., Jeon, G., Kumar, R. (eds.) Data Analytics for Smart Grids Applications—A Key to Smart City Development. Intelligent Systems Reference Library, vol. 247. Springer, Cham (2023). https://doi.org/10.1007/978-3-031-46092-0_11

G. Arun Sampaul Thomas M.E., Ph.D., M.I.S.T.E. M.I.E.T. Received **B.E.** degree in **Information Technology**, and **M.E.** and **Ph.D.** degrees in **Computer Science and Engineering** from Anna University, Chennai, in 2006, 2010, and 2018, respectively. He is currently an **Associate Professor and HOD** in the department of **Artificial Intelligence and Machine Learning, J.B. Institute of Engineering and Technology, Hyderabad**, India. He has overall teaching experience more than **ten years**. His research areas include **Computer networks, Data Science, Big Data Analytics, and IoT**. He Presented papers in five international conferences and various national conferences. His papers were published in various international journals. He attended several seminars, and workshops**.** He is an active **life member of ISTE & IET**.

S. Muthukaruppasamy M.E., Ph.D., Dr. S. Muthukaruppasamy is working as an Associate Professor and HOD, Department of Electrical and Electronics Engineering at Velammal Institute of Technology, Chennai, Tamil Nadu He received **B.E degree in Electrical and Electronics Engineering** from Manonmaniam Sundaranar University and **M.E degree in power electronics and drives** from Sathyabama University in 2000 and 2005 respectively. He completed **Ph.D in Electrical Engineering from Anna University**, Chennai, India. He has published 12 research papers in international journals.

S. Sathish Kumar M.E., (Ph.D.) Received **B.E.** degree in **Computer Science and Engineering**, and **M.E.** degree in **Software Engineering** from Anna University, Chennai, in 2010 and 2014, respectively. He is currently an **Assistant Professor** in the department of **Artificial Intelligence and Machine Learning, J.B. Institute of Engineering and Technology, Hyderabad**, India. He has overall teaching experience more than **five years**. His research areas include **Machine Learning, Deep Learning, Computer Vision, Big Data Analytics, and IoT**. He Presented papers in four international conferences and various national conferences. His papers were published in various international journals.

Beulah J. Karthikeyan M.E Received **B.E. degree in Computer Science and Engineering**, and **M.E. degree in Computer Science and Engineering** from Anna University, Chennai, in 2017 and 2020, respectively. She is currently an **Assistant Professor** in the department of **Artificial Intelligence and Machine Learning,J.B. Institute of Engineering and Technology, Hyderabad**, India. Her research areas include **Machine Learning, Deep Learning, Computer Vision, Big Data Analytics, and IoT**. She attended several seminars, and workshops.

K. Saravanan M.E., Ph.D., is working as an **Associate Professor, Department of Computer Science and Engineering** at ***College of Engineering, Guindy, Anna University***, Chennai-25, Tamil Nadu. He has done M.E Software Engineering and Ph.D. in Computer Science engineering. His research interest includes Cloud Computing, Software Engineering, Internet of Things, Smart cities. He has published papers in 20 international conferences and 40 international journals. He has also written 20 book chapters and edited 12 books.

Artificial Intelligence Algorithm Models for Agents of Embodiment for Drone Applications

Sateesh Kourav, Kirti Verma, and M. Sundararajan

Abstract AI algorithm models for embodiment in drone applications require designing algorithms that let drones appropriately perceive and interact with their surroundings. Embodiment refers to an agent's capacity to sense and respond to its surroundings, in this case, a drone. One component of tasks in embodied artificial intelligence is a natural language query to an intelligent agent. As a result, natural language processing technologies will be required to convert the inquiry into a format suited for generating an effective action plan. There are two fundamental techniques to solving this problem. The first is based on custom models that have been trained with specific instances of instructions that have been transformed into the agent-executable format. The other way relies on enormous language models trained with a significant quantity of unlabeled input to store common sense information. Creating realistic and communicative virtual beings is a difficult challenge. In this video, we demonstrate a functioning version of our modular system of lifelike, conversational. As a consequence, such models may be used to create an agent's action plan in plain English without prior knowledge. This work delves further into models based on the second method and their application to embodied artificial intelligence issues. Over the last decade, generative AI has progressed, enabling AI systems to synthesize new knowledge based on previously obtained models. The creative outputs of generative AI take numerous forms, including essays, code, artwork, poetry, and music. Google Brain was critical in the development of Generative AI. The requirement for natural

S. Kourav
Department of Electronics and Communication Engineering, Indian Institute of Information Technology, Allahabad, UP, India
e-mail: kourav530@gmail.com

K. Verma (✉)
Department of Engineering Mathematics, Gyan Ganga Institute of Technology and Sciences, Jabalpur, MP, India
e-mail: kirtivrm3@gmail.com

M. Sundararajan
Department of Mathematics and Computer Science, Mizoram University, Aizawl, Mizoram, India
e-mail: dmsrajan.mzu@gmail.com

P. Raj et al. (eds.), *Building Embodied AI Systems: The Agents, the Architecture Principles, Challenges, and Application Domains*, Information Systems Engineering and Management 14, https://doi.org/10.1007/978-3-031-68256-8_4

language processing technology in the creation of effective action plans demonstrates AI's current and future usefulness in practical problem-solving.

Keywords Artificial Intelligence (AI) · Embodied AI · Machine intelligence · CNN

1 Introduction

Computation and cognition have advanced dramatically throughout history, and this advancement is currently occurring. It is influenced by society's awareness, demands, and issues, as well as advances in our knowledge of intelligence, its origins, and intelligent behaviours or interactions with technology. AI technology advancements have resulted in the creation of general-purpose computer systems capable of doing tasks quickly and effectively [1]. Recognizes significant historical advancements in computers and intelligence. The emphasis is on the idea that computer and cognitive progress is a continuous and ongoing process. Understands that societal awareness, needs, and difficulties impact the path of mathematical and intellectual advancement. The author observes that improvements in understanding intelligence, its origins, and intelligent behaviours or interactions with technology all contribute to progress. Highlights the present significance of achievements, particularly in the context of addressing global health, climate, and social challenges. It underlines the critical importance of developing the capabilities of intelligent embodied systems in intelligible, fair, and moral technological growth has led to the creation of general-purpose computer systems capable of doing tasks swiftly and efficiently scalability. AI is also used in a variety of physical applications. It has a great deal of promise for the globe. AI, on the other hand, falls short when compared to biological systems. While interacting with their environment, based robots continue to struggle with unexpected or previously experienced tasks. Artificial intelligence-enabled devices, for example, can no longer Animals, on the other hand, can recover from setbacks swiftly and vigorously. Physical responses. Natural embodied interactions, according to pioneers in the field, provide "physical controllers," "cognitive processes," and "raw materials" for computer systems.

New manufacturing technologies, including multi-material 3D printing and micro-fabrication, have advanced EI research by enabling the creation of innovative designs, materials, and systems [2]. These technologies provide additional issues in modelling and understanding interactions because of the scale, complexity, or soft nature of the machines. The progress of artificial intelligence was linked to the direction of major philosophers such as Alan Turing and the use of the embodied philosophy of living systems. The reference to early cybernetic-based robots underlines the historical history of AI systems, highlighting milestones such as goal-seeking behaviour and learning capabilities. This point of view highlights the relevance of combining physical touch and embodied cognition in the evolution of intelligent creatures. Early cybernetic-based robots displayed goal-seeking conduct, homeostasis, and learning capacities, which is regarded as a watershed moment in the evolution of artificial intelligence. Refers to Francisco Varela's embodied philosophy, which emphasizes the position that human cognition can only be understood via the human body and its link to the physical environment. These were the first manifestations of embodied intelligence. In recent years, artificial intelligence has advanced tremendously, with improvements in machine learning allowing the creation of intelligent computers capable of performing a wide range of tasks. However, there is still a significant gap between the capabilities of these systems and those of people, who have total knowledge of their surroundings and can navigate and interact with them flexibly and intelligently. Humans can perceive and reason about their surroundings by using common sense, social conventions, and the ability to reuse previously acquired information. We'd want to create AI bots that can mimic such behaviour. Notes the significant advancement in AI, particularly in recent years, as a result of advances in machine learning. Demonstrates the enormous gap between AI systems and human abilities, particularly in terms of environmental awareness, navigation, and flexible and intelligent interaction. Human Abilities Human abilities like as an entire understanding of the environment, navigation, and flexible interaction are identified as distinct characteristics that current AI systems lack. Human Cognitive Capability Humans can reason about their environment using common sense, social conventions, and the ability to reuse previously acquired information. Desire to Model Human Behavior Expresses a desire to construct AI bots that can replicate human behaviour, emphasizing a goal of creating AI systems that can emulate the cognitive talents and flexibility inherent in human intelligence. Embedded communicators are computer-generated animated figures with human-like qualities such as the capacity to produce and respond to spoken and nonverbal communication. It is (a) a multimodal interface that employs natural communication techniques such as speech, facial expressions, hand gestures, and body posture; (b) a software client that represents a computer in human–computer interaction or a human as a user in a computer environment; and (c) a communication system that creates and manages verbal and nonverbal communication tools in a conversation between a user and a computer. Installed narrators must interact with the animated figures on the screen face-to-face. Alan Turing proposed training robots equipped with sensors and actuators to do reasoning when artificial intelligence is intended to build abstract intelligent entities capable of thinking. Human knowledge, according to Francisco Varela and the concept of living systems, can only be understood in terms of the human body and the physical world with which it interacts. Early cybernetic robots displayed goal-seeking behaviour, homeostasis, and learning ability, which proved to be a predecessor. As answers to social, climate, and global health concerns emerge, there is an urgent need to extend the capabilities of intelligent immune systems and understand how to employ them wisely, justly, and properly. Many of the difficulties in creating or completely implementing AI and capabilities are related to the need to manage systems with processing capacity and physical implementation. As a result, the notion of "Embodied Intelligence" was established. (EI) encompasses all physical systems, materials, and designs.

2 Embodied AI

Embodied AI is the study and development of intelligent robots that can interact with their surroundings in the same way that people do. It is a modern AI direction distinguished by the progress of its synthetic examination of natural cognitive processes. Embodied AI refers to artificial intelligence that operates on a tangible "thing," such as a robot arm or a self-driving car. It, like humans, can move around the world and influence its physical environment. Sensors are built into embodied AI agents. The embodiment theory states that "intelligence emerges as a result of an agent's interaction with its environment and sensorimotor activity." The goal of Embodied AI is to bridge the gap between digital AI and real-world applications. The study of artificial systems that exhibit intelligent activity by interacting with their environments through their bodies is known as embodied AI. The first generation of embodied AI researchers focused on robotic embodiments, arguing that robots must interact with their noisy environments using a diverse set of sensors and effectors, resulting in high-bandwidth interaction that challenges classical AI approaches' fundamental assumptions of clean inputs, clean outputs, and static world states. Rich simulation frameworks, typically derived from scans of actual buildings and models of real robots, have enabled more recent embodied AI research to mimic circumstances that are closer to the real world than those previously available. Passive AI tasks rely on predictions based on independent samples of the environment, such as photographs obtained without the involvement of a decision-making agent. Embodied AI tasks, on the other hand, necessitate the use of an active artificial agent, such as a robot, to observe and interact with the outside world on purpose to achieve its goals, even if the environment is unstructured or even hostile. Embodied AI, enabled by developments in computer vision and robotics, is the next frontier of difficulty for exploring and measuring intelligent models and algorithms for the physical environment.

3 History Artificial Intelligence (AI)

AI may be traced back to ancient stories and philosophical studies in which talented artisans and scholars examined the notion of imbuing artificial entities with consciousness. In the 1940s, the emergence of programmable digital computers was a watershed moment, laying the framework for AI research. The growing interest in developing electronic brains sparked a large workshop at Dartmouth College in 1956, which kicked off organized AI research. A workshop at Dartmouth College in the summer of 1956 is credited with beginning the formal field of artificial intelligence research, marking a one-of-a-kind event that started the creation of AI as a distinct topic. A small group of scientists grew intrigued by the programmable digital computer and its underlying theories, initiating a serious investigation into the possibility of constructing an electronic brain. Has sparked debate about the future of writing and the role of AI in the creative process. Some believe that AI-generated text lacks the depth and complexity of human writing, but others view technology as a tool that may help humans be more creative by offering new ideas and perspectives. Generative AI can transform a wide range of things. The notion of a self-contained machine is not new, but for the sake of this article, we shall limit ourselves to the twentieth century. 1900-1950 Much information was created in the early 1900s. The phrase "artificial intelligence" may be traced back to ancient thinkers debating life and death thousands of years ago. In ancient times, mechanical "automatons" were built, devices that moved without direct human input. Etymology of "Automaton". The term "automaton" is derived from ancient Greek and means acting freely. Leonardo da Vinci's Automaton (1495) It is worth noting that in 1495, Leonardo da Vinci constructed one of the most renowned automatons, which served to influence the history of autonomous machines. While debates over the nature of creativity and the ethics of utilizing AI in these fields remain, it is evident that generative AI is a potent instrument that will shape the future of technology and the arts. The term "artificial intelligence" was coined thousands of years ago by ancient philosophers discussing the meaning of life and death [1]. Ancient inventors constructed mechanical "automatons" that moved without the aid of humans. The term "automaton" is derived from Greek and indicates "freely acting." Focus on the Twentieth Century While acknowledging that the notion of robots acting independently is not new, the article's concentration is on the twentieth century. Acknowledges that experts were first dubious about the notion of developing an artificial brain, voicing concerns about its practicality. Mentions the construction of early prototypes of steam-powered robots capable of making facial expressions and walking, even if they were primitive. Contribution of Karel Apek (1921) The term "robot" was initially used to denote artificial individuals in Karel Apek's science fiction drama "Rossum's Universal Robots," which popularized the notion. Makoto Nishimura's Japanese Robot (1929) Notes that Makoto Nishimura created the first Japanese robot, Gakutensoku, in 1929, sparking international interest in robotic development. Mentions Edmund Callis Berkley's 1949 essay "Giant Brains, or Machines that Think," in which he linked contemporary computer models to human brains, reflecting the growing interest in artificial intelligence. The Birth of AI (1950–1956): Describes the period between 1950 and 1956 as a period of increased interest in artificial intelligence, exemplified by the development of Alan Turing's Turing Test. Cites Alan Turing's influential paper "Computer Machinery and Intelligence" (1950) and the proposal of The Imitation Game as a test for computer intelligence, leading to the coining and widespread use of the term "artificial intelligence" effectively captures the evolution of AI, from early scepticism to the field's birth, highlighting key events.

3.1 Definition of Artificial Intelligence (AI)

AI is defined as computers' mimicry of human intelligence processes, with a focus on computer systems' capacity to accomplish tasks related to learning, reasoning, and perception. Covers AI goals, such as computer-assisted learning, reasoning, and perception. Notes the varied uses of AI in industries such as finance and healthcare. Expert systems, natural language processing, speech recognition, machine vision, decision-making, pattern identification, and problem-solving are among the AI techniques and applications listed. Mentions that, because of artificial intelligence, robots may already emulate human characteristics such as learning and problem-solving [3]. Describes how AI learns by scanning huge quantities of data for patterns and then guiding the learning process through reward or discouragement based on outcomes. Regarding the Definition of "Artificial Intelligence," John McCarthy invented the phrase "artificial intelligence" in 1956 while arranging the first academic conference on the subject. Historical context acknowledges that research into whether robots can think predates the term "artificial intelligence."

3.2 Motivation

Many AI developments have been attributed to machine learning and deep learning during the last decade. Machine learning and deep learning have proven useful as data quantities and computing capacity have increased. This type of "Internet data" does not, however, originate in reality from a first-person human perspective. Data from satellites, selfies, and Twitter feeds is jumbled and randomized, and none of this is how a person sees the world. This data is used by machine learning algorithms to solve NLP, CV, and navigation challenges. While "Internet data" and "Internet AI" have helped certain organizations expand, they are not the ideal data or approach for the task. Machine learning approaches do not correspond to how people learn. Humans learn through observation, movement, interaction, and communication with others. Humans learn best when encounters are sequential rather than chaotic and surprising. Embodied AI is based on the idea that embodied beings should be able to learn in the same way that humans do. That is why professionals in Cognitive Science and Psychology are so vital. Sight, movement, speech, and interaction should be used to teach virtual robotics. While "Embodied AI" differs from "Internet AI," it can benefit from many of Internet AI's triumphs. Some jobs can presently be performed pretty successfully by computer vision and natural language processing [4]. These advancements in CV and NLP considerably boost the chance of Embodied AI success. In addition, various realistic 3D sceneries are available for use as simulated environments for Embodied AI training. SUNCG, Matterport3D, Gibson, Replica, Habitat, and DART are some of the alternatives. These conditions are significantly more realistic than those utilized in prior simulations of the study. The abundance and open availability of these datasets increases the likelihood of Embodied AI success.

4 Current Trends Embodiment and Intelligence

Embodied intelligence is a method of judging intelligent behavior in computer-assisted agents. It addresses the agent's interaction with its surroundings as well as the agent's physical limitations. Embodied intelligence may also refer to one's capacity to relate to and learn from hard situations via bodily experience. It might suggest becoming more conscious of your body's inherent tendencies for responding, attacking, or connecting throughout the day. Embodied intelligence focuses on the physical entity which makes up the human body. It looks at how our minds and bodies change as a result of physical interactions with the outer environment [2]. According to several research, embodiment plays an important role in intellectual development. The physical forms of virtual animals, for example, influenced their capacity to learn new behaviours. Morphologies that learn and grow in more challenging environments or while performing more demanding jobs learn faster and more effectively. Embodiment can also refer to the idea that intelligence cannot exist just as an abstract algorithm, but must have a physical manifestation, a body. The word refers to the fact that a certain agent is realized as a physical robot or as a simulation agent in artificial systems. Intelligence cannot develop without embodiment or involvement with its surroundings. Intelligent beings use embodiment to perform motor activities and impact their environment. Sensors in the vehicle record the response of the surroundings. A changing environment also includes embodiment. Understanding, modeling, and learning are all capabilities of intelligence. Based on motor and sensor characteristics, intelligent agents may assess, simulate, and grasp states and constraints. The manifestation of the intelligence core interacts with the cosmos, as seen in. Figure 1 depicts a sensory-motor coordination connection that is closed-loop. This specification might alter the controlled mechanism. The functionality of the embodiment and the intelligent agent's interaction with the environment expand as the embodiment expands. Second, the embodiment does not need to be constantly linked to the intellectual core to continue sensory-motor contact with the outer environment. We could, for example, train and link a Motivation in Embodied Intelligence machine.

5 Creating Embodied Intelligence

Learning is an ongoing process. Sensors tell AI about its surroundings, and sensory-motor coordination allows it to interact with it. Motor neurons activate in response to excitations linked with intended actions in response to observable situations. Learning which acts are desirable and which are not increases the learning agent's ability to survive in a hazardous environment. An organism may adapt to its surroundings in two ways: evolutionary adaptation and cognitive adaptation. Describes learning as a dynamic process in which AI gathers knowledge about its surroundings via sensors and interacts with its environment via sensory-motor coordination. How

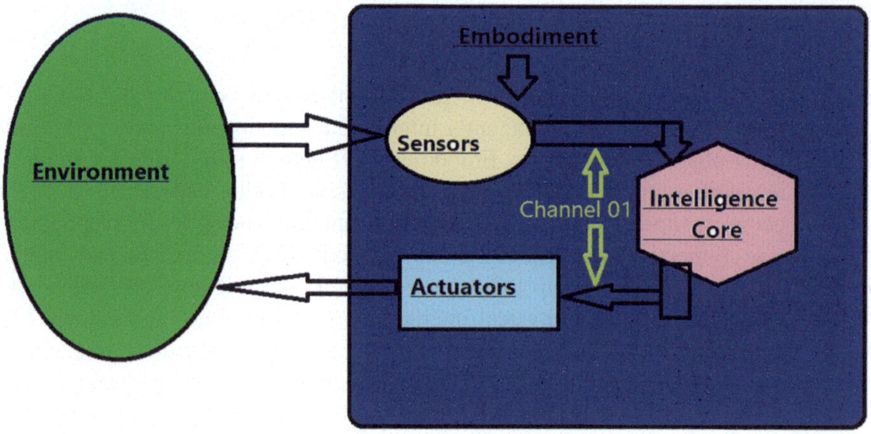

Fig. 1 Intelligence embodiment

motor neurons activate in response to excitations associated with desired actions and observed conditions, suggesting a relationship between sensory information and motor responses. The necessity of learning which acts are wanted and which are not is emphasized, adding to the agent's capacity to live in a dangerous environment. Describes the interaction of memory components in real-time. Memory's numerous components are linked and interact in real-time, underlining memory's dynamic nature. Because neurons quickly form memories of earlier patterns, learn to interact with their environment, and provide predictions about future occurrences, self-organization is stressed as a critical component in human brain development [4]. The brain of a six-year-old is thought to be densely packed with redundant and pliable connections, ready to absorb a wide range of information. Years of learning produce a decrease in brain connections, retaining just the most important knowledge and building linked memories and talents. Mentions how an agent may adapt to its environment through a variety of techniques, such as evolutionary adaptation via natural selection of fit agents and cognitive adaptation via learning, memory, associations, pattern recognition, representation building, and goal execution. Differentiates between evolutionary adaptation and cognitive adaptation as it relates to intelligence [5]. Because of its relationship to an agent's intellect, the description concentrates on the second form of adaptation cognitive adaptation for the development of AI.

Collective intelligence, as well as the functions of spatiotemporal patterns, mental representations, and memory in shaping our worldview. Group intelligence is described as a type of intelligence that is based on the individual intelligence of group members. According to this study, all spatiotemporal trends encountered throughout life contribute to knowledge and mental representations of the universe. Patterns are made up of references of attributes at various levels of abstraction, and the relationships between these components are learned and remembered. Highlights the abstract representation of motor actions and skills, demonstrating a greater level of representation than direct physical motions. Memory is defined as episodic and associative, with components, connections, and skills captured. Introduces the concept that sparse connections, as seen in a rat's neural network, may increase synaptic storage capacity while decreasing network energy costs. The significance of developing a neural network memory for intelligent robots, stressing properties such as consistency, hierarchy, and sparse connections. It also presents the Goal Generation System (GCS) as an essential component for creating targets and promoting sensory-motor coordination. Suggests that building a neural network memory, similar to human learning, is good for the goal of creating intelligent robots. Emphasizes the significance of organizing memory consistently, hierarchically, and with sparse connections. The author observes that such a memory structure enables Embodied Intelligence (EI) to learn mainly unattended by responding to environmental signals, stressing a conscious and continuing learning process. Environment Consistency The learning process is directly linked to the machine's environmental goals, allowing efficient resource organization to acquire new information and develop important talents. Introduces the GCS as a built-in framework for specifying goals in intelligent machine behaviour. GCS functions The GCS promotes sensory-motor coordination, goal-directed learning of perceptions and actions, and acts as a stimulant for involvement with the environment, according to the definition. Hierarchical Structure that is Consistent The fact that both machine memory and GCS have a consistent hierarchical and self-organizing structure is highlighted. Complexity with Goal Hierarchy: As the goal hierarchy grows, so does the structure. In the Goal Creation Process, Hetero Hierarchy Goal creation is described as supporting the development of a hetero hierarchy reflecting sensory inputs and a matching hetero hierarchy expressing.

6 Embodied Intelligence Goal Creation

One of the primary goals of AI is to create systems that can analyze large amounts of data, uncover patterns, and make data-driven decisions [6]. This capacity to solve issues and make judgments quickly is useful in a variety of sectors, including healthcare, banking, transportation, and manufacturing. This sort of artificial intelligence can do any intellectual work as well as a person. The objective of artificial intelligence in this discipline is to develop a system that can think for itself in the same way that humans do.

- Making Decisions and Solving Problems—One of the primary goals of AI is to create systems capable of analyzing enormous amounts of data, detecting patterns, and making data-driven judgments. This capacity to solve issues and make judgments quickly is useful in a variety of industries, including healthcare, finance, transportation, and manufacturing.

- Deep Learning and Machine Learning—Deep learning and machine learning are AI subsets that enable computers to learn from data without the need for explicit programming. These technologies have resulted in substantial advances in disciplines such as computer vision, speech recognition, and recommendation systems, among others.
- Automation and robotics—Integrating AI with robotics has resulted in intelligent devices capable of performing physical tasks with precision and accuracy. AI-powered automation is transforming sectors globally, from assembly line robots in manufacturing facilities to self-driving cars.
- Improving Healthcare and Medicine—The purpose of artificial intelligence in healthcare is to enhance diagnosis, treatment planning, and patient care. AI algorithms may be used by medical practitioners to assess medical imaging, forecast illness outcomes, and design individualized treatment regimens for patients.
- Foster Creativity and Innovation—In addition to practical applications, AI has the potential to drive creativity and innovation. Artificial intelligence-powered technology can assist artists, authors, and designers in developing novel ideas and pushing the boundaries of human creativity.

7 Future Trends of Artificial Intelligence

The research organization highlighted four emerging technical trend topics: developer experience, pervasive cloud, human-centric privacy, and security [7]. Artificial intelligence isn't just for home and business usage; the medical and aviation sectors are already embracing AI to improve their services. If artificial intelligence outperforms human efforts, then choosing. There has been an increase in data quantities 2.5 quintillion bytes of data produced each day and very high processing capacities in recent years. Chat GPT includes 175 billion parameters that were trained using 570 GB of data acquired from books, Wikipedia, and other textual content on the Internet—such scales are only conceivable because of high computing power and the widespread availability of inexpensive, off-the-shelf hardware and open-source libraries. The reduction in image classification system training expenditures (63.6%) and improvement in training time (94%), for example, since only 2018 offer a strong indicator of how far humans have advanced in AI technology. As a result, AI has advanced significantly as more people use it. Artificial intelligence (AI) has been at the forefront of technical innovation for quite some time and is predicted to gain importance in the future. This article provides an outline of how this area has progressed over time and discusses some of the major advancements and uses of AI in recent years. I am a key enabler of the Fourth Industrial Revolution. The modern industrial era's ever-increasing data quantities, fast-changing social trends, more interconnection, and cross-pollination of varied technologies rely on machine intelligence to draw insights and automate reactions [8].

- AI Analyst/Specialist—The function of an AI Analyst or Specialist. Your tasks are similar to those of an AI engineer, with a focus on building and implementing AI-oriented solutions to improve services in certain areas. AI analysts or specialists utilize data-analysis abilities to look for trends and patterns in datasets, with the objective of improving healthcare, finance, geology, cybersecurity, and other fields.

  The salary range you provided for AI analysts in India, ranging from INR 3–10 lakhs per year, depicts the pay scale depending on characteristics such as experience and employment circumstances.
- Data Scientist—Data scientists play a significant role in data analysis and machine learning, performing tasks such as discovering relevant data sources, cooperating with data engineers to automate data collection, and analyzing large datasets to reveal patterns and trends. They usually employ a variety of tools and computer languages to do their duties, including SQL, Python, Scala, SAS, SSAS, and R. mentioned some popular frameworks used by data scientists, such as Amazon ML, Azure ML Studio, and Spark MLlib, which are used to build and deploy machine learning models at scale.
- Research Scientist—To work as an AI research scientist, you must have a master's or PhD degree. An AI research scientist in India currently earns at least INR 35 lakhs per year, according to current data. Research Scientist A Research Scientist is another fascinating artificial intelligence career. As we all know, basic research skills are required for every topic.
- Product Manager—Machine learning is critical in the position of product manager in any major organization. A product manager's role is to tackle challenging circumstances by meticulously obtaining data. You must be able to identify critical business obstacles as well as obtain supporting datasets for data analysis. Following data interpretation, the product manager employs effective artificial intelligence (AI) approaches to investigate the business implications of data interpretation results. Given the significance of the position, every company need a qualified product manager.
- Robotics Scientist—In our fast-paced, technologically sophisticated culture, robots are undoubtedly snatching the jobs of those who perform physical, repetitive, and boring labor. On the contrary, it employs robotics experts. These self-driving vehicles must be created and managed by a robotics engineer.

8 Artificial Intelligence Enhanced Models for the Drone

AI algorithms allow drones to execute tasks like object identification, navigation, and data processing more quickly and accurately. This technology has expanded the possibilities for drone applications, resulting in increased acceptance across a variety of sectors. Nowadays, technology is ubiquitous. Orders placed from home and shop closures have cemented modern civilization as a digital period. Artificial intelligence has facilitated increased automation and digitalization. Artificial intelligence is all about teaching computers and robots to make judgments in the same way that people do. By teaching computers to replicate human thought patterns, they can execute elements of our work while this is a frightening concept cue the sci-fi movies about robots taking over the world AI may make processes far more productive and, in many cases, more accurate models are tools and methodologies for training computers to read and evaluate data in the same way that people do. Machine learning is a large class of artificial intelligence models in which computers are taught to think for themselves and build algorithms based on massive quantities of data [9]. Other artificial intelligence models need the computer to be programmed with an algorithm, and the code can be developed and altered based on experience. Finally, some models cannot learn on their own; they must be trained and interact with humans. The current inquiry has been fascinated by artificial intelligence (AI), which has been assigned to several sectors such as pavement distress detection, pothole identification, and healthcare. The detection of distress on the streets and pavements, as well as the delivery of healthcare and medical services, must be monitored utilizing cutting-edge technologies such as drone technology. Construction site improvements and healthcare delivery are two important challenges. A 3D time-of-flight (TOF) camera observes the reflected light after lighting the scene with a modulated light source. The phase shift between light and reflection is calculated and converted to distance.

Using an LED light source, a structured light-based depth-sensing camera produces light patterns (typically striped ones) on the target object. The distortions produced can be used to calculate the distance to the item. A structured light 3D scanner is widely used to reproduce the 3D representation of an object. This review research provides appropriate monitoring methodologies for academics worldwide who use AI-enabled drones. There are two sorts of techniques based on principles: direct and indirect approaches. Figure 2 shows how laser emitters directly generate light pulses. When light interacts with an object, it bounces back and is detected by the sensor, and the distance between the item and the sensor is calculated using the light term.

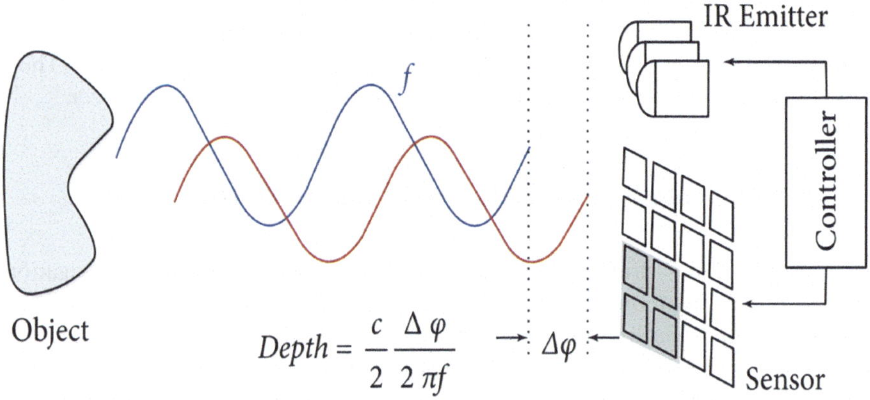

Fig. 2 Working principle for enhanced models for the drone

The pattern is captured by the 3D scanner's cameras, and the distortion allows the software to produce the 3D model. Record how the object deforms the light pattern using a single camera (preferably two). Vision systems may identify the depth and surface information of objects in an image by studying how patterns morph as individuals on surfaces. A pair of binoculars has four prisms. One prism is put horizontally and one prism is inserted vertically in each tube. The binocular camera's structure and physical depiction may be observed. It also exposes the fundamental operating mindset. This strategy is made up of three separate parts. The first is that Integral Image may be used to determine the characteristics of the detector. Because it is calculated quickly, its computational complexity is decreased. The second method is to construct the classifier by selecting a few important qualities and then using a learning process to develop a highly structured light scanner's primary function is to create a structured light pattern and project it onto an object. The light beams are guided by the lenses and subsequently by the Porro prisms. According to prior research, while selecting a depth camera, one should take caution [10, 11]. The third and fastest strategy is a cascaded collection of more complex classifiers. The basic objective of a structured light scanner is straightforward: you project a structured light pattern onto an item, then film it with at least one camera (usually two cameras) to capture how the light pattern deforms as a result of the object's deformation (Fig. 3).

Its capacity to extract features is used to evaluate its ability to extract abstract features. It can be processed with increased accuracy and speed since it is based on supervised learning [12]. A CNN that has been systematically improved is known

Fig. 3 Structured light scanning

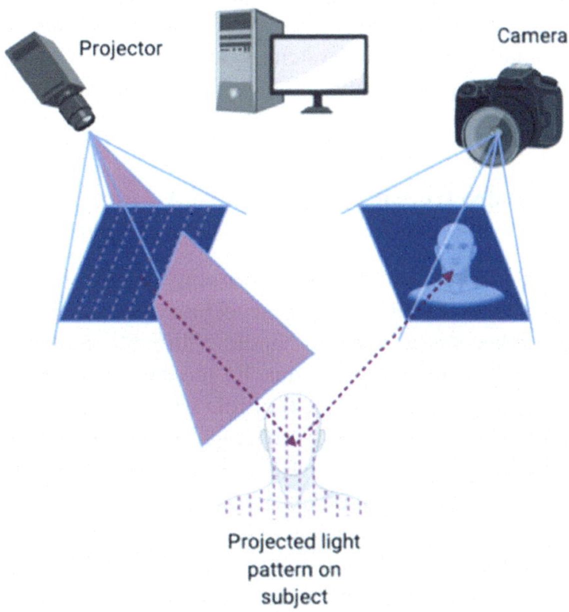

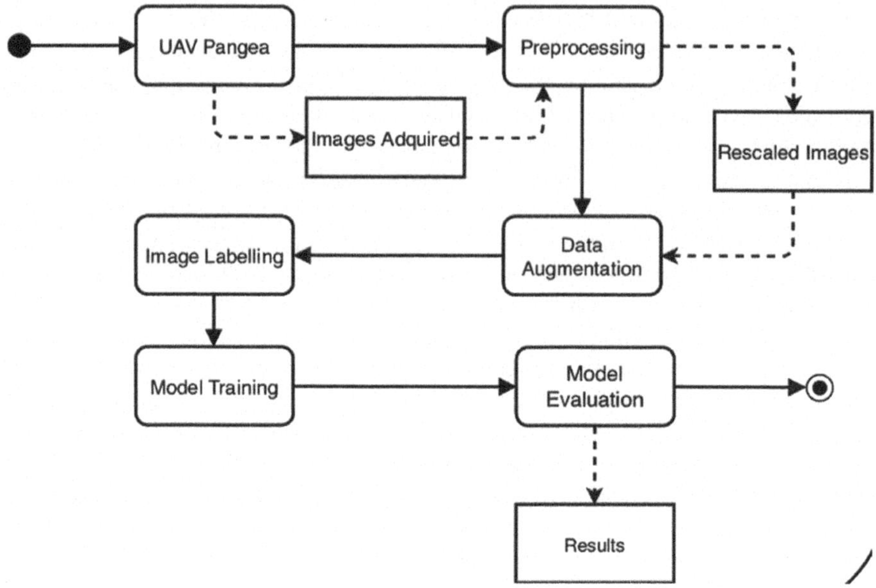

Fig. 4 Models of artificial intelligence for drones

as supervised learning. Its popularity arises from its ability to thrive in unsupervised learning. We picked CNN to increase speed and efficiency based on the study presented above. The two components of this system necessitate image-based learning. Because the system is heavily reliant on machine learning, it can perform better with proper calibration and sufficient learning (Fig. 4).

9 CNN Based Drone Control

Deep learning and machine learning are being used successfully in academics, industry, and healthcare. These technologies are well known and renowned for their versatility and efficacy in handling difficult challenges. The surge in popularity of IoD can be linked to its capacity to adapt to unexpected events. UAVs, or drones, are an important component of IoD because they have practical properties including minimal mobility, wireless coverage extension, and access to difficult-to-reach areas. UAVs may be employed efficiently in a variety of settings such as farming, mission-critical services, surveillance systems, and rescue missions. In emergencies, their capacity to reach unreachable regions makes them invaluable. Several studies use the words IoD and UAVs interchangeably to emphasize the tight link between the Internet of Drones and the deployment of unmanned aerial vehicles. Drones contribute to network topology efficiency by enhancing latency, throughput, connection, and reliability. High degrees of mobility, the unpredictability of the wireless medium, and the possibility of rapid topological changes owing to battery life have all been noted as obstacles in drone deployment. These elements add to the difficulty of efficiently operating and commanding UAVs. Within the IoD-UAV framework, the research describes the Internet of Things (IoT) in the context of potential uses and contrasting operational scenarios. This implies that IoT technology integration is being studied in the management and optimization of UAV operations. Under the context of IoD-UAVs, machine learning is classified depending on its applications. This holistic approach implies a thorough examination of the IoD-UAV landscape. According to the findings, Convolutional Neural Networks (CNN) are the most commonly utilized machine learning approach in IoD-UAV articles. CNNs excel at tasks like picture recognition and object identification, which are critical in UAV applications. According to study findings, the bulk of studies under the IoD-UAV framework have focused on resource and mobility management. This demonstrates a strong interest in optimizing resource allocation and improving drone mobility. CNNs have been employed in a range of computer vision applications such as localization, detection, and segmentation. CNNs surpassed typical computer vision algorithms in various tests [13, 14]. CNNs scored well in the Image Net Large Scale Visual Recognition Competition (ILSVRC), particularly in image classification. This breakthrough has solidified CNNs as a powerful tool for occupations requiring the interpretation and understanding of visual data [15, 16]. CNNs are used to generate autonomous navigation behaviour in a range of mobile devices, including cars, robotics, and aerial drones. Specifically, in automobile navigation, an end-to-end trained deep CNN can map front-facing camera input images to steering angles. This enables the machine to navigate freely by associating visual input [17, 18]. A deep neural network training method that combines Reinforcement Learning (RL) with model predictive control. This integration proposes a hybrid strategy that use both RL techniques and predictive control methods to improve the performance of a deep neural network in the context of drone obstacle avoidance [19]. This entails employing RL principles, which let the system learn optimum strategies through interactions with its environment, as well as model predictive control, which is forecasting future system behaviour to optimize control actions [20]. A drone navigation model based on Reinforcement Learning (RL). The RL framework states that the drone learns the optimum navigation tactics through trial-and-error interactions with its environment.

The primary focus of this RL-based model is on visually searching for target items. This indicates that the drone has been trained to independently explore within spaces while actively searching for predefined target items using its visual perception skills [21]. Active Neural Localization is a system that predicts a probability map for an agent's location on a map. This data is then used to forecast navigation policy. The active component indicates that the system is actively participating in determining the location of the agent on the map. The fundamental benefit of Reinforcement Learning (RL) in the Active Neural Localization system is the elimination of manual data labeling. Without depending on pre-labeled data, RL allows the system to learn optimum navigation algorithms through contact with the environment [22]. The usage of an Imitation Learning (IL) algorithm, a common strategy in machine learning, is mentioned in reference [23]. For navigation, Kelchtermans et al. [24] used an

IL algorithm in combination with a Gated Recurrent Unit (GRU) recurrent neural network. The model includes not just visual characteristics, but also other types of features, such as a low pass filtered history of prior orders.

Giusti et al. [25] created a deep learning-based algorithm for drone navigation along forest routes. This paradigm makes use of distinct actions such as "go right," "go straight," and "go left." The actions are most likely selected based on a study of input data, including visual information.

Smolyanskiy et al. [26] used similar models for trail following and added an extra entropy incentive to stabilize drone navigation. The use of entropy as a reward proposes a way to encourage exploration while keeping the model from becoming overly predictable.

Kim et al. [27] developed a CNN model particularly optimized for indoor target search. This means that the model is designed to travel and search for specific objectives within indoor surroundings. The researchers used a transfer learning strategy, beginning the training with a pre-trained model. Transfer learning entails using information obtained from training on one task (pre-trained model) to improve performance on another activity (indoor target search). This can improve the model's generalization capabilities. Kim et al. used extra training data to increase the model's generalization. This shows an effort to improve the model's capacity to detect and traverse interior situations successfully, perhaps allowing a greater range of circumstances (Fig. 5).

As it follows a specified course with predefined waypoints, the drone captures its current front view and yaw. It is designed to fly inside an envelope around the

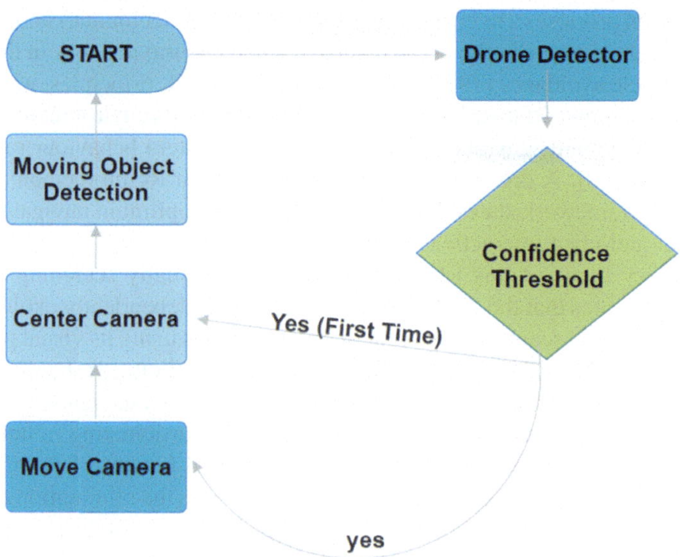

Fig. 5 Flow chart of CNN based drone movement control

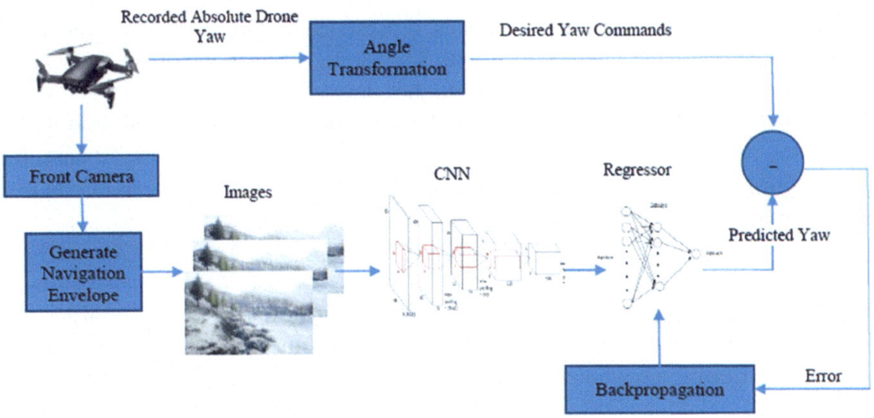

Fig. 6 CNN based drone movement control

waypoints in order to collect a variety of data for the purpose of constructing a dependable system. The recorded yaw command is used to generate relative yaw instructions. These directives are most likely the intended changes in orientation (yaw) that the drone should follow depending on its current condition. The neural network architecture is made up of a CNN and a regressor. A trained VGG-16 network trained for the ImageNet Large Scale Visual Recognition Competition (ILSVRC) classification challenge is used as a feature extractor in CNN. Following the feature extractor, the regressor employs either a Fully Connected Neural Network (FCNN) or a Recurrent Gated Neural Network (GRU). The regressor is trained from start to finish to learn the mapping from extracted features to yaw commands. The training technique comprises modifying the neural network parameters to lower the mean squared error (MSE) between the real and predicted yaw angles. The convolutional layers used for feature extraction (VGG-16) are frozen to prevent the extraction of features biased toward simulated situations. This ensures that when the model is utilized for visual route following in real-world scenarios, the learned properties are more relevant and generalize effectively. The loss function during training is the mean squared error (MSE) between the true and predicted yaw angles. This statistic assesses the difference between the predicted and actual yaw angles and directs the training method to reduce it. The proposed model for drone navigation is based on visual cues, using a pre-trained CNN (VGG-16) for feature extraction and a regressor (FCNN or GRU) for predicting yaw angles. The whole training process is focused on minimizing the mean squared error to create accurate yaw predictions, and strategies are utilized to increase generalization to real-world settings (Fig. 6).

$$MSE = \frac{1}{n}\sum_{i}(y_i - p_i)^2$$

where n is the number of training instances, represents the true yaw for sample i, and represents the predicted yaw for example. The angle between the drone's direction and the next waypoint, or the angle by which it deviates from its position, is used to determine the true yaw. The network links the deviation angle, waypoints, and visual input using the loss function. A path is divided into many waypoint-based parts for training purposes. It is expected that the entire navigation path has been taught and that the waypoints do not overlap.

Evaluation Metrics

This statistic quantifies path following model performance by computing the total difference between each waypoint location and the closest position reached by the drone during the whole journey. The total number of waypoints is then averaged. Cross-Track Mean Distance. The Mean Cross Track Distance, which computes the shortest distance between the drone's present position and the next two nearest waypoints, is the second error metric. This statistic adds to the model's accuracy in following the planned route. To quantify the complexity of a journey based on the change in navigation angles, a third metric, the Sum of Angle Change, is added. This statistic sheds light on the path's complexity from a navigation standpoint. In the future, the authors propose using Generative Adversarial Networks (GANs) to achieve style transfer between synthetic and genuine training images. This program attempts to increase system performance while perhaps lowering the size of the training dataset. Style transfer can help the model generalize from synthetic to real-world scenarios.

Another area for future research is implementing the proposed end-to-end navigation system into UAV middleware. Based on prior work in these domains, this integration might position the system as an independent component or as part of a more comprehensive framework for target detection and tracking. The current work focuses on quantitative evaluation using defined error metrics, and future work includes additions such as leveraging GANs for style transfer and integrating the navigation system into UAV.

10 Artificial Intelligence Applications

Embodied AI is a subfield of artificial intelligence that focuses on enabling agents to interact autonomously with humans and their surroundings. Artificial intelligence has various applications in modern civilization. It is becoming a need in our day and age because to its ability to handle complex situations. AI benefits a wide range of industries, including healthcare, entertainment, banking, and education. AI has accelerated and enhanced the ease of our daily lives. Modern artificial intelligence-based technologies are the ideal fit for every intelligent endeavor [28, 29]. The list of AI applications is growing by the day. Some of the uses of artificial intelligence are listed here.

1. **Education**—AI has shown to be an excellent educational technology. AI allows for the automation of basic activities in educational institutions. AI-powered systems aid in the generation of timely and regular feedback for students and teachers. AI technologies are now developed to provide pupils with tailored learning. AI can automate grading to give the lecturer more time to lecture. AI chatbots can communicate with students as a teaching tool. AI may one day act as a handy, on-demand virtual educator for pupils. This tutor would be available at all times and from any location.
2. **Healthcare**—AI in healthcare is a life-saving medical assistance. AI algorithms can detect potentially fatal conditions and even inform emergency medical services. Doctors rely on AI systems to help them perform more efficiently. AI might potentially be used to enhance medication accuracy. AI has advanced and will have a substantial impact on the healthcare industry in the next five to ten years. In the healthcare field, AI is being utilized to diagnose patients faster than humans. AI can assist physicians in making diagnoses and notifying them. They can identify changes in a patient's health. In order for the patient to be treated before being admitted to the hospital.
3. **Finance**—AI applications have helped financial services over the past two decades. AI is used in finance to anticipate stock prices, handle bookkeeping, and identify fraud in banks. The finance and artificial intelligence businesses work well together. Machine learning, algorithm trading, and chatbots are all being used in the financial business. It utilizes adaptive intelligence and automation in financial processes [32, 33].
4. **Agriculture**—AI is being used in agriculture to help cultivate better crops. It helps farmers with insect control, soil and growth monitoring, and data organization. AI reduces workloads while improving agricultural jobs. It benefits farmers and traders all along the food supply chain. Agriculture is a field that necessitates a large amount of resources, effort, money, and time to provide the best outcomes. As AI becomes more ubiquitous, it becomes more computerized. Artificial intelligence may help agriculture robotics, solid and crop monitoring, and predictive analysis. Agriculture artificial intelligence has the potential to significantly benefit farmers [34, 35].
5. **Business**—AI may increase the productivity, efficiency, and accuracy of business goals. It aids businesses in boosting their income and return on investment. My and Ml's effect on businesses will last for many years. By introducing AI into corporate environments, you may be able to save time on repetitive procedures. It boosts employee productivity and the overall consumer experience. Businesses may profit in areas ranging from marketing to operations to sales [36, 37].

11 Conclusion and Future Aspect

Business AI may increase the productivity, efficiency, and accuracy of business goals. It aids businesses in boosting their income and return on investment. My and Ml's effect on businesses will last for many years. By introducing AI into corporate environments, you may be able to save time on repetitive procedures. It boosts employee productivity and the overall consumer experience. Businesses may profit in areas ranging from marketing to operations to sales. Recent advancements in Internet AI have aided Embodied AI's early success. Once Embodied AI exceeds the successes of Internet AI, it will be a big step forward in enabling robots to correctly learn how to interact with the physical environment. Embodied AI's early success has been fueled by recent advancements in Internet Artificial Intelligence. Technique Employed Mentions the use of artificial curiosity and the formation of abstract objectives as system-level strategies for facilitating efficient and deliberate learning. Abstract Objectives at a Higher Level The system produces higher-level abstract objectives, which contribute to a growth in the internal complexity of memory representations and skills. Learning Efficiency indicates that this system has shown the capacity to learn better and quicker than traditional reinforcement learning systems. Internal Reward System Highlights a crucial contrast from standard reinforcement learning by emphasizing that the GCS creates an internal reward according to the abstract goal fulfilled by the computer. Transparency in Reinforcement This internal reward system makes the reinforcement process transparent, hence limiting external control over the machine. Within the GCS, a more dynamic and adaptable learning process in which the agent gradually builds its grasp of objectives, actions, and values. The integration of a value system appears to be more sophisticated, including the connecting of basic desires with internal moods in order to produce higher-level goals. This may result in more flexible and intricate learning as compared to standard reinforcement learning approaches. Abstract ambitions and External Influences: Shows how external demands such as buying food or paying bills may impose abstract ambitions such as earning money, and how accomplishing these higher-level objectives may not always alleviate lower-level needs such as hunger or warmth. basic Pain and External Rewards: Introduces the notion of "basic pain" as a fundamental, palpable desire and differentiates externally measuring and Optimizing Rewards: Highlights the difficulty in measuring and optimizing rewards, particularly for secondary objectives connected to higher-level goals, as opposed to the entire external benefit acquired to fulfil core objectives. This explanation adds layers to the goal-setting and learning processes, emphasizing the hierarchical structure of objectives as well as the delicate interaction between external incentives and the accomplishment of distinct goals at various levels. It denotes a more refined understanding of how the agent navigates and prioritizes goals inside the GCS.

Future Aspect—Digitalization and technology are prevalent in today's environment. Every other day, we humans come into contact with new technology. AI is one of the most rapidly growing and beneficial technologies in computer science. AI is a fascinating and all-encompassing topic in computer science. It's past time to talk about some of the benefits of artificial intelligence. You must weigh these benefits and choose if AI has a bright future or not Artificial intelligence is a fluid field. AI provides several prospects for the development of intelligent machines. They advocate for a positive shift in computer science. Intelligent technologies have the potential to reduce human errors while increasing the efficiency of specialized occupations. The economy of a country benefits from artificial intelligence. It facilitates task advancement. When man-made intelligence is joined with AI intelligence, occupations and projects may be completed more efficiently. AI systems can now perform challenging tasks that humans cannot. AI-based apps are widely used by humans. AI algorithms are used in Apple's Siri and Microsoft's Cortana, for example. Developers employ AI technology to create digital assistants. AI assistants decrease labour and do tough tasks with astonishing efficiency. Artificial intelligence is most beneficial to the medical business. AI technologies in the medical business help to remove manual tasks, allowing physicians to focus on more vital tasks. Artificial intelligence is poised to usher in a new age of global transformation through intelligent systems. It may not only aid in the automation of the labour process, but it may also aid in the making of better business decisions. You gain critical insights with minimum effort. AI is also being utilized to enhance the capabilities of these tools, allowing them to do increasingly complex tasks. Soldering and assembling electrical components are two examples. Complex jobs may be completed with more control. More precise and less invasive medications will benefit the patient. You may also utilize Big Data approaches in combination with AI and ML to collect and analyze massive volumes of data.

References

1. Nguyen, P.T., Di Rocco, J., Iovino, L., Di Ruscio, D., Pierantonio, A.: Evaluation of a machine learning classifier for metamodels. Softw. Syst. Model. **20**(6), 1797 (2021)
2. Chowdhury, M.E.H., Rahman, T., Khandakar, A., et al.: Can AI help in screening viral and COVID-19 pneumonia? IEEE Access **8**, 132665–132676 (2020)
3. Nguyen, P.T., Di Ruscio, D., Pierantonio, A., Di Rocco, J., Iovino, L.: Convolutional neural networks for enhanced classification mechanisms of metamodels. J. Syst. Softw. **172**, 110860 (2021)
4. Silva, L.A., Sanchez San Blas, H., PeralGarcía, D., Sales Mendes, A., Villarubia González, G.: An architectural multiagent system for a pavement monitoring system with pothole recognition in UAV images. Sensors **20**(21), 6205 (2020)
5. Wang, J., Chen, S., Liu, Y., Lau, R.: Intelligent metaverse scene content construction. IEEE Access **11**, 76222–76241 (2023)
6. Gao, M., Wang, X., Zhu, S., Guan, P.: Detection and segmentation of cement concrete pavement pothole based on image processing technology. Math. Probl. Eng. **2020**(Article ID 1360832), 13 (2020)
7. Redmon, J., Divvala, S., Girshick, R., Farhadi, A.: You only look once: unified, real-time object detection. In: Proceedings of the IEEE Conference on Computer Vision and Pattern Recognition, pp. 779–788 (2016)
8. Wazid, M., Bera, B., Mitra, A., Das, A.K., Ali, R.: Private blockchain-envisioned security framework for AI-enabled IoT-based drone-aided healthcare services. In: Proceedings of the

2nd ACM MobiCom Workshop on Drone Assisted Wireless Communications for 5G and Beyond, London, UK, September 2020, pp. 37–42

9. Zhang, H.-T., Zhang, J.-S., Zhang, H.-H., et al.: Automated detection and quantification of COVID-19 pneumonia: CT imaging analysis by a deep learning-based software. Eur. J. Nucl. Med. Mol. Imaging **47**(11), 2525–2532 (2020)
10. Li, L.: Evaluation of regional economic development quality in Hunan Province based on projection pursuit method. J. Hunan Univ. **33**(03), 48 (2019)
11. Wu, H.-D.: Artificial intelligence era. J. Inst. Arrange. Leg. Regul. Leg. Sci. 35(5), 1674–5205 (2017). (Journal of Northwest University of Political Science and Law)
12. Negrescu, C., Ciobanu, A., Ilie, M.D.: Specific acoustic unit processing in concatenative Romanian speech synthesis used for talking agents. In: 2013 7th Conference on Speech Technology and Human-Computer Dialogue (SpeD), pp. 1–8 (2013)
13. Deng, J., Dong, W., Socher, R., Li, L.-J., Li, K., Fei-Fei, L.: Imagenet: a large-scale hierarchical image database. In: Computer Vision and Pattern Recognition, 2009. CVPR 2009. IEEE Conference on, pp. 248–255. IEEE (2009)
14. Sermanet, P., Eigen, D., Zhang, X., Mathieu, M., Fergus, R., LeCun, Y.: OverFeat: integrated recognition, localization and detection using convolutional networks, pp. 1–16. https://doi.org/10.48550/arXiv.1312.6229 (2013)
15. Girshick, R.: Fast R-CNN. In: Proceedings of the IEEE International Conference on Computer Vision, pp. 1440–1448 (2015)
16. Badrinarayanan, V., Kendall, A., Cipolla, R.: SegNet: a deep convolutional encoder-decoder architecture for image segmentation. https://doi.org/10.48550/arXiv.1511.00561 (2015)
17. Ronneberger, O., Fischer, P., Brox, T.: U-net: convolutional networks for biomedical image segmentation. In: International Conference on Medical Image Computing and Computer-Assisted Intervention, pp. 234–241. Springer (2015)
18. Bojarski, M., Del Testa, D., Dworakowski, D., Firner, B., Flepp, B., Goyal, P., Jackel, L.D., Monfort, M., Muller, U., Zhang, J., et al.: End to end learning for self-driving cars, pp. 1–9. https://doi.org/10.48550/arXiv.1604.0731 (2016)
19. Ross, S., Melik-Barkhudarov, N., Shankar, K.S., Wendel, A., Dey, D., Bagnell, J.A., Hebert, M.: Learning monocular reactive UAV control in cluttered natural environments, pp. 1–9. https://doi.org/10.48550/arXiv.1211.1690 (2012)
20. Zhang, T., Kahn, G., Levine, S., Abbeel, P.: Learning deep control policies for autonomous aerial vehicles with MPC-guided policy search, pp. 1–11. https://doi.org/10.48550/arXiv.1509.06791 (2015)
21. Zhu, Y., Mottaghi, R., Kolve, E., Lim, J.J., Gupta, A., Fei-Fei, L., Farhadi, A.: Target-driven visual navigation in indoor scenes using deep reinforcement learning. In: IEEE International Conference on Robotics and Automation (ICRA), pp. 3357–3364. IEEE (2017)
22. Chaplot, D.S., Parisotto, E., Salakhutdinov, R.: Active neural localization, pp. 1–16. https://doi.org/10.48550/arXiv.1801.08214 (2018)
23. Ross, S., Gordon, G., Bagnell, D.: A reduction of imitation learning and structured prediction to no-regret online learning. In: Proceedings of the Fourteenth International Conference on Artificial Intelligence and Statistics, pp. 627–635 (2011)
24. Kelchtermans, K., Tuytelaars, T.: How hard is it to cross the room?–Training (recurrent) neural networks to steer a UAV, pp. 1–19. https://doi.org/10.48550/arXiv.1702.07600 (2017)
25. Giusti, A., Guzzi, J., Ciresan, D.C., He, F.-L., Rodríguez, J.P., Fontana, F., Faessler, M., Forster, C., Schmidhuber, J., Di Caro, G., et al.: A machine learning approach to visual perception of forest trails for mobile robots. IEEE Robot. Autom. Lett. 1(2), 661–667 (2016)
26. Smolyanskiy, N., Kamenev, A., Smith, J., Birchfield, S.: Toward low-flying autonomous MAV trail navigation using deep neural networks for environmental awareness, pp. 1–7. https://doi.org/10.48550/arXiv.1705.02550 (2017)
27. Kim, D.K., Chen, T.: Deep neural network for real-time autonomous indoor navigation. https://doi.org/10.48550/arXiv.1511.04668 (2015)
28. Wang, P., Hu, Y., Dai, Y., Tian, M.: Asphalt pavement pothole detection and segmentation based on wavelet energy field. Math. Probl. Eng. **2017**(Article ID 1604130), 13 (2017)

29. Ilie, M.D., Ciobanu, A., Negrescu, C., Stanomir, D.: Towards expressive Romanian speaking 3D avatars for multimedia interfaces. In: 2013 20th International Conference on Systems, Signals and Image Processing (IWSSIP), pp. 47–50 (2013)
30. Gupta, R., Kumari, A., Tanwar, S.: Fusion of blockchain and artificial intelligence for secure drone networking underlying 5G communications. Trans. Emerg. Telecommun. Technol. **32**(1), Article ID e4176 (2021)
31. Amer, K., Samy, M., Shaker, M., ElHelw, M.: Deep convolutional neural network-based autonomous drone navigation, pp. 1–9. https://doi.org/10.48550/arXiv.1905.01657 (2019)
32. Chandnani N., Khairnar C.N.: Quality of Service (QoS) Enhancement of IoT WSNs Using an Efficient Hybrid Protocol for Data Aggregation and Routing SN Computer Science **4**(6) https://doi.org/10.1007/s42979-023-02165-6 (2023)
33. Chandnani N., Khairnar C.N.: A Novel Hybrid Protocol in Achieving QoS Regarding Data Aggregation and Dynamic Traffic Routing in IoT WSNs Wireless Personal Communications **131**(1) 295–335 https://doi.org/10.1007/s11277-023-10429-w (2023)
34. Chandnani N., Khairnar C.N.: A Reliable Protocol for Data Aggregation and Optimized Routing in IoT WSNs based on Machine Learning Wireless Personal Communications **130**(4) 2589–2622 https://doi.org/10.1007/s11277-023-10393-5 (2023)
35. An analysis of architecture framework security and challenging aspects for data aggregation and routing techniques in IoT WSNs Theoretical Computer Science 92995–113 https://doi.org/10.1016/j.tcs.2022.06.032 (2022)
36. Chandnani N., Khairnar C.N .: Bio-Inspired Multilevel Security Protocol for Data Aggregation and Routing in IoT WSNs Mobile Networks and Applications **27**(3) 1030–1049 https://doi.org/10.1007/s11036-021-01859-6 (2022)
37. Chandnani N., Khairnar C.N .: A Comprehensive Review and Performance Evaluation of Recent Trends for Data Aggregation and Routing Techniques in IoT Networks. Lecture Notes in Networks and Systems, Springer, **100** 467–484, 2020, https://doi.org/10.1007/978-981-15-2071-6_37

Artificial Intelligence Algorithms and Models for Embodied Agents: Enhancing Autonomy in Drones and Robots

Gnanasankaran Natarajan, Elakkiya Elango, B. Sundaravadivazhagan, and Sandha Rethinam

Abstract The integration of artificial intelligence (AI) algorithms and models in the realm of embodied agents, such as drones and robots, has emerged as a transformative field with profound implications for various industries. This chapter provides a comprehensive overview of the state-of-the-art AI techniques employed to augment the autonomy, adaptability, and efficiency of these autonomous systems. The study begins by examining the fundamental principles that underpin AI algorithms for embodied agents, emphasizing the importance of sensor fusion, perception, and decision-making processes. It explores the symbiotic relationship between machine learning and embodied agents, showcasing how models evolve over time to enhance adaptability in dynamic environments. One of the key focal points is the application of deep learning architectures, including convolutional neural networks (CNNs) and recurrent neural networks (RNNs), to bolster the perceptual capabilities of drones and robots. These architectures enable real-time object recognition, scene understanding, and spatial mapping, facilitating improved navigation and interaction within complex surroundings. The chapter delves into the significance of reinforcement learning (RL) in training embodied agents, elucidating how these algorithms enable agents to learn from experiences and optimize their decision-making processes. Specifically, it investigates the challenges of implementing RL in the context of drones and robots, addressing issues related to sample efficiency, exploration–exploitation trade-offs, and safety concerns. Furthermore, the integration of AI algorithms for collaborative multi-agent systems is explored, emphasizing the coordination and communication strategies essential for achieving collective goals. The research investigates swarm

G. Natarajan (✉) · S. Rethinam
Department of Computer Science, Thiagarajar College, Madurai, Tamil Nadu, India
e-mail: sankarn.iisc@gmail.com

E. Elango
Department of Computer Science, Government Arts College for Women, Sivagangai, Tamil Nadu, India

B. Sundaravadivazhagan
Department of Information Technology, University of Technology and Applied Sciences, Al Mussanah, Oman

P. Raj et al. (eds.), *Building Embodied AI Systems: The Agents, the Architecture Principles, Challenges, and Application Domains*, Information Systems Engineering and Management 14, https://doi.org/10.1007/978-3-031-68256-8_5

intelligence, where groups of embodied agents work together seamlessly, inspired by natural systems like ant colonies and flocking behaviour in birds. The ethical considerations and societal implications of deploying AI-powered embodied agents are also discussed. As these systems become increasingly autonomous, concerns surrounding accountability, transparency, and the potential impact on employment and privacy are scrutinized. Finally, the chapter highlights emerging trends and future directions in AI for embodied agents, including the fusion of symbolic reasoning and deep learning, the integration of edge computing for real-time processing, and the continual evolution of AI models through online learning paradigms. This chapter provides a comprehensive synthesis of the current landscape of AI algorithms and models for embodied agents, shedding light on the advancements, challenges, and potential future developments in this rapidly evolving field.

Keywords Artificial Intelligence · Embodied agents · Autonomous systems · Deep learning · Reinforcement learning · Neural networks · Swarm intelligence

1 Brief Introduction About Artificial Intelligence (AI)

Artificial intelligence (AI) is at the cutting edge of advancements in technology progress, encompassing a discipline of software engineering devoted to the development of intelligent computers that can simulate human cognitive capabilities. The main objective of AI is to give robots the capacity to learn, reason, solve problems, and adapt on their own, allowing them to accomplish activities that would otherwise need human intellect. At its foundation, AI covers a wide range of methods, algorithms, and approaches that aid in machine learning, natural language processing, computer vision, and robotics. Machine learning, a subset of artificial intelligence, focuses on giving robots the ability to understand patterns from data and make intelligent decisions requiring being explicitly programmed.

AI has progressed through important milestones, through initial rule-based systems to modern deep learning architectures that are capable of digesting massive datasets and reaching human-like performance in specialised tasks. AI applications have invaded many aspects of our everyday lives, ranging from virtual assistants and personalised recommendations to sophisticated medical diagnostics and self-driving cars. As AI advances, it poses critical challenges about ethics, transparency, and the social consequences of intelligent computers. Combining the promise for remarkable gains with ethical concerns represents an important component of AI development. This overview delivers an insight into the vast and dynamic world of AI, where continuing research and innovation are transforming the field of technology, delivering novel perspectives and revolutionary potential for the future [8].

2 An Overview on the Artificial Intelligence Algorithms for Model Building and Its Use Cases

First, let us study about the many forms of learning, and then we will look at each of the AI algorithms.

2.1 *Supervised Learning*

Supervised learning is a machine learning paradigm that occurs when an algorithm is refined on a dataset that has been labelled in order to anticipate or learn mappings between input data and output labels. The word "supervised" denotes that the algorithm adapts from a set of labelled samples supplied over the training phase, with each input connected with the right output. Here are the key elements of supervised learning:

- **Labelled Dataset**: In supervised learning, the training dataset includes input samples and output labels. The labels represent the fundamental truth or intended outcome that the algorithm seeks to forecast.
- **Training Process**: During training, the algorithm modifies internal parameters to minimise the discrepancy between predicted and real labels. This entails iteratively modifying the model based on mistakes or inconsistencies between predicted and actual results.
- **Prediction or Inference**: The trained model can generate predictions or inferences from fresh data. The objective is to generalise from the training data in order to properly predict output labels for previously unidentified inputs.

2.1.1 Types of Supervised Learning

- **Regression**: The algorithms anticipate continuous output variables. For example, projecting property values based on factors such as square footage and the number of bedrooms.
- **Classification**: The algorithm classifies input data into discrete groups or classes. Examples include spam detection in emails (binary classification) and picture recognition (multiclass classification).

Supervised learning is widely used in a variety of fields; include natural language processing, picture identification, speech recognition, and medical diagnosis. It serves as the foundation for developing intelligent systems capable of learning and making decisions based on past data, allowing for automation and pattern recognition in a wide range of applications [21].

2.2 Unsupervised Learning

Unsupervised learning is a type of machine learning in which algorithms are trained on datasets without explicitly labelled outcomes or goal variables. Unsupervised learning, as opposed to supervised learning, in which the algorithm learns from labelled examples, involves extracting patterns, structures, or connections from input data in the absence of preset outputs. The main objective is to investigate the intrinsic structure of the data or to identify hidden patterns. The main properties of unsupervised learning are:

- **No Labelled Outputs**: In unsupervised learning, training data solely includes input characteristics without output labels. The algorithm has to determine the fundamental trends or connections in the input without explicit instructions on what the result should be.
- **Clustering**: It is a frequent unsupervised learning problem that groups related data points based on intrinsic features. For instance, K-Means clustering, hierarchical clustering, and DBSCAN.
- **Dimensionality Reduction**: Unsupervised learning employs dimensionality reduction techniques. Common approaches for dimensionality reduction include Principal Component Analysis (PCA) and t-distributed Stochastic Neighbour Embedding (t-SNE).
- **Association Rule Learning**: Identifies intriguing links connecting variables in data. Apriori and Eclat are two instances of algorithms for detecting common item sets in transaction data.
- **Generative Models**: Some unsupervised learning techniques use models that are generative to simulate the data's probability distribution. Examples are Gaussian Mixture Models (GMMs) and Variational Auto encoders (VAEs).
- **Anomaly Detection**: Unsupervised learning may find abnormalities and outliers in data. Anomaly detection methods include one-class SVMs (Support Vector Machines) and isolation forests.

Unsupervised learning is very beneficial when the purpose is to investigate and gain insights from data, discover hidden structures, or prepare data for subsequent tasks. It is widely used in disciplines including data clustering, exploratory data analysis, and feature engineering. Unsupervised learning is well-suited for cases in which labelled data is rare or expensive to get.

2.3 Reinforcement Learning

Reinforcement learning is a machine learning paradigm that describes how an agent trains to make decisions through interaction with its environment. The agent receives input in the form of incentives or penalties, helping it to learn the best conduct via trial and error. The lack of labelled training data is a critical feature of reinforcement learning; instead, the agent investigates the world, makes actions, and trains via the results of those activities.

Here are the basic aspects of reinforcement learning:

- **Agent**: The agent is an entity or system that learns and makes decisions in its surroundings. The agent acts based on its present knowledge of the environment, aiming to maximise its cumulative reward over time.
- **Environment**: It refers to the external system that occurs when the agent interacts. The environment reacts to the agent's behaviour by offering response in the way of rewards or punishments. The condition of the environment varies dependent on the agent's activities.
- **Actions**: The agent's probable motions or decisions based on the current environment. The agent makes decisions depending on its current policy or strategy.
- **State**: It refers to the current status or arrangement of the environment. The agent's decision-making process is frequently reliant on the present state, which changes as the agent performs activities.
- **Rewards**: The agent receives numerical values as feedback from the environment for their activities. The rewards represent how desirable the agent's behaviours are. The agent's goal is to acquire a policy that maximises the total reward over time.
- **Policy**: It refers to the agent's learned technique for mapping states to actions. The objective is to find a policy that result in the most beneficial cumulative rewards.

Reinforcement learning algorithms seek an ideal policy by iteratively examining the environment, performing actions, obtaining feedback, and modifying the policy depending on observed rewards. Q-Learning, Deep Q Networks (DQN), Policy Gradient Methods, and Actor-Critic Methods are examples of commonly used reinforcement learning algorithms. Reinforcement learning has found applications in a wide range of fields, involving game playing (e.g., AlphaGo), robots, autonomous systems, finance, and others, where decision-making in dynamic and unpredictable settings is a critical task [6].

2.4 Ensemble Learning

Heterogeneous learning is a machine learning approach that combines predictions from numerous different models to increase both performance and accuracy. The concept is that by combining the outputs of several models, the ensemble may typically outperform any one model functioning alone. Ensemble approaches are very useful for managing complex and varied datasets or models. The most important components of ensemble learning are:

- **Ensemble learning**: It includes training multiple base models separately. According to the ensemble approach, these basis models might be homogeneous (same type) or heterogeneous (different kind).
- **Ensemble learning's**: It power comes on its many base models. If the multiple models contribute distinct views or capture different parts of the basic trends in the data, the whole system will perform better. Diversity may be accomplished by utilising various techniques, data subsets, or sets of features for each base model.
- **Combining Predictions**: Base models' predictions are merged to get a final forecast. The aggregation procedure differs according to the ensemble approach. Common approaches include averaging (for regression problems) and voting (for categorization).
- **Numerous ensembles**: It approaches exist, each with a unique approach to mixing models. Some common ensemble approaches are:
  - **Bagging (Bootstrap Aggregating)**: Creates many models using distinct bootstrap samples from training data and combines their predictions (e.g., Random Forest).
  - **Boosting**: Trains models consecutively, with each model focused on rectifying the faults of its predecessor (for example, AdaBoost and Gradient Boosting).
  - **Stacking**: This involves developing a meta-model that incorporates the predictions of numerous base models, resulting in a more complicated and perhaps more accurate final forecast.
- **Benefits**: Ensemble learning includes better generalisation, robustness, and performance relative to individual models. It can lessen the effect of overfitting, decrease variance, and improve the model's capacity to deal with various features of the data.

Ensemble learning is frequently utilised in many machine learning applications, such as as classification, regression, and anomaly detection. Random Forest, AdaBoost, Gradient Boosting Machines (GBM), and neural network-based ensemble approaches are some frequent instances of ensemble methods [19].

3 Types of AI Algorithms for Model Building

Now, let us investigate the sorts of AI algorithms.

3.1 Supervised Learning Algorithms

Generating models in artificial intelligence (AI) entails using algorithms to teach machines to do tasks or make choices without explicit programming. The algorithms used are determined by the nature of the issue, the type of data provided, and the intended conclusion. Here's a full discussion of some widely used AI methods for model building:

(a) **Linear Regression:** Linear regression is a fundamental technique for performing regression problems. It establishes a linear link between input data and a continuous target variable, making it appropriate for numerical prediction.

Use Case: Predicting property values based on factors such as square footage and number of bedrooms.

(b) **Logistic Regression:** It is used for binary classification issues. It models the likelihood of an instance belonging to a given class.

Use Case: Spam detection in emails, with the purpose of classifying emails as spam or not.

(c) **Support Vector Machines (SVM):** These techniques are useful for combining regression and classification. They seek to identify a hyperplane that optimally divides distinct classes in the feature space.

Use Case: Image categorization and handwriting recognition.

(d) **Decision Trees and Random Forests:** Decision trees recursively partition data based on characteristics, whereas random forests combine predictions from numerous decision trees to enhance accuracy and resilience.

Use Cases: predicting customer attrition and detecting fraud.

(e) **Neural Networks:** Neural Networks architectures use linked layers of neurons to discover complicated patterns and represent data.

Use Cases: Image recognition, natural language processing, and speech recognition [17].

3.2 Unsupervised Learning Algorithms

(a) **K-Means Clustering**: It divides data into clusters based on similarity. It allocates all the points to the cluster whose centroid is closest.

Use Cases: Customer segmentation and anomaly detection.

(b) **Hierarchical Clustering**: This method generates a tree of clusters, with the root being a single cluster that contains all data points and the leaves being individual points.

Use Cases: Taxonomy construction and gene expression analysis.

(c) **Principal Component Analysis (PCA)**: PCA is a dimensionality reduction approach that determines the most essential characteristics and maps the data to a lower-dimensional space.

Use case: Feature reduction and visualisation [20].

3.3 Reinforcement Learning Algorithms

(a) **Q-Learning**: Q-learning is a model-free reinforcement learning algorithm that trains to make judgements by assigned a value (Q-value) to each behaviour in a certain state.

    Use Case: Game play and robotic control.

(b) **Deep Q Networks (DQN)**: DQN builds on Q-learning by employing a deep neural network to approximate Q-values, allowing the technique to handle more complicated state-action spaces.

    Use Case: DeepMind's AlphaGo for robotic control in difficult situations.

3.4 Ensemble Learning Algorithms

(a) **Gradient Boosting**: Gradient boosting creates a strong predictive model by merging the predictions of numerous weak models in a sequential order.

    Use Cases: Predicting stock prices and click-through rates.

(b) **AdaBoost**: AdaBoost focuses on progressively training weak models, giving larger weights to classified incorrectly cases in order to increase overall accuracy.

    Use Case: Face detection and text classification.

These algorithms are a subset of the many tools accessible for developing AI models. The method used is determined by the unique qualities of the data as well as the aims of the modelling endeavour. To attain optimal performance, successful model development frequently requires experimenting and fine-tuning [1, 11].

4 A Brief Introduction to Embodied Agents (Drones and Robots)

Embodied agents, including drones and robots, are a type of autonomous system meant to interact with the physical world. Figure 1 depicts a rudimentary robotic model. Embodied agents, as opposed to virtual agents, have a physical presence and are typically powered by artificial intelligence (AI) algorithms. These agents are able to see their surroundings, make decisions, and carry out actions in real-world scenarios.

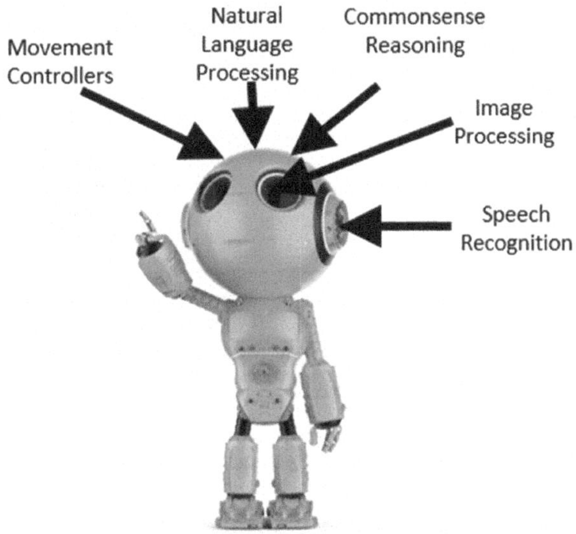

Fig. 1 Representation of a robot an example of an embodied agent

4.1 Advantages of Embodied Agents

- **Autonomy**: Embodied agents can navigate and make decisions autonomously, even in unexpected surroundings. This autonomy enables them to complete activities without requiring continual human assistance.
- **Adaptability**: These agents are able to adapt to varying situations and make real-time judgements based on sensor data. This flexibility is essential for jobs requiring interaction with a broad and dynamic physical environment.
- **Versatility**: Embodied agents are versatile, including drones, ground robots, and undersea vehicles. Their adaptability allows them to handle an extensive variety of applications across sectors.
- **Efficiency**: Embodied agents improve efficiency in several jobs. They may conduct repetitive, risky, or time-consuming jobs with accuracy and consistency, lowering the workload for human operators.
- **Data Collection**: Embodied agents have sensors that enable data collecting in various contexts. Drones, for example, can take aerial pictures, and ground robots can penetrate difficult terrains to acquire data [22].

Fig. 2 A drone in a rescue operation

4.2 *Applications of Embodied Agents*

4.2.1 Search and Rescue

Drones and robots play an important role in search and rescue efforts, particularly in disaster-prone areas, as seen in Fig. 2. They can explore hard-to-reach regions, assess damage, and identify survivors.

4.2.2 Surveillance and Security

Embodied agents are deployed for monitoring in both civilian and military settings. Drones can monitor huge regions from the air, whereas ground robots may patrol and give a physical presence in safety operations, as seen in Fig. 3.

Fig. 3 Drone in security surveillance

Fig. 4 Drone equipped in smart farming

4.2.3 Precision Agriculture

Figure 4 shows how drones equipped with sensors and cameras help to precision agriculture by monitoring crops, analysing soil conditions, and optimising resource utilisation. This increases agricultural yield while reducing environmental effect [2].

4.2.4 Logistics and Delivery

Companies are looking at the utilisation of drones and ground robots during last-mile delivery, which would provide a more rapid and effective conveyance of products. This application has become popular in the e-commerce market. Figure 5 depicts the use of robots in packaging items for shipment and logistics.

4.2.5 Environmental Monitoring

Embodied agents are used in environmental research and monitoring, as seen in Fig. 6. Drones can scan forests, follow wildlife, and collect climatic data, all of which help with ecological research and conservation.

4.2.6 Manufacturing and Warehousing

Robots are used in industrial operations for activities such as assembling, welding, and quality monitoring, as seen in Fig. 7. In warehouses, autonomous robots can roam shelves, retrieve things, and improve inventory control.

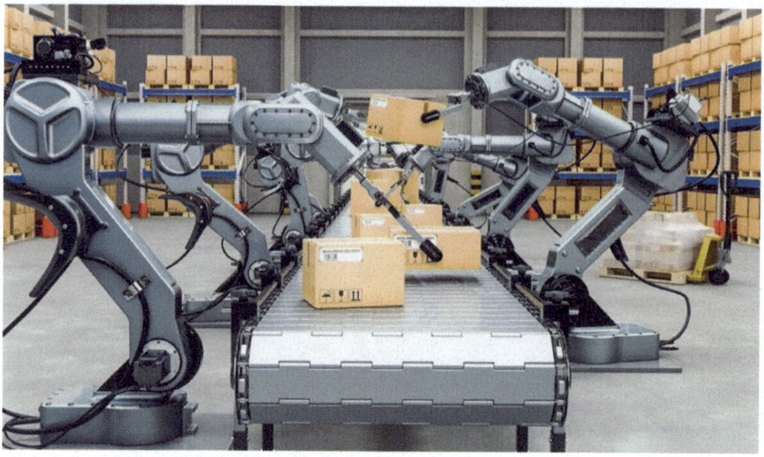

Fig. 5 Robots equipped for packing in a logistics company

Fig. 6 Drone utilised for pollution control monitoring

4.2.7 Healthcare Assistance

Robots are rapidly being employed in healthcare to perform activities such as patient care, pharmaceutical distribution, and surgical aid. Drones may also be used to provide medical supplies in distant or emergency circumstances, as seen in Fig. 8.

Fig. 7 Robots working in Industry and warehouse

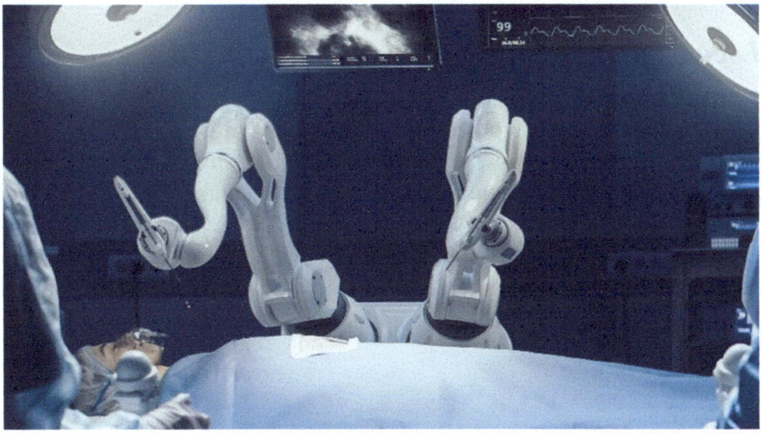

Fig. 8 Robots equipped for surgery process in a hospital

5 Role of Artificial Intelligence Algorithms on Embodied Agents to Maintain the Autonomy and Efficiency of the Autonomous Systems

Artificial intelligence (AI) algorithms serve an important role in improving the powers and autonomy of embodied agents like drones and robots. These algorithms enable these agents to sense their surroundings, make intelligent judgements, and carry out actions, allowing them to function independently in the real world. AI approaches are critical in ensuring the autonomy and efficiency of autonomous systems across several domains. These strategies enable these systems to observe, learn, adapt, and make decisions in real time, allowing them to function autonomously and successfully. Here's a full analysis of the important functions that AI algorithms play in embodied agents:

5.1 Perception and Sensing

AI algorithms allow embodied entities to understand data from a variety of sensors, such as cameras, LiDAR, radar, and other environmental sensors. These algorithms analyse raw sensor data to find relevant traits, identify objects, and comprehend the agent's environment. Computer vision algorithms, sensor fusion techniques, and deep learning models (including Convolutional Neural Networks) are frequently used for perception tasks. These techniques enable agents to recognise objects, estimate distances, and construct a cohesive picture of their surroundings.

5.2 Localization and Mapping

To successfully navigate, embodied agents must first grasp where they are in the surroundings and develop maps. AI algorithms aid in simultaneous mapping and localization (SLAM), which enables agents to create and updated maps while recognising their present position. SLAM methods, which frequently combine probabilistic methodologies and sensor fusion, allow agents to generate accurate maps of their environment. Particle filters, Advanced Kalman Filters, and graph-based algorithms are widely employed for localization and mapping [9].

5.3 Decision-Making and Planning

AI algorithms enable embodied entities to make intelligent decisions based on their perceptions of their surroundings and desired outcomes. These algorithms establish the appropriate activities the agent must perform to attain its goals while taking into account the environment's dynamic character. Reinforcement learning is commonly used in decision-making, in which agents learn optimum policies via trial and error. Markov Decision Processes (MDPs) and Monte Carlo Tree Search (MCTS) are also used to plan and make decisions in complicated and unpredictable situations.

5.4 Path Planning and Navigation

AI techniques let embodied agents explore their surroundings more efficiently. These algorithms make sure agents may go from their present place to their destination while avoiding impediments and optimising for time, energy, and safety. Common path planning techniques include the A* algorithm (A-star), Dijkstra's algorithm, and quickly expanding random trees (RRT). Machine learning techniques can additionally be used to adjust and update navigation tactics depending on experience [14].

5.5 Learning and Adaptation

AI algorithms enable embodied beings to learn from their experiences and adapt to changing situations. This flexibility is critical for agents to deal with changing circumstances, uncertainties, and unexpected problems. Reinforcement learning is a very effective model for learning adaptive actions. Agents acquire knowledge through the rewards and punishments they experience when interacting with the environment. Online learning approaches, transfer learning, and meta-learning are all used to improve adaptability.

5.6 Communication and Collaboration

AI algorithms make communication and cooperation easier when numerous embodied beings are working together. These algorithms allow agents to communicate, coordinate activities, and reach common goals. Collaboration is facilitated by the use of communication protocols, consensus methods, and coordination mechanisms. Swarm intelligence and multi-agent reinforcement learning are used to facilitate decentralised coordination among numerous actors [7].

5.7 Human–Robot Interaction

AI algorithms help to create natural and user-friendly interfaces for human–robot interaction. This involves recognising human gestures, recognising voice, and responding appropriately to orders or requests. Natural Language Processing (NLP) algorithms, gesture recognition models, and human-aware planning approaches are used to improve the interactions of embodied creatures with humans.

5.8 Fault Detection and Diagnostics

AI approaches aid in the identification of errors or abnormalities in autonomous systems, as well as the facilitation of diagnostic processes for effective troubleshooting. Anomaly detection methods, machine learning classifiers, and failure prediction models help to monitor the system's health. These strategies allow for early discovery of faults, which improves overall dependability.

5.9 Energy Optimization

AI approaches help to optimise energy usage in autonomous systems, resulting in more effective resource use. Machine learning algorithms may forecast energy requirements based on past data, allowing systems to optimise their operations for energy efficiency. Optimisation algorithms assist in route planning and work scheduling to reduce energy usage.

Thus, AI algorithms provide the foundation of embodied agents, allowing them to feel, see, plan, learn, and communicate with the physical world. These algorithms let embodied entities achieve the autonomy and intelligence needed to accomplish different tasks across several domains. The collaboration of AI with embodied agents is constantly evolving, resulting in novel applications and advances in sectors such as robotics, self-driving cars, and intelligent infrastructure. AI approaches help autonomous systems maintain their autonomy and efficiency by allowing them to detect, learn, make decisions, and deal with dynamic and complicated situations. As these approaches continue to progress, autonomous systems grow increasingly competent, adaptable, and dependable to handle a wide range of applications, including autonomous cars and robotic systems.

6 Deep Learning: An In-Depth Overview

Deep learning is a subtype of machine learning that entails teaching artificial neural networks to complete tasks without explicit programming. The term "deep" relates to the usage of deep neural networks, which include numerous layers (hence the name) and process data to create predictions or classifications. Deep learning has been extremely popular and successful in a variety of fields, notably computer vision, natural language processing, speech recognition, and others.

6.1 Key Concepts in Deep Learning

- **Neural Networks**: It relies heavily on artificial neural networks, which are modelled after the human brain. These networks are made up of linked nodes organised in layers.
- **Layers**: Neural networks generally consist of three layers: input, hidden, and output. Each layer comprises nodes (neurons), and the connections between nodes are weighted.
- **Activation functions**: Activation functions add nonlinearity to the network, allowing it to simulate complicated interactions. Rectified Linear Unit (ReLU), Sigmoid, and Hyperbolic Tangent (tanh) are examples of commonly used activation functions.
- **Weights and Biases**: During training, the network changes weights and biases to reduce the gap between projected and actual results. Stochastic gradient descent is a common optimisation approach used for this procedure.
- **Loss Function**: It measures the difference between expected and actual values. The purpose of training is to minimise this loss, allowing the model to generate more accurate predictions [15].

6.2 Deep Learning Architectures

Deep learning architectures are neural network topologies created to process and learn from complicated data. These designs use numerous layers (deep layers) to automatically learn hierarchical representations of incoming data, allowing them to capture complex patterns and characteristics. Figure 9 depicts a popular deep learning architecture.

Here's an in-depth look at several popular deep learning architectures.

6.2.1 Feedforward Neural Networks (FNN)

It represents perhaps the most basic type of neural network, with information flowing in a single direction from the input layer to the output layer via hidden layers. Its typical applications include image and audio recognition, as well as classification jobs.

6.2.2 Convolutional Neural Networks (CNN)

This network is designed specifically for processing grid-structured data, such as pictures. CNNs employ convolutional layers to analyse input data for local patterns, which makes them ideal for applications that need spatial connections. CNN has a

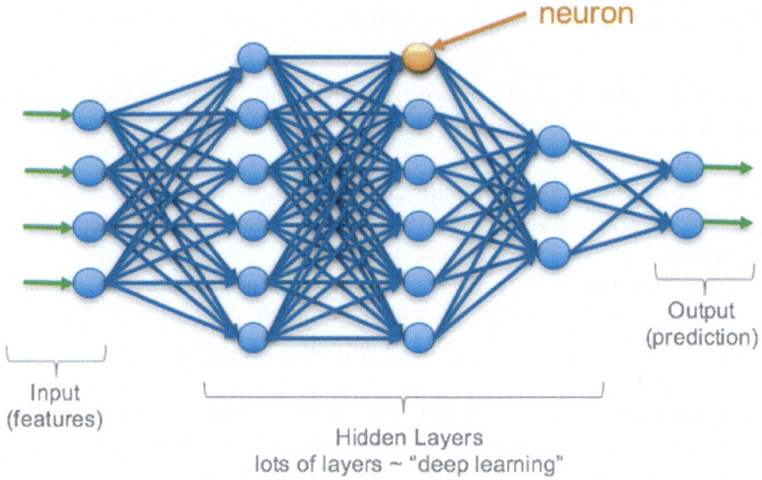

Fig. 9 Architecture of a common deep learning network

few popular applications, including image categorization, object identification, and facial recognition.

6.2.3 Recurrent Neural Networks (RNN)

RNN was created for sequential data processing, including feedback loops that allow information to remain over time. Traditional RNNs are prone to vanishing gradient issues. RNN's uses include natural language processing and time series prediction.

6.2.4 Long Short-Term Memory (LSTM) Networks

Unlike typical RNNs, LSTMs employ memory cells and gates to overcome vanishing gradient concerns. They can identify long-term dependencies in sequential data. LSTM uses include speech recognition, language translation, and sentiment analysis.

6.2.5 Gated Recurrent Units (GRU)

Like LSTMs, GRUs is additionally computationally effective and have a simpler design. They employ gating mechanisms to regulate the rapid transfer of information. Its applications are comparable to LSTMs: natural language processing and sequence modelling.

6.2.6 Autoencoders

It is used for unsupervised learning and feature learning. Autoencoders consist of an encoder that maps input data to a latent space and a decoder that reconstructs the input from the latent space. Data denoising, dimensionality reduction, anomaly detection are few applications of autoencoders.

6.2.7 Generative Adversarial Networks (GAN)

GANs, which consist of a generator and a discriminator, are used to produce fresh data that is indistinct from the training set. The generator generates data, while the discriminator verifies its legitimacy. GAN has a few uses, including image production, style transfer, and data augmentation.

6.2.8 Transformers

The network is Transformers, which were first introduced for natural language processing problems, employ self-attention techniques to make predictions based on distinct sections of the input sequence. They are currently essential in language comprehension tasks. Transformers have a few uses, including machine translation, text summarization, and language modelling [5].

6.2.9 Capsule Networks (CapsNets)

The network is CapsNets, proposed as a replacement to convolutional neural networks, are intended to overcome difficulties such as perspective variance. They employ capsules to indicate the hierarchical connections between pieces of an object. CapNets uses include image recognition, particularly when dealing with variations in object poses.

6.2.10 Deep Reinforcement Learning (DRL) Networks

These are a combination of deep learning and reinforcement learning. DRL networks educate agents how to make decisions in a situation by engaging with it and getting rewards or punishments. DRL applications include game playing (such as AlphaGo), robotic control, and autonomous systems. These architectures are a subset of the broad range of deep learning models. The architecture used is determined by the nature of the work, the qualities of the data, and the expected consequences. Continued research and innovation in deep learning architectures help to drive the field's rapid expansion and applicability across several fields.

7 Training Deep Learning Models

- **Forward Pass**: The network receives input data and generates predictions.
- **Loss Computation**: A loss function measures the difference between expected and actual values.
- **Backward Pass (Backpropagation)**: It minimises loss by computing gradients with regard to model parameters and updating weights accordingly.
- **Iterative optimisation**: The method is performed across numerous epochs until the model achieves suitable performance.
- **Computational Resources**: Deep learning model training requires sophisticated hardware, such as GPUs or TPUs, due to its computational demands.
- **Deep learning models**: It requires significant volumes of labelled data for efficient training. Insufficient data may result in overfitting.
- **Deep learning models**: These are typically termed "black boxes" because to their complexity, making it difficult to understand their decision-making processes.

Despite hurdles, deep learning has transformed several industries, showing extraordinary success in tasks which were previously thought to be incredibly difficult. Continued research and breakthroughs in architectures, training techniques, and interpretability are propelling the progress of deep learning [13].

8 Significance of Reinforcement Learning in Training the Embodied Agents

Reinforcement learning (RL) is a machine learning paradigm that is especially useful in the development of embodied agents like drones and robots. Unlike traditional programming techniques that rely on explicit instructions, reinforcement learning allows agents to acquire knowledge by communicating with their surroundings and getting feedback in the way of rewards or penalties. Below are few main reasons why reinforcement learning is important in training embodied agents.

8.1 Learning from Interaction

Embodied agents work in dynamic, unpredictable contexts. RL enables children to acquire knowledge by communicating with their surroundings, making decisions, and experiencing the results of their actions. This is critical for adjusting to real-world scenarios in which the agent may face unexpected events.

8.2 *Adaptability to Shifting Environments*

RL enables agents to react to changes in their environment. Drones and robots must be able to constantly learn and adapt how they act since they work in a variety of and frequently unpredictable environments. RL algorithms enable agents to learn new abilities and change current ones as required [4].

8.3 *Self-Directed Decision-Making*

Reinforcement learning allows embodied entities to arrive at autonomous decisions based on previously learned information. This is particularly important in situations when continual human interaction is impractical or not possible, such as autonomous drones or robots deployed in remote or hazardous settings.

8.4 *Optimizing Long-Term Performance*

RL is designed to optimize cumulative rewards over time, encouraging agents to make decisions that lead to favourable outcomes in the long run. This is essential for embodied agents that may need to achieve complex goals or navigate challenging environments, as they can learn strategies that lead to better overall performance.

8.5 *Handling High-Dimensional and Continuous Input Spaces*

RL is intended to optimise cumulative rewards over time, motivating agents to make actions that result in positive long-term results. This is critical for embodied agents who need to attain complicated goals or negotiate difficult surroundings, since they can learn tactics that lead to improved overall performance.

8.6 *Transferability of Learned Skills*

Many real-world issues need high-dimensional and continuous input spaces, which might be difficult for standard programming approaches. RL techniques, particularly deep reinforcement learning (DRL), have shown useful in dealing with such complicated input fields, allowing agents to process and react to a wide variety of sensory data.

8.7 Real-Time Decision-Making

Embodied agents frequently need to make judgements in real time. RL algorithms, particularly those optimised for efficiency, can aid in quick decision-making by letting agents to swiftly learn and alter their behaviour in response to instant feedback from the environment. Reinforcement learning is important in training embodied agents because it offers a framework for learning through interaction, adapting to changing environments, making autonomous decisions, optimising long-term performance, dealing with complex input spaces, sending learned skills, and providing real-time decision-making. These variables are critical to the successful deployment of drones and robots in real-world applications [18].

9 Integration of AI Algorithms for Collaborative Multi Agent Systems

The addition of artificial intelligence (AI) algorithms into interactive multi-agent systems is an important field of study and application, wherein AI techniques serve to improve interaction, communication, and decision-making among numerous agents. Collaborative multi-agent systems are composed of several intelligent creatures that collaborate to achieve shared goals or solve challenging challenges. Here's a summary of how to make use of AI algorithms in this context.

9.1 Cooperative Decision-Making

AI algorithms allow agents to reach decisions which contribute towards the group's overall success. Cooperative decision-making entails evaluating the behaviours of other agents and adopting choices that maximise collective performance. Game theory, for example, is frequently used to simulate strategic interactions and decision-making in multiagent systems.

9.2 Communication and Information Sharing

Effective communication is crucial for collaboration among agents. AI algorithms can be used to design communication protocols that allow agents to share relevant information, coordinate actions, and negotiate decisions. Natural language processing (NLP) and communication protocols based on signalling or shared beliefs are examples of AI techniques used in this context.

9.3 Distributed Learning

Agent cooperation relies heavily on effective communication. AI algorithms may be utilised to create communication protocols that enable agents to communicate important information, coordinate activities, and reach agreements. Artificial intelligence techniques employed in this area include natural language processing (NLP) and communication protocols based on signalling or shared ideas.

9.4 Task Allocation and Resource Management

Agents in a collaborative multi-agent system can use AI algorithms to facilitate distributed learning. Each agent learns via how it interacts with the environment and other agents, which adds to the system's aggregate knowledge. Reinforcement learning, transferable learning, and federated education are all strategies used to promote distributed learning in multi-agent systems.

9.5 Coordination and Synchronization

Coordinating the behaviours of several agents presents a challenging task. AI algorithms contribute to the development of coordination mechanisms that enable agents to synchronise their activity. Coordination and conflict resolution in multi-agent systems are achieved using approaches such as consensus algorithms, auction-based procedures, and market-based processes.

9.6 Adaptability and Dynamic Environments

Multi-agent systems frequently work in dynamic, uncertain contexts. AI algorithms improve the system's flexibility by allowing agents to learn and update their strategies as they react to environmental changes. Agents' adaptability may be enhanced using machine learning approaches such as online learning and adaptive algorithms.

9.7 Swarm Intelligence

Swarm intelligence is a type of artificial intelligence that is inspired by the collective behaviour of social insects. It entails creating algorithms that allow a group of agents to engage in intelligent behaviour via local interactions. Ant colony optimisation, particle swarm optimisation, and bee-inspired algorithms are some examples of swarm intelligence approaches used in collaborative multi-agent systems.

9.8 Ethical Considerations and Norms

AI algorithms can be used to impose ethical principles and standards inside a collaborative multi-agent system. This entails implementing rules and principles that regulate agent activity to promote fairness, transparency, and ethical compliance. The addition of AI algorithms into interactive multi-agent systems is a multidisciplinary endeavour that employs a variety of AI approaches to improve intelligent agents' communication, decision-making, coordination, and adaptability. These systems have applications in a variety of fields, including robots, autonomous cars, smart grids, and decentralised networks [10].

10 Challenges and Considerations

The development and implementation of artificial intelligence (AI) algorithms as well as models for embodied agents such as drones and robots presents a number of obstacles. Here are some of the primary problems connected with increasing autonomy in drones and robots using AI:

- **Complexity of Real-World Environments**: Drones and robots work in complicated and dynamic real-world contexts. Designing AI algorithms that can explore and make judgements in the face of uncertainties, impediments, and unanticipated events is still a serious task.
- **Sensor Limitations and Integration**: Embodied agents rely on sensors to interact with their surroundings. Integrating input from multiple sensors (e.g., cameras, lidar, and accelerometers) and dealing with constraints like as noise, occlusions, and sensor failures provide obstacles in constructing strong AI models that can make correct judgements.
- **Autonomous Decision-Making under Uncertainty**: Real-world unpredictability is a challenge for autonomous decision-making. AI systems must be able to handle ambiguity and make intelligent judgements even when presented with incomplete or noisy data.
- **Adaptability to Dynamic Environments**: Embodied agents must adapt to changing environments in order to function effectively. Creating AI models that can learn and adapt in real time to dynamic changes in their environment is difficult, especially when training data does not completely represent all conceivable scenarios.

- **Human–Robot Interaction and Collaboration**: Understanding and responding to human cues, intentions, and social conventions is a complicated task for drones and robots. Designing AI models that enable seamless human–robot collaboration while maintaining safety and trust is a continuing area of research.
- **Energy Efficiency and Computational Resources**: Deep learning methods, in particular, can need significant computer resources. Drones, in particular, have limited onboard processing resources and energy requirements. Creating AI models that match the demand for processing capacity with the platform's resource constraints is critical for effective implementation.
- **Explain ability and Trust**: Embodied agents must be explainable and trustworthy to end-users as they gain autonomy. Ensuring AI models' openness and interpretability is critical for increasing trust in their abilities and allowing human monitoring.
- **Robustness to Adversarial Attacks**: AI models, particularly those using machine learning, are vulnerable to adversarial assaults. Adversaries may modify input data to confuse the model and lead to inaccurate judgements. Ensuring the resilience of AI systems against adversarial assaults is a key problem, particularly in security-sensitive applications.
- **Legal and Ethical Considerations**: The utilisation of autonomous embodied beings presents legal and ethical considerations, including as responsibility, accountability, and privacy. To address these challenges, ethical and legal rules for developing and deploying AI for embodied beings must be established alongside technical solutions.
- **Scalability and Generalization**: AI models for embodied beings must be scalable and generalizable across several contexts and circumstances. Scalability and making sure models work well in a variety of scenarios are difficult tasks, especially when training data is restricted or biassed.

Addressing these difficulties needs a multidisciplinary strategy that includes skills in AI, robotics, computer vision, human–computer interaction, and other domains. Continuous research and developments in AI algorithms and models are required to overcome these obstacles and realise the full potential of autonomous drones and robots in a variety of applications [16].

11 Emerging Trends and Future Directions in AI for Embodied Agents

Several trends and directions have emerged in the discipline of AI for embodied entities. While I am unable to offer real-time updates, here are some basic patterns and likely future directions based on the trends as of my previous update:

11.1 Embodied AI and Robotics Integration

There is an increasing trend of closer integration among AI and robots. Researchers are working on constructing embodied agents that may communicate with the physical environment in more complex ways, which will improve applications in fields like as industrial automation, healthcare, and autonomous cars.

11.2 Explainable AI (XAI) for Trust and Transparency

As embodied agents get increasingly sophisticated, there is a greater emphasis on generating their decision-making processes understandable and visible. Explainable AI (XAI) is a growing field that aims to give insights into how AI models make decisions, hence building trust and understanding in human-AI interactions.

11.3 Sim-To-Real Transfer Learning

Because real-world training is expensive and dangerous, simulated settings are frequently employed to teach embodied agents. Transfer learning approaches are becoming more important in adapting agents taught in simulated settings to the complexity of the real world.

11.4 Multi-Modal Sensing and Perception

Future embodied agents are likely to use a variety of sensory mechanisms, such as vision, audition, and touch. Integrating information from many sensory modalities enables agents to have a better awareness of their surroundings while making more educated judgements.

11.5 Continual Learning and Adaptability

Embodied agents that can constantly acquire knowledge and adjust to new circumstances are gaining popularity. This includes creating algorithms and structures that enable agents to learn new abilities, adapt to novel tasks, and enhance their performance over time [12].

11.6 Social Intelligence and Human–Robot Interaction

Improving social intelligence is crucial for embodied creatures designed for human interaction. Future themes include creating agents that are capable of responding to human emotions, gestures, and social cues, resulting in them being better suited for collaborative activities in a variety of situations.

11.7 Edge AI for Real-Time Processing

The integration of AI at the network's edge, rather than depending purely on centralised processing, is a development that improves embodied agents' real-time performance. This is especially significant in applications involving autonomous cars and drones.

11.8 Ethics and Bias Mitigation

There is a growing recognition of the ethical concerns and possible biases in AI systems. Future AI advancements for embodied beings are anticipated to feature more sophisticated efforts to eliminate biases, maintain fairness, and follow ethical principles throughout design and deployment.

11.9 Neuromorphic Computing

Neuromorphic computing, which was inspired by the design and operation of the human brain, is a field of study that potentially influence the creation of more efficient and brain-like AI systems. This might result in more energy-efficient and adaptable embodied agents.

11.9.1 Quantum Computing for AI

As quantum computing technology progresses, it has the potential to revolutionise artificial intelligence. Quantum algorithms might greatly accelerate some calculations related to embodied agents, resulting in stronger and more effective AI systems [3].

12 Conclusion

In order to integrating artificial intelligence algorithms and models for embodied agents, with an emphasis on improving autonomy in drones and robots, marks a significant step forward in the discipline. These intelligent systems use AI to learn from their surroundings, react to changing situations, and make educated judgements in real time. The combination of AI with embodied agents enables autonomous navigation, collaborative work, and effective communication with the physical world. As technology advances, ongoing study in explainable AI, multi-modal sensing, continuous learning, and ethical considerations is expected to improve the functions of these agents, allowing for seamless integration into a wide range of applications such as industrial automation, healthcare, and beyond. The future promises hopeful prospects for the continuous progress of AI-driven autonomy in embodied entities, paving the way.

References

1. Asadi, E., Da Silva, M.G., Antunes, C.H., Dias, L., Glicksman, L.: Multi-objective optimization for building retrofit: a model using genetic algorithm and artificial neural network and an application. Energy Build. **81**, 444–456 (2014)
2. Chen, X., Hosseini, R., Panetta, K., Sinapov, J.: A framework for multisensory foresight for embodied agents. In: 2021 IEEE International Conference on Robotics and Automation (ICRA), pp. 10927–10933. IEEE (2021)
3. Duan, J., Yu, S., Tan, H.L., Zhu, H., Tan, C.: A survey of embodied AI: from simulators to research tasks. IEEE Trans. Emerg. Top. Comput. Intell. **6**(2), 230–244 (2022)
4. Erbas, M.D., Winfield, A.F., Bull, L.: Embodied imitation-enhanced reinforcement learning in multi-agent systems. Adapt. Behav. **22**(1), 31–50 (2014)
5. Iphofen, R., Kritikos, M.: Regulating artificial intelligence and robotics: ethics by design in a digital society. Contemp. Soc. Sci. **16**(2), 170–184 (2021)
6. Kallem, S.R.: Artificial intelligence algorithms. IOSR J. Comput. Eng. (IOSRJCE) **6**(3), 1–8 (2012)
7. Larson, D.A.: Artificial Intelligence: robots, avatars, and the demise of the human mediator. Ohio St. J. Disp. Resol. **25**, 105 (2010)
8. Norvig, P.R., Intelligence, S.A.: A modern approach. Prentice hall upper saddle river, NJ, USA. In: Rani, M., Nayak, R., Vyas, O.P. (2015) (eds.) An Ontology-Based Adaptive Personalized e-Learning System, Assisted by Software Agents on Cloud Storage, vol. 90, pp. 33–48. Knowledge-Based Systems (2002)
9. Obaid, M., Johal, W., Mubin, O.: Domestic drones: context of use in research literature. In: Proceedings of the 8th International Conference on Human-Agent Interaction, pp. 196–203 (2020)
10. Oliveira, E., Fischer, K., Stepankova, O.: Multi-agent systems: which research for which applications. Robot. Auton. Syst. **27**(1–2), 91–106 (1999)
11. Pelikan, M.: Probabilistic model-building genetic algorithms. In: Proceedings of the 13th Annual Conference Companion on Genetic and Evolutionary Computation, pp. 913–940 (2011)
12. Pfeifer, R., Iida, F.: Embodied artificial intelligence: trends and challenges. Lect. Notes Comput. Sci. 1–26 (2004)
13. Pouyanfar, S., Sadiq, S., Yan, Y., Tian, H., Tao, Y., Reyes, M.P., Iyengar, S.S.: A survey on deep learning: algorithms, techniques, and applications. ACM Comput. Surv. (CSUR) **51**(5), 1–36 (2018)

14. Roy, N., Posner, I., Barfoot, T., Beaudoin, P., Bengio, Y., Bohg, J., Van de Panne, M.: From machine learning to robotics: challenges and opportunities for embodied intelligence. https://doi.org/10.48550/arXiv.2110.15245 (2021)
15. Shrestha, A., Mahmood, A.: Review of deep learning algorithms and architectures. IEEE Access **7**, 53040–53065 (2019)
16. Sikora, R., Shaw, M.J.: A multi-agent framework for the coordination and integration of information systems. Manag. Sci. **44**(11-part-2), S65–S78 (1998)
17. Smith, A., Director, F.: Using Artificial Intelligence and Algorithms. FTC (2020)
18. Tellols, D., Lopez-Sanchez, M., Rodríguez, I., Almajano, P., Puig, A.: Enhancing sentient embodied conversational agents with machine learning. Pattern Recogn. Lett. **129**, 317–323 (2020)
19. Tõugu, E.: Algorithms and Architectures of Artificial Intelligence, vol. 159. IOS Press (2007)
20. Wang, Z., Srinivasan, R.S.: A review of artificial intelligence based building energy use prediction: contrasting the capabilities of single and ensemble prediction models. Renew. Sustain. Energy Rev. **75**, 796–808 (2017)
21. Weiss, G. (ed.): Multiagent Systems: A Modern Approach to Distributed Artificial Intelligence. MIT press (1999)
22. Xia, F., Zamir, A. R., He, Z., Sax, A., Malik, J., Savarese, S.: Gibson Env: real-world perception for embodied agents. In: Proceedings of the IEEE Conference on Computer Vision and Pattern Recognition, pp. 9068–9079 (2018)

Dr.N. Gnanasankaran MCA., Ph.D., RUSA PDF,. is presently working as an Assistant Professor in the Department of Computer Science, Thiagarajar College, Madurai, India. He has 15 Years of Teaching and 14 years of Research Experience. His areas of research specialization include Software Engineering, SQA, Data Science, Machine Learning, Big Data and IOT. He pursued his Masters in Computer Application (2008) and Doctorate in Computer Science (2014) from Madurai Kamaraj University, Madurai, Tamilnadu, India and Completed his Post Doc in Computer Science (2020) from Alagappa University, Karaikudi, Tamilnadu, India. He has contributed more than 35 research articles in Scopus and Web of Science Indexed journals/ conferences from IEEE, Elsevier, Springer and Tech Science Press etc. He has contributed One Text Book publication, Two Edited Book Volume in Springer Nature (UK) and Apple Academic Press (Canada). He has published Ten Book Chapters in prominent Publishers such as Taylor and Francis, CRC Press, Routledge, De Gruyter, IEEE River publishers, IGI Global, Springer Nature etc. He holds eight patent work (two Indian design and one utility patent/ three UK design patent/ two German utility patent) in the field of Cyber Security, IoT, Drone Technology and Healthcare. He has secured major and minor funded projects from UGC, India. He is also a review committee member in many Scopus and SCI indexed Journals such as Tech Science Press, Inder Science Publishers, CRC Press, IGI Global Publishers, ASEAN Journal, Malaysia and acted as a review committee member and advisory committee member in IEEE and Springer Conferences. He has organized many seminars, workshops, and faculty development programs in various institutions he served. He has delivered several Guest Lectures and represented as a key note speaker and resource person in International Conferences and Faculty Development Programs. He is a life member in Indian Society of Technical Education and Institution Innovation Council and Yukti, India.

Scopus ID: 57215985065
ORCID: 0000-0001-9486-6515
Vidwan ID: 338870
Google Scholar ID: b0tx6GIAAAAJ
LinkedIn: https://www.linkedin.com/in/dr-n-gnanasankaran-natarajan-49b4b851
ResearchGate: https://www.researchgate.net/profile/NatarajanGnanasankaran
Google Scholar: https://scholar.google.com/citations/user=b0tx6GIAAAAJ&hl=en
Vidwan: https://vidwan.inflibnet.ac.in/profile/338870

Researcher ID: JFS-1817-2023.

Dr. Elakkiya Elango M.Sc., M.phil., Ph.D., is currently working as a Guest Lecturer in the Department of Computer Science under the Government Arts College for Women, Sivagangai. She has 8 Years of Teaching and 11 years of Research Experience. His areas of research specialization include Data Mining, Machine Learning, Artificial Intelligence, NLP and IoT. She has completed his Bachelor's in Information Technology from 2004 to 2007, Master's in Information Technology from 2007 and 2009, and Master of Philosophy in Computer Science from 2009 to 2011and PhD in Computer science from 2014 to 2018 respectively. She has contributed more than 8 research articles in Scopus Indexed journals from Elsevier, Thomsan Router etc. and published 4 Book Chapters in CRC Press, Elsevier and also communicates and approved in more than 8 chapters in CRC Press, Elsevier, and Taylor Francis.etc. She holds 7 UK patents, 2 German utility and 1 Indian Design patent. She is the life member in Professional Member of ICSES, Life Membership of International Association of Engineers (IAENG) and Lifetime Membership of MathTech Thinking Foundation (MTTF). She is also a potential Reviewer in ETASR journal and other international publishers.

Google scholar ID: https://scholar.google.com/citations?user=faGCPFQAAAAJ&hl=en
Research Gate ID: https://www.researchgate.net/profile/Elakkiya-Elango-2
Orchid ID: https://orcid.org/0000-0002-5373-7116
Scopus ID: https://www.scopus.com/freelookup/form/author.uri
Linkedin ID: https://www.linkedin.com/in/elakkiya-e-775a291a8/.

B. Sundaravadivazhagan has obtained PhD in computer science from Anna University, Chennai India. Currently, he is faculty in the Department of Information Technology at the University of Technology, and Applied Science-AL Mussanah in Oman He has professional memberships in ISACA and ISTE and is a senior member of IEEE. His academic and research background spans more than 23 years at many institutions. Working on two financed research projects for the Ministry of Higher Education Research Innovation in Oman via TRC. His research interests include the Cyber Security, Artificial Intelligence and machine learning, deep learning, and cloud computing. He has published more than 55 technical articles in journals and conferences throughout the world.

Orcid Id: https://orcid.org/my-orcid?orcid=0000-0002-5515-5769
Academic URL: https://www.act.edu.om/staffdirectory/index.aspx?XY=496
LinkedIn: https://www.linkedin.com/in/dr-b-sundaravadivazhagan-a3562b136/
Google Scholar: https://scholar.google.com/citations?user=gWbzTjMAAAAJ&hl=en.

Dr. R. Sandha MCA Ph.D. is presently working as an Assistant professor in the department of computer science, Thiagarajar College, Madurai, India. She has 10 years of teaching and 9 years of research experience. Her areas of research specialization include Cloud computing, image processing, cryptography and datamining. She pursued her master in computer application (2009) and doctorate in computer science in (2020) From Bharathiar University, coimbatore, Tamil Nadu, India. She has contributed more than 5 research articles in Scopus and conferences from Elsevier,Springer and IEEE. She is also a review committee member. She has organized many seminars, workshops and faculty development programs in various institutions she served. She has delivered several guest lectures and represented as a keynote speaker and chairperson in international conferences.

Enhanced Security and Privacy from Industry 4.0 and 5.0 Vision

Tarun Kumar Vashishth, Vikas Sharma, Kewal Krishan Sharma, and Bhupendra Kumar

Abstract Advent of Industry-4.0 and envisioned future of Industry-5.0 bring unprecedented advancements in technology, connectivity, and automation. This paper explores the critical aspects of enhanced security and privacy within the context of these industrial revolutions. Industry 4.0, characterized by incorporation of cyber-physical systems, IoT, and data analytics, introduces new challenges and opportunities for securing sensitive information and ensuring user privacy. As the landscape evolves towards Industry 5.0, emphasizing human–machine collaboration, the need for robust security measures becomes even more pronounced. This research investigates cutting-edge technologies, encryption methods, and privacy-preserving protocols that play a pivotal role in fortifying the security posture of Industry 4.0 and anticipating the requirements of Industry 5.0. The analysis includes a comprehensive examination of potential threats, risk mitigation strategies, and the ethical considerations surrounding the implementation of enhanced security and privacy measures in smart manufacturing environments. Through a forward-looking lens, the paper underscores the imperative for proactive approaches, collaboration between stakeholders, and ongoing research to address the evolving challenges of cybersecurity and privacy in the Industry 4.0 and Industry 5.0 era.

Keywords Industry 4.0 · Industry 5.0 · Cybersecurity · Privacy · Enhanced security · Smart manufacturing · Internet of Things (IoT) · Data analytics · Human–machine collaboration · Encryption · Risk mitigation · Ethical considerations · Cyber-physical systems · Technology integration · Automation

---

T. K. Vashishth (✉) · V. Sharma · K. K. Sharma · B. Kumar
School of Computer Science and Applications, IIMT University, Meerut, UP, India
e-mail: tarunvashishth@gmail.com

P. Raj et al. (eds.), *Building Embodied AI Systems: The Agents, the Architecture Principles, Challenges, and Application Domains*, Information Systems Engineering and Management 14, https://doi.org/10.1007/978-3-031-68256-8_6

1 Introduction

The advent of Industry 4.0 and its subsequent evolution into Industry 5.0 have ushered in a transformative era, marked by the convergence of digital technologies, intelligent systems, and interconnected devices. As organizations increasingly embrace the benefits of these industrial revolutions, the paramount concern becomes the assurance of enhanced security and privacy in the dynamically evolving digital landscape. Industry 4.0, categorized by mixing of cyber-physical systems, IoT, and data analytics, brings forth unprecedented opportunities for efficiency and innovation. However, this digital interconnectedness also introduces new vulnerabilities and risks, demanding a robust security infrastructure to safeguard critical assets, intellectual property, and sensitive information.

In the context of Industry 5.0, which areas a stronger emphasis on human–machine collaboration and socio-technical systems, significance of security and privacy is further heightened. As human workers interact seamlessly with advanced technologies and AI-driven systems, ensuring the protection of personal data and preserving individual privacy becomes a paramount concern. The vision of Industry 5.0 underscores necessity for a holistic approach to security, encompassing not only technological solutions but also policy frameworks, ethical considerations, and proactive risk management strategies. The amalgamation of cutting-edge technologies in Industry 4.0 and the human-centric approach of Industry 5.0 calls for a paradigm shift in security and privacy practices, necessitating a comprehensive and adaptive framework to fortify the foundations of the next industrial revolution. Such introduction sets stage for a deeper exploration of the multifaceted challenges and innovative solutions that arise in the pursuit of improved security and privacy in the Industry 4.0 and 5.0 landscape.

1.1 Background

The roots of enhanced security and privacy concerns within context of Industry 4.0 and its evolutionary successor, Industry 5.0, trace back to the fundamental shifts in industrial paradigms. Industry 4.0, often referred to as the Fourth Industrial Revolution, represents a convergence of physical and digital realms, marked by the integration of smart technologies into industrial processes. With the widespread adoption of cyber-physical systems, IoT, and bigdata analytics, organizations found themselves at the forefront of a technological revolution, unlocking unprecedented efficiencies and capabilities. However, this rapid digitization and interconnectivity brought forth a new frontier of cybersecurity challenges. The vulnerabilities exposed by the interconnectedness of devices, machines, and networks highlighted the critical importance of fortifying digital infrastructures against potential threats.

As Industry 4.0 matured into Industry 5.0, the narrative evolved to include a more profound human element, emphasizing collaborative interactions between humans and intelligent systems. While this human–machine synergy promised increased productivity and innovation, it also underscored the need for a heightened focus on privacy concerns. Industry 5.0 envisions a future where technology and human expertise coalesce, necessitating a delicate balance between innovations and safeguarding individual privacy. The evolving landscape demands a comprehensive understanding of the historical trajectory of Industry 4.0, coupled with a foresight into the nuanced challenges posed by Industry 5.0, to effectively address and enhance security and privacy measures in this dynamic and evolving industrial ecosystem.

1.2 Objectives of the Chapter

The chapter on enhanced security and privacy within the context of Industry 4.0 and 5.0 envisions several key objectives, each aimed at comprehensively addressing the multifaceted challenges and opportunities arising in the evolving industrial landscape:

- Understanding the Technological Landscape: Provide a thorough analysis of the technological advancements characterizing Industry 4.0, exploring incorporation of cyber-physical systems, IoTs, artificial intelligence, and data analytics. Additionally, elucidate the human centrical approach of Industry 5.0, emphasizing the need for a harmonious collaboration between humans and intelligent systems.
- Identifying Security Risks and Vulnerabilities: Systematically examine the security risks and vulnerabilities inherent in the interconnected ecosystems of Industry 4.0 and 5.0. This involves a comprehensive assessment of potential threats to cyber-physical systems, IoT devices, and the intricate networks that underpin the industrial processes.
- Evaluating Privacy Concerns: Delve into the evolving landscape of privacy concerns, particularly in the context of Industry 5.0, where human–machine collaboration is a central theme. Explore the challenges related to the collection, processing, and safeguarding of personal data, emphasizing the importance of ethical considerations and regulatory compliance.
- Developing Robust Security Frameworks: Propose and discuss state of the art security frameworks and methodologies tailored to the unique challenges posed by Industry 4.0 and 5.0. This includes the integration of proactive actions such as encryption, anomaly discovery, and controls of access to mitigate potential threats and vulnerabilities.
- Addressing Ethical and Policy Implications: Explore the ethical considerations surrounding enhanced security and privacy, considering the socio-technical implications of Industry 5.0. Additionally, examine existing and potential policy frameworks to ensure a harmonious balance between innovation and the protection of individual rights.
- Promoting Industry Best Practices: Highlight industry best practices and case studies that showcase successful approaches to enhancing security and privacy in

Industry 4.0 and 5.0. Draw insights from real-world implementations to provide practical guidance for organizations navigating the complexities of the digital industrial landscape.

By addressing these objectives, the chapter aims to offer a comprehensive and insightful exploration of opportunity and challenge associated with enhanced security and privacy in the transformative contexts of Industry 4.0 and 5.0.

2 Scope and Significance

Scope:

The scope of addressing enhanced security and privacy within the ambit of Industry 4.0 and 5.0 is expansive and multi-dimensional. Primarily, it encompasses the examination of technological landscapes, scrutinizing the intricate interplay of cyber-physical systems, IoT devices, artificial intelligence, and data analytics in Industry 4.0. Moreover, it extends into the human-centric paradigm of Industry 5.0, investigating collaborative dynamics among humans and intelligent systems. The scope also involves identifying and assessing security risks and vulnerabilities, evaluating privacy concerns associated with data collection and processing, and proposing robust frameworks to fortify digital infrastructures. Additionally, ethical considerations and policy implications, integral components of the scope, necessitate an exploration of the socio-technical aspects and the establishment of guidelines that balance innovation with individual rights.

Significance:

The significance of addressing heightened security and privacy with context of Industry 4.0 and 5.0 cannot be overstated. It aligns with the imperative of safeguarding critical assets, intellectual property, and sensitive information in an era defined by unprecedented digital connectivity. Significantly, as Industry 5.0 areas a heightened emphasis on human and machine collaboration, ensuring the protection of personal data becomes paramount. Beyond mere technological considerations, the significance extends into the ethical and regulatory domains, demanding a responsible and compliant approach to data management. Enhanced security and privacy measures not only mitigate the risks associated with cyber threats but also foster trust among stakeholders, promoting the sustainable development of industries. Moreover, by delving into best practices and case studies, this exploration contributes to the collective knowledge base, empowering organizations to navigate the complexities of the digital industrial landscape with resilience and foresight. In essence, the significance lies in fortifying the foundations of Industry 4.0 and 5.0, ensuring a secure, ethical, and privacy-conscious evolution of the industrial landscape.

2.1 *Overview of Industry 4.0 and 5.0*

Industry 4.0 Overview:

Industry 4.0, often referred to as the 4th Industrial Revolution, represents a paradigm shift in manufacturing and industry, driven by the integration of advanced digital technologies. It builds upon the foundations laid by previous industrial revolutions but distinguishes itself by the convergence of physical and digital systems. Key mechanisms of Industry 4.0 include:

- Cyber-Physical Systems (CPS): Integration of physical processes with digital technologies, enabling real-time monitoring, control, and optimization.
- Internet of Things (IoT): Interconnected devices and sensors that facilitate seamless communication and data exchange between machines, enabling intelligent decision-making.
- Big Data and Analytics: Utilization of vast amounts of data generated by IoT devices and other sources to derive insights, optimizes processes, and enhances decision-making.
- Artificial Intelligence (AI): Deployment of ML algorithms and AI to automate tasks, predict outcomes, and improve overall efficiency.
- Additive Manufacturing (3D Printing): The ability to create complex and customized products through layer-by-layer additive processes, transforming traditional manufacturing methods.
- Cloud Computing: Centralized data storage and processing capabilities, enabling accessibility and collaboration across different locations.
- Augmented Reality (AR) & Virtual Reality (VR): Addition of AR & VR technologies to enhance training, maintenance, and overall human–machine interaction.

The ultimate goal of Industry 4.0 is to create "smart factories" that are flexible, efficient, and responsive to market demands through the seamless integration of digital technologies (Fig. 1).

Industry 5.0 Overview:

Industry 5.0 represents the evolution beyond Industry 4.0, with a notable emphasis on human–machine collaboration and a return to a more people-centric approach. While Industry 4.0 focused on automation and data-driven processes, Industry 5.0 seeks to integrate human skills, creativity, and intuition with advanced technologies. Key features of Industry 5.0 include:

- Human–Machine Collaboration: A shift towards collaborative environments where humans and machines work together to leverage the strengths of both, emphasizing the importance of human skills and creativity.
- Socio-Technical Systems: Integration of social and technical elements, acknowledging the impact of technology on society and emphasizing ethical considerations in the development and deployment of advanced technologies.

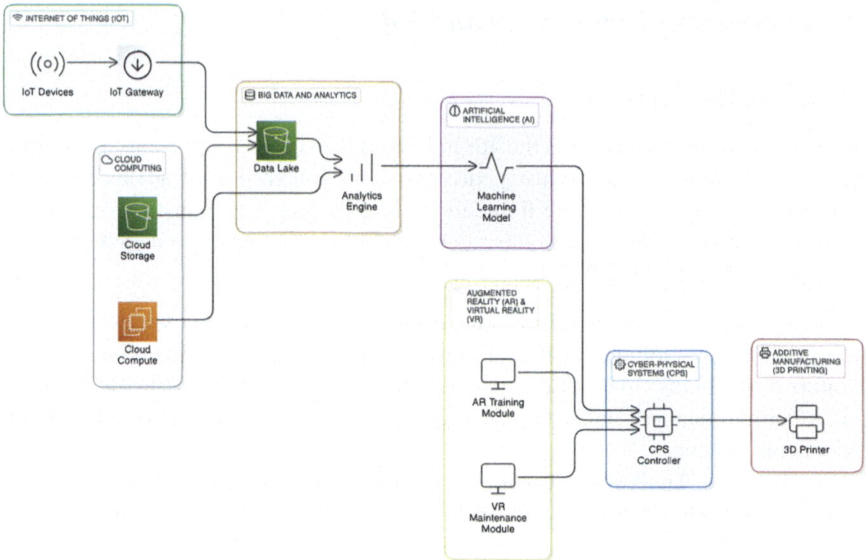

Fig. 1 Industry 4.0

- Customization and Personalization: The ability to produce highly customized and personalized products at scale, responding to individual customer needs and preferences.
- Decentralized Decision-Making: Empowering workers at various levels with decision-making authority, moving away from centralized control towards distributed decision-making.
- Holistic Approach: Consideration of broader socio-economic and environmental factors, aligning industry goals with sustainable and responsible practices.

Industry 5.0 envisions a future where technology complements human abilities, fostering a more inclusive, creative, and socially responsible industrial ecosystem. The transition 4.0 to 5.0 signifies a maturation of digital transformation efforts towards a more balanced and harmonious integration of technology and human elements.

3 Literature Review

In Pant et al. [1] addressed the issue of security in Industry 5.0. With increased use of interconnected devices, the risk of cyber-attacks has also risen. The authors proposed a security model that using ML to detect and prevent potential cyber threats, ensuring the safety of the production process. In Baz et al. [2] emphasized the importance of continuous monitoring and updating of the framework to ensure its effectiveness in addressing emerging security threats. They also suggested that the framework could be used as a benchmark for healthcare organizations to compare their security risk management practices with industry standards. In Wang et al. [3] focused on developing a user authentication scheme that addresses the challenges of industry 4.0. Their proposed scheme not only ensures the privacy of user data but also guarantees forward secrecy, meaning that even if a user's private key is compromised, previous and future communications remain secure. In Garg et al. [4] emphasize the need for secure and provable authentication in the IoT ecosystem to ensure the success of Industry 4.0. They also highlight the importance of collaboration between researchers and industry experts to develop practical and effective solutions for authentication in the IoT. With the rapid growth of IoT and Industry 4.0, it is crucial to address these security concerns and make authentication a top priority. In Joshi et al. [5] sheds light on the adoption of blockchain technology as a means to enhance security and privacy in framework of Industry 4.0. With its potential to revolutionize data management and address the challenges posed by 4th Industrial Revolution, blockchain technology has emerged as a talented solution for businesses and industries. It is crucial for organizations to consider incorporating blockchain into their digital strategies to stay ahead in the ever changing technological scene. In Czeczot et al. [6] presents a compelling case for the integration of AI in IIoT cybersecurity management. With the increasing complexity and interconnectivity in industries, it is crucial to leverage AI to strengthen security measures and safeguard against potential cyber attacks. In Bayanati [7] highlighted the potential risks associated with the use of IoT in various industries such as healthcare, transportation, and manufacturing. As more and more devices were connected to the internet, the vulnerability to cyber-attacks and data breaches increased. In Qahtan et al. [8] explored the potential of Industry 4.0 in revolutionizing the healthcare sector through the implementation of blockchain technology. The team examined various aspects of this integration, including security and privacy concerns, and proposed solutions to address them. In Jhanjhi et al. [9] highlight that major worry with IIoT is the potential for cyber-attacks. As IIoT devices are connected to the internet, they become vulnerable to hacking and other malicious activities. This can result in the compromise of sensitive data and disruption of critical operations. In the past, there have been numerous incidents of cyber attacks on IIoT systems, causing significant financial and reputational damage to organizations. In Velliangiri et al. [10] proposed mechanism utilizes the advantages of ECC, which is known for its high level of security and efficiency. The researchers have designed a lightweight ECC-based scheme that can be easily integrated into existing Industry 4.0 systems without significant modifications. This makes it a cost-effective solution for industries of all sizes.

In Pradhan et al. [11] explored potential of blockchain technology in revolutionizing the industrial sector, while also highlighting the challenges and risks associated with its implementation. In Saura et al. [12] discussing the increasing use of social networks in Industry 4.0, as these platforms have become essential for businesses to connect with customers and collaborate with other organizations. However, this reliance on social networks also brings about potential security and privacy concerns that need to be addressed. In Pedreira et al. [13] provide a comprehensive review of the different types of attacks and vulnerabilities that have been identified in Industry 4.0. These include attacks on smart devices, cloud computing, and wireless networks. They also discuss the potential impact of these attacks, which can range from financial losses to reputational damage. In Vashishth et al. [14] proposed the use of AI algorithms to analyze and forecast resource demand, as well as blockchain technology to securely and transparently allocate resources. This combination would not only optimize resource utilization but also enable real-time monitoring and tracking of resources. In Kumar and Sharma [15] shed light on the pivotal role of AI in enhancing the security and privacy of data in smart cities. Various techniques and advancements discussed in the conference prove that AI has the potential to make smart cities safer and more private for its citizens. It is now up to the governments and city planners to incorporate these technologies in the development of smart cities, ensuring a secure and private future for its citizens. In Salam et al. [16] proposed the use of deep learning techniques, specifically Convolutional Neural Networks (CNNs), to detect and prevent web-based attacks in real-time. They collected a large dataset of web traffic from various industrial systems and trained the CNN model to identify patterns and anomalies associated with malicious activities. In Hammad et al. [17] highlighted growing use of network-based systems in manufacturing, which has led to a rise in cyber threats. With the rise of Industry 4.0, where machines and systems are interconnected and communicate with each other, the potential for cyber attacks has also increased. In Jaisan et al. [18] presents a valuable contribution to the field of M2M communication in Industry 4.0. The Secured SECS/GEM mechanism offers a much-needed layer of security to protect the sensitive data and operations of smart factories. It is a promising step towards creating a safer and more efficient Industry 4.0 ecosystem. In Laghari et al. [19] sheds light on the potential dangers of cyberattacks on SECS/GEM communications in the Industry 4.0 ecosystem. It serves as a wake-up call for companies to take proactive measures to secure their systems and safeguard the smooth functioning of the manufacturing industry. With the continuous evolution of technology, it is critical to stay vigilant and adapt to new security threats to protect our critical infrastructure. In Leng et al. [20] provides valuable insights into the potential of blockchain knowhow in securing smart industrial processes in Industry 4.0.

4 Security Challenges in Industry 4.0

4.1 Integration of Cyber-Physical Arrangements

Incorporation of Cyber-Physical Systems (CPS) in Industry 4.0 introduces numerous security challenges, reflecting the convergence of digital and physical elements. While Industry 4.0 promises improved efficiency and innovation, interconnected nature of CPS creates vulnerabilities that must be addressed to ensure a secure industrial landscape. Here are key security challenges associated with the integration of CPS in Industry 4.0:

- Interconnected Networks and Devices: The seamless connectivity of CPS components across networks increases the attack surface, providing malicious actors with more entry points. The interdependence of devices also raises the risk of cascading failures if one component is compromised, potentially leading to widespread disruptions.
- Data Integrity and Privacy Concerns: The massive volume of data generated by CPS, including sensitive operational data and intellectual property, poses challenges related to data integrity and privacy. Unauthorized access or manipulation of data can have severe consequences, impacting decision-making processes and compromising the confidentiality of critical information.
- Lack of Standardized Security Protocols: Absence of standardized security protocols across various CPS components and manufacturers can lead to inconsistencies in security implementations. It lack of uniformity makes it issues to establish a cohesive security framework, leaving potential vulnerabilities unaddressed.
- Real-Time Threats and Attacks: CPS operates in real-time, and any delay in response to security threats can result in immediate and significant consequences. Traditional security measures designed for slower, less dynamic environments may struggle to keep pace with the rapid pace of cyber threats targeting CPS.
- Physical Consequences of Cyber Attacks: Unlike purely digital systems, CPS controls physical processes. Cyber attacks on CPS can have direct physical consequences, such as equipment malfunctions, production disruptions, or even safety hazards. This intersection of digital and physical realms requires a comprehensive approach to security.
- Supply Chain Vulnerabilities: The interconnected nature of supply chains in Industry 4.0 introduces additional security challenges. Cyber attacks on suppliers or third-party components can propagate through the supply chain, impacting the overall integrity and security of CPS within an industrial ecosystem.
- Insufficient Authentication and Access Controls: Less adequate authentication machineries and lax access controls can result in not authorized access to CPS components. Malicious actors gaining unauthorized control over critical systems may manipulate processes, cause disruptions, or extract sensitive information.
- Legacy System Integration: The integration of CPS often involves connecting with existing legacy systems. Legacy systems may have outdated security measures and protocols, creating potential vulnerabilities that can be exploited by cyber threats.

Addressing these security challenges requires a holistic and proactive approach. Implementing robust authentication protocols, encryption mechanisms, continuous monitoring, and regular updates to security practices are essential to fortify mixing of Cyber-Physical Systems in Industry 4.0 against evolving cyber threats. Additionally, collaboration within the industry to establish standardized security practices and

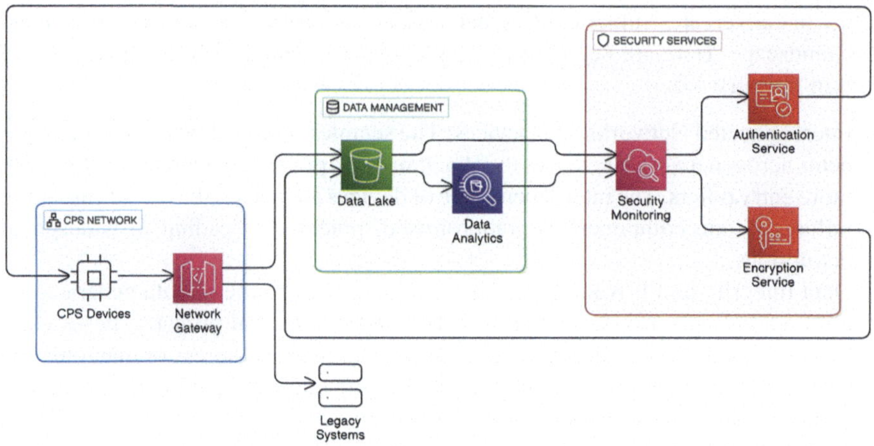

Fig. 2 Integration of cyber physical systems in Industry 4.0

frameworks can contribute to an additional secure and resilient industrial landscape (Fig. 2).

4.2 *IoT in Industrial Settings*

Mixing of the IoT in manufacturing settings within the framework of Industry 4.0 presents a complex tapestry of security challenges, stemming from the proliferation of interconnected devices and the convergence of physical and digital realms. The sheer diversity and abundance of IoT devices introduce a heterogeneous landscape, complicating the management of security protocols across different platforms. Authentication and authorization mechanisms become particularly challenging, as resource-constrained IoT devices may struggle to implement robust credentialing practices. The extensive data generated by these devices, crucial for operational insights, becomes a double-edged sword, posing risks to data integrity and confidentiality if not adequately protected. The absence of standardized security protocols exacerbates these concerns, leading to inconsistencies in security implementations and potential vulnerabilities left unaddressed. Networks connecting these devices become potential vectors for cyber threats, and the supply chain risks associated with compromised IoT components pose additional challenges. The limited capability for over-the-air updates raises concerns about patching vulnerabilities promptly, while physical access to IoT devices introduces new dimensions of risk. As these devices integrate with legacy systems, security challenges are compounded, requiring a delicate balance between technological advancement and the preservation of a robust security posture. Tackling these multifaceted challenges demands a comprehensive strategy encompassing encryption, authentication measures, regular audits, and collaborative efforts to establish and adhere to industry-wide security standards, ensuring the resilient and secure evolution of industrial environments in the Industry 4.0 era.

4.3 Data Analytics and Industry

In the era of Industry 4.0, where information analytics plays a crucial role in driving operational efficiencies, innovation, and informed decision-making, several security challenges emerge. Vast amounts of statistics generated by interconnected devices and systems are a prime target for cyber threats. Data integrity and confidentiality become critical concerns, particularly as organizations leverage analytics to gain valuable insights from sensitive operational and business data. The sheer volume of data exacerbates challenges related to secure storage, transmission, and processing, demanding robust encryption and access control measures. Additionally, the reliance on cloud-based analytics platforms introduces new attack vectors, as data traverses external networks. Supply chain vulnerabilities also come to the forefront, as data analytics solutions often involve third-party services or tools. Furthermore, the interconnected nature of Industry 4.0 systems poses a risk of cascading failures, where a security breach in one component can compromise the entire data analytics ecosystem. The implementation of machine learning algorithms for predictive analysis introduces concerns about adversarial attacks and biased data, which can undermine the reliability of analytical insights. As organizations strive to harness the power of data analysis in Industry 4.0, comprehensive and adaptive security framework is imperative, encompassing encryption, regular audits, and proactive measures to counter evolving cyber threats, ensuring the resilience and trustworthiness of data-driven industrial processes.

4.4 Implications for Information Security

Advent of Industry 4.0 has steered in transformative advancements, but with these technological leaps come profound implications for information security. Integration of cyber-physical systems, the IoT, data analytics, and artificial intelligence has exponentially expanded the attack surface for malicious actors. Interconnected nature of Industry 4.0 systems, where devices, machines, and networks communicate seamlessly, introduces vulnerabilities that, if exploited, can have far-reaching consequences. Information security faces heightened challenges in preserving the confidentiality, integrity, and availability of sensitive data crucial for decision processes. The reliance on cloud computing for storage and processing amplifies concerns regarding data privacy and the potential exposure of intellectual property. Additionally, the use of machine learning algorithms in autonomous decision-making processes raises questions about the robustness of these systems against adversarial attacks. Supply chain risks become more pronounced as organizations collaborate with external partners, and the convergence of IT (Information Technology) and OT (Operational Technology) blurs traditional security boundaries, necessitating a unified and adaptive security strategy. Industry 4.0's drive for efficiency and innovation must be coupled with a strategic and proactive approach to information security, encompassing robust encryption, continuous monitoring, employee training, and collaborative efforts to establish industry-wide standards, thereby fortifying the foundations of the fourth industrial revolution against evolving cyber threats.

4.5 Privacy Concerns in Industry 4.0

Concerns about Privacy in Industry 4.0 arise due to the pervasive integration of advanced technologies, interconnected devices, and the extensive collection and utilization of data. Several factors contribute to these privacy challenges:

- Data Proliferation and Sensitivity: Industry 4.0 relies heavily on the generation and analysis of vast amounts of data. This data often includes sensitive information related to operations, employee details, and sometimes customer data. The sheer volume and sensitivity of the data raise concerns about how it is handled, stored, and shared.
- IoT and Real-Time Tracking: The widespread deployment of IoT devices enables on time monitoring and tracking of physical assets, including machinery and personnel. While this enhances operational efficiency, it also raises privacy concerns, especially in relation to employee monitoring and location tracking.
- Personalization and Customer Data: Industry 4.0 facilitates the customization and personalization of products and services based on individual preferences. However, this involves the collection and analysis of consumer behavior data, which can lead to concerns about the privacy of customer information and potential misuse.
- Employee Privacy: The integration of smart technologies in the workplace, such as wearable devices and sensors, can lead to continuous monitoring of employees. This constant surveillance raises concerns about employee privacy, as organizations need to strike a balance between monitoring for efficiency and respecting individual privacy rights.
- Cross-Organizational Data Sharing: Collaborative efforts within supply chains and across industries involve the sharing of data between organizations. Ensuring that this data exchange is conducted securely and with due consideration for privacy regulations becomes a critical challenge.
- Regulatory Compliance: Industry 4.0 operations necessity to adhere to privacy regulations and standards, like General Data Protection Regulation (GDPR) in Europe. Navigating complex landscape of compliance requirements across different regions and industries poses a significant privacy challenge.

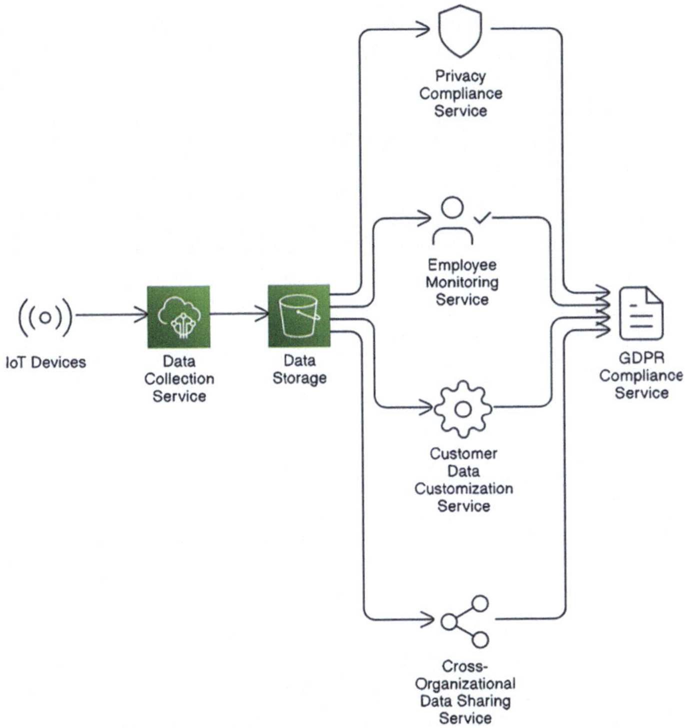

Fig. 3 Privacy concerns in Industry 4.0 architecture

To address these privacy concerns, organizations need to implement privacy via design principles, conduct thorough privacy impact assessments, prioritize encryption and information anonymization, and stay abreast of evolving privacy regulations. Striking a balance between leveraging the benefits of Industry 4.0 technologies and safeguarding individual privacy is crucial for building trust among stakeholders and ensuring the responsible evolution of the industrial landscape (Fig. 3).

5 Technologies and Methods for Enhanced Security

5.1 Cutting-Edge Security Technologies

Cutting-edge security technologies are pivotal in fortifying digital infrastructures against evolving cyber threats. Quantum Key Distribution (QKD) stands at the forefront, leveraging principles of quantum-mechanics to safe message channels by transmitting cryptographic keys in a quantum state. This ensures unprecedented levels of security as any attempt to intercept the quantum key would alter its state, immediately alerting the parties involved. Artificial Intelligence (AI) and Machine Learning (ML) play pivotal roles in predictive threat analysis, enabling the identification of patterns and anomalies in real-time. Behavioral biometrics, encompassing advanced techniques like keystroke dynamics and gait analysis, provides an additional layer of identity verification based on unique behavioral traits. Secure Multiparty Computation allows many participants to jointly compute a function over their inputs while caring inputs private. Blockchain technology ensures the integrity of data through decentralized and tamper-resistant ledgers. The integration of these cutting-edge technologies, along with continuous advancements, reflects a proactive stance in addressing contemporary security challenges and fostering resilient defenses in the digital landscape.

5.2 Encryption Methods

Encryption methods play a pivotal role in enhancing security across various digital applications, defending sensitive data from unlawful access and safeguarding confidentiality and integrity of information. Advanced encryption methods leverage mathematical algorithms to transform plaintext into ciphertext, rendering it unreadable to anyone without the corresponding decryption key. One of most widely used encryption methods is Advanced-Encryption-Standard (AES), a symmetric key algorithm that employs varying key lengths for robust protection. Public Key Infrastructure (PKI) utilizes asymmetric encryption, employing a pair of public and private keys to facilitate secure communication and digital signatures. Transport Layer Security (TLS) and its predecessor, Secure Sockets Layer (SSL), apply encryption protocols to secure data transmitted over networks, commonly used for securing web communications. Homomorphic-encryption permits computation on encrypted data short-of decrypting it, ensuring privacy during data processing. Quantum Key Distribution, based on the principles of quantum mechanics, enables the secure exchange of cryptographic keys, leveraging attributes of quantum-states. These encryption methods collectively contribute to a layered and comprehensive security strategy, safeguarding information in transit, at rest, and during processing in diverse digital environments.

5.3 Privacy-Preserving Protocols

Privacy-preserving protocols are crucial technologies and methods that focus on safeguarding sensitive information while allowing for secure interactions and data sharing. Homomorphic Encryption is a pioneering approach that permits computations on encrypted data short of need for decryption, ensuring the privacy of sensitive information during processing. Secure Multi-Party Computation is another privacy preserving protocol that allows multiple parties to jointly compute a function over their inputs while keeping those inputs private. This technique is valuable in collaborative scenarios where organizations need to analyze data collectively without exposing individual contributions. Differential Privacy introduces noise or randomness into data before analysis to protect individual privacy while still providing accurate aggregate results. Zero-Knowledge Proofs let one party to verify the authenticity of certain information to another party without revealing the information itself. It is particularly useful in authentication and verification processes without disclosing sensitive data. Privacy-preserving protocols are instrumental in navigating the delicate balance between the need for data-driven insights and the imperative to protect individuals' privacy in the digital landscape, fostering trust and compliance in various sectors.

5.4 Role of Artificial Intelligence in Security

Role of AI in security is transformative, contributing significantly to the development of advanced methods to enhance cybersecurity measures. AI augments traditional security approaches by providing real-time threat detection, response, and adaptive defense mechanisms. ML, a subset of AI, plays a crucial role in predictive analysis, identifying patterns, and anomalies in massive datasets to recognize potential security threats. AI-powered systems can nonstop learn and adapt to evolving attack techniques, refining their ability to detect sophisticated cyber threats. Behavioral Analytics is another key application of AI in security, focusing on understanding normal user behavior to detect deviations that might indicate a security breach. AI-driven systems can analyze user activities, network traffic, and application usage, identifying unusual patterns that may signify a security incident. Additionally, AI is instrumental in automating and orchestrating responses to security incidents, enabling faster and more efficient threat mitigation. Security orchestration, automation, and response (SOAR) platforms leverage AI to streamline incident response workflows, dropping the time required to identify, analyze, and remediate security incidents. Furthermore, AI is utilized in the development of deception technologies, where fake data or decoy systems are strategically placed to mislead and divert attackers, providing organizations with valuable insights into potential threats.

In summary, AI plays a multifaceted role in security, offering predictive capabilities, behavioral analysis, automated incident response, and innovative approaches like deception technologies. Its adaptive nature positions AI as a critical component in fortifying digital infrastructures and staying ahead of evolving cyber threats in an increasingly complex and interconnected digital landscape.

6 Transition to Industry 5.0

6.1 Human–Machine Collaboration

The transition to Industry 5.0 presents a paradigm shift from largely automated and data-driven processes of Industry 4.0 to a more collaborative and human-centric model. At the core of Industry 5.0 is idea of human and machine collaboration, emphasizing the integration of human skills, creativity, and intuition with advanced technologies. This evolution recognizes that while automation and technology have driven efficiency gains, human expertise is irreplaceable, particularly in complex decision-making, problem-solving, and innovation. One key aspect of Industry 5.0 is development of socio-technical systems where humans and machines work together synergistically. Advanced technologies such as AI, robotics, and the IoT are designed to augment human capabilities rather than replace them. For example, collaborative robots or cobots are employed to work alongside human workers, improving productivity with security.

In Industry 5.0, focus extends beyond efficiency gains to customization and personalization. Human input becomes critical in tailoring products and services to meet each customer requirements. This shift requires a high degree of adaptability in manufacturing processes, with technology serving as a tool that amplifies human creativity and decision-making. The transition to Industry 5.0 also necessitates a reevaluation of workplace dynamics. Human-machine collaboration implies changes in job roles, skill requirements, and skilling programs. Organizations need to foster a culture that encourages continuous learning and adaptability among the workforce to thrive in this collaborative environment. Ethical considerations become paramount in Industry 5.0, requiring careful management of data privacy, transparency in AI decision-making, and addressing societal concerns about job displacement. Striking the right balance between the advantages of technology and the ethical implications of its implementation is crucial for successful human-machine collaboration.

The transition to Industry 5.0 marks a parting from automation-centric approach of Industry 4.0 to a more inclusive, human-centered industrial paradigm. Human-machine collaboration is not just a technological shift but a cultural and organizational transformation that emphasizes the unique strengths of both humans as well as machines, driving innovation and creating a more adaptable and sustainable industrial ecosystem.

6.2 Evolving Landscape of Industry 5.0

Transition to Industry 5.0 signifies a profound evolution in the industrial landscape, marked by the convergence of digital technologies, advanced automation, and a renewed focus on human-machine collaboration. This evolution builds-up foundations of Industry 4.0 but introduces a more holistic and human-centric approach to industrial processes. Several key elements define the evolving landscape of Industry 5.0:

- Human–Machine Synergy: Industry 5.0 emphasizes a symbiotic relationship between humans and machines. Rather than changing human workers with automation, the focus is on augmenting human capabilities with advanced technologies. Collaborative robots, wearable devices, and augmented reality interfaces enable seamless interactions between humans as well as machines.
- Customization with Personalization: A prominent feature of Industry 5.0 is the ability to provide highly customized and personalized products and services. Advanced technologies facilitate flexible and agile manufacturing processes that can respond rapidly to individual customer needs and preferences.
- Decentralized Decision-Making: Industry 5.0 moves away from centralized control structures, empowering workers at various levels with decision-making authority. This decentralization enhances agility and responsiveness in industrial processes, allowing for quicker adaptations to changing conditions.
- Socio-Technical Systems: The evolution to Industry 5.0 recognizes the impact of technology on society. It emphasizes the development of socio-technical systems that integrate not only advanced technologies but also social and ethical considerations. The interplay between technological innovation and societal values is a critical aspect of this evolving landscape.
- Dynamic and Adaptive Manufacturing: Industry 5.0 envisions manufacturing processes that are dynamic and adaptive. Smart factories equipped with intelligent systems can reconfigure themselves based on real-time data and demand fluctuations. This adaptability enhances efficiency and resource utilization.
- Data-Centric Operations: The role of data becomes even more central in Industry 5.0. Advanced analytics, machine learning, and AI contribute to data-driven decision-making processes. The ability to extract meaningful insights from vast datasets enables continuous optimization and improvement.
- Ethical and Responsible AI: As AI technologies show an important part in Industry 5.0, there is heightened emphasis on ethical and responsible AI practices. Transparency, fairness, and accountability in AI decision-making are critical considerations to ensure the ethical deployment of advanced technologies.
- Workplace Transformation: The evolving landscape of Industry 5.0 necessitates a transformation in the workplace. The workforce is expected to acquire new skills related to collaboration with intelligent systems, adaptability, and problem-solving. Continuous learning and development become integral to staying relevant in this dynamic environment.

The evolving landscape of Industry 5.0 presents a shift towards a more inclusive, adaptable, and human-centric industrial ecosystem. The integration of advanced technologies with a focus on ethical considerations positions Industry 5.0 as a transformative phase that goes beyond automation to embrace the collaborative potential of humans and machines (Fig. 4).

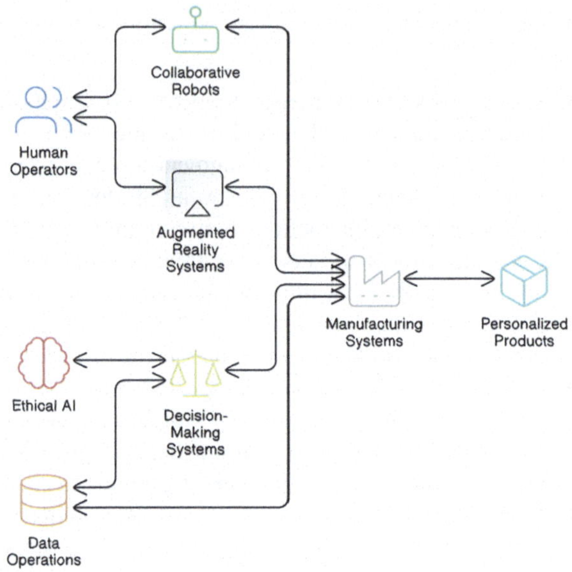

Fig. 4 Industry 5.0 evolution

6.3 Security Requirements in Industry 5.0

Transition to Industry 5.0 necessitates a comprehensive revaluation of security requirements to accommodate the dynamic fusion of human-machine collaboration, interconnected systems, and ethical considerations. With a heightened focus on the human element, security measures must prioritize individual privacy, emphasizing robust access controls and identity management. Real-time threat detection and response mechanisms become imperative in Industry 5.0's dynamic and adaptive environment, requiring advanced analytics and artificial intelligence to promptly identify and mitigate evolving security threats. Secure connectivity, underpinned by strong encryption protocols, is essential to protect against unauthorized access and data tampering in the interconnected ecosystem. Adaptive access controls dynamically adjusting based on contextual factors ensure that access privileges align with the evolving operational context. Security measures must also extend to the ethical deployment of artificial intelligence, addressing biases and ensuring transparent, fair, and accountable decision-making processes. With collaborative efforts across supply chains, supply chain security assumes heightened importance, necessitating comprehensive risk management strategies. Employee training, regulatory compliance, and robust incident response planning round out the multifaceted security requirements, aiming to create a resilient and trustworthy foundation for Industry 5.0's transformative industrial landscape.

7 Threat Analysis and Risk Mitigation

7.1 Potential Threats in Smart Manufacturing

Smart manufacturing, key components of Industry 4.0/5.0, introduces transformative technologies that enhance efficiency but also brings forth potential threats. Cybersecurity vulnerabilities loom large as smart manufacturing heavily relies on interconnected systems, IoT devices, and data exchange. One of the primary threats is risk of unsanctioned access and data breach, where malicious actors exploit vulnerabilities to gain entry into the network, potentially compromising sensitive information and intellectual property. The interconnected nature of smart manufacturing introduces the risk of supply chain attacks, where compromised components or software can propagate vulnerabilities throughout the ecosystem. The reliance on data analytics and machine learning also opens the door to threats such as adversarial attacks, where manipulated data inputs can deceive AI systems. Disruption of operations through ransomware attacks poses a significant threat, with the potential to halt production processes. Suring security of edge devices and endpoints is critical, as these devices serve as entry points for cyber threats. Robust threat analysis and risk mitigation strategies, including encryption, network segmentation, continuous monitoring, and employee training, are imperative to safeguard smart manufacturing environments from these multifaceted threats.

7.2 Risk Mitigation Strategies

Effective risk mitigation strategies are paramount in safeguarding organizations from the diverse array of threats present in today's dynamic digital landscape. First and foremost, comprehensive cybersecurity measures are critical, encompassing robust network security protocols, firewalls, and intrusion detection systems to thwart unauthorized access and cyber attacks. Regular and thorough risk assessments aid in identifying vulnerabilities, allowing organizations to proactively address potential weaknesses before they are exploited. Encryption of delicate data, both in transit and at rest, ensures an additional layer of protection, rendering intercepted information unreadable to unlawful entities. Establishing a strong incident response plan enables organizations to swiftly detect, contain, and recover from security incidents, minimizing potential damage. Employee training and awareness programs are vital in building a security-aware culture, reducing the likelihood of human error leading to security breaches. Additionally, diversifying technology solutions and employing a defense-in-depth approach, incorporating multi-layers of security actions, enhances overall resilience. Collaborative efforts, information sharing within the industry and adherence to regulatory standards contribute to a holistic approach, creating a fortified defense against evolving threats.

7.3 Ethical Considerations in Cybersecurity

Ethical considerations in cybersecurity are integral to navigating the complex landscape of digital defense, ensuring that protective measures align with principles of fairness, transparency, and respect for privacy. A key ethical concern involves the responsible use of technology, ensuring that cybersecurity practices prioritize the well-being of individuals and organizations while avoiding undue harm (Fig. 5).

Transparency in the deployment of cybersecurity measures is crucial, particularly in contexts involving surveillance or data collection. Respecting user privacy is a paramount ethical principle, emphasizing the need for clear consent mechanisms and secure handling of personal information. Ethical cybersecurity also encompasses considerations of equity and justice, ensuring that protective measures do not disproportionately impact certain groups or contribute to digital divides. Cybersecurity professionals must adhere to a code of ethics, emphasizing integrity, honesty, and a commitment to continuous learning to stay abreast of evolving threats responsibly. Balancing the imperative for robust defense with ethical principles is essential for cultivating trust and accountability in the cybersecurity domain, promoting a secure digital environment that respects the rights and dignity of individuals.

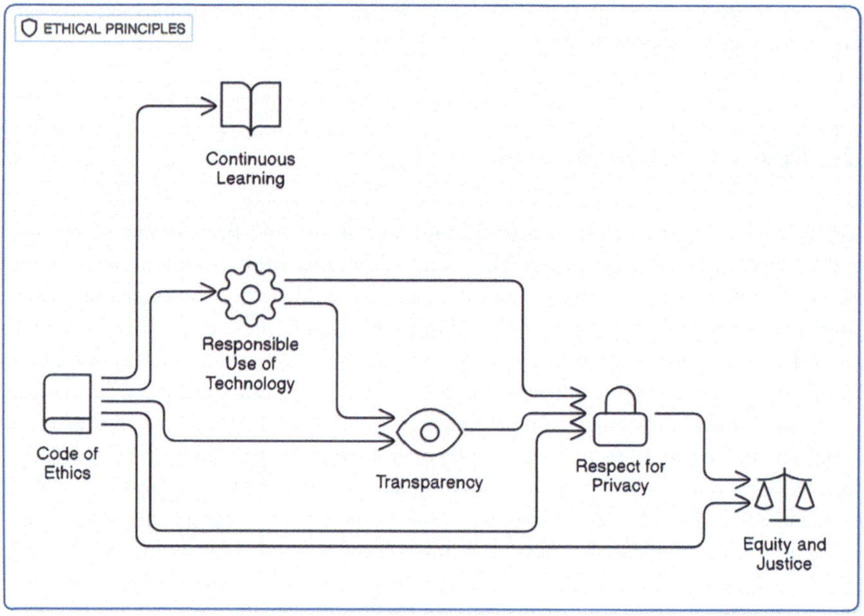

Fig. 5 Ethical considerations in cybersecurity

8 Proactive Approaches and Collaboration

8.1 Need for Proactive Security Measures

The need for proactive security measures has become imperative in the face of ever-evolving and sophisticated cyber threats. Reactive approaches, relying solely on responding to incidents after they occur, are insufficient in the rapidly changing digital landscape. Proactive security measures involve anticipating and mitigating potential risks before they manifest, providing a preemptive defense against cyber threats. This approach involves continuous monitoring of network activities, vulnerability assessments, and the implementation of robust access controls. Regularly updating and patching systems, conducting security audits, and staying informed about emerging threats are integral components of proactive cybersecurity. The growing interconnectedness of systems, the rise of advanced persistent threats, and the increasing sophistication of cybercriminal tactics underscore the importance of a proactive stance. Collaborative efforts, data sharing, and industry wide initiatives contribute to a collective protection against cyber threats. By adopting proactive security measures, organizations can strengthen their resilience, minimize the impact of potential breaches, and foster a cybersecurity posture that is adaptive and forward-looking in addressing the challenges of the digital era.

8.2 Stakeholder Collaboration

Stakeholder collaboration is a linchpin in fostering robust cybersecurity defenses through proactive approaches. In an era where cyber threats transcend organizational boundaries, collaboration among stakeholders such as government entities, private industries, academia, and cybersecurity professionals is essential. Sharing threat intelligence, best practices, and lessons learned creates a collective defense mechanism against evolving cyber risks. Public-private partnerships enable the pooling of resources, expertise, and insights, enhancing the ability to detect and mitigate cyber threats effectively. Collaborative steps can include joint research with development, information-sharing forums, and coordinated incident response efforts. Engaging in regular dialogues allows stakeholders to align their cybersecurity strategies, address common challenges, and collectively raise the bar for industry-wide resilience. Additionally, collaboration fosters a culture of awareness and proactive information exchange, reinforcing the significance of cybersecurity across diverse sectors. By working collaboratively, stakeholders can stay ahead of emerging threats, respond more effectively to incidents, and collectively contribute to a safer digital environment for individuals, organizations, and society at large.

8.3 Industry-Government Collaboration

Industry-government collaboration stands as a cornerstone in establishing proactive cybersecurity approaches that effectively mitigate and respond to evolving cyber threats. The synergy between private industries and government entities creates a robust framework for sharing threat intelligence, expertise, and resources. Governments play a pivotal role in setting regulatory standards, establishing best practices, and providing a legal and policy framework for cybersecurity. Meanwhile, private industries bring real-world insights, innovation, and agility to the table. Joint initiatives, such as information-sharing platforms and collaborative research projects, enable a more comprehensive understanding of cyber threats and effective countermeasures. Governments can incentivize private sectors to adopt and adhere to cybersecurity standards, while industries provide crucial insights into emerging threats and vulnerabilities. This collaborative effort not only fortifies national cybersecurity resilience but also fosters a proactive and adaptive approach to countering cyber threats on a global scale. By combining strengths and resources, industry-government collaboration becomes a linchpin in creating a cyber-resilient ecosystem that can effectively navigate the complex and rapidly evolving cybersecurity landscape.

8.4 Ongoing Research Initiatives

Ongoing research initiatives play a pivotal role in advancing proactive approaches to cybersecurity by staying at the forefront of emerging threats, vulnerabilities, and innovative solutions. Collaborative efforts among academic institutions, research organizations, and industry partners contribute to a deep understanding of evolving cyber landscapes. Researchers explore cutting-edge technologies, such as AI, ML and quantum cryptography, to bolster proactive threat detection and response capabilities. They investigate new attack vectors, analyze sophisticated malware, and devise countermeasures to enhance the resilience of digital systems. Industry-academic partnerships facilitate the practical application of research findings, allowing for the development of proactive cybersecurity tools and methodologies. Furthermore, interdisciplinary research initiatives bring together experts from diverse fields, including computer science, psychology, and law, to address multifaceted aspects of cybersecurity. Continuous exploration and dissemination of research outcomes provide a knowledge foundation for cybersecurity professionals, policymakers, and organizations, fostering a proactive mindset that adapts to the evolving nature of cyber threats and ensures the development of effective countermeasures.

9 Future Trends in Cybersecurity for Industry 4.0 and 5.0

Future trends in cybersecurity for Industry 4.0 and 5.0 anticipate the evolving nature of cyber threats and the dynamic technological landscape. As industrial systems become more interconnected and reliant on advanced technologies, AI and ML are expected to play an gradually prominent role in cybersecurity. These technologies will empower proactive threat detection and response by analyzing vast datasets for anomalous patterns and potential vulnerabilities. Quantum-resistant cryptography is gaining attention as a future trend, preparing systems for the era of quantum computing and ensuring the resilience of cryptographic protocols. Zero Trust Architecture (ZTA) is likely to become a standard, emphasizing the continuous verification of identities and restricting access based on least privilege principles. Automation and orchestration of cybersecurity processes will become more prevalent, streamlining incident response and allowing for faster mitigation of cyber threats. Additionally, the integration of cybersecurity into the fabric of digital transformation initiatives and the development of industry-specific standards and regulations will shape the future cybersecurity landscape for Industry 4.0 and 5.0, promoting a secure and resilient industrial ecosystem.

9.1 Innovations and Emerging Technologies

The future of cybersecurity for Industry 4.0 and 5.0 is intricately tied to innovations and emerging technologies that are poised to shape the digital landscape. Quantum computing, despite posing a threat to traditional encryption, is expected to catalyze expansion of quantum-resistant cryptographic methods, ensuring the security of sensitive data in the quantum era. AI and machine learning (ML) will witness further advancements, enabling more sophisticated threat detection capabilities and proactive defense mechanisms. The rise of Extended Detection and Response (XDR) solutions will integrate various security components, providing a unified and comprehensive approach to threat management. As IoT proliferates, integration of Blockchain technology is anticipated to enhance the security and integrity of IoT ecosystems by establishing decentralized and tamper-resistant ledgers. Edge computing will demand new security paradigms to protect distributed systems, giving rise to innovative approaches such as Confidential Computing that ensures data privacy even during processing. Moreover, conjunction of Operational Technology (OT) and Information Technology (IT) in Industry 4.0 and 5.0 will drive the development of specialized security solutions tailored to industrial environments, addressing the unique challenges posed by cyber threats in critical infrastructure and manufacturing systems. These emerging technologies collectively signal a transformative era in cybersecurity, paving the way for resilient and adaptive defense mechanisms against the evolving threat landscape.

9.2 Adaptive Security Measures

Adaptive security measures are set to define the future of cybersecurity in Industry 4.0 and 5.0, reflecting a shift from traditional, static defense mechanisms to dynamic, responsive strategies. As the threat landscape becomes more sophisticated and dynamic, adaptive security relies on real-time threat intelligence, machine learning, and behavioral analytics to continuously assess and adapt to evolving risks. Context-aware security, leveraging information about user behavior, device posture, and network conditions, will enable more precise and proactive threat detection. Zero Trust Architecture (ZTA), characterized by continuous verification and least privilege access, is gaining prominence as an adaptive approach to network security. Additionally, self-healing systems equipped with automated response capabilities will become integral, allowing organizations to autonomously identify and mitigate security incidents. The integration of risk-based decision-making and predictive analytics into adaptive security models will enhance organizations' ability to anticipate and thwart emerging threats. Ultimately, adaptive security measures are poised to provide a dynamic and resilient defense framework, aligning with the fluid and interconnected nature of Industry 4.0 and 5.0 environments.

9.3 Impact of Regulatory Changes

The future trends in cybersecurity for Industry 4.0 and 5.0 will be significantly influenced by the impact of regulatory changes as government's worldwide grapple with the evolving cyber threat landscape. With the increasing integration of digital technologies in critical infrastructures and industrial processes, regulatory bodies are expected to enact stringent cybersecurity standards and frameworks to ensure the resilience of these systems. Industry-specific regulations will likely be refined and expanded, addressing the unique challenges posed by the interconnectivity of devices, data-driven decision-making, and the reliance on artificial intelligence. Compliance requirements related to data protection, privacy, and incident reporting will play a crucial role in shaping organizations' cybersecurity strategies. Cross-border data transfer regulations and international collaborations in setting cybersecurity standards are also anticipated. Compliance with these evolving regulations will not only be a legal imperative but a critical factor in building trust among stakeholders. Organizations will need to adopt agile and adaptive cybersecurity frameworks that align with emerging regulatory landscapes, fostering a proactive and collaborative approach to cybersecurity governance in time of Industry 4.0 and 5.0.

10 Conclusion

In conclusion, the trajectory of cybersecurity for Industry 4.0 and 5.0 reflects a dynamic landscape characterized by technological advancements, evolving threats, and the imperative for proactive defense mechanisms. The future of cybersecurity in these industrial revolutions is intricately tied to innovations such as quantum-resistant cryptography, artificial intelligence, and machine learning, which empower organizations to anticipate and mitigate sophisticated cyber threats. Adaptive security measures, including context-aware security and Zero Trust Architecture, are poised to become pivotal in addressing the fluid and interconnected nature of these industrial environments. The impact of regulatory changes is an influential factor, as governments worldwide seek to establish stringent cybersecurity standards to safeguard critical infrastructures and data integrity. Industry-government collaboration and ongoing research initiatives contribute to a collective defense against emerging threats, fostering a resilient and adaptive cybersecurity ecosystem. As organizations navigate this complex landscape, the convergence of ethical considerations, technological innovations, and regulatory compliance will shape a cybersecurity paradigm that not only fortifies digital infrastructures but also fosters trust, collaboration, and resilience in face of evolving cyber threat landscape.

References

1. Pant, P., Rajawat, A.S., Goyal, S.B., Singh, D., Constantin, N.B., Raboaca, M.S., Verma, C.: Using machine learning for industry 5.0 efficiency prediction based on security and proposing models to enhance efficiency. In: 2022 11th International Conference on System Modeling & Advancement in Research Trends (SMART), pp. 909–914. IEEE (2022)
2. Baz, A., Ahmed, R., Khan, S.A., Kumar, S.: Security risk assessment framework for the healthcare Industry 5.0. Sustainability **15**(23), 16519 (2023)
3. Wang, C., Wang, D., Xu, G., He, D.: Efficient privacy-preserving user authentication scheme with forward secrecy for industry 4.0. Sci. China Inf. Sci. **65**(1), 112301 (2022)
4. Garg, S., Kaur, K., Kaddoum, G., Choo, K.K.R.: Toward secure and provable authentication for Internet of Things: realizing industry 4.0. IEEE Internet Things J **7**(5), 4598–4606 (2019)
5. Joshi, S., Pise, A.A., Shrivastava, M., Revathy, C., Kumar, H., Alsetoohy, O., Akwafo, R.: Adoption of blockchain technology for privacy and security in the context of industry 4.0. Wirel. Commun. Mob. Comput. **2022**, (2022)
6. Czeczot, G., Rojek, I., Mikołajewski, D., Sangho, B.: AI in IIoT management of cybersecurity for Industry 4.0 and Industry 5.0 purposes. Electronics **12**(18), 3800 (2023)
7. Bayanati, M.: Security and privacy analysis based on Internet of Things in the fourth industrial generation (Industry 4.0). Int. J. Innov. Manag., Econ. Soc. Sci. **3**(2), 1–12 (2023)
8. Qahtan, S., Yatim, K., Zulzalil, H., Osman, M.H., Zaidan, A.A., Alsattar, H.A.: Review of healthcare Industry 4.0 application-based blockchain in terms of security and privacy development attributes: comprehensive taxonomy, open issues and challenges and recommended solution. J. Netw. Comput. Appl. **209**, 103529 (2023)
9. Jhanjhi, N.Z., Humayun, M., Almuayqil, S.N.: Cyber security and privacy issues in industrial Internet of Things. Comput. Syst. Sci. & Eng. **37**(3), (2021)

10. Velliangiri, S., Manoharn, R., Ramachandran, S., Venkatesan, K., Rajasekar, V., Karthikeyan, P., Kumar, P., Kumar, A., Dhanabalan, S.S.: An efficient lightweight privacy-preserving mechanism for Industry 4.0 based on elliptic curve cryptography. IEEE Trans. Ind. Inform. **18**(9), 6494–6502 (2021)
11. Pradhan, D., Agarwal, A., Tun, H.M., Naing, Z.M., & Oo, T.: Critical security & privacy issue in blockchain technology intended to Industry 4.0. Middle East Res. J. Eng. Technol. **2**(1), 1–7 (2022)
12. Saura, J.R., Ribeiro-Soriano, D., Palacios-Marqués, D.: Evaluating security and privacy issues of social networks based information systems in Industry 4.0. Enterp. Inf. Syst. **16**(10–11), 1694–1710 (2022)
13. Pedreira, V., Barros, D., Pinto, P.: A review of attacks, vulnerabilities, and defenses in Industry 4.0 with new challenges on data sovereignty ahead. Sensors **21**(15), 5189 (2021)
14. Vashishth, T.K., Sharma, V., Sharma, K.K., Kumar, B., Chaudhary, S., Panwar, R.: Intelligent resource allocation and optimization for industrial robotics using AI and blockchain. In: AI and Blockchain Applications in Industrial Robotics, pp. 82–110. IGI Global (2024)
15. Sharma, V., Kumar, S.: Role of Artificial Intelligence (AI) to enhance the security and privacy of data in smart cities. In: 2023 3rd International Conference on Advance Computing and Innovative Technologies in Engineering (ICACITE), pp. 596–599. IEEE (2023)
16. Salam, A., Ullah, F., Amin, F., Abrar, M.: Deep learning techniques for web-based attack detection in Industry 5.0: a novel approach. Technologies **11**(4), 107 (2023)
17. Hammad, M., Jillani, R.M., Ullah, S., Namoun, A., Tufail, A., Kim, K.H., Shah, H.: Security framework for network-based manufacturing systems with personalized customization: an Industry 4.0 approach. Sensors **23**(17), 7555 (2023)
18. Jaisan, A., Manickam, S., Laghari, S., Rehman, S.U., Karuppayah, S.: Secured SECS/GEM: a security mechanism for M2M communication in Industry 4.0 ecosystem. Int. J. Adv. Comput. Sci. Appl. **12**(8), (2021)
19. Laghari, S., Manickam, S., Karuppayah, S., Al-Ani, A., Rehman, S.U.: Cyberattacks and vociferous implications on SECS/GEM communications in Industry 4.0 ecosystem. Int. J. Adv. Comput. Sci. Appl. **12**(7), (2021)
20. Leng, J., Ye, S., Zhou, M., Zhao, J.L., Liu, Q., Guo, W., Cao, W., Fu, L.: Blockchain-secured smart manufacturing in Industry 4.0: a survey. IEEE Trans. Syst., Man, Cybern. Syst. **51**(1), 237–252 (2020)

Exploring Applications: Intelligent Drones and Robots in Industrial Settings

Divya Bansal, Naboshree Bhattacharya, and Priyanka Shandilya

Abstract The integration of embodied AI systems, encompassing drones and robots, has revolutionized various industrial sectors. In manufacturing, embodied AI systems streamline operations through automated inspection, precise material handling, and efficient assembly processes. These systems enhance productivity, quality control, and safety measures. Additionally, they play a vital role in surveillance and security, conducting monitoring, patrolling, and intruder detection tasks with heightened accuracy and efficiency. Maintenance and repair procedures benefit from predictive maintenance capabilities and autonomous repair functionalities, minimizing downtime and optimizing equipment performance. Furthermore, embodied AI systems contribute to environmental monitoring by assessing air and water quality and exploring hazardous environments, ensuring workplace safety and sustainability. Data analytics and integration play a crucial role in leveraging the potential of embodied AI systems, facilitating big data analytics and seamless integration with the Internet of Things (IoT) and other AI technologies. Despite their transformative potential, challenges such as regulatory compliance and technical limitations persist. However, ongoing advancements in emerging technologies and increasing industry adoption promise a future where embodied AI systems continue to redefine industrial processes, driving efficiency, innovation, and economic growth. This current chapter provides an overview of their applications, challenges, and future trends.

D. Bansal (✉) · N. Bhattacharya · P. Shandilya
Amity University, Noida, Jharkhand, India
e-mail: divyadogra26@gmail.com

P. Raj et al. (eds.), *Building Embodied AI Systems: The Agents, the Architecture Principles, Challenges, and Application Domains*, Information Systems Engineering and Management 14, https://doi.org/10.1007/978-3-031-68256-8_7

1 Introduction

1.1 Background

The integration of intelligent drones and robots in industrial operations signifies a substantial advancement. Utilizing AI, machine learning, and advanced sensors, these technologies enhance productivity and address operational challenges across sectors. In automotive manufacturing, robots with machine vision systems optimize assembly precision and speed. In e-commerce, drones expedite order fulfillment in warehouses. Beyond, in agriculture, drones monitor crops for health and optimize irrigation while healthcare robots aid in patient care and sterilization. The convergence of IoT, cloud computing, and 5G further accelerates integration, fostering automation and adaptability for sustainable growth and competitiveness. Ultimately, this integration heralds a future of seamless human–machine collaboration and enhanced industrial efficiency.

1.2 Overview of Drones and Robots in Industrial Settings

Drones and robots have swiftly transitioned from innovative concepts to essential components within various industrial domains, reshaping conventional manufacturing, logistics, and operational methodologies across a spectrum of sectors [1, 2].

1. **Manufacturing Automation:**
   - Robots revolutionize manufacturing tasks, ensuring precision and efficiency.
   - Collaborative robots, or cobots, enhance human capabilities in safer work environments (Rethink Robotics).
2. **Logistics and Warehousing:**
   - Drones monitor inventory and perform surveillance in warehouses [3].
   - Autonomous mobile robots (AMRs) optimize storage and fulfillment processes (Kiva Systems).
3. **Supply Chain Optimization:**
   - Drones provide real-time insights into supply chain activities.
   - Robotic arms streamline goods movement in logistics hubs (Fanuc America Corporation).
4. **Inspection and Maintenance:**
   - Drones conduct aerial inspections of industrial infrastructure.
   - Robotic crawlers and UGVs ensure asset integrity in hazardous environments).

5. **Precision Agriculture:**
   - Drones monitor crop health and optimize agricultural practices.
   - Autonomous tractors and harvesters improve agricultural efficiency.
6. **Healthcare Assistance:**
   - Robots assist in medical tasks and enable remote patient monitoring.
   - Telepresence robots facilitate virtual consultations in healthcare setting.
   - Drones and robots drive automation, productivity, and safety in industries, promising further advancements in intelligent manufacturing and logistics [4].

1.3 Importance of Intelligent Systems in Enhancing Industrial Processes

Intelligent systems, integrating AI, ML, robotics, and automation, revolutionize industrial processes, enhancing efficiency, quality, and safety. They optimize production schedules and ensure quality control, reducing errors and downtime. In supply chain management, they aid in forecasting demand, optimizing inventory, and improving logistics. Moreover, they mitigate risks, enhance customization, and promote sustainability by minimizing waste and conserving resources. Leveraging these technologies provides organizations with a competitive edge in today's dynamic market environment [5, 6].

2 Applications in Manufacturing

2.1 Automated Inspection

2.1.1 Quality Control Through Drone Inspections

Drones transform quality control inspections, providing precision and efficiency (Nguyen et al. 2019). With high-resolution cameras and thermal sensors, they swiftly detect defects (Chen et al. 2018), accessing inaccessible areas for enhanced safety (Petrov et al. 2021). AI-driven analysis identifies deviations from standards [58], reducing downtime and optimizing operations [59]. Their adoption increases across sectors due to their ability to promote workplace safety and streamline processes [7, 8].

2.1.2 Detection of Defects and Anomalies

Drone inspections revolutionize industrial quality control, employing UAVs with advanced sensors. High-resolution cameras, thermal imaging, and LiDAR enable precise defect detection. Machine learning enhances automated defect recognition. Thermal sensors identify anomalies like overheating, while LiDAR constructs 3D models for structural analysis. Real-time inspections enable proactive maintenance, reducing downtime and safety risks [9–11].

2.2 *Material Handling*

2.2.1 Robotic Systems for Efficient Material Movement

Robotic systems streamline material movement, leveraging advanced technologies like robotics, automation, and AI across logistics, manufacturing, and healthcare sectors. Let's delve into some essential elements and technologies involved, while referencing pertinent sources where applicable [12, 13].

Autonomous Navigation: Key for drones and robots in material transportation, enabled by SLAM technology integrating LiDAR and cameras for dynamic mapping.

Manipulation and Grasping: Essential for efficient handling, robots utilize grippers with sensors and actuators for precise manipulation, enabled by advancements in soft robotics and tactile sensing.

Machine Learning and AI: Algorithms enhance material handling approaches, employing reinforcement learning for adaptive decision-making, notably improving warehouse operations.

Integration of Drones and Ground Robots: Combining capabilities expands material handling functions, utilizing drones for aerial tasks and ground robots for indoor manipulation, fostering collaborative solutions.

Safety and Human–Robot Collaboration: Prioritizing safety, collaborative robots equipped with advanced sensors and interfaces ensure safe human–robot interaction in shared workspaces.

2.2.2 Integration with Warehouse Management Systems

Integrating warehouse management systems (WMS) with robotic systems in warehouses and distribution centers is essential for maximizing efficiency and functionality. Acting as central control hubs, WMS manage inventory, orders, and logistics. This integration ensures a smooth flow of real-time data, improving decision-making and warehouse productivity [14, 15].

- **Real-Time Data Exchange**: Integration with Warehouse Management Systems (WMS) facilitates dynamic decision-making based on inventory and order status.
- **Optimized Task Allocation**: Direct task assignment from WMS enhances collaboration, optimizing task prioritization and improving throughput.
- **Inventory Accuracy and Traceability**: WMS integration ensures real-time updates, enhancing inventory control and minimizing stockouts.
- **Dynamic Path Planning and Coordination**: Robots utilize warehouse data for efficient route planning, boosting throughput and efficiency.
- **Scalability and Flexibility**: WMS integration allows adaptability to changing warehouse needs and demand fluctuations, ensuring sustained operational efficiency [16, 17].

2.3 Assembly and Production

2.3.1 Robot-Assisted Assembly Lines

Robot-assisted assembly lines represent a transformative shift in manufacturing, employing robotic technology to elevate efficiency, accuracy, and adaptability. These systems integrate industrial robots, streamlining tasks from basic to complex assembly processes. By automating repetitive tasks, they ensure high throughput and minimize errors, ensuring consistent product quality. Flexible and collaborative robots enhance safety and teamwork, while data-driven optimization drives efficiency and predictive maintenance, fostering operational excellence.

3 Surveillance and Security

3.1 Monitoring and Patrolling

3.1.1 Drones for Real-Time Surveillance

Drones have transformed surveillance practices by providing real-time aerial monitoring capabilities in various scenarios, including security, law enforcement, disaster management, and environmental monitoring. Equipped with advanced sensors, cameras, and communication systems, these unmanned aerial vehicles (UAVs) collect and transmit data rapidly from vantage points often inaccessible to ground-based surveillance. Drones provide powerful capabilities for real-time surveillance, enabling operators to monitor vast areas, detect threats, and gather actionable intelligence from elevated perspectives. By leveraging mobility, cost-effectiveness, and advanced sensor technologies, drones are reshaping surveillance operations across various domains, prompting careful consideration of privacy and ethical considerations in their deployment [18, 19].

3.1.2 Enhanced Security Measures in Industrial Facilities

Enhanced security measures in industrial facilities using drones represent a proactive strategy to protect critical infrastructure, assets, and personnel from potential threats. Drones offer unique capabilities for perimeter surveillance, intruder detection, and incident response, making them valuable tools for enhancing security in industrial settings. Below, I'll explore the key aspects and benefits of using drones for security in industrial facilities, citing relevant sources where applicable [20].

Drones are instrumental in industrial security, offering comprehensive surveillance, rapid threat response, cost-effective solutions, and enhanced situational awareness. Equipped with advanced sensors, they detect breaches and hazards swiftly, aiding risk mitigation. While cost-effective, regulatory compliance and privacy safeguards are crucial to ensure responsible deployment [21].

3.2 Intruder Detection

3.2.1 Intelligent Robots for Identifying Unauthorized Access

Intelligent robots equipped with advanced sensors and artificial intelligence (AI) algorithms are pivotal in identifying unauthorized access and bolstering security measures across various environments like industrial facilities, warehouses, and sensitive installations. These robots utilize sensing technologies, machine learning algorithms, and autonomous navigation to detect and respond to security breaches in real-time. Here's a detailed exploration of the key aspects and benefits of using intelligent robots for identifying unauthorized access, along with relevant sources [22, 23]:

i. **Continuous Surveillance and Patrol:**

Intelligent robots conduct continuous surveillance and patrol duties, monitoring designated areas for unauthorized access or suspicious activities. Equipped with sensors like cameras, LiDAR, thermal imagers, and motion detectors, these robots detect anomalies and alert security personnel promptly. By autonomously patrolling predefined routes, robots ensure comprehensive coverage and timely detection of unauthorized access attempts.

ii. **Real-time Threat Detection and Response:**

With AI-driven algorithms, intelligent robots analyze sensor data in real-time to identify patterns indicative of unauthorized access or security threats. Machine learning enables robots to differentiate between normal and abnormal behavior, recognize potential intruders, and prioritize alerts. By processing data locally or transmitting to centralized security operations centers, robots facilitate rapid threat detection and response coordination, allowing security personnel to take timely actions.

iii. **Enhanced Situational Awareness:**

Intelligent robots contribute to enhanced situational awareness by providing real-time views of the operational environment. By streaming live video feeds, sensor data, and analytical insights, robots help operators assess security incidents, track unauthorized movement, and coordinate response efforts effectively. This situational awareness enables informed decision-making, efficient resource allocation, and proactive risk mitigation by security teams.

iv. **Adaptive Security Measures:**

Intelligent robots dynamically adapt surveillance and response strategies to evolving threats and operational conditions. Through continuous learning, robots refine detection algorithms, adjust patrol routes, and optimize resource allocation based on historical data and real-time feedback. This adaptability helps robots anticipate emerging threats, prevent security breaches, and enhance security system resilience.

v. **Integration with Security Ecosystems:**

Intelligent robots integrate into broader security ecosystems, complementing surveillance cameras, access control systems, and intrusion detection technologies. Seamless integration with centralized management platforms and security information and event management (SIEM) systems enables unified command and control, automated incident response workflows, and data-driven analytics. This integration enhances security operations' effectiveness, facilitates information sharing, and strengthens overall security posture.

Intelligent robots are pivotal in identifying unauthorized access and enhancing security measures by providing continuous surveillance, real-time threat detection, enhanced situational awareness, adaptive security measures, and integration with broader security ecosystems. Leveraging intelligent robots' capabilities strengthens security infrastructure, mitigates risks, and safeguards critical assets and personnel against unauthorized access and security threats.

3.2.2 Integration with Alarm and Response Systems

Integration with alarm and response systems significantly improves the effectiveness of intelligent robots in identifying unauthorized access and promptly addressing security threats. These integrations facilitate seamless communication and coordination between intelligent robots and alarm systems, enabling rapid response and mitigation of security incidents. Here are the key aspects and benefits of integrating intelligent robots with alarm and response systems, supported by relevant sources [24–26]:

i. **Real-time Alert Generation:**

Intelligent robots integrate with alarm systems to promptly detect security breaches or unauthorized access attempts, triggering real-time alerts.

ii. **Automated Incident Response:**

Upon receiving alerts, robots autonomously assess situations and initiate response actions, such as analyzing video feeds or deploying countermeasures.

iii. **Coordination with Human Operators:**

Robots coordinate with human operators in security operations centers, providing critical information and situational updates for informed decision-making.

iv. **Enhanced Situational Awareness:**

Integration with alarm systems enhances situational awareness by sharing real-time data between robots and centralized platforms, enabling better threat visualization (Chowdhury et al. 2020).

v. **Streamlined Incident Management:**

Robots streamline incident management by automating tasks, logging event details, and facilitating seamless information sharing in centralized platforms.

4 Maintenance and Repairs

4.1 Predictive Maintenance

4.1.1 Use of Drones for Preventive Maintenance

Drones revolutionize preventive maintenance in industries, enhancing efficiency, reducing downtime, and optimizing asset performance through timely and cost-effective inspections using specialized sensors and cameras [27–30].

Utilizing drones for preventive maintenance offers numerous advantages in industrial upkeep. Aerial inspections enable swift and precise monitoring of critical infrastructure, aiding in early issue detection. Drones excel in hazardous environments, ensuring safety and efficiency in remote areas. Data analytics provide actionable insights for predictive maintenance, optimizing asset performance. Drones offer cost-effective solutions, reducing expenses and downtime. Furthermore, they aid in regulatory compliance and documentation, furnishing visual records for audits. Integrating drones into maintenance strategies enhances asset reliability and operational efficiency.

4.1.2 Monitoring Equipment Health and Performance

Monitoring equipment health and performance is integral to ensuring operational efficiency, minimizing downtime, and prolonging the lifespan of critical assets across industries like manufacturing, energy, transportation, and healthcare. Advanced monitoring techniques such as sensor-based monitoring, predictive analytics, and condition-based maintenance empower organizations to proactively identify equipment issues, optimize performance, and prevent costly failures. Let's explore the key aspects and benefits of monitoring equipment health and performance, citing relevant sources where applicable [31–33].

i. **Sensor-based Monitoring Systems**: Sensor-based monitoring systems, utilizing vibration, temperature, pressure, and flow sensors, enable real-time equipment health assessment.
ii. **Predictive Analytics and Machine Learning**: Predictive analytics and machine learning analyze sensor data to predict equipment failures accurately, guiding preventive maintenance.
iii. **Condition-based Maintenance Strategies**: This uses real-time data and predictive analytics to optimize schedules, reducing costs and downtime.
iv. **Remote Monitoring and Telemetry**: Remote monitoring and telemetry provide global real-time visibility into equipment health, enabling proactive maintenance.
v. **Integration with Asset Management Systems**: Integration with asset management systems centralizes data for comprehensive asset management, enabling informed decision-making and regulatory compliance.

Monitoring equipment health and performance through sensor-based monitoring, predictive analytics, condition-based maintenance strategies, remote monitoring, and integration with asset management systems enables organizations to optimize maintenance practices, reduce costs, and enhance operational reliability across industries.

4.2 Autonomous Repairs

4.2.1 Robotic Systems Capable of Minor Repairs

Robotic systems designed for minor repairs represent a significant advancement in industrial automation, promising to enhance maintenance efficiency, minimize downtime, and boost overall operational productivity. These systems integrate cutting-edge technologies such as robotics, artificial intelligence (AI), and computer vision to autonomously tackle various minor repair tasks.

Robotic systems equipped with advanced sensors and algorithms autonomously detect anomalies and execute repairs with precision, streamlining maintenance processes. They efficiently address maintenance needs, preventing downtime and extending equipment lifespan. In hazardous industries, robots enhance safety by handling risky tasks. Integrating robots optimizes resource allocation, enhances efficiency, and enables continuous improvement through AI [32, 34].

5 Environmental Monitoring

5.1 *Air and Water Quality*

5.1.1 Drones for Monitoring Industrial Emissions

Drones are increasingly employed for monitoring industrial emissions, providing a versatile and efficient method for assessing air quality, identifying pollutant sources, and ensuring compliance with environmental regulations. Equipped with specialized sensors and imaging technologies, drones can gather real-time data on air pollutants, greenhouse gases, and particulate matter emissions from industrial facilities. Let's explore the key aspects and benefits of using drones for monitoring industrial emissions, with relevant sources cited where applicable [35–38].

i. **Remote Sensing Capabilities**: Drones equipped with specialized sensors monitor industrial emissions, detecting pollutants like SO_2, NOx, VOCs, and PM, providing real-time data on air quality.
ii. **High Spatial Resolution**: Drones offer high spatial resolution for mapping emission sources in industrial complexes, aiding regulators and operators in prioritizing mitigation.
iii. **Real-time Monitoring and Data Analytics**: Drones offer real-time data analysis for industrial emissions monitoring, aiding in compliance detection and proactive air quality management.
iv. **Cost-effective and Flexible Deployment**: Drones provide cost-effective, flexible emission monitoring, swiftly adapting to changing conditions, regulations, and concerns, maximizing coverage while minimizing costs.
v. **Improved Safety and Accessibility**: Drones provide cost-effective, flexible emission monitoring, swiftly adapting to changing conditions, regulations, and concerns, maximizing coverage while minimizing costs.

Drones offer significant advantages for monitoring industrial emissions, including remote sensing capabilities, high spatial resolution, real-time monitoring and data analytics, cost-effective deployment, and improved safety and accessibility. Leveraging drone capabilities, environmental agencies, regulatory authorities, and industrial stakeholders can enhance their understanding of air quality dynamics, mitigate environmental impacts, and promote sustainable industrial practices.

5.1.2 Robotics in Water Quality Assessment

Robotics is pivotal in water quality assessment, providing innovative solutions for monitoring, sampling, and analyzing water bodies. With advancements in robotics, sensors, and autonomous navigation systems, robotic platforms can gather real-time data, detect pollutants, and evaluate water quality parameters with precision and efficiency. Let's delve into the key aspects and benefits of using robotics in water quality assessment, with relevant sources cited where applicable [39–41].

i. **Remote Sensing and Monitoring**: Robotic platforms equipped with sensors enable remote sensing and continuous monitoring of water quality parameters.
ii. **High-resolution Mapping and Sampling**: AUVs and USVs equipped with sonar and imaging systems facilitate high-resolution mapping and targeted assessment of pollution hotspots.
iii. **Real-time Data Collection and Analysis**: Robotic platforms provide real-time data collection and analysis, aiding in timely responses to environmental threats.
iv. **Cost-effective and Scalable Solutions**: Robotic systems offer cost-effective and scalable solutions for water quality assessment, reducing labor-intensive fieldwork.
v. **Integration with Environmental Monitoring Networks**: Robotics seamlessly integrates with environmental monitoring networks, enhancing data interoperability and accessibility.

Robotics in water quality assessment offer substantial advantages, including remote sensing and monitoring capabilities, high-resolution mapping and sampling, real-time data collection and analysis, cost-effective and scalable solutions, and integration with environmental monitoring networks. By leveraging robotics, stakeholders can deepen their understanding of water quality dynamics, pinpoint pollution sources, and implement targeted interventions to safeguard aquatic ecosystems for future generations.

5.2 *Hazardous Environment Exploration*

5.2.1 Robots for Inspecting and Managing Hazardous Areas

Robots are increasingly utilized for inspecting and managing hazardous areas, where human access is limited or poses significant risks to safety and health. These robots, equipped with specialized sensors, manipulators, and mobility systems, navigate through hazardous environments, conduct inspections, and perform maintenance tasks with precision and efficiency. Let's explore the key aspects and benefits of using robots for inspecting and managing hazardous areas, with relevant sources cited where applicable [42–45].

i. **Remote Inspection and Surveillance:**

Robots enable remote inspection and surveillance of hazardous areas, including industrial facilities, nuclear sites, and disaster zones, without endangering humans. Equipped with cameras, sensors, and imaging systems, these robots capture high-resolution images, videos, and sensor data in real-time, providing operators with comprehensive situational awareness and diagnostic information. By navigating through confined spaces, elevated platforms, or contaminated areas, robots identify anomalies, monitor environmental conditions, and detect potential safety hazards more effectively than traditional inspection methods.

ii. **Detection and Monitoring of Hazardous Substances:**

Robots are equipped with sensors capable of detecting various hazardous substances, such as toxic chemicals, radiation, and biological agents. Gas sensors, spectrometers, and radiation detectors allow robots to identify and quantify hazardous materials, assess exposure risks, and establish exclusion zones to protect personnel. Additionally, robots collect air samples, soil samples, and surface swabs for laboratory analysis, providing valuable data for risk assessment, emergency response planning, and environmental remediation efforts.

iii. **Safe and Precise Handling of Materials:**

Robots feature manipulators and end-effectors designed to handle hazardous materials safely and precisely. Whether removing radioactive debris, dismantling chemical weapons, or containing oil spills, robots perform delicate tasks with dexterity and control, minimizing human exposure and environmental contamination. Advanced grippers, suction devices, and cutting tools enable robots to manipulate objects, seal leaks, and mitigate hazards without direct human intervention, ensuring operational safety and regulatory compliance in hazardous environments.

iv. **Emergency Response and Disaster Management:**

Robots are crucial in emergency response and disaster management scenarios, where rapid assessment and intervention are vital for saving lives and minimizing damage. Search and rescue robots equipped with cameras, sensors, and thermal imaging capabilities locate survivors in collapsed buildings, hazardous terrain, or debris-filled environments, providing first responders with real-time intelligence and situational awareness. Drones and unmanned aerial vehicles (UAVs) survey disaster zones from the air, assess infrastructure damage, and support decision-making processes for resource allocation and evacuation planning.

v. **Continuous Monitoring and Maintenance:**

Robots enable continuous monitoring and maintenance of hazardous areas, ensuring the integrity of critical infrastructure and industrial facilities. Autonomous inspection robots equipped with sensors and diagnostic tools detect structural defects, corrosion, and wear in pipelines, storage tanks, and chemical reactors, enabling proactive maintenance interventions to prevent catastrophic failures. By implementing predictive maintenance strategies, organizations optimize asset performance, extend equipment lifespan, and minimize downtime, while reducing risks to personnel and the environment.

Robots for inspecting and managing hazardous areas offer numerous benefits, including remote inspection and surveillance capabilities, detection and monitoring of hazardous substances, safe handling of materials, emergency response and disaster management support, and continuous monitoring and maintenance of critical infrastructure. By leveraging robotics technology, organizations enhance safety, efficiency, and resilience in hazardous environments, protecting human lives and safeguarding the environment.

5.2.2 Applications in Nuclear, Chemical, and Other High-Risk Industries

Applications of robotics in industries such as nuclear, chemical, and other high-risk sectors are instrumental in improving safety, efficiency, and productivity while reducing human exposure to hazardous conditions. These industries rely on advanced robotics for tasks like inspection, maintenance, decontamination, and emergency response, especially in environments marked by radiation, toxic chemicals, and other risks. Let's explore the key applications of robotics in these sectors, with relevant sources cited where applicable [1, 46, 47].

i. **Nuclear Industry:**

In nuclear facilities, robots are vital for tasks such as facility inspection, reactor maintenance, and handling radioactive waste. Robots with radiation-resistant materials and sensors can access areas with high radiation levels, conduct visual inspections, and measure radiation intensity to assess facility conditions. Remote-controlled manipulators and robotic arms facilitate precise handling of radioactive materials, fuel rods, and contaminated components during decommissioning and waste management processes.

ii. **Chemical Industry:**

Robots are crucial in managing hazardous materials, preventing chemical spills, and ensuring process safety in the chemical industry. Automated systems equipped with chemical sensors and monitoring devices detect leaks, spills, and abnormal chemical concentrations in storage tanks, pipelines, and processing units. Robotic platforms perform chemical sampling, analysis, and decontamination tasks, reducing human exposure to toxic substances and ensuring compliance with safety regulations.

iii. **Oil and Gas Industry:**

In the oil and gas sector, robots are employed for offshore exploration, pipeline inspection, and maintenance operations in harsh environments. Subsea robots with cameras, sonar, and manipulation systems inspect underwater infrastructure, detect pipeline defects, and perform repairs without human intervention. Unmanned aerial vehicles (UAVs) and drones conduct aerial surveys, flare stack inspections, and emergency response activities, enhancing operational efficiency and safety.

iv. **Mining Industry:**

Robots are utilized for exploration, excavation, and safety monitoring in the mining industry. Systems with LiDAR scanners and 3D mapping technology survey mine tunnels, assess geological formations, and identify hazards like unstable rock formations and gas leaks (Kam et al. 2021). Autonomous haulage trucks and mining vehicles enhance productivity and safety by reducing accidents and injuries associated with manual operation [60].

v. **Emergency Response and Disaster Management:**

Robotics play a critical role in emergency response and disaster management across various industries. Search and rescue robots equipped with cameras, sensors, and communication systems locate survivors in collapsed buildings, industrial accidents, and natural disasters, aiding timely rescue operations. UAVs and drones perform aerial reconnaissance, mapping, and damage assessment in disaster zones, supporting decision-making and resource allocation for emergency response teams.

Robotics find diverse applications in nuclear, chemical, and other high-risk industries, including inspection, maintenance, decontamination, and emergency response. By harnessing robotics technology, these sectors enhance safety, efficiency, and resilience in hazardous environments while safeguarding human lives and minimizing environmental impacts.

6 Data Analytics and Integration

In today's data-driven world, businesses and organizations are recognizing the importance of harnessing the power of data analytics to gain valuable insights, make informed decisions, and drive business success. The availability of large and feature-rich datasets has enabled organizations to analyze complex patterns, trends, and relationships that were previously not possible. Furthermore, advancements in computing and technology have opened up opportunities for integrating various sources of data, including spatial and hydrologic data, to provide a more comprehensive understanding of water resources and hydrologic engineering problems. By integrating various sources of data, including spatial and hydrologic data, organizations can obtain a more holistic view of water resources and hydrologic engineering problems. This allows for better decision-making and the development of more efficient management strategies [47]. For example, in the field of water resources engineering, the use of spatial data such as terrain, aerial imagery, digital elevation models, and infrastructure data can provide valuable information for flood plain analysis and flood management. Additionally, the availability of water consumption data with spatial information, such as household address or census block identification, has allowed municipalities to analyze patterns of water demand and identify areas of high usage or potential water conservation opportunities. The integration of data analytics and spatial analysis techniques, such as geographic information systems, has also proven to be beneficial in water resource mapping, satellite rainfall measurements, runoff simulation, and flood risk management [21]. Overall, the use of data analytics and integration techniques in water resources and hydrologic engineering offers numerous benefits. It can enhance decision-making, improve understanding of system behavior, optimize operation strategies, and facilitate policy evaluation. Overall, data analytics and integration techniques are valuable tools for addressing water resources and hydrologic engineering problems. Overall, data analytics and integration techniques are valuable tools for addressing water resources and hydrologic engineering problems. Overall, data analytics and integration techniques are valuable tools for addressing water resources and hydrologic engineering problems.

6.1 Big Data Analytics

One of the significant advancements in recent years, big data analytics, has emerged as a powerful tool for addressing water resources and hydrologic engineering problems. Through the analysis of large and complex datasets, big data analytics allows for a deeper understanding of water resources systems and enables more informed decision-making. This can include analyzing data related to rainfall patterns, streamflow levels, groundwater levels, and water usage. By identifying patterns and trends in the data, big data analytics can help identify areas of high water demand, predict potential water shortages or flooding events, and optimize water allocation and management strategies. In addition, big data analytics can also integrate various sources of data, such as spatial data, hydrologic simulation models, and real-time sensor data, to provide a comprehensive and holistic view of water resources system. This comprehensive view can lead to more effective planning, mitigation, and response strategies for water resource management. In the field of water resources engineering, system dynamics concepts have been applied to model a variety of water resource systems. These models can be used to simulate and evaluate different scenarios, allowing decision-makers to assess the effectiveness of various management strategies. Furthermore, data analytics and integration techniques have immense potential in the field of hydrologic rainfall-runoff modeling. By analyzing rainfall data and other relevant variables, data analytics can be used to develop accurate models for predicting floods and streamflow. This can greatly contribute to improved flood forecasting and better understanding of water flow patterns in rivers [48, 49].

6.1.1 Utilizing Data Collected by Drones and Robots

One source that can be used appropriately for addressing water resources and hydrologic engineering problems is the utilization of data collected by drones and robots. Drones and robots equipped with sensors can collect data on various hydrologic variables such as water levels, flow rates, and water quality parameters. This data can then be integrated and analyzed using data analytics techniques to gain valuable insights into water resources systems and make informed decisions for water management.

6.1.2 Extracting Actionable Insights for Decision-Making

Big data analytics can extract actionable insights from the vast amount of data available in water resources and hydrologic engineering. These insights can then be used for informed decision-making related to water resource management, such as identifying areas of high water demand, predicting potential water shortages or water scarcity, and optimizing water allocation strategies to ensure efficient use of resources.

Integration with IoT and AI: Integrating data analytics with IoT and AI technologies can further enhance the effectiveness of water resource management. For example, IoT devices can collect real-time data on various aspects of water resources systems, such as water levels, precipitation, and soil moisture. This data, combined with AI algorithms, can provide real-time analysis and predictions for efficient water allocation and conservation.

1. **Interconnected systems for seamless operation**: Interconnected systems can be implemented to ensure seamless operation in water resource management. These systems can integrate data from various sources, such as weather forecasts, river flow measurements, and groundwater levels, to provide a comprehensive understanding of the water resources system. This comprehensive understanding can then be used to optimize water allocation, manage flood events, and make proactive decisions for sustainable water resource management [50].
2. **AI-driven decision support for industrial processes**: AI-driven decision support systems can be implemented in industrial processes related to water resources to optimize operations and minimize environmental impacts (Sit et al. 2020). These systems can use data analytics techniques to analyze the impact of industrial activities on water resources, identify potential risks or issues, and suggest strategies for mitigation or improvement. In addition to data analytics, the integration of IoT and AI technologies can further enhance the effectiveness of water resource management. For example, IoT devices can collect real-time data on water usage, chemical levels, and equipment performance in industrial processes. This data, combined with AI algorithms, can provide real-time analysis and insights to optimize water usage, detect anomalies or inefficiencies in the processes, and make proactive decisions for sustainable water management [50, 51].

7 Challenges and Solutions

Some challenges in the integration of data analytics and IoT for water resources management include:—Data quality and reliability: Ensuring that the data collected from IoT devices is accurate, reliable, and comprehensive can be a challenge. However, this can be mitigated by implementing quality control measures, regularly calibrating and maintaining the IoT devices, and using data validation techniques.—Data privacy and security: As the integration of IoT devices increases the amount of data being collected and transmitted, ensuring data privacy and security becomes crucial. One solution to address this challenge is to implement strong encryption protocols and data access controls to protect sensitive information.

- **Data integration and interoperability**: Integration of data from various sources, such as different sensors or databases, can be challenging due to differences in formats and structures. However, this challenge can be overcome by using standardized data formats, implementing data integration platforms or APIs, and establishing interoperability protocols between different systems.
- **Scalability**: As the number of IoT devices and data streams increases, scalability becomes a challenge. However, this can be addressed by using cloud-based platforms that can handle large volumes of data and implementing distributed computing solutions.—Data analysis and decision-making: Analyzing large volumes of data collected from IoT devices can be overwhelming. However, machine learning algorithms and artificial intelligence can be used to automate the analysis process, identify patterns and trends, and make data-driven decisions to efficiently manage water resources. In summary, integrating data analytics with IoT for water resources management presents numerous opportunities for optimizing water usage and improving sustainability. However, it also comes with challenges such as data quality and reliability, data privacy and security, data integration and interoperability, scalability, and data analysis
- **Regulatory Compliance**: Ensuring compliance with regulatory standards and guidelines related to data collection, storage, and usage is another challenge in the integration of data analytics and IoT for water resources management. Overall, integrating data analytics with IoT for water resources management presents both opportunities and challenges. Proper data analytics and integration is crucial for effectively managing and utilizing the vast amount of data generated by IoT devices in water resources management [52].

Using data analytics and integration solutions in the field of IoT for water resources management presents numerous opportunities for optimizing water usage and improving sustainability. Integration of spatial and hydrologic data, along with the use of hydrologic simulation models and virtual reality environments, can provide valuable insights for students studying water resources and hydrologic engineering. The use of 3D VR simulation environments in education can enhance student understanding of floodplain management by allowing them to explore and evaluate different engineering options for flood management and assess the severity of flooding in specific locations. Additionally, the combination of system dynamics modeling and IoT data analytics can be applied in various water resource systems, such as environmental policy simulation studies, groundwater modeling applications, irrigation planning, and policy evaluation. The integration of data analytics and IoT for water resources management presents numerous opportunities for optimizing water usage, improving sustainability, and enhancing student understanding of floodplain management and water resource systems. In summary, the integration of data analytics and IoT for water resources management offers opportunities for optimizing water usage, improving sustainability, and enhancing decision-making. Furthermore, the use of IoT data for water resources management requires a proper analytics platform or infrastructure to effectively analyze and extract insights from the data. Furthermore, the heterogeneity, noise, variety, and rapid growth of IoT data present challenges in processing and analyzing this data. Therefore, organizations must develop robust data analytics and integration solutions to effectively manage and utilize IoT data in water resources management.

7.1 Adhering to Aviation and Safety Regulations

Adhering to aviation and safety regulations is essential for ensuring the safety and security of air travel. Organizations in the aviation industry must comply with stringent regulations and standards set by governing bodies such as the International Civil Aviation Organization and the Federal Aviation Administration. They must implement robust systems and processes to effectively manage safety risks, conduct regular inspections and audits, and ensure that aircraft, equipment, and facilities meet the required safety standards. Additionally, organizations must train and certify personnel in accordance with regulatory requirements, maintain accurate and up-to-date records, and establish a culture of continuous improvement to continuously enhance safety practices and procedures. In the context of aviation and safety regulations, organizations must devote significant resources to monitoring compliance, conducting regular inspections and audits, and maintaining accurate records. The integration of data analytics and IoT for water resources management presents numerous opportunities for optimizing water usage, improving sustainability, and enhancing student understanding of floodplain management and water resource systems. To effectively manage and utilize IoT data in water resources management, organizations must develop robust data analytics and integration solutions that can handle the unique characteristics of IoT data such as heterogeneity, noise, variety, and rapid growth [53–57].

8 Conclusion

In conclusion, the future of industrial applications of drones and robots is poised for significant growth and innovation. By addressing challenges and embracing emerging technologies, industries can harness the full potential of intelligent systems to enhance productivity, safety, and sustainability in manufacturing, surveillance, maintenance, and environmental monitoring.

References

1. Javaid, M., et al.: Exploring contributions of drones towards Industry 4.0. Ind. Robot. Int. J. Robot. Res. Appl. **49**(3), 476–490 (2022)
2. Suzuki, S.: Recent researches on innovative drone technologies in robotics field. Adv. Robot. **32**(19), 1008–1022 (2018)
3. Galar, D., Kumar, U., Seneviratne, D.: Robots, Drones, UAVs and UGVs for Operation and Maintenance. CRC Press (2020)
4. Dzedzickis, A., et al.: Advanced applications of industrial robotics: new trends and possibilities. Appl. Sci. **12**(1), 135 (2021)
5. Yang, T., et al.: Intelligent manufacturing for the process industry driven by industrial artificial intelligence. Engineering **7**(9), 1224–1230 (2021)
6. Tyagi, A.K., et al.: Intelligent automation systems at the core of industry 4.0. In: International Conference on Intelligent Systems Design and Applications. Springer International Publishing, Cham (2020)
7. Memon, Q.A.: Intelligent network system for process control: applications, challenges, approaches. In: Rodic, A.D. (ed.) Automation Control-Theory and Practice, pp. 177–197 (2009)
8. Dagnaw, G.: Artificial Intelligence Towards Future Industrial Opportunities and Challenges (2020)
9. Pires, J.N., et al. The role of robotics in additive manufacturing: review of the AM processes and introduction of an intelligent system. Ind. Robot. Int. J. Robot. Res. Appl. **49**(2), 311–331 (2021)
10. Uysal, M.P., Mergen, A.E.: Smart manufacturing in intelligent digital mesh: Integration of enterprise architecture and software product line engineering. J. Ind. Inf. Integr. **22**, 100202 (2021)
11. Teerasoponpong, S., Sugunnasil, P.: Review on artificial intelligence applications in manufacturing industrial supply chain–industry 4.0's perspective. In: 2022 Joint International Conference on Digital Arts, Media and Technology with ECTI Northern Section Conference on Electrical, Electronics, Computer and Telecommunications Engineering (ECTI DAMT & NCON). IEEE (2022)
12. Jenett, B., et al.: Material–robot system for assembly of discrete cellular structures. IEEE Robot. Autom. Lett. **4**(4), 4019–4026 (2019)
13. Skibniewski, M.J., Wooldridge, S.C.: Robotic materials handling for automated building construction technology. Autom. Constr. **1**(3), 251–266 (1992)
14. Zhao, N., et al.: Trajectory predictions with details in a robotic twin-crane system. Complex Syst. Model. Simul. **2**(1), 1–17 (2022)
15. Sharma, N., Shamkuwar, M., Ramdasi, P.: Digital dimensions of industry 4.0: Opportunities for autonomic computing and applications. Auton. Comput. Cloud Resour. Manag. Ind. **4**, 347–383 (2021)
16. Lahmar, M. (ed.): Facility Logistics: Approaches and Solutions to Next Generation Challenges. CRC Press (2007)
17. Liu, S., Sun, C.: Master–slave control system for virtual-physical interactions using hands. Sensors **23**(16), 7107 (2023)
18. Mohsan, S.A.H., et al.: Unmanned aerial vehicles (UAVs): practical aspects, applications, open challenges, security issues, and future trends. Intell. Serv. Robot. **16**(1), 109–137 (2023)
19. Hildmann, H., Kovacs, E.: Using unmanned aerial vehicles (UAVs) as mobile sensing platforms (MSPs) for disaster response, civil security and public safety. Drones **3**(3), 59 (2019)
20. Balci, H.: The Use of Drones in Airport Security Operations (2023)
21. Pinto, L.R., et al.: Radiological scouting, monitoring and inspection using drones. Sensors **21**(9), 3143 (2021)
22. Guo, P.: Detection and Prevention: Toward Secure Mobile Robotic Systems. The Pennsylvania State University (2018)

23. AlNaaji, H.M.M.: Automating unauthorized access attempts detection and handling using robotic process automation. Dissertation, Princess Sumaya University for Technology (Jordan) (2022)
24. Casper, J.L., Mark M., Robin R.M.: Issues in intelligent robots for search and rescue, vol. 4024. In: Unmanned Ground Vehicle Technology II. SPIE (2000)
25. Luo, R.C., et al.: The development of intelligent home security robot. In: IEEE International Conference on Mechatronics, ICM'05. IEEE (2005)
26. Ko, A.W.Y., Henry, Y.K.L.: Intelligent robot-assisted humanitarian search and rescue system. Int. J. Adv. Robot. Syst. **6**(2), 12 (2009)
27. Anandan, R., et al., (eds.): Industrial Internet of Things (IIoT): Intelligent Analytics for Predictive Maintenance. Wiley & Sons (2022)
28. Gbadamosi, A.-Q.O.: An internet of things enabled system for real-time monitoring and predictive maintenance of railway infrastructure. Dissertation, University of the West of England, Bristol (2023)
29. Askarzadeh, T.: Quantitative Evaluation of Drone-Based Linear Asset Condition Monitoring. North Dakota State University, Diss (2023)
30. Weyers, E.: The use of drones to improve downtime management on South African mines. Dissertation, University of Johannesburg
31. Lozano-Ramírez, N.E., et al.: Digitalization and Sustainability in Linear Projects Trends: A Bibliometric Analysis. Sustainability **15**(22), 15962 (2023)
32. Hovsha, J., Daragh O.B.: Drone use cases and their privacy impacts: a taxonomy. Tilburg Law Rev. **27**(1), (2022)
33. Roach, D.P.: FAA research program webinar series on structural health monitoring-module 1: introduction to SHM and implementation, no. SAND2016-8003PE. Sandia National Lab (SNL-NM), Albuquerque, NM (United States) (2016)
34. Noor-A-Rahim, M., et al.: Wireless communications for smart manufacturing and industrial IoT: existing technologies, 5G and beyond. Sensors **23**(1), 73 (2022)
35. Schuyler, T.J.: Applications of Drones in Atmospheric Chemistry (2020)
36. Soubans, G.: Formation and mitigation of toxic emissions from blasting in mining: experimental investigation and development of aerial sampling methods. Doctoral dissertation, Murdoch University (2021)
37. Prisacariu, V., Piticar, A., Stoiculete, A.: Aspects regarding environmental and atmospheric monitoring with unmanned aerial vehicles. Sci. Bulletin'Mircea Cel Batran'Naval Acad. **25**(2), (2022)
38. Trnka, D.: Policies, Regulatory Framework and Enforcement for Air Quality Management: The Case of Korea (2020)
39. Sharma, U., Sharma, M., Rana, R.: Artificial Intelligence for Surface Water Quality Monitoring (2023)
40. Lloyd-Davies, C.: Water Quality Monitoring Drone (2019)
41. Pearce, R.H.: 'Do-it-yourself': evaluating the potential of Arduino technology in monitoring water quality, 10 p. Unpublished Bachelor's dissertation (2018). https://doi.org/10.13140/RG.2.2.31105.33127
42. Yinka-Banjo, C., Bagula, A., Osunmakinde, I.O.: Autonomous multi-robot behaviours for safety inspection under the constraints of underground mine terrains. Ubiquitous Comput. Commun. J. **7**(5), 1316 (2012)
43. Neto, J.A.B.: Using a mobile robot for hazardous substances detection in a factory environment. Dissertation, Instituto Politecnico de Braganca (Portugal) (2019)
44. Zhao, J., et al.: A search-and-rescue robot system for remotely sensing the underground coal mine environment. Sensors **17**(10), 2426 (2017)
45. Smyth, D., Glavin, F.G.: A Multi-Agent System to aid the Automation of Search and Examination in Hazardous Environments
46. Nayan, R.: Nuclear mission of drones. J. Def. Stud. **16**(4), 141–158 (2022)
47. Restas, A.: Drone applications for preventing and responding HAZMAT disaster. World J. Eng. Technol. **4**, 76–84 (2016)

48. Веселівський, Р.Б.: Rationale Use of Unmanned Aircraft Technology as a Means of Detecting Accidents and Emergencies Situations (2017)
49. Abusohyon, I.A.S., Aiello, G., Certa, A.: Drone Based Inspection of Above Ground Storage Tanks. A Multi Criteria Evaluation Approach
50. Yigit, K.A., Dalkiran, A., Karakoc, T.H.: Applications of Drones in the Health Industry, pp. 69–93. Unmanned Aerial Vehicle Design and Technology (2023)
51. Popov, O., et al.: Features of the modern UAV-based complexes use to solve radiation control problems. In: Systems, Decision and Control in Energy IV: Volume II, pp. 35–57. Nuclear and Environmental Safety. Cham: Springer Nature Switzerland (2023)
52. Calandrillo, S., Oh, J., Webb, A.: Deadly drones: why FAA regulations miss the mark on drone safety. Stan. Tech. L. Rev. **23**, 182 (2020)
53. Cunliffe, A.M., et al.: A UK Civil Aviation Authority (CAA)-approved operations manual for safe deployment of lightweight drones in research. Int. J. Remote. Sens. 38(8–10), 2737–2744 (2017)
54. Elsayed, M., Mohamed, M.: The impact of airspace regulations on unmanned aerial vehicles in last-mile operation. Transp. Res. Part D Transp. Environ. **87**, 102480 (2020)
55. Ejaz, W., et al.: Unmanned aerial vehicles enabled IoT platform for disaster management. Energies **12**(14), 2706 (2019)
56. Niu, C., et al.: An embarrassingly simple approach for visual navigation of forest environments. Front. Robot. AI **10**, (2023)
57. Lockhart, A., et al.: Making space for drones: the contested reregulation of airspace in Tanzania and Rwanda. Trans. Inst. Br. Geogr. **46**(4), 850–865 (2021)
58. Shi, Z., Chang, X., Yang, C., Wu, Z., & Wu, J.: An acoustic-based surveillance system for amateur drones detection and localization. IEEE transactions on vehicular technology. **69**(3), 2731–2739 (2020)
59. Yang, G., Shi, X., Feng, L., He, S., Shi, Z., & Chen, J.: CEDAR: A cost-effective crowdsensing system for detecting and localizing drones. IEEE Transactions on Mobile Computing, **19**(9), 2028–2043 (2019)
60. Jin, R., Jiang, J., Qi, Y., Lin, D., & Song, T.: Drone detection and pose estimation using relational graph networks. Sensors, **19**(6), 1479 (2019)

Dr. Divya Bansal Associate Professor at Amity University, Noida has more than 19 years of experience in teaching Business Management students at UG and PG level. Passion for learning new things and sharing knowledge has led her to a conscious decision of opting teaching as a profession. She has come a long way, both in terms of academics and research with 19+ years of experience. She has published more than thirty research articles in various reputed journals and presented many papers in National and International conferences. Her research interests lies in the diverse areas of marketing, Universal Banking, customer satisfaction, and entrepreneurship. She has keen interest in learning new packages and analytical tools in Statistics. She has organised various guest lectures, conferences and workshops.

Dr. Naboshree Bhattacharya is an economist with a specialization in healthcare economics and a broader research focus on development economics. With a rich academic background, she holds alumni status from prestigious institutions including St. Xavier's College, Ranchi University, and Jadavpur University. Over the course of her career spanning 12 years, Naboshree has garnered extensive experience in both academic and industrial settings. Currently serving as an Assistant Professor at Amity University Jharkhand, she brings a wealth of knowledge and expertise to her role. Naboshree's contributions to the field extend beyond the classroom. She is a prolific author, having penned six books on various topics within economics, collaborating with both national and international authors. Additionally, her research output includes numerous papers and book chapters published in esteemed journals and publications, both domestically and internationally.

Her dedication to advancing economic theory and practice, particularly in the realm of healthcare economics, underscores her commitment to making meaningful contributions to the field. Naboshree Bhattacharya's multifaceted background, scholarly achievements, and commitment to education position her as a respected figure in the academic community.

Priyanka Shandilya a dynamic force in the realm of academia, serves as the esteemed Course Coordinator for the Department of Humanities at Amity University Online. Currently, on the journey towards earning her Ph.D. in Economics, her scholarly pursuits delve into the intricate realms of Quantitative Techniques and EdTech.

With a rich tapestry of experience spanning over 9 years in the education industry, Priyanka Shandilya epitomizes innovation and determination in her academic endeavors. Her keen expertise extends beyond traditional pedagogy, embracing the transformative power of digital tools within the landscape of Education 2.0.

The Industrial Revolution: Harnessing Embodied AI Systems

B. Sundaravadivazhagan, S. Lakshmisridevi, N. A. Natraj, M. Ramkumar, and K. Madhusudhana Rao

Abstract Embodied AI systems are causing a significant transformation in numerous industrial sectors by integrating physical robotics with artificial intelligence. This chapter delves into the pragmatic implementations and significant ramifications of embodied AI within the industrial domain, encompassing precision agriculture, healthcare, construction, retail, environmental monitoring, quality control and inspection, warehousing and logistics, and manufacturing and automation. The chapter demonstrates, through the use of real-world illustrations and case studies, how embodied AI improves productivity, safety, and efficiency. Complex tasks are being automated by robots that are integrated with AI algorithms in collaboration with humans. Additionally, quality control standards are being elevated by AI-driven vision systems. The importance of embodied AI in mitigating labour shortages, addressing formerly hazardous tasks, and increasing the efficiency and productivity of industries is emphasised. Through an examination of the various industrial sectors that can implement embodied AI systems—from manufacturing to healthcare, environmental monitoring, and conservation—the reader will acquire valuable knowledge. Additionally, the chapter alludes to the prospective development of embodied

B. Sundaravadivazhagan
Department of Information and Technology, University of Technology and Applied Sciences-Al Mussana, Muladdah, Oman

S. Lakshmisridevi
Department of CSE, Chennai Institute of Technology, Chennai, Tamil Nadu, India

N. A. Natraj (✉)
Symbiosis Institute of Digital and Telecom and Management, Symbiosis International (Deemed University), Pune, Maharashtra, India
e-mail: natraj@sidtm.edu.in

M. Ramkumar
Department of ECE, Sri Krishna College of Engineering and Technology, Coimbatore, Tamil Nadu, India

K. Madhusudhana Rao
Department of ECE, KKR and KSR Institute of Technology and Science, Guntur, Andhra Pradesh, India

P. Raj et al. (eds.), *Building Embodied AI Systems: The Agents, the Architecture Principles, Challenges, and Application Domains*, Information Systems Engineering and Management 14, https://doi.org/10.1007/978-3-031-68256-8_8

AI, which promises to incorporate enhanced human–robot collaboration, improved sensory capabilities, and the ability to adapt to various environments.

Keywords Embodied AI · Industry 4.0 · AI · Healthcare · IR · Manufacturing · Automation · Quality control · Smart agriculture

1 Introduction

Artificial intelligence (AI) has evolved dramatically in recent years, pushing the boundaries of what machines are capable of completing and producing. As a result of these breakthroughs, artificial intelligence can now accomplish jobs that were previously regarded to be only the domain of human intelligence. Traditional AI systems, on the other hand, are sometimes referred to as disembodied AI since they primarily function in the digital realm and have no direct relationship to the physical world. As a result of this constraint, they are unable to completely engage with and grasp their surroundings, limiting their ability to acquire human levels of intellect [1].

Embodied artificial intelligence, also known as embodied cognition, is a method of bridging the gap between the mental and physical worlds. This contrasts sharply with classical AI, which takes a more traditional approach. This is performed by integrating AI systems into physical bodies or robotic platforms. The human brain, which is inextricably tied to the human body and continually interacts with its surroundings through sensory input and motor responses, spurred this paradigm change. As a result, in this approach, the brain of a human was considered. Incorporating embodiment into Artificial Intelligence is intended to develop a profound comprehension of intelligence while advancing AI entities to work in the real-world space, such as operating and exploring independently. It is an input component that enables the AI systems to freely impart the same sensory and motor capabilities.

Unlike disembodied AI which means an emphasis on the ability of the computer system to emulate or mimic human behaviour through virtual presence, embodied AI highlights the role of the physical environment in which the system is designed. This in-depth and long assessment will shed light on the core concepts and how it can affect artificial intelligence as a whole. It promises to transform the artificial intelligence field. Here we shall put our emphasis over how AI faces a change in the way it communicates with humans.

a. Key Benefits of Embodied AI

In the following section, I outline the three vital merits of AI through the integrated body:

Interaction with the physical world: Implemented AI systems are able to conduct experimental learning procedure and proceed in error and fact situations by practices their own. This process can be done by interacting with a simulated environment providing physical capabilities that may be difficult or otherwise impossible to have on their own.

Common-sense reasoning: The therapy for understanding of and development of common-sense reasoning abilities foster for embodied AI systems whose foundation is entirely based on their relation to concrete experiences. In this case, algorithms acquire a capability to make more psychological predictions and inferences about people or the surrounding world [2].

Transferability of skills: Learning in a certain type of surroundings can be usefully transferred to a similar type of places, and thus, leads to the minimization of retraining demands and demonstrates the flexibility of AI systems.

Embodiment effects: The development of cognitive processes and possible emergence of new behavioural patterns in AI systems is largely affected by the physical structure, which results in achievements the complete realisation of which would be hardly possible without the physical form.

b. Embodied AI: A Conceptual Shift

The facility of embodied AI is of the topmost tier in artificial intelligence that is referred as the quantum leap. It poses a challenge to the dogma of intelligence focusing in separating it with physical interaction. AI systems are widely used today in the real world under the form of a conventional AI (artificial intelligence) and a "disembodied AI"—a kind of an AI that is not related to the tissue. It implies that data and algorithms are used to process information and make decisions. However, these technologies can't establish a base layer of direct rapport with the physical world which is perceived and interacted physically. Nevertheless, these approaches focus only on matching AI capabilities with the human ones while we confront a disparity between these two domains that is apparent in the fact that AI can only replicate the cognitive functions of humans, while being designed with the logic mechanisms alone and without the capability to perceive their environment and to match this data with the commands intended to be delivered. It is mainly this knowledge that it is our physical embodiment and human intelligence are not separate, but constitutes an integrated whole [3]. Our motor actions facilitate our environment interaction, and at almost the same time sensory apparatus are sending our brain the external stimuli. In addition, the external world plays an extremely important part in the act of cognition, which is a trait that is typical of humans. Through an embodied AI the proactive agents would be integrated into the physical environment which could emulate the same relationship between anticipating and responding to the conditions of that environment [4].

The essence of this outlook alteration lies in ability to growth the AI field. If we scratch the surface, we may find that region of AI, known widely as disembodied AI, is very much capable of improving effectiveness in many spheres, but unfortunately it is far from being actually applicable for the issues of this world that demand physical interaction or sensory perception. Autonomous AI systems feature among the most innovative and disruptive technologies as they not only have the motion capability, but also the ability to modify the physical environment, making it possible to explore infinite sectors that have not been tapped into. The feature of embodied AI that makes it distinct, similar to other robotics, from other AI technologies is the dependency on the physical presence. AI algorithms, actuators, and sensors are actually the third components that constitute the mechanism that enables these systems to interact with the world. Sensors enable this AI system to take in and acquire data from its environment and process it in a way that is similar to its sensory organs imitation [5]. The essential feature of the sensors here is that they can efficiently read wide sets of data including microphones for audio, cameras for video, touch-sensitive input, and various sensor types to deal with other data ranges like temperature or pressure. Actuators, which enable the AI's interaction with the outside world, constitute the other layer in AI mechanics. Such organizations may be designed as acting through robotic arms, wheels, motors and other devices that can be mechanically operated smoothly. AI algorithms, constitutes the core block of embodied AI, identify, interpret sensory input, then provide meanings and/or actions in response. The algorithms they involve include various techniques which can be a deep neural network, reinforcement learning, and even computer vision and natural language processing. These foundations of the AI system make it possible for it to think, comprehend and make decisions based upon and in contrary to received data.

Also, there are what you might call decision-making skills that are responsible for analysis of information coming from sensors and the data analysis and then carrying out actions or making decisions guided by pre-set targets. Frequently computational algorithms replace decision making processes in the system, and so, AI grows in full capacity with every new decision. An embedded AI's ability of learning from experience further and being changeable accordingly, would contribute to its performance and flexibility [6]. The feedback mechanisms help the embodied AI systems to have good learning and adaptive abilities by making them learn in response to environments information. A considerable number of these systems feature mechanisms that are able to enable the acquisition of knowledge through assessing the implementation of the actions and their outcomes. Reinforcement learning which is a pertinent machine learning technology can help the system to refine its behaviour once it gets the feedback concerning the outcomes of its actions. Moreover, influence by humans in both preparation of the AI system model and its progressive improvement could be considered as a determinant factor of the old system complexity and compliance with the human purposes.

c. Significance of AI and Robotics in Industrial Applications

AI and robotics in industry affect several industries. Technological advances have optimized industrial processes and solved difficulties. Embodied AI has a wide-ranging influence, and its implementations encompass a variety of sectors and fields [7]. The Fig. 1 shows the industrial applications of Embodied AI.

The following are the most notable domains in which incorporated AI has exhibited substantial promise:

1. Manufacturing and Automation: The manufacturing and automation sector is undergoing a paradigm shift due to the implementation of embodied AI. Algorithms powered by artificial intelligence enable robots to execute complex

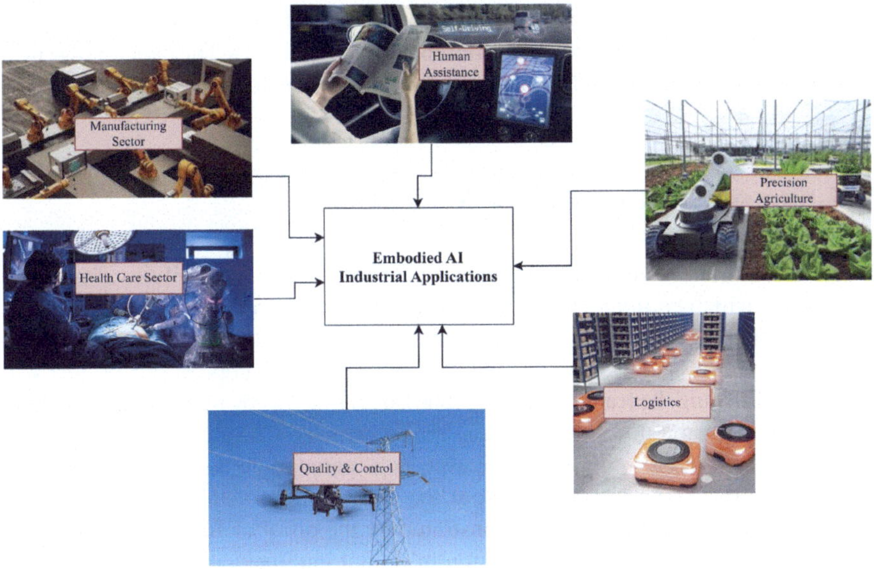

Fig. 1 Embodied AI applications in various industrial sectors

assembly tasks with pinpoint accuracy and to adjust to alterations in the manufacturing process; this results in increased output, decreased errors, and enhanced product quality. The level of productivity and seamlessness in the collaboration between human labourers and AI-driven robots is increasing.

2. Quality Control and Inspection: Through digital images recognition, quality control systems are able to identify deficiencies, non-codified situations and components non-conforming to the quality guidelines. Thus, the final goods that emerge into the market are top notch. This introduces the ubiquitous AI technology that leads to marked improvements in product quality and uniform across the productions.
3. Warehouse and Logistics: The conclusion of AI embodiment the variable in the warehousing and logistics industry has caused a prominent paradigm shift. Through adoption of machine learning algorithmic path planning, smart robots that are infused with AI attributes advance material handling, inventory management and delivery services, thereby enable error free operations which are also effective. These AI-based systems guarantee precision to the order, cost-effectiveness and also raise the level of performance of all the operations [8].
4. Precision Agriculture: The field of agriculture got the herald boost in the precision farming aspect via the functionality of the embodied AI. With AI-augmented drones, ground-based robots, farming tasks can be completed like monitoring production, precisely applying fertilisers and pesticides, and therefore completing various operations.

This Chapter will be all about the disruption affect of the embodied AI as a representative technology in several different industrial sectors. The paper will reflect the practical sides ad benefits of AI-robot fusion for industrial applications.

2 Understanding Embodied AI Systems

Recent advances in Artificial Intelligence (AI) are boosting its capabilities. Embodied AI systems, which combine digital intelligence with the physical world, are a major achievement in the field. AI systems apply remotes and identifies the things they feel, and make decisions solely based on the data available. The scope of this project will involve establishing the framework of embodied AI systems by listing and unpacking their distinctive parts that include actuators, sensors, and AI algorithms execution. With this, we would also highlight the aspects of perception, decision-making, and action in embodied AI to strike right at the demarcation of this technology. In a nutshell, what AI is all about is demonstrated by the quest of AI into embodiment. Tangram AI machines through the fusion of digital intelligence and physical being empower AI agents with the capacity to perceive and grip that are analogous to those of the people surrounding them. Artificial intelligence (AI) that is commonly discussed refers to an AI that works with numbers and machines and is away from real people, therefore called digital or "disembodied AI." Because of the deficiency in the sense and motor operations command, the systems remain virtually disconnected from the physical world.

There is a problem of distant sensing, which is solved by the use of AI in the field. This brings the AI mechanism into a real human environment to feel. This theory—that mind and body are connected—holds that people's intellects and bodies are similar. Our epidermis, ears, and eyes send us signals that our nervous system integrates into messages and then sends to our brains for analysis and due information. We tend not only to feel the surroundings but to connect with our environment actively, strategic, and purposefully. The intricate relationship between conscious process and action has a significant role in the human brain operations.

2.1 Fundamental Elements of Embodied AI Systems

Embodied AI systems are comprised of a number of essential components that facilitate their efficient interaction with the physical environment:

Sensors: The sensory instruments of embodied AI systems are sensors. Their duty entails the reception and accumulation of data originating from the external environment. Sensors are available in a multitude of modalities; they include contact sensors, specialised sensors for detecting particular environmental data (e.g., temperature, humidity, pressure), and visual sensors (cameras). Sensors function as the AI system's visual and auditory organs, facilitating the acquisition of vital information from its environment. Actuators furnish the mechanisms by which the artificial intelligence system executes tangible actions in accordance with its perception. Actuators may manifest in diverse physical configurations, encompassing robotic appendages, wheels, motors, or other such entities. By performing operations or tasks, these latter components convert the signals that come to them from the AI system to the physically executing component, as human appendages and muscles do.

AI Algorithms: Embodied AI involves AI models or systems that collaborate with the human team at its decision-making nucleus. The algorithms are saddled with the duty of to analyze the data the sensors pick up, to interpret it, and to formulate a suitable response to action getting from the information. AI algorithms include a variety of approaches like Deep Neural Networks, Computer Vision, Natural Language processing and Reinforcement Learning algorithms. They give the knowledge for the logic, deduction and understanding functions to the artificial intelligence system, the same way you have these functions in the human brain.

a. Significance of Actuators, Sensors, and AI Algorithms

By cooperating, the three essential components of embodied AI systems enable these systems to actively interact with and comprehend the physical environment.

1. Sensors: The critical input that empowers the AI system to perceive its surroundings is supplied by the sensors. Beside different types of sensory components such as seismic, acoustic, and optical, a set of devices is composed of a number of sensors that measure various physical properties and parameters. Simultaneously, visual sensors capture images and video data while the auditory sensors record sound and speech. Sensor data becomes the vehicle through which AI can sense its environment and thereby understand its surroundings.
2. AI Algorithms: The data gathered by the sensors is judged and determined by Artificial Intelligence algorithms. For example, machine learning algorithms can tell abnormal situations, and matrix and object being in the visual data. Additionally, algorithms for natural language processing can do text mining that not only comprehends the semantics of the written or spoken language but also hears data. Algorithms for reinforcement learning gives AI system to establish it own knowledge through its actions and subsequently adjust its practice to approached. The responsibility of tackling this hurdle of interpreting and comprehending the sensory data into actionable knowledge remains on the shoulders of these algorithms only.
3. Actuators: Actuators are the work- Horse of AI system. They are capable of carrying out instructions that have been analyzed and processed by the system. AI follow-up algorithms specified above malfunctions on the servos that implement commands and orders. Actuators bring mobility and make the robotic arms or the wheels in a robot system move as task demands; they help the AI systems operate in a physical environment, hence interaction and manipulation of the physical environment is possible.

b. Imperative of Perception, Action, and Decision-Making

Deploying AI systems to interact effectively with the physical world is contingent upon the following three fundamental components of embodied AI: attitude, preference, choice, and activity. Perception is the fundamental source of the foundational entity of embodied AI. it is the informational conduit that the AI system utilizes to collect data from its environment using various sensors. Perception, by its end, is capable of the system to amass environmental data, object identification, change detection, and situational awareness systems. There will be problems if the artificial intelligence system cannot observe the external environment, which leads to this artificial intelligence system does not make a good plan and decisions.

Determination: Finally, based upon either of the two previous neural networks, the AI system is expected to make decisions which should be based on this knowledge. The analytical process covers the steps of consequently, making sense of the data, determining the patterns, and finally the best way forward. Algorithms assisted by artificial intelligence hence play a meaningful role here as they facilitate the system to determine the impact of the sensor data and pass instantaneously the reactions or decree commands. The main element involved in the manifested AI is actions. Implementing the decisions is the final position of the intelligent robot which after going through analysis and processing produces results through function. The executions of it actions are performed by actuators due to the fact that they are perceived as the material expansions of AI system, which allows the system to touch its surroundings, process data and react to it in the environment of its existence. The adaption feature of AI systems to developing situations and accomplishing predetermined objectives or aims using their intended operations is shown through action.

Correspondingly, embodied AI systems imply how a new paradigm is attained by the emergence of the synergy between digital intelligence and tactile interactions. Intelligent systems operate by sensing their surroundings, having a knowledge based to carry out an informed judgement, and later accomplishing the required actions. Thus, through their interpretation on the adopted sensors, actuators, and AI algorithms, they generate the necessary reactions. It is the integration of perceptions, decisions, and activities which gives embodied AI its unique signature, and which makes its view go beyond the Dichotomy of matter and mind and offer a greater spectrum of opportunities for different fields of work. As these systems roll on, their engagement options with "real-world" will be incessantly expanded, and this will lead to the advent of novel engineering's know-how's and a potential for novel achievements.

2.2 Key Technologies in Embodied AI: Unlocking the Potential of Physical Intelligence

The invented methodology called Embodied AI (EAI) entails the combination of actual robotics and AI, which gives the machines a capacity of perceiving, interacting with, and modifying their environment. The emerging perception of the world takes advantage of sophisticated software and hardware solutions as its associates, enabling a seamless transfer between virtual intelligence and physical environment [9]. The examination will cover the two major technologies serving as the foundation upon which embodied AI is built, that is, the software and hardware devices collectively popular as the powerhouse of embodied AI. Other advanced technologies in robotics, vision, natural language processing, and machine learning, as part of an embodied AI framework and tools, are vital for AI system design and deployment, thus it's imperative also that they will be addressed.

a. Technologies of Hardware in Embodied AI:

Sensors: Sensors, as the members of an AI embodied systems act as machines' sensory extension and are the apparatus that allows them to capture their reality. Sensors act as cameras in the vision machine learning world as they take pictures, and microphones are usable on the speaker side to record audio. Similarly, in addition to GPS (Global Positioning System) and the inertia measurement system provided, specialised sensors such as LiDAR for depth perception, contact sensors to provide tactile feedback, and environment sensors to gather data on temperature, humidity and other points, may be incorporated. The Fig. 2 shows the major elements of the embodied AI system.

Actuators: Actuators are the tangible constituents that enable embodied artificial intelligence systems to execute actions in accordance with their perception. These may manifest in diverse ways, including robotic appendages, wheels, motors, and other manipulable and mobile mechanisms. The selection of actuators is application-specific, with robots being engineered to perform a variety of tasks, including laborious environments, warehouse automation, and exploration.

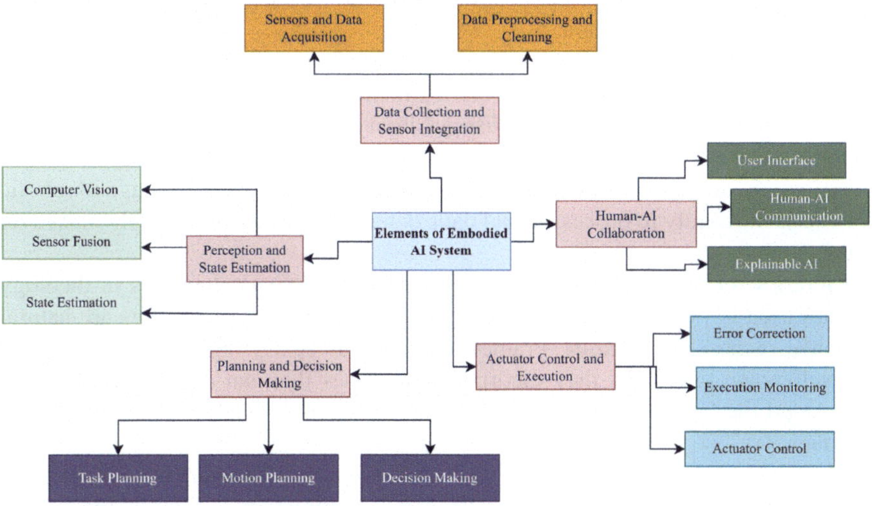

Fig. 2 Elements of embodied AI system

Hardware platforms: Hardware platforms are of paramount importance in the advancement of embodied AI systems. These platforms frequently consist of high-performance computational units, including GPUs (Graphics Processing Units) and TPUs (Tensor Processing Units), which are employed for the rapid processing and analysis of data. In addition, robotic hardware platforms serve as the fundamental building blocks for tangible robotics, integrating essential components such as control systems, actuators, and sensors.

Communication Systems: Communication is one of the most important part of effective relationships in AI embodiments. Through networking hardware, protocols and a unified central processing unit (CPU), actuators and sensors can transfer the data in a smooth stream of data. This kind of cooperation between robots or among devices working interclaiming duties.

b. Technologies of Software in Embodied AI:

AI Algorithms: Embodied AI systems are powered by AI algorithms. It is the responsibility of these algorithms to process sensory data, interpret it, and generate responses or actions. A multitude of AI methodologies are implemented, encompassing computer vision, reinforcement learning, deep learning, machine learning, and natural language processing [10]. By utilising these algorithms, embodied AI systems are capable of deciphering data, identifying patterns, and reaching well-informed conclusions.

Robot Operating Systems (ROS): An extensively utilised and adaptable software framework, ROS facilitates the development of robotic applications.

The digital platform provides a lot of versatile tools and utilities, used with the purposes of robotics, that cover data dashboard, device abstraction, remote control unit and communication protocols. ROS can robotic intelligence systems with great complexities which are created with the provision of the efficient integration and administration of the software.

Computer Vision Libraries: Computer vision which is a critical AI component for visual perception products is an embodiment AI technology. Libraries such as OpenCV (Open-source Computer Vision Library), as well as tens of other decent options, cover an enormous ground of tools and features for managing image and video content analysis. Of these, the one that gives the greatest advantage to robotic systems is the ability to process visual data, detect objects, and effectively maneuver in the environment.

Libraries for Natural Language Processing (NLP): Critical factors in theology that are used to perceive conversations between the humans and robots are natural language processing libraries. Frameworks such as spaCy and NLTK are among the wellknown ones, and they incorporate the ones that can detect speech input, understand the language there is, and produce responses that have a sound natural language.

Machine learning frameworks: It acts as an essential component of embodied AI, being the feature that is concerned with data accumulation as well as the system that adapts and acquires knowledge. With frameworks like scikit-learn, and TensorFlow machine learning models and tools such as, classification, regression and reinforcement learning just to name out some, are offered.

Simulators: The simulations environments are of crucial importance for testing and training the artificial intelligence systems that are planned to be used as embodied ones. This way, robots or systems' virtual environments can be created in the Gazebo or Unity3D simulators that enable developers to validate and update their software on their own virtual machines without the necessity of purchasing physical hardware.

c. Developments in Robotics:

Advanced Robot Kinematics: The field of robot kinematics, which is responsible for the location and orientation of mobile robots in the physical world, has seen tremendous advancement and is now able to program complicated manoeuvres and the improved agility of robots systems. The advancements in end-effector technology and serial and parallel manipulators have significantly broadened the functionalities of embodied AI systems across multiple sectors.

Humanoid Robotics: Humanoid robotics is the study of creating machines that mimic the movements and interactions of humans. These robotic systems find utility in the healthcare, service, and research sectors, facilitating organic collaboration between humans and robotics [11].

Soft robotics: Soft robotics refers to the development of flexible and deformable structures that empower soft robots to engage in secure human interactions and adjust to ever-changing environments. Particularly suited for applications in which rigid robots could pose a hazard.

Bio-Inspired Robotics: The field of biomimicry has yielded robotic systems that draw inspiration from the movements and conduct of invertebrates and animals. Bio-inspired robots possess distinctive functionalities that render them well-suited for missions involving tough terrain exploration or search and rescue.

d. Transformations in Computer Vision:

Deep Learning for Vision: In terms of the field of computer vision, deep learning has dramatically changed everything thanks to convolutional neural networks, (CNNs) that is. However, these channelizations manifest superior performance on image segmentation, object detection and classification that latter makes them the essential components of visual perception of embodied AI systems.

Object Recognition: The steady improvement of object recognition algorithms has led to more sophisticated and detailed embodied AI systems which can now detect and follow objects around the surveillance area. These technologies turn out to be invaluable in industries for example production, distribution and self guiding machines.

Visual SLAM (Simultaneous Localization and Mapping): Visual SLAM algorithms (Simultaneous Localization and Mapping): Robots can do this concurrently to generate visual maps of their surroundings based on where they are placed and determine their position in relation to the created maps. This is a fair enough ground for research and self driving.

e. Advances in Natural Language Processing:

Reinforcement Learning: Such technologies as chatbots, virtual assistants and voice recognition systems are developing rapidly nowadays, and being able to understand and react to the context without a knowledge of person's mood or topic of discussion becomes more and more natural. Through the doing of these developments, there is a great improvement in the kind and level of reception from the human users of embodied AI technology.

Transfer Learning: BERT(Bidirectional Encoder Representations from Transformers)and GPT (Generative Pre-trained Transformer) are the examples of NLP modules which have taken over language understanding with NLP models such as BERT (Bidirectional Encoder Representations from Transformers) and GPT (Generative Pre-trained Transformer). These models together accelerated the comprehension and increased the effectiveness of using embodied AI systems to deal competently with human speech.

f. Advances in Machine Learning:

Reinforcement Learning: Advances in reinforcement learning, that is, what allows robots to develop knowledge through the interaction with their environment, have now been greatly improved. The capacity of these programs to constantly learn and correct as they gather data infers that they could be really useful as robotic controls and in autonomous decision-making.

g. Notable AI tools and frameworks include:

TensorFlow: This is a comprehensive, open source machine learning framework by Google, which mainly focuses on machine learning model training and deployment. The bundle is a comprehensive package that can include a variety of reinforcement learning and deep learning tools.

PyTorch: It is a deep learning framework that focuses on dynamic computation graph it provides GPU acceleration and capabilities it's is the hottest in embodied AI development research and experimentation.

OpenAI Gym: OpenAI Gym is a broadly used platform that is grasped in the course of training and testing reinforcement learning methods. It comprises a realistic variety of training and evaluation for agents' AI.

ROS (Robot Operating System): The core technology built around ROS or the Robot Operating System plays a crucial role in robotic applications development. It serves as the link between software and hardware so that embodied AI systems can be used through many development tools targeted to robots.

To conclude, the sector of embodied AI beholds a myriad of technologies to incorporate diverse hardware and software to provide AI systems the ability to transact the physical reality. Advances in robotics, computer vision, and machine learning constitute the basis for the extension of the artificial intelligence technology where its application is focused on the industrial sector. The state-of-the-art AI frameworks and tools that act as a platform greatly simplify the construction of autonomous AI mechanisms that are complicated, pliable, and adaptive [12]. Being evolving nature of these devices, one can expect more cutting-edge aspects of these technologies as well as new areas of their application in this highly dynamic discipline, and in the years to come.

3 Implementation of Embodied AI for Different Industrial Sectors

3.1 Embodied AI for Automation and Industrial Robotics: Revolutionising Manufacturing and More

Consequently, we see the implementation of AI systems in industrial robotics and automation leading to an era of a sharp contrast to the models of production and assembly processes across various industries. Among other major achievements, these systems have made possible the improvement of productivity and accuracy. Additionally, they are the foundation of brand new automation solutions developed for complicated and monotonous jobs. This research covers how AI that is embodied or machine-to-machine communications have an effect on assembly and manufacturing lines, and explains specific instances when robots have been put to work on the factory floor, ending it with the possibility that these kinds of robots can now be applied to perform tasks such as those previously observed as too laborious or repetitive to human workers.

a. Embodied AI's Impact on Manufacturing and Assembly Transformation: Embodied AI's Impact on Manufacturing and Assembly Transformation:

AI, through serial and specialized production, creates clear transformation in manufacturing and assembly and redefines the processes hampering production and assembly of goods. The subsequent are the effects of these systems: The subsequent are the effects of these systems:

Enhanced Efficiency: Here we mean physical AI systems which are fully equipped with sensors of the highest performance and AI algorithms. These systems turn out to be annoyingly quick and accurate when it comes to performing tasks. Consequently, production rates are faster, while at the same production time is maximized.

Quality Control: Implementing of computer vision and machine learning can allow robots to perform precise inspection of relevant products and put aside any poor quality ones. Therefore, this activity upgrades the manufacture standards and it also lowers the likelihood of flaw products being sacrificed to the market.

Flexibility and Adaptability: Industrial robots capability for learning, which includes embodied AI, to a great degree employs flexibility and dexterity. The most useful traits for this is their ability to be reprogramed or reconfigured for a wide range of duties and they are applicable to more extensive products and applications. In industries that have support for dynamic customized product demands or changes this responsiveness is of key importance.

Safety Enhancement: The robots may be protected by AI to work in environment with is risky for humans. The scope of tasks for this equipment might include, among others, the toxic material handling as well as jobs requiring that one survive in a dangerously hot environment. This safety acts as a measure that aims at the goal of providing a safe working environment by minimizing the opportunities for accidents and injuries.

Collaboration with Humans: Usually called "cobots," collaborative machines are created in such a way that they can be used efficiently along with humans. They are the ones who can undertake risky, fatiguing, or boring work, thereby freeing up human coworkers to engage with more creative parts of their jobs. Both human and machine partake on production with the sum of their efforts exceeding the result of a partial effort of either side.

b. Illustrations of robots in industrial environments: Illustrations of robots in industrial environments:

Welding Robots: Because of many reasons, welding robots are popular in metal cutting, for instance, in an automotive factories. These automated bots enable to perform welds on components precisely and consistently thus offer an escalation in the speed and quality of these welding processes.

Pick-and-Place Robots: Rulleyo machines can be used in various fields of industrial manufacturing and logistics. Their efficiency is expressed in successful conveyance and positioning of objects in a planned area of work, which makes them irreplaceable machine tools for work like assembly, materials handling, and packing.

Painting Robots: Bracing of painting robot is played a significant role in different industries, including automotive manufacture. They not only ensure that precisely and consistently enough paint is applied but also serve to limit paint wastage and improve the quality of the overall finish.

Inspection and Quality Control Robots: These machines are equipped with state of the art cameras so that they can search products for any kind of defects and undefined quality standards. They are equipped with a feature which allows them to recognize differences in numerous areas of industries like pharmaceuticals and electronics, accurate and fast.

Assembly Line Robots: The staff of robots on the assembly line is programmed to perform a mix of part assembly and process control functions that result in the finished piece. These robots have a high-precision and are configurable to a dramatically low extent so that diverse product variants can be produced.

Collaboration Robots (Cobots): The design of cobots is focused on security of the human robot communication, and therefore, they are considered to be collaborative robots. These devices are employed to make the assembly and packaging tasks and those which require a high degree of dexterity for example more streamlined for human labourer while never posing any safety issues.

Palletizing Robots: Robotic palletizing machines are a foremost situation in inventory du and logistics sites. While they use pallet dacking and ordering products on pallets in a proper manner, they attempt to optimize the packing and shipping process.

c. Implementations for the Automation of Complex and Repetitive Tasks:

Material Handling: Material handling refers to one of the settings where embodied AI systems carry out tasks which are either too much work for their human counterparts, or impossible for them to take part in, such as loading and unloading of heavy-duty objects. While best suit various domains like logistics and manufacture.

Packaging: The traditional human-centric packing operation is turning to robotics progressively. Picking, sorting as well as labeling products are jobs they easily do. And they can also place products into boxes and packages. The introduction of robotization leads to repetition and speed up of the packaging operation.

Machine Tending: Machine tending is an advanced process in manufacturing that takes effect with robotic undertakings. Being completely independent, they take care of charging tools and materials into machines, are in charge of the manufacturing process and keep it running.

Assembly: Robotic arm is considered to be the best in such assemblies which require fixity and uniformity. They do this without error. The ability of these robots to be instantly reprogramed for use in different assembly operations that enable them to handle changes in the production volume illustrate their adaptability to a dynamic production environment.

Quality Inspection: The fast-paced operations of the industries are facilitated by the robots working in symbiosis with computer vision systems to do quality inspections with almost perfect accuracy. This helps to identify products with even the tiniest flaws or abnormalities which get removed to ensure only top-notch quality products are dispatched.

Metal Fabrication and Welding: Automated welding is achieved by the welding robots, as they allow creation of strong common welds. They are broadly used in the marine and automotive industry, including the majority others.

Food Processing: Angry robot automize many kinds of food making process like chopping, slicing and utilizing packing machine. Style: Informal The use of these technologies in the production process also helps in getting safe and consistent products, and decreasing need of manual work.

Pharmaceutical Manufacturing: While robots are performing the tasks of filling, labeling and packaging medications, they will still be doing these in the pharmaceutical manufacturing industry. The tests help to ensure the sterility and accuracy that are necessary to manufacture pharmaceuticals.

Agriculture: A diverse range of agricultural machines that do weeding, planting, and harvesting tasks are being used today to ensure minimal errors and wastage during the development of agriculture. The machines will replace workers, therefore there will be less cost of labour and high production rate on the agricultural side.

3D printing: Additive manufacturing is a multifaceted application of 3D printing technology among other robotics disciplines. The 3D printing technology not only offers the possibility of quick and custom-based fabrication of components and structures by making it move from one layer to another.

This transformation stems from an increasing number of intelligent automation systems, which are creating a joint concept for industrial processes such as assembly and production within every industry. Automation enables these systems to diminish safety risks, ramp up the processes and improve quality management for a multitude of operations within the scope of material handling, quality inspection, welding, and packaging. As the developments in robotics and AI are on the rise, industrial automation is likely to go through even greater improvements and increase of creative applications which brings about the fast developing of ways assembly and manufacturing of goods technology. Such systems are installed with the cameras and sensors that are enabled to fetch and analyze the visual data at the same time. By detecting any defects, incompatibilities, or inconsistencies among information, products, or practice, they only ensure that the highest quality products/information are the only ones which will be recommended (put on the market).

3.2 Embodied AI for Quality Control and Inspection

Visual Perception: AI vision systems possess the capability to analyse visual data through the identification and classification of patterns, shapes, colours, and text, rendering them adaptable across diverse sectors.

Machine learning algorithms, including convolutional neural networks (CNNs) and deep learning, are utilised in the training of vision systems to detect defects in the presence of annotated data sets. The algorithms consistently enhance the precision of the system by means of learning and adaptation.

Real-Time Analysis: Vision systems are well-suited for high-speed production lines and dynamic manufacturing processes due to their ability to process visual data in real time.

Integration: The incorporation of vision systems into automation systems, conveyors, and industrial robotics enables immediate action in response to the detection of defects.

Quality Metrics: The generation of exhaustive quality metrics and reports that offer manufacturers insights into the production process facilitates the implementation of data-driven decision making.

3.3 Utilization of Embodied AI Quality and Inspection of Particular Sectors:

The Automotive Sector:

Defect Detection: Vision systems propelled by artificial intelligence examine automotive assemblies and components, including electrical systems, engines, and body panels, to detect flaws such as scratches, dents, and misalignments. Error prevention systems are of paramount importance as they guarantee the absence of defects in safety–critical components such as brake systems and airbags [13].

Vehicle paint inspection: This involves the utilisation of vision systems to identify flaws, colour inconsistencies, and irregularities in the coating.

Electronics Production:

Printed Circuit Board (PCB) Inspection: To ensure the functionality and dependability of electronic devices, vision systems inspect PCBs for soldering defects, missing components, and circuit integrity.

Component Placement: These integration systems ensure satisfactory soldering and alignment through quality control to check the components are fixed on the PCBs accurately.

Food Manufacturing: Vision system helps in monitoring food processing lines by identifying poor quality items, dirt particles, and contaminant items instituting quality standards while achieving food safety.

Label Inspection: Detecting exactly what kind of information has been put in the food packaging labels (for example the list of ingredients, nutrition facts, and expiration note) is one of their responsibilities.

The Pharmaceutical Sector: Together with the use of vision systems for the purpose of inspection of pharmaceutical packaging, the manufacturers try to prevent mislabelling, substitution, and other cases of product manipulation. Due to this, qualities of the final product are also maintained. For the purpose of the inspections of capsules and tablets in Capsule and Tablet Inspection, these systems carefully look for flaws, discolorations, or any possible contaminations that may have happened in the production process so as to ensure that these pharmaceutical products are of good quality.

The Textile Sector: Various vision systems are used for making sure that the fabric quality is maintained. They mostly spot across the fabric stains, rips, and displaced patterns. This guarantees that the produced fabrics are only of the highest quality. One product ingredient administrator uses a marker importance for ensuring that the garments do not have any embroidery flaw, uneven size, and irregular fabric; they guarantee that the fashion industry's product quality is maintained.

3.4 Benefits of Artificial Intelligence for Consistency and Defect Reduction

Vision systems built on artificial intelligence do inspections faster and more reliable, which leads to the reduction of the number any mistakes that would be made due to the tiredness or inattentiveness of a person. Rather than unreliable and inconsistent kind of quality control process these systems give a guarantee of consistent criteria that when applied to every item outcome uniform quality product. AI vision systems help industrialists manage lower defect rates through early detection and get the value of the resources which would otherwise be spent on rectification and product recalls during the production phase.

Enhanced Efficiency: Instead, autonomy, which is the better method, comes up with automated quality control and inspection where efficiency and time are considered and processed quickly through faster production cycle and lesser lead time.

Vision system providers deliver significant volumes of data and reports that improve many business processes, enabling manufacturers to optimise their production and produce decisions based on factual knowledge.

Labour Savings: By automating inspection duties with AI, employees will be able to shift the time and energy from routine repetitive tasks towards more meaningful work or more high added value responsibilities.

Error Prevention: In respect to quality control vision systems, they have low error rate in contrast to those conducted by human checkers, which create a zilch error labelling case.

Adaptability: One of the outstanding features of these systems is that, they can be improved in iterations or re-configuration to incorporate product design or quality standards updates, which ensures its ability to adjust to any production amendments.

Synthetically, the vision systems that are driven by the AI technology have become crucially important tools for densitometry and defect checking in different industries. The capacity to recognize the faults and errors accurately, especially their position, size and circumstance and detect them with speed, accuracy and consistency lead to the increase of the quality of the product itself and a decrease in the expenses associated with violation of quality standards. With a huge number of applications in various industries, like automotive, electronics and food production, these systems help manufacturers in meeting the highest quality standards in conformance with the pressure of the market advantage and improving the efficiency of the process. Continuous evolution in AI and computer vision field provides wonder opportunities to uphold quality control standards and surveillance, thus these processes will enable getting the best products possible.

3.5 Embodied AI Agriculture and Precision Farming: Cultivating the Future

Digitalisation in agriculture sector is an ongoing process driven by digital technologies that accomplish triple objective of increasing crops yields, perfect the administrative operations, and enhance sustainability. As we shed the traditional manual way, the use of robotics and AI is crucial in this transfiguration, computerizing not only cultivation but stock monitoring and lumbering as well. AI-based crop inspecting systems involve the seamless operation of cameras and sensors through drones to scan and proffer precise information on the status of plants, analyze signs of infections and stresses, and make actionable recommendations to the farmers. Such real-time data allows for the quick intervention of early stage that will ensure bigger success in their yield and minor loses. Moreover, AI algorithms decide optimum value of plantation depth and position for uniform growth of their crops. Robot-based technologies make use of the computer vision to distinguish between crops and weeds which helps in controllable destruction of weeds through herbicide reduction thereby saving on spraying chemcials. Unmanned harvesting and picking robots that are equipped with robotic arms and AI vision systems grasp and arrange ripe produce to reduce error and labor costs, while increasing productivity, and the crops are not damaged.

AI-based fertilization systems perform soil analysis and nutrient levels which provides optimal ways of fertilizer dissipation thus reduce more groundwater contamination and environment pollution. Auto irrigation management performed by AI overweighs weather data, moisture sensors, and growth stage information in order to make water input precise, thus, saving on water resources. AI and robots open up numerous economic and environmental possibilities that can largely be attributed to the precision farming. This helps in enhancing resource efficiency that decreases the amount of water, fertilizers, and pesticides. This carries out cost saving and less environmental footprint. The applications of AI such as predictions and interventions are empowering to farmers for better growths in crop yields and quality. Thus, it will bring more money and profits. In addition to this, practices that are sustainable including results-based interventions, water management and dynamic utilization of chemical inputs are transforming the sector into an environmentally friendly and sustainable one. Mechanized processes with autonomy are decreasing the labour demand and leading to the unemployment of humans particularly in situations where lots of people are working at one time-harvesting. Accuracy of the agriculture results from precision has implications for the planning, resource allocation and crop management decisions often over a long-term.

3.6 The Revolutionising Impact of Embodied AI on Healthcare and Medical Robotics on Patient Care

The integration of artificial intelligence (AI) and robotics in healthcare is driving transformative advances, revolutionizing patient care, diagnostics, and surgical procedures. AI-powered medical robots are facilitating minimally invasive surgeries with enhanced precision, stability, and reduced patient trauma, leading to faster recovery times and improved surgical outcomes. AI algorithms are greatly important in the area of diagnostic imaging, by speedily decoding MRIs, X-rays, and CT scans and spotting tine differences which improve the patient's prognosis and helps in early disease detection. AI provides also has an unique feature in individualizing well for different patients by benefitting from combination of groups of data in order to build an individual treatment solution unique for each and every person, which improves the effectiveness and reduces the undesirable effects. Furthermore, the area of medicine that involves searching for new molecules, known as drug discovery, is being boosted by AI, where the learning machine models are used to analyze biological data in order to make predictions concerning drug interactions and in the short term, speed up the identification of new drug candidates. Soft tissues and internal organs damage induced by open surgeries are now being prevented by minimally invasive procedures made possible through use of surgical robots. This has led to the utilization of fewer incisions, minimum pain and short recovery times. Being precise and not damaging the healthy tissues is a surgery advantage with robotic systems. This contributes to dealing with the complex procedures that require more precision. Additionally, reduced size of incision gives rise to lesser infection rates, which open up doorways for patients to stay safer and witness high outcomes.

The application of AI-enabled robots in rehabilitation sector mainly involves helping the patients in regaining their strength and mobility via personalized exercises, tracking the progression and the gamified approach. This results in the encouragement of the patients to work on and engage with the treatment. Robots-gadgets like exoskeleton powered by AI make even paralyzed people independent. The autonomy is added and the general well-being is improved. These advancements go beyond clinical and rehabilitation stage. They cover multiple areas of healthcare. Virtual assistants and chatbots brighten healthcare by offering continuous maintenance and answering medical questions, allowing patients to be engaged and giving them reminders of prescription medications and appointments, thus relieving healthcare workers from overwork. AI-powered mental health apps and chatbots provide comforting and accessible mood monitoring as well as distorted emotional support, especially for people that experience mental healt difficulties. AI enabled remote patient monitoring devices which monitor the health metrics and vital signs through sensors can spot the changes on time and lead to the better management of chronic conditions. AI also assists in the performance of healthcare administrative procedures like patient record management, invoicing and appointment scheduling and prevails in investments of productive time and administration related tasks burden. Despite the fact that the utilisation of AI in healthcare definitely has its challenges—as such, data privacy, regulatory compliance, and preventing algorithmic biased—the advantages are undoubtedly significant. The development will happen further in sections such as robotic engineering whenever medicines will be combined, telemedicine, and AI-assisted diagnostics. The healthcare field will be greatly benefitted with the further developments of AI and robotics that will come up with the innovations which will engage in more improved patient outcomes, higher level of accessibility and large-scale healthcare delivery. To sum up, the humanized Artificial Intelligence and medical robotics are the main contributors of healthcare industry that provide better patient care, accurate diagnostics, minimal invasion through surgery, and the new treatment strategies during the rehab. The health system is by far the most affected sector from the ongoing digital transformation in which AI is set to bring in even more improvements in this domain. Along this way by the end, the patients care is enhanced, and the healthcare system improved in quality.

4 Challenges and Future Directions for Embodied AI: Navigating the Path Forward

Although the implementation of embodied AI systems in industry signifies a bright future of technological progress, it is not without its share of obstacles. The sets of problems are technical challenges, ethical discourses, data privacy protection, and labor force changes. In addition to that, tantalizing progresses and the colossal development opportunities promised for embodied AI are abundant in front of us as we move to the future. The research part of this article will focus on who is the limit of embodiment AI in the industrial realm and what are some of the concerns when it comes to ethics and data privacy and see what the future trends in these areas are.

4.1 Implementation Obstacles for Embodied AI Systems in Industry

The use of embodied systems may be a promising approach to industrial AI, but there are a number of substantial challenges that accompany this. Integration of such systems into an existing industrial environment and processes, however, implies a high degree of complexity that entails conformance with protocols for corresponding machinery communication and coordination with human workers. Besides, the creation, application, and conducting of these systems are time consuming as well as resource-consuming which could scare off the SMEs from such technology in the long term. The technical challenges, however, imply the requirement of highly specialized hardware and software together, with algorithms, that is able to perform a variety of industrial jobs. The creation of artificial intelligence systems with the features of being robust, reliable, scalable always of work and continuously even in the harsh working conditions of the industries, adaptation to changes in the environment and demands is important as well as complicated. Overcoming these problems is no less essential for an AI technique embedded in the industrial applications to be integrated successfully.

4.2 Ethical Considerations

The choice of embodied AI systems together with the ethics committee is sufficiently critical ground. The eradication of biases and fairness in the process of decision-making has to be monitored, because the same way as human beings AI systems may be biased in their preferences from the training data. The issue of transparency is yet another ethical concern that is brought forth by this nature of the "black box", which implies a difficulty in understanding the decision-making process, making it impracticable to hold accountability and build the desired levels of confidence.

In addition, there is the problem of the growing displacement of workers by AI-engineered robots. At the risk of those industries which are dependent on manual labour, this question is of ethics in prioritizing technological growth over job security. However, using data from sensors and cameras for the embodied AI systems is an added value to the rising privacy issue, reminding the ethical responsibility to secure individuals' confidentiality and avoid the undesirable consequence of private data misuse.

4.3 Regarding Data Privacy

Ethics is a major issue that there is a need to consider for AI systems that store data. Preserving data safe via such strong actions, as prevention of any unauthorized access and data breaches, in which the process of collected data handling and storage is of vital significance, has to be the utmost priority. It is paramount to note that getting the consent of subjects and maintaining transparency in the collection of such data are the basic steps in safeguarding privacy. The complex difficulties like data possession, whose hands may also the embodied AI systems nan have to their disposal needs to be highly weighed. Ensuring proper data retention periods and putting in place specific guidelines or rules aimed at data deletion also are vital in this regard.

4.4 Future Developments and Trends

Embodied AI systems represents the next step in the evolution of this technology, offering various exciting updates and trends. More and more humans are going to get involved in a teamwork where AI is going to support and enhance human capabilities and prospect to boost productivity and provide an innovative output. Applications of embodied AI, including construction operation, logistics, healthcare, and agriculture will witness an increase. A crucial stage of this process is development of robots that run independently with no intervention from the human in dynamic and unstructured environment. Further, sensory achievement of virtual metaphors would permit AI systems to obtain more universal awareness of surrounding life. However, the ethical framework and a regulatory foundation will consistently develop. Their responsibility system is to ensure the safe development and deployment of the AI-based technology. In a few decades, there will be broader integration of AI in manufacturing, necessitating adaptability, customization, and sustainability. Sustainability will be seen as an implementation of AI, creating systems that are tailored to industry needs while promoting resource efficiency and environmental protection. Providing retraining and adapting programs will stand out as the important solution to anticipate existing job displacement worries that could emerge in those industries where AI integration proceeds.

In summary, incorporated AI systems possess the capacity to revolutionise sectors and provide resolutions to a number of the most critical global issues. However, they are not without their fair share of difficulties, such as data privacy concerns, ethical dilemmas, and technical complexities. In order to surmount these obstacles, it will be necessary for technology developers, regulators, and industry stakeholders to work in tandem. Embodied AI exhibits a bright future, as evidenced by developments that prioritise responsible AI development, autonomous robotics, improved sensory capabilities, and diverse applications involving AI and humans [14]. Its continued development will profoundly affect societies, industries, and the manner in which we interact with AI systems on a daily basis.

5 Conclusion

In summary, this chapter has presented an exhaustive examination of the profound impact that embodied AI can have across diverse sectors. By bridging the divide between the digital and physical worlds, embodied AI systems are reshaping numerous industries, including manufacturing, healthcare, and agriculture. This chapter explored the implementation of various verticals of Embodied AI in various sectors. The chapter explored the responsibility of Embodied AI systems bringing about a paradigm shift in the manufacturing sector through the automation of intricate procedures, the improvement of accuracy, and the guarantee of uniform quality. The incorporation of embodied AI in various sectors is explained in this chapter. This chapter promotes the investigation of AI technologies for industrial applications in greater depth. The incorporation of embodied AI systems holds the potential to enhance the safety and efficiency of industrial processes, thereby potentially improving the overall quality of life in addition to yielding economic benefits. Sectors can sustain their innovative prowess and effectively tackle the intricate challenges of the twenty-first century by remaining at the vanguard of AI technology.

References

1. Hughes, J., Abdulali, A., Hashem, R., Iida, F.: Embodied artificial intelligence: enabling the next intelligence revolution. IOP Conf. Ser.: Mater. Sci. Eng. **1261**(1), 012001. IOP Publishing (2022)
2. Downing, K. L.: The predictive basis of situated and embodied artificial intelligence. In: Proceedings of the 7th Annual Conference on Genetic and Evolutionary Computation, June 2005, pp. 43–50 (2005)
3. Steels, L., Brooks, R. (eds.): The Artificial Life Route to Artificial Intelligence: Building Embodied, Situated Agents. Routledge (2018)
4. Anderson, M.L.: Embodied cognition: a field guide. Artif. Intell. **149**(1), 91–130 (2003)
5. Floreano, D., Mondada, F., Perez-Uribe, A., Roggen, D.: Evolution of embodied intelligence. In: Embodied Artificial Intelligence: International Seminar, Dagstuhl Castle, Germany, 7–11 July 2003. Revised Papers, pp. 293–311. Springer, Berlin, Heidelberg (2004)

6. Pfeifer, R.: Embodied artificial intelligence 10 years back, 10 years forward. In: Informatics: 10 Years Back, 10 Years Ahead, pp. 294–310. Springer, Berlin, Heidelberg (2001)
7. Pfeifer, R., Iida, F.: Embodied Artificial Intelligence: Trends and Challenges. Lecture Notes in Computer Science, pp. 1–26 (2004)
8. Abulibdeh, A., Zaidan, E., Abulibdeh, R.: Navigating the confluence of artificial intelligence and education for sustainable development in the era of industry 4.0: challenges, opportunities, and ethical dimensions. J. Clean. Prod. 140527 (2024)
9. Savva, M., Kadian, A., Maksymets, O., Zhao, Y., Wijmans, E., Jain, B., Straub, J., Liu, J., Koltun, V., Malik, J., Parikh, D., Batra, D.: Habitat: a platform for embodied AI research. In: Proceedings of the IEEE/CVF International Conference on Computer Vision, pp. 9339–9347 (2019)
10. Chrisley, R.: Embodied artificial intelligence. Artif. Intell. **149**(1), 131–150 (2003)
11. Pee, L.G., Pan, S.L., Cui, L.: Artificial intelligence in healthcare robots: a social informatics study of knowledge embodiment. J. Am. Soc. Inf. Sci. **70**(4), 351–369 (2019)
12. Abrahamson, D., Worsley, M., Pardos, Z.A., Ou, L.: Learning analytics of embodied design: enhancing synergy. Int. J. Child-Comput. Interact. **32**, 100409 (2022)
13. Brunette, E.S., Flemmer, R.C., Flemmer, C.L.: A review of artificial intelligence. In: 2009 4th International Conference on Autonomous Robots and Agents, February 2009, pp. 385–392. IEEE (2009)
14. Holland, O.: (2004, July). The future of embodied artificial intelligence: Machine consciousness?. In Embodied Artificial Intelligence: International Seminar, Dagstuhl Castle, Germany, July 7–11, 2003. Revised Papers (pp. 37–53). Berlin, Heidelberg: Springer Berlin Heidelberg.

Synergistic Fusion: Vision-Language Models in Advancing Autonomous Driving and Intelligent Transportation Systems

Abha Kiran Rajpoot and Gaurav Agrawal

Abstract In recent years, combining visualization and language models has opened many possibilities. Control and Intelligent Traffic Investigation (ITS) is a very useful application. This article examines the role and importance of integrating visual and linguistic models to increase the efficiency and safety of traffic management and improve traffic outcomes. Integrating vision and language models has been proven to increase the performance and safety of autonomous vehicles and improve traffic and transportation. Perception and decision processes in autonomous vehicles. Leveraging the power of both methods, these models help gain a deeper understanding of the environment by interpreting multimodal information such as images, videos, and text documents. Thanks to image recognition, object recognition and natural language understanding, these machines can understand traffic conditions, road signs, pedestrian behavior, technology, and contextual information, enabling safe and reliable autonomous navigation. In addition, the use of visual language models includes a wide range of intelligent traffic in addition to the intelligent car. By analyzing and interpreting various streams of data from various sources, including data collected from cameras, sensors and city information systems, social media and databases, these models help with real-time traffic management, congestion forecasting and monitoring and optimization methods. Their ability to transmit information from different components allows vehicles to respond to changing conditions, improving safety and efficiency. This article provides an overview of the current advances, approaches and challenges in the development and application of visual modeling in the automotive industry and intelligent transportation systems. Explores the integration of computer vision and powerful language processing techniques and discusses their impact on improving comprehension, decision making, and overall performance. The need to be strong, meaningful, and able to improve the nature of driving and traffic capital according to principles is also important in future guidance and research. In short, it can be stated that the fusion model of vision and language has a lot of promise in the development of driving and intelligent driving. These models combine the understanding and interpretation of multimodal information, resulting in

A. K. Rajpoot (✉) · G. Agrawal
Department of Computer Science & Engineering, Galgotias University, Greater Noida, India
e-mail: akrajpoot@gmail.com

P. Raj et al. (eds.), *Building Embodied AI Systems: The Agents, the Architecture Principles, Challenges, and Application Domains*, Information Systems Engineering and Management 14, https://doi.org/10.1007/978-3-031-68256-8_9

safer, more efficient, and more flexible transportation, bringing us closer to a future of intelligent and more connected people. Finally, we discuss issues and research gaps in depth paper the goal is to provide researchers with current work and the future Trends in VLM in AD and ITS.

Keywords Vision-language model · Large language model · Autonomous driving · Intelligent transportation systems

1 Introduction

In the ever-evolving panorama of self-sufficient and intelligent transport structures (ITS), an innovative synergy has emerged at the intersection of imaginative, visionary, and linguistic modes. This union, regularly referred to as a "Synergistic Fusion", indicates a giant leap forward in the competences of these structures. It embodies the convergence of contemporary technology, connecting imaginative and foresight-based total faith with the contextual expertise of linguistic mode skills, reshaping the very material of the way self-sufficient engines navigate and engage in wise transportation networks.

Traditionally, motor autonomy has relied largely on visible belief structures, using an array of sensors, cameras, and mapping technologies to decipher and navigate the complex tapestry around them. While these structures have shown huge improvements, they have encountered obstacles in filming nuanced contexts and interpreting multifaceted scenarios that have provided challenging situations in terms of adaptability and resilience. Conversely, language modes rooted in the domain names of herbal language processing (NLP) excel at interpreting the complex semantic systems [1] and contextual nuances found in textual information, reflecting human expertise in language.

The union of imaginative, predictable, and linguistic modes represents a key juncture, an unlocking paradigm in which the powers of these modalities converge and transcend the boundaries of character competencies. This amalgamation enables self-sufficient engines and intelligent traffic structures to decode and respond to a symphony of multimodal stimuli encompassing every visible and linguistic information. This integration enhances their cognitive abilities and gives them an extraordinary ability to recognize various environmental stimuli and make informed decisions in real time.

The core of this chapter lies in exploring the transformative power embedded in the synergy of imaginative, visionary, and linguistic modes, catapulting the evolution of self-sufficient and wise transport structures into uncharted territories. By delving into the architecture, mechanisms, and trend enhancements in imaginative and prescient language fashions, this article seeks to illuminate the complex pathways by which these fashions facilitate a holistic interpretation of an amalgam of visible and linguistic cues. Critical to better navigation, interaction and decision-making within these structures, this complete expertise paves the way for a whole new generation of smart mobility.

Furthermore, beyond contemporary nation-states of self-sufficiency, the results of imaginative and insightful language modes extend far and wide, permeating the material of wise transport structures. These methods provide extraordinary capabilities, from optimizing site visitor flows and improving infrastructure control to supporting seamless and intuitive human-device interfaces. The integration of imaginative and insightful modes and language already guarantees not only strengthening, but also redefining the very nature of the transport ecosystem.

The emergence of LLM [2–6] and VLM [7–12] provides a potential solution to the inherent current limitation of autonomous driving and intelligent transportation systems. These new technologies synthesize linguistic and visual data and promise a future where vehicles and systems are deeply embedded to understand your surroundings. This heralds a new era of intelligent, efficient, and explainable transport. In addition to improving traditional tasks in AD or ITS, such as object detection or traffic flow prediction, emerging domains include null shot perception and accident analysis, as shown in Fig. 1. Due to the increase in research applying language models to autonomous driving and intelligent systems, a systematic and comprehensive survey is of great importance to the research community. However, existing surveys [12–16] also focus on LLM, VLM, AD or ITS individually. To our knowledge, there is still no systematic review dealing with VLM applications in AD and ITS. To this end, we present an overview of existing visual language algorithm models in autonomous driving systems and intelligent transportation systems, with an emphasis on the latest technological trends in the research community. We will show the taxonomy of this.

The main contributions of this chapter can be summarized as follows:

- Presentation of the first comprehensive application survey of visual language models in autonomous driving and intelligent transport systems.
- We systematically summarize and analyse existing work and datasets.

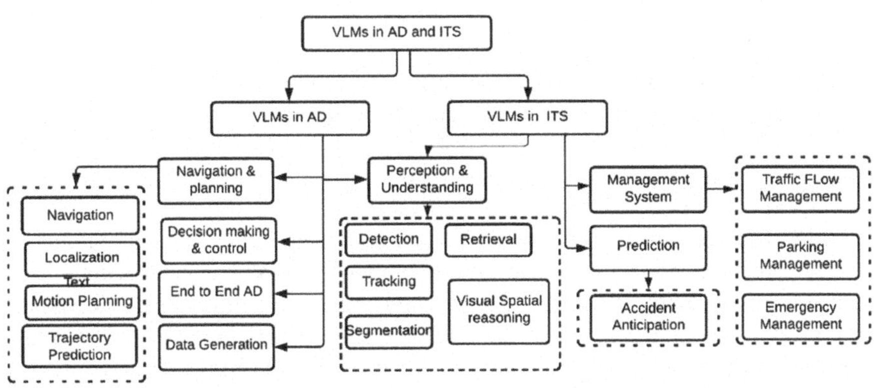

Fig. 1 Overview of taxonomy of VLMS in autonomous driving and intelligent traffic systems

- We explore potential applications and technological advances in AD and ITS.
- We provide an in-depth discussion of the challenges and research gaps in this area.

1.1 Scope and Purpose

The integration of vision and language models stands as a transformative force in revolutionizing autonomous driving and intelligent transportation systems (ITS). By fusing the strengths of computer vision, which processes visual data, and natural language processing, which interprets textual information, these models elevate vehicles' understanding of their surroundings. Their scope extends beyond mere perception to contextual awareness, incorporating linguistic cues with visual data for a more comprehensive understanding of the environment. This integration empowers vehicles to make informed decisions by comprehending complex scenarios on the road, thereby enhancing safety and efficiency. Moreover, these models facilitate more natural human-vehicle interactions, enabling passengers to communicate with vehicles and contributing to a smoother, more intuitive transportation experience. Ultimately, the purpose lies in advancing the safety, reliability, and adaptability of autonomous vehicles and transportation systems through the harmonious synergy of vision and language technologies.

1.2 Application Scenario and Key Tasks for the Embodied Agent to Perform

An exemplary application scenario for synergistic fusion of vision-language models in advancing autonomous driving and intelligent transportation systems involves an embodied agent, such as an autonomous vehicle, navigating urban environments. The agent aims to perform several key tasks by amalgamating vision and language capabilities:

Navigation and Route Understanding: The embodied agent utilizes vision models to perceive and interpret visual cues from the environment, such as traffic signs, lane markings, and pedestrians. Simultaneously, language models process textual information from maps, GPS, and road signs to understand route instructions and contextual details, assisting in precise navigation.

Traffic Interpretation and Decision-Making: Combining vision and language, the agent comprehends traffic situations comprehensively. Vision models detect and analyze the movement of vehicles, while language models interpret traffic signals and textual instructions (e.g., stop signs, speed limits). This fusion assists in making informed decisions about speed adjustments, lane changes, and interaction with other vehicles.

Human Interaction and Assistance: The agent interacts with passengers and pedestrians through language capabilities, responding to verbal or textual queries. For instance, it may provide information about the route, nearby landmarks, or estimated time of arrival. This interaction relies on the fusion of natural language understanding and contextual visual perception.

Adaptive Responses to Dynamic Environments: Vision-language fusion enables the agent to adapt to dynamic scenarios. For instance, if the road conditions change due to construction, the agent can interpret both visual changes (detour signs, altered lanes) and textual information (roadwork instructions) to modify its route accordingly.

Safety Protocols and Emergency Situations: By integrating vision and language, the agent can swiftly respond to safety protocols and emergency signals. It can detect visual cues indicating emergencies (such as accident scenes) and understand textual warnings (e.g., emergency detour instructions) to take appropriate actions, ensuring passenger safety.

Overall, the application scenario involves an embodied agent (autonomous vehicle) utilizing synergistic fusion of vision and language models to navigate, understand, interact with, and adapt to complex and dynamic urban environments, ensuring safe and efficient transportation experiences.

2 Literature Survey

Research in autonomous driving and intelligent transportation systems (ITS) has witnessed a significant shift towards leveraging vision-language models to enhance vehicle perception and decision-making capabilities. Integrating vision and language has become a focal point, aiming to improve scene understanding and interpret complex environments encountered by autonomous vehicles. This synergy aims to fuse natural language descriptions with visual data obtained through various sensors such as cameras, lidar, and radar, enabling a more comprehensive perception system.

One key area of exploration involves semantic understanding and scene interpretation. Researchers are striving to develop models that effectively combine vision and language [12] to enable vehicles to recognize objects, comprehend road conditions, and respond adeptly to diverse traffic scenarios. By integrating these two modalities, autonomous systems are anticipated to gain better context awareness, leading to more informed decision-making processes.

Another important facet revolves around human-AI interaction and communication. Vision-language models not only aid in better understanding the environment but also facilitate more intuitive interactions between autonomous vehicles and humans. This could include enabling the vehicle to communicate its intentions or understanding human instructions, thereby enhancing the overall user experience and safety.

Ethical considerations and safety implications remain paramount. Researchers are actively addressing how these models handle unforeseen situations and ensuring their alignment with ethical and legal standards. Additionally, creating comprehensive datasets that combine vision and language for training and evaluating these models has emerged as a critical area. Benchmarking against standardized datasets aids in assessing performance and tracking progress.

However, challenges persist. Implementing these advanced models in real-time autonomous driving systems poses computational complexity, latency, and hardware constraints. Addressing these challenges is crucial to achieve practical and efficient integration of vision-language models in autonomous vehicles, paving the way for safer and more intelligent transportation systems.

2.1 Limitations of Current Sensor Technologies

The advancement of autonomous driving and intelligent transportation systems heavily relies on sensor technologies for data collection and perception. However, these sensors come with certain limitations that can impact the effectiveness of synergistic fusion with vision-language models:

Resolution Limitations: Sensor technologies like cameras, LiDAR (Light Detection and Ranging), and radar systems have specific resolutions that might not capture fine details crucial for precise decision-making. Lower-resolution sensors may struggle to capture intricate visual data, potentially impacting the accuracy of object recognition and scene understanding.

Accuracy and Noise: Sensors can be susceptible to noise, inaccuracies, and occlusions. For instance, adverse weather conditions like heavy rain, fog, or snow can affect the accuracy of LiDAR and radar systems, leading to distorted or incomplete data. This inconsistency can hinder the accuracy of perception and, consequently, the decisions made by autonomous vehicles.

Robustness to Environmental Conditions: Different sensor technologies exhibit varying degrees of robustness to environmental factors. While LiDAR might struggle with fog or heavy rain due to scattering, cameras might face challenges [13] with low-light conditions or glare. Such limitations in specific environmental conditions can affect the reliability of data captured, impacting the vehicle's ability to perceive and navigate effectively.

Cost and Scalability: High-quality sensors that offer better resolution and robustness often come at a higher cost. This factor poses challenges in scaling up autonomous systems for mass deployment, especially in scenarios where a large number of sensors are needed for comprehensive coverage.

Sensor Fusion Complexity: Integrating data from multiple sensors (fusion of camera, LiDAR, radar, etc.) adds complexity to the perception process. Aligning data from disparate sources while accounting for discrepancies in their measurements and characteristics requires sophisticated algorithms and computational resources, which can pose challenges in real-time decision-making.

Addressing these limitations involves ongoing research and development in sensor technology, such as enhancing resolution, improving robustness to environmental conditions, reducing noise, and developing more cost-effective solutions. Moreover, advancements in AI algorithms that can effectively [14] handle noisy or incomplete sensor data are crucial for mitigating these limitations and improving the overall reliability of autonomous systems relying on vision-language fusion.

2.2 Autonomous Driving

Autonomous operation is an interesting term that allows the car to travel and operate without human intervention. It requires a combination of sensors, cameras, radar, lidar and advanced software algorithms to understand the environment.

Explain the different levels of automation, from level 0 (no automation) to level 5 (full automation), with help from SAE International. Most cars on the road today have a variety of driver aids such as cruise control or lane departure warning (Level 1 or 2 autonomy). However, driverless cars that can perform all tasks without human intervention are still being developed. The huge potential of self-sufficiency includes increased [15] safety (reducing injuries caused by human error), improved mobility for elderly or disabled enterprises, reduced visitor traffic and greener use of transportation.

However, there are still significant issues and challenges that need to be addressed, including governance and crime, public accountability, cybersecurity, moral responsibility, and need a strong and reliable time to resolve.

Despite these regulations, great progress has been made in self-satisfaction over time by using different organizations and research [16] centers, and we think that the development will continue in the coming years.

2.3 Intelligent Transportation System

Autonomous operation is an interesting term that allows the car to travel and operate without human intervention. It requires a combination of sensors, cameras, radar, lidar and advanced software algorithms to understand the environment.

Intelligent Transportation Systems (ITS) include ideas and the use of various technologies designed to improve transportation, safety, and security. These systems integrate advanced communications, sensors and information technology into transportation and vehicles to improve the movement of people and goods. Traffic management and control: ITS collects real-time data from sensors, cameras, and [17] other monitoring systems for good traffic management. This includes signal transition control, collision control and event detection to reduce delays and improve overall control. Advanced passenger information system: It is important to provide accurate and accurate information to passengers. ITS enables users to make informed travel [18] decisions by providing updates on traffic times, alternatives, Vehicle to Infrastructure (V2I) and Vehicle to Vehicle (V2V) Communications: V2I and V2V communications [19] enable vehicles to communicate with each [20] other and with each other. This communication helps improve safety and efficiency by facilitating the exchange of information about roads [21]. Autonomous and Connected Vehicles: ITS combines autonomous and connected vehicles to increase safety and efficiency. Powered by artificial intelligence and sensor technologies, autonomous vehicles can navigate and operate without human intervention, while connected vehicles communicate with each other and with infrastructure to avoid accidents and optimize traffic flow. Planning and management of transport systems: ITS helps in the planning and management of transport by analyzing data collected from various sources to optimize routes, infrastructure development and resource allocation, resulting in more sustainable efficient transportation system. Intelligent transport systems continue to evolve, driven by advances in technologies such as artificial intelligence, machine learning and the Internet of Things (IoT). The goal is to create smarter, safer, and more efficient transportation networks that increase mobility, reduce congestion, improve environmental sustainability, and ultimately improve the overall quality of transportation experiences for individuals and businesses.

3 Large Language Models

Large language models, such as GPT-3, are advanced AI systems designed to understand and generate human-like text based on the vast amount of data they have been trained on. These models are built on deep learning architectures, using neural networks with millions or even billions of parameters.

Key characteristics of large language models include:

Scale and Complexity: These models are massive in scale, comprising numerous interconnected layers of artificial neurons. GPT-3, for example, consists of 175 billion parameters, allowing it to capture and understand complex patterns in text data.

Natural Language Understanding and Generation: These models excel in natural language processing tasks, understanding context, semantics, and syntax in human language. They can generate coherent and contextually relevant text based on prompts provided to them.

Adaptability and Versatility: Large language models demonstrate adaptability across a wide range of tasks, from text completion and summarization to translation and question answering. They can be fine-tuned on specific datasets to perform more specialized tasks.

Training and Data Requirements: Training such models requires vast amounts of data and substantial computational resources. They learn from diverse sources of text available on the internet, learning to predict the next word in a sentence based on the context provided by the preceding words.

Ethical Considerations: There are ethical considerations regarding the use of large language models, such as concerns about bias, misinformation, and potential misuse. Researchers and developers continually work on mitigating these issues through responsible AI practices and model improvements.

These models have diverse applications across industries, including language translation, content generation, customer service automation, healthcare diagnostics, and more. However, they also raise questions about data privacy, accountability, and the societal impacts of AI-powered systems, prompting ongoing discussions on their responsible use and regulation.

4 VLMS in Autonomous Driving

An increasing number of initiatives are endeavoring to implement VLMs across various aspects of AD. In this section, we present the existing work of VLMs within AD consisting of perception and understanding (III-A), navigation and planning (III-B), decision-making and control (III-C), end-2-end AD (III-D), and data generation (III-E) The summarized current methods are shown in Table 1.

Table 1 Overview of LLMS and VLMS in autonomous driving

Method	Year	Tasks	Type	Contribution
NIC [28]	2020	IC	V2T, VTM	Introduce image captioning task to autonomous driving for traffic scene understanding. VLPD [24]
UMPD [22]	2023	PD	VTM	Propose a fully-unsupervised multi-view pedestrian detector without manual annotations
MSSG [25]	2023	MOR-T	VTM	Propose referring multi-object tracking task, dataset, and benchmark based on KITTI
Prompt Track [27]	2023	MOR-T	VTM	Propose multi-view 3D referring multi-object tracking dataset based on scenes
Open Scene [23]	2023	OV-3DSS	VTM	Distil CLIP knowledge to a 3D network for 3D scene understanding
UP-VL [35]	2023	OV-3DOD-T	VTM	Introduce semantic-aware unsupervised detection for objects in any motion state
Zelda [36]	2023	VQA	M2T, VTF	Introduce visual question-answering task and baseline model in autonomous driving based on scenes. Talk2BEV [29]

TASKS: **IC**: IMAGE CAPTIONING **PD**: PEDESTRIAN DETECTION **SOR**: SINGLE-OBJECT REFERRING **MOR-T**: MULTI-OBJECT REFERRING AND TRACKING **OV-3DSS**: OPEN-VOCABULARY 3D SEMANTIC SEGMENTATION **OV-3DOD-T**: OPEN-VOCABULARY 3D OBJECT DETECTION AND TRACKING **LVR**: LANGUAGE-GUIDED VIDEO RETRIEVAL **VQA**: VISUAL QUESTION ANSWERING

4.1 Perception and Understanding

In autonomous driving perception, VLMs, especially those models pre-trained on large-scale datasets with image-text-matching like, has spurred numerous new studies. These studies [22, 23] leverage the substantial prior knowledge of pre-trained large VLMs to boost the performance of perception and understanding, and further introduce many novel tasks in the field.

Pedestrian Detection. Human-like object confusion and insufficient border case samples are the inherent challenges in pedestrian detection. To this end, VLPD [24] first proposes a vision-language extra-annotation-free method to enhance model's capability of distinguishing confusing human-like objects. It employs CLIP to acquire the pixel-wise explicit semantic contexts, and distance pedestrian features from the features of other categories through contrastive learning, improving the detection capabilities for broader cases. UMPD [22] also utilizes the zero-shot semantic classes from the CLIP, and proposes a fully-unsupervised multi-view Pedestrian Detection approach without the need of manual annotations. Object Referring. Compared to traditional perception tasks in AD, such as object detection, tracking, and semantic segmentation, the introduction of language enables the model to attain a more finely-grained and comprehensively unconstrained capability to understand the surrounding environment.

The use of a container or mask to identify specified equipment is often based on questions. MSSG [25] proposed a multi-modal 3D single object (SOR) project in standalone use. Bird's Eye View (BEV) imaging combines lidar and language capabilities to train the multi-purpose single-lens terrestrial version and estimate position analysis by the detector without delay and without any subsequent processing. Trans RMOT [26] has been extended from the SOR project to Multiple Object Reference and Tracking (MOR-T) and compiled into the Guide-KITTI benchmark based mainly on the KITTI dataset. When asked a question, Trans RMOT can find and score all the equipment used in the movie. Similarly, Prompt Track [27] prepared prompts to run on the scenes dataset and created the prompt benchmark. Compared to Refer KITTI, Prompt leverages more views of Scenes, making it more relevant to MOR-T views.

Open-Vocabulary 3D Object Detection and Semantic Segmentation: Turn on 3D language object exploration and semantic segmentation. Thanks to the powerful zero-shot transform and cross-modal mapping capabilities of large VLMs, it has provided the ability to perform object search and semantic segmentation, explicit word search, and segmentation of invisible structures. The UP-VL framework [35] initially proposed an unsupervised, multi-level automatic labeling pipeline to generate content-level features and object-level bounding boxes and trajectories for an explicit language and class-independent 3D detection monitor, and added usage to specify 3D bounding. inference time boxes. The UP-VL framework, together with semantic labels derived from matching, enables unsupervised open-source 3D detection and tracking (OV-3DOD-T) of participants in AD scenarios. Open Sene [23] leverages 2D-3D projection correspondence to maintain consistency between 3D point cloud embeddings and composited 2D image features. Essentially, it combines

Fig. 2 Evaluation of GPT-4 V [30] on traffic anomaly scene interpretability and reasoning. The traffic scenario A is sourced from Providential The traffic accident is highlighted in the red rectangle dotted box at the left-top of this figure. In the traffic scenario B from a video, where a car's door is open while the car is in the middle of the lane. The red dotted boxes show the opened door and the traffic light

3D point cloud representation with CLIP image representation to achieve zero-correction perception of dense 3D point features. Open Scene is primarily evaluated in an indoor environment but also shows interest in its Open Word 3D Semantic Segmentation (OV-3DSS) capabilities.

Traffic Scene Understanding: It is important to have a good understanding and level of traffic in terms of driving safety. Figure 2 shows an example of understanding the capability of GPT-4 V [30] in the event of a traffic accident and the appearance of the road that may be at risk in the city. There are some research studies that use VLM to understand traffic conditions resulting from specific activities. Reference [28] attempted to understand traffic using image signatures (ICs) to identify the scene image. Reference [23] converted visual data into descriptive language and then used the ability of LLM. for visual impairment (SAD). Zelda [27] used VLM's semantic similarity function in video search (VR) and achieved better performance than other high-end video analytics. Scenes-QA provides benchmarks for visual response (VQA) tasks in AD scenes based on the scene's dataset, laying the foundation for further research. Speak to the map model and the understanding model through understanding and understanding of FMC... OV-AD We note that most of the above projects are still in their early stages, including 3DOD, OV-3DSS, SOR, MOR-T, but we believe there is promise and we hope it is there. In the future, there will be many interesting projects where more and more jobs will be created.

5 VLMS in Intelligent Transportation Systems

Intelligent transportation systems are becoming more and more prevalent and work as an important step to achieve fully self-driving capabilities [75]. ITS usually consists of a multisensory setup in which cameras are almost always present due to its rich and dense semantic information. Like AD, VLMs can also facilitate the development of ITS in many aspects including ITS perception and understanding (IV-A), and ITS management system.

ITS Perception and Understanding: The rapid increase in the number of electronic devices installed in transportation has greatly improved the ability of intelligent transportation. This allows for better monitoring and understanding of the traffic environment, allowing ITS to clearly identify and interpret complex situations. Search for cars by language. Traffic monitoring is an important part of understanding and understanding smart transportation. To guide researchers to focus on this aspect, the AI City Challenge [23] establishes a reverse vehicle pursuit using narrative as a challenge. As part of this collaboration, proposed a multi-layer search engine for language-based vehicle detection. A multi-question search based on additional words is used as the main topic. The idea behind this mod is to use various missing descriptions to get a better and more realistic look. Reference [30] proposed an interesting method to deal with inaccessible data for vehicle detection. The main task of this work is to introduce the transfer of information from recorded information to invisible information through the creation of fake texts. MLVR proposed a multilingual transport system that uses text and image separately to encode features and then generates a vector image sequence from video recognition. By integrating different traffic patterns, MLVR creates rich traffic data to manage relationships and fulfill search queries. Determine traffic visibility. Another emerging area of VLM in ITS is the visual understanding of events, often in the form of question answering tasks (VQA) proposed the use of automatic captioning method for video query response in the weak control process network domain. The main program is to provide online subtitle to city car films using the created files. Then automatically created video images are used for optimization by adding local information to the training model. Proposed a different reasoning problem to analyze the context of the time that led to the effect of problematic thinking at the level. Care -based modules can be studied on the nature of the body and problem pairs. He learned an adaptable concept to reduce the temporal difference between images and videos by learning over time. It performs excellently in video field studies. Anomaly CLIP uses the CLIP model to detect video anomalies. By identifying the abnormal group and optimizing the content, it can distinguish normal and abnormal patterns, thus detecting anomalies. Anomaly CLIP achieves good results on a wide range of data, including anomaly detection. VLM's advanced semantic anomaly detection algorithms can be extended to ITS disaster or emergency response.

Potential Applications: Although many approaches have explored leveraging the capabilities of VLMs in various transportation scenarios, the full potential of VLMs in ITS remains untapped. Accident Anticipation and Detection. The capability to promptly detect and anticipate accidents is critical in ensuring road safety. As points out, current vision-based Traffic Accident Detection (TAD) and Traffic Accident Anticipation (TAA) still face various challenges, including its long-tailed and safe-critical properties, complex scene evolution, harsh environments, and determination uncertainty. Large VLMs, thanks to their exceptional abilities in zero-shot generalization, profound scene comprehension, border case recognition, and multi-step reasoning, serve as a promising solution to address current challenges. Crash

Reporting. Rapid crash analysis enhances traffic efficiency and prevents further congestion. M2V conditional generative VLMs represent a potential solution. By utilizing video footage recorded by drivers, along with their descriptions, the model can instantly generate narratives of the accident scene.

6 Comparative Study Between VLMS in ADS and ITS

Here is a comparative table outlining the differences between Vision-Language Models (VLMS) in Autonomous Driving Systems (ADS) and VLMS in Intelligent Transportation Systems (ITS):

Table 2 highlights the key differences between VLMS used in ADS and VLMS applied in ITS, outlining their specific application focuses, functionalities, utilization of data, objectives, technological roles, challenges, impact, and integration into transportation systems.

Table 2 Overview of comparison between VLMS in ADS and ITS

Aspect	VLMS in Autonomous Driving Systems (ADS)	VLMS in Intelligent Transportation Systems (ITS)
Application focus	Integration into perception systems for autonomous vehicles	Deployment for traffic analysis, management, and infrastructure planning
Functionality	Enhancing environment perception for autonomous vehicles	Analyzing traffic patterns, providing insights for traffic management
Data utilization	Utilizes vision-language fusion for scene understanding	Utilizes fusion for interpreting traffic-related visual and textual data
Objectives	Enhancing vehicle perception, aiding decision-making	Optimizing traffic flow, reducing congestion, enhancing safety
Technological role	Supporting autonomous vehicle navigation and decision-making	Facilitating data-driven decision-making for transportation networks
Challenges	Ensuring accurate scene understanding in varied conditions	Interpreting complex traffic scenarios, data standardization challenges
Impact	Improved autonomous vehicle capabilities	Enhanced traffic management, reduced congestion, safer transportation
Integration	Enhancing perception systems within individual vehicles	System-wide integration to optimize transportation networks

6.1 VLMS in Autonomous Driving Systems (ADS)

Focus: Perception for Autonomous Vehicles: VLMS in ADS primarily aid in enhancing the perception capabilities of autonomous vehicles. They interpret visual data and linguistic context to understand the environment for safe navigation.

Functionality: Scene Understanding: These models fuse vision and language to interpret scenes, identify objects, understand road conditions, and aid in decision-making within the context of autonomous driving.

Objectives: Enhanced Vehicle Autonomy: VLMS in ADS aim to improve the autonomy of vehicles by providing a comprehensive understanding of the surroundings, supporting navigation and decision-making processes.

Challenges: Varied Environmental Conditions: Addressing challenges related to accurately interpreting diverse environmental conditions, ensuring robustness in complex scenarios, and handling real-time decision-making.

6.2 VLMS in Intelligent Transportation Systems (ITS)

Focus: Traffic Analysis and Management: VLMS in ITS contribute to analyzing traffic-related data, interpreting visual and textual information to optimize traffic flow, and enhance overall transportation systems.

Functionality: Data-Driven Decision-Making: These models integrate vision and language to interpret traffic-related data, aid in predicting traffic patterns, and assist in making informed decisions for traffic management and infrastructure planning.

Objectives: System-Wide Optimization: VLMS in ITS aim to optimize transportation networks, reduce congestion, improve safety, and enhance the overall efficiency of transportation systems.

Challenges: Complex Traffic Scenarios: Addressing challenges related to interpreting complex traffic scenarios, data standardization across diverse sources, and ensuring the accuracy of insights for effective traffic management.

6.3 Integration and Impact

ADS Integration: VLMS in ADS are integrated into the perception systems of autonomous vehicles, enhancing their ability to understand and navigate their surroundings safely.

ITS Integration: VLMS in ITS are integrated into the broader transportation infrastructure, aiding in data-driven decision-making for traffic analysis, management, and infrastructure planning.

7 Conclusion

Fusion of Vision-Language Models (VLMS) is a transformative force shaping the future of both autonomous driving and intelligent transportation systems (ITS). This approach promises to raise awareness, enable comprehensive decision-making, and create safer and more efficient transport networks. However, this success comes with huge challenges, such as the complexity of actual implementation, the ethical aspects of decision-making, and the need for seamless integration with existing infrastructure. The collaboration of cross-disciplinary innovations, regulatory frameworks and ethical guidelines is essential to harness the true potential of VLMS to ensure a harmonious balance between technological development and societal needs in shaping the world of transport for a safer and more efficient future.

References

1. Vaswani, A., Shazeer, N., Parmar, N., Uszkoreit, J., Jones, L., Gomez, A.N., Kaiser, Ł., Polosukhin, I.: Attention is all you need **30** (2017)
2. Touvron, H., Martin, L., Stone, K., Albert, P., Almahairi, A., Babaei, Y., Bashlykov, N., Batra, S., Bhargava, P., Bhosale, S., et al.: Llama 2: open foundation and fine-tuned chat models (2023). arXiv:2307.09288
3. Touvron, H., Lavril, T., Izacard, G. Martinet, X., Lachaux, M.-A., Lacroix, T. Rozie‘re, B., Goyal, N., Hambro, E., Azhar, F., et al.: Llama: open and efficient foundation language models (2023)
4. Scao, T.L., Fan, A., Akiki, C., Pavlick, E., Ilić, S., Hesslow, D., Castagne´, R., Luccioni, A.S., Yvon, F., Galle´, M., et al.: Bloom: a 176b-parameter open-access multilingual language model (2022). arXiv:2211.05100
5. Chowdhery, A., Narang, S., Devlin, J., Bosma, M., Mishra, G., Roberts, A., Barham, P., Chung, H.W., Sutton, C., Gehrmann, S., et al.: Palm: scaling language modeling with pathways (2022). arXiv:2204.02311
6. Brown, T., Mann, B., Ryder, N., Subbiah, M., Kaplan, J.D., Dhariwal, P., Neelakantan, A., Shyam, P., Sastry, G., Askell, A., et al.: Language models are few-shot learners. Adv. Neural. Inf. Process. Syst. **33**, 1877–1901 (2020)
7. Zhu, D., Chen, J., Shen, X., Li, X., Elhoseiny, M.: Minigpt-4: enhancing vision-language understanding with advanced large language models (2023). arXiv:2304.10592
8. Peng, Z., Wang, W., Dong, L., Hao, Y., Huang, S., Ma, S., Wei, F.: Kosmos-2: grounding multimodal large language models to the world (2023). arXiv:2306.14824
9. Gong, T., Lyu, C., Zhang, S., Wang, Y., Zheng, M., Zhao, Q., Liu, K., Zhang, W., Luo, P., Chen, K.: Multimodal-GPT: a vision and language model for dialogue with humans," (2023). arXiv:2305.04790
10. Liu, S., Fan, L., Johns, E., Yu, Z., Xiao, C., Anandkumar, A.: Prismer: a vision-language model with an ensemble of experts (2023). arXiv:2303.02506
11. Wu, C., Yin, S., Qi, W., Wang, X., Tang, Z., Duan, N.: Visual chatgpt: talking, drawing and editing with visual foundation models (2023). arXiv:2303.04671
12. Zhang, J., Huang, J., Jin, S., Lu, S.: Vision-language models for vision tasks: a survey (2023). arXiv:2304.00685
13. Chen, L., Wu, P., Chitta, K., Jaeger, B., Geiger, A., Li, H.: End-to-end autonomous driving: challenges and frontiers (2023). arXiv:2306.16927

14. Gu, J., Han, Z., Chen, S., Beirami, A., He, B., Zhang, G., Liao, R., Qin, Y., Tresp, V., Torr, P.: A systematic survey of prompt engineering on vision-language foundation models (2023). arXiv:2307.12980
15. Zhao, W.X., Zhou, K., Li, J., Tang, T., Wang, X., Hou, Y., Min, Y., Zhang, B., Zhang, J., Dong, Z., et al.: A survey of large language models (2023). arXiv:2303.18223
16. Yuan, T., Neto, W.D.R., Rothenberg, C.E., Obraczka, K., Barakat, C., Turletti, T.: Machine learning for next-generation intelligent transportation systems: a survey. Trans. Emerg. Telecommun. Technol. **33**, 4 (2022)
17. S. International: Taxonomy and definitions for terms related to driving automation systems for on-road motor vehicles (2021)
18. Xu, Z., Zhang, Y., Xie, E., Zhao, Z., Guo, Y., Wong, K.K., Li, Z., Zhao, H.: Drivegpt4: interpretable end-to-end autonomous driving via large language model (2023). arXiv:2310.01412
19. Qian, T., Chen, J., Zhuo, L., Jiao, Y., Jiang, Y.-G.: Nuscenes-qa: a multi-modal visual question answering benchmark for autonomous driving scenario (2023). arXiv:2305.14836
20. Jin, B., Liu, X., Zheng, Y., Li, P., Zhao, H., Zhang, T., Zheng, Y., Zhou, G., Liu, J.: Adapt: action-aware driving caption transformer (2023). arXiv:2302.00673
21. Fu, D., Li, X., Wen, L., Dou, M., Cai, P., Shi, B., Qiao, Y.: Drive like a human: Rethinking autonomous driving with large language models," (2023). arXiv:2307.07162
22. Liu, M., Zhu, C., Ren, S., Yin, X.-C.: Unsupervised multi-view pedestrian detection (2023). arXiv:2305.12457
23. Peng, S., Genova, K., Jiang, C., Tagliasacchi, A., Pollefeys, M., Funkhouser, T., et al.: Open-scene: 3d scene understanding with open vocabularies. In: Proceedings of the IEEE/CVF Conference on Computer Vision and Pattern Recognition, pp. 815–824 (2023)
24. Liu, M., Jiang, J., Zhu, C., Yin, X.-C.: VLPD: context-aware pedestrian detection via vision-language semantic self-supervision. In: Proceedings of the IEEE/CVF Conference on Computer Vision and Pattern Recognition, pp. 6662–6671 (2023)
25. Cheng, W., Yin, J., Li, W., Yang, R., Shen, J.: Language-guided 3d object detection in point cloud for autonomous driving (2023). arXiv:2305.15765
26. Wu, D., Han, W., Wang, T., Dong, X., Zhang, and J. Shen: Referring multi-object tracking," in *Proceedings of the IEEE/CVF Conference on Computer Vision and Pattern Recognition*, pp. 14633–14642 (2023)
27. Wu, D., Han, W., Wang, T., Liu, Y., Zhang, X., Shen, J.: Language prompt for autonomous driving (2023). arXiv:2309.04379
28. Li, W., Qu, Z., Song, H., Wang, P., Xue, B.: The traffic scene understanding and prediction based on image captioning. IEEE Access **9**, 1420–1427 (2020)
29. Dewangan, V., Choudhary, T., Chandhok, S., Priyadarshan, S., Jain, A., Singh, A. K., Srivastava, S., Jatavallabhula, K.M., Krishna, K.M.: Talk2bev: language-enhanced bird's-eye view maps for autonomous driving (2023). arXiv:2310.02251
30. Chen, R., Liu, Y., Kong, L., Zhu, X., Ma, Y., Li, Y., Hou, Y., Qiao, Y., Wang, W.: Clip2scene: towards label-efficient 3d scene understanding by clip. In: Proceedings of the IEEE/CVF Conference on Computer Vision and Pattern Recognition, pp. 7020–7030 (2023)

Dr. Abha Kiran Rajpoot is presently working as Professor in the Department of Computer Science & Engineering at Galgotias University, Greater Noida. She holds Ph.D., M. Tech, and B. Tech (all in Computer Science). She has 16 years of teaching, research, and administrative experience at various accredited institutes and universities like Sharda University (NAAC A+) and KIET Group of Institutions. Her field of interest includes Machine Learning, Deep Learning, Business Intelligence, Computer Networks, Operating Systems, Human-Computer Interaction, and Theoretical Computer Science. She has published/presented more than 30 papers in peer-reviewed journals of repute and conferences held in India. Fifteen patents are also to her credit. She has chaired some sessions at international conferences. She has been a member of the organizing committee of 02 international conferences held in India. She is an active member of the Society for Research Development (SRD) and a member of IEEE. She is the reviewer of some reputed conferences.

Dr. Gaurav Agrawal is presently working as Professor in the Department of Computer Science & Engineering at Galgotias University, Greater Noida. He holds Ph.D., M. Tech, and B. Tech (all in Computer Science). He has 20 years of teaching, research, and administrative experience at various accredited institutes of AKTU universities. His field of interest includes Machine Learning, Deep Learning, Business Intelligence, Computer Networks, web accessibility, Operating Systems, and Human-Computer Interaction. He has published/presented many papers in SCI/SCOPUS peer-reviewed journals of repute and conferences held in India. He also has five patents in his credit. He has been a member of the organizing committee of many international conferences held in India and has chaired sessions at international conferences. He is also serving as the reviewer in SCI/SCOPUS journals and conferences.

Health Care Industry Use Cases of Embodied AI

S. N. Kumar, Jomin Joy, Alen J. James, and Andrew Dixen

Abstract Embodied AI systems, also known as embodied artificial intelligence, represent a category of AI systems that are physically integrated into tangible entities. These systems can interact with and manipulate the physical world, utilizing sensors to perceive their surroundings, making decisions based on gathered information, and often possessing the capacity for physical interaction. They can adapt and learn from their interactions and may operate autonomously or in conjunction with humans. Additionally, embodied AI systems can combine their physical attributes with digital AI components for tasks like language understanding and complex decision-making. In the healthcare industry, embodied AI is gaining traction as it is applied to various use cases to enhance patient care, streamline operational processes, and advance medical research. The robotic surgical platform assists surgeons in performing minimally invasive procedures with precision and dexterity. Surgeons control the robot to execute surgeries, resulting in smaller incisions, reduced pain, and faster recovery times. Embodied AI-driven exoskeletons aid patients with mobility challenges in regaining their ability to walk, providing support and facilitating recovery after injuries or surgical procedures. The utilization of embodied AI in healthcare holds the promise of improving patient outcomes, reducing medical errors, and elevating the overall quality of care. The outcome of this chapter paves towards the applications of embodied AI system in healthcare system.

Keywords Artificial Intelligence · Robotics · Machine learning · Deep learning · IoT

S. N. Kumar (✉) · J. Joy
Department of EEE, Amal Jyothi College of Engineering, Kanjirappally 686518, Kerala, India
e-mail: appu123kumar@gmail.com

A. J. James · A. Dixen
Tata Elxsi, Thiruvananthapuram, Kerala, India

P. Raj et al. (eds.), *Building Embodied AI Systems: The Agents, the Architecture Principles, Challenges, and Application Domains*, Information Systems Engineering and Management 14, https://doi.org/10.1007/978-3-031-68256-8_10

1 Introduction to Embodied Artificial Intelligence

The term "Embodied agent" refers to an artificial intelligence system that interacts with the natural world. This could be a robot, a virtual assistant, or any other type of intelligent system. An embodied agent typically has a body or a physical form that allows it to interact with its environment through sensors and actuators. What makes an embodied agent unique is that they are placed in the environment, which means that they can interact naturally with their environment [1]. For instance, a robot can pick up objects or move around the room. An embodied agent also has a body, which could be a physical object such as a robot or a virtual object such as an avatar in virtual reality systems. The body provides the agent with a means of interacting with the world. Embodied artificial intelligence (AI) is nowadays going up in clinical relevance for therapeutic applications in healthcare and mental health services. Numerous aspects of embodied AI exist in the healthcare sector, including the key feature that it can help create simulations that are more realistic and credible. This is because embodied agents can interact with their environment more realistically, which allows for more realistic scenarios to be created. It can also improve the efficiency of learning. This is because embodied agents can acquire data from the surrounding stimuli. This helps us to reduce the resources we spend and the time to train AI is saved. Manifestation, AI systems that are embodied have a mind–body relation. This implies that they can communicate and express themselves through their bodies, as well as detect and interact with the world around them. The trends and challenges of embodied AI are highlighted in Pfeifer and Iida [1].

Additionally, embodiment suggests that the mind is shaped by the body as the former learns experiences from its history and responds to its circumstances. In the context of embodied AI systems, the notion of situatedness relates to the concept because such devices are situated or placed in a specific setting, which may be defined by elements like task, location, or social context. This feature suggests that these AI systems can understand and react to pertinent aspects of their environment. Furthermore, they can consider the objectives and guidelines that are set for them by the AI system itself as well as by other parties like human operators or users. For a better understanding, we can further classify situatedness based on place, task, social context etc. The AI operates on physical and virtual platforms. They include rooms, industry space or even digital space. For better navigation, the system must have a thorough idea about the location and its place in which present. The range of AI tasks is wide and includes a multitude of functions to meet various applications and demands. AI systems do exceptionally well at fundamental tasks such as data analysis, language processing, image processing and picture recognition. As we move up the complexity ladder, AI systems may be able to perform activities that are ever more sophisticated, such as speaking human languages, playing strategic video games, and even coming to their own conclusions in unpredictable situations. Object recognition and manipulation tasks are pivotal components of AI systems, contributing significantly to their functionality and versatility. Object recognition, a foundational task, enables AI systems to comprehend their environment by identifying and categorizing objects through visual features. This process, facilitated by techniques like computer vision and neural networks, is essential for applications ranging from autonomous vehicles navigating traffic to surveillance systems detecting potential threats. The categorization of visual elements provides a structured understanding of the surroundings, laying the groundwork for more advanced tasks as AI systems learn to associate objects with specific attributes and behaviors. Complementing object recognition, manipulation tasks involve the physical interaction of AI systems with objects, expanding their capabilities beyond passive observation to active engagement. This capability, seen in scenarios such as robotic arms assembling products or virtual assistants performing user interface actions, highlights the adaptability and versatility of AI systems. The classical machine learning models when coupled with the deep learning models were found to exhibit proficient results [2].

Manipulation tasks facilitate physical movement, picking up, and manipulation of objects, demonstrating the system's dynamic responsiveness to its environment. Together, recognizing objects and manipulation tasks serve as key building blocks for sophisticated functionality within AI systems. The foundation of higher-level cognitive processes is formed by the knowledge obtained by identifying items and the abilities acquired via their manipulation. Artificial intelligence systems may learn and adapt to various contexts by performing these fundamental functions. For instance, a robot that has been taught to identify and use tools in a controlled environment can use these newly acquired abilities in a range of situations. Furthermore, the data gathered from these assignments enhances the system's general contextual awareness, which helps make complicated decisions. The past features and future futures of embodied AI were put forward in Pfeifer [3] and Holland [4].

The AI system's ability to make well-informed decisions is demonstrated by the integration of visual signals and physical encounters, demonstrating the comprehensive nature of its cognitive powers. On the other hand, AI systems demonstrate outstanding skills in intricate decision-making procedures. This might include resource allocation, strategic planning, or even moral issues. The systems' intelligence and flexibility can be seen by their capacity to navigate across complex decision spaces. The ability of an AI system to understand the nuances of the tasks that are given to it as well as the environment in which these activities take place is critical to its efficacy. Understanding the local surroundings is simply one aspect of contextual knowledge; another involves analyzing the larger conditions that might impact the way a task is carried out systems constantly collect data from their environment because they have been fitted with sensors and input instruments. The system's capacity to quickly evaluate this data and modify its decisions and actions in response to modifications in the environment is essential for real-time adaptation. Acceptance tasks require not only simply performing things out; additionally, they involve linking operations against broader goals. Irrespective of have significance of whether these objectives have been selected by developers, users, or machine learning systems, the AI system's potential for producing substantial outcomes is enhanced whenever it is aware of its targets. The interactions and connections that an AI system has in a

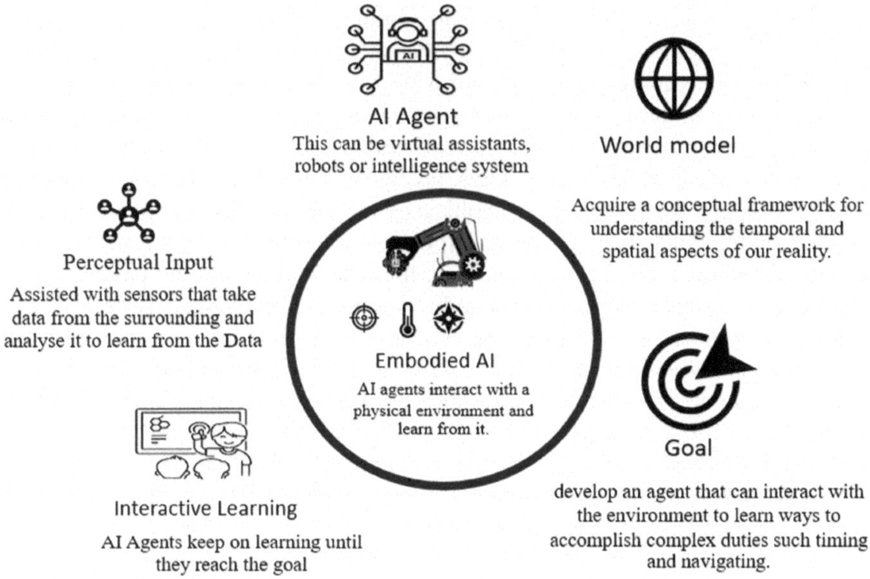

Fig. 1 AI-powered robot learn from data and sensor inputs along with the physical surrounding

social environment. This includes communicating with humans or other AI systems, detecting signs of society, and adjusting their behavior appropriately. In cooperative and collaborative AI circumstances social context is crucial. Embodied AI systems employ their situatedness to improve their effectiveness and agility [4]. Here are additional details on how they utilize this capacity. The AI system utilizes imaging equipment, mics, and other input equipment to process information collected from its immediate surroundings. It applies the analysis of this data to generate a contextual awareness of the world around it. The characteristics of AI-powered robot are depicted in Fig. 1.

The AI system's ability to adjust instantly to stimuli or changes in the world around it derives from its awareness of it. To enable it to perform its allocated duty, this may include making decisions, modifying its actions, or completing specific operations. Goal and Standard Alignment: The AI system attempts to accomplish the objectives and benchmarks specified for it. These objectives may be specified by developers, changed by users, or even eventually picked up by the AI system employing machine learning techniques. In certain circumstances, an AI system might demonstrate signs of self-awareness, grasping its abilities, shortcomings, and ambitions. The artificial intelligence's aptitude for self-awareness permits it adapt and work at its greatest potential under different conditions. However, there are risks and obstacles linked to embodied AI, including technological, social, legal, and moral problems. We have to therefore develop and apply embodied AI to health care with caution, cooperation, and consciousness [4]. A few instances of these issues include how to oversee displayed AI's duty and transparency, protect the security and confidentiality of the information it uses, strike a balance among the obligations of humans and computers in the field of healthcare, and figure out the possibility of biases, errors or harms put on by embodied AI.

2 Embodied Robots for Surgery in Health Care

Embodied AI plays a vital part in the healthcare industry. In the healthcare industry, embodied AI has the potential to transform patient care, diagnostics, and therapy. Embodied AI is based not just on information processing but also on interaction and manipulation of the physical world, allowing them to observe their surroundings, move within their environment, and engage in activities similar to humans and other living organisms. With the use of surgical robots like the da Vinci Surgical System, surgeons may carry out minimally invasive treatments with more control and precision. These instruments translate the surgeon's hand movements into more accurate actions, resulting in smaller incisions, less scarring, and speedier recovery for patients. Researchers hope to give the AI system a physical shape to develop a more adaptable and adaptive machine system capable of moving and understanding the world in the same way that humans can. As a result, tasks like object recognition and manipulation are being built to respond to the external world. Embodied AI systems frequently learn from their fundamental interactions with their environment, and this learning process can be facilitated by a feedback mechanism that allows them to adapt and improve over time. Advancements in robotics, computer vision, and machine learning would aid in the development of the healthcare industry. Embodied AI in surgery involves integrating artificial intelligence technologies with physical instruments and robotic systems to improve surgical processes. The goal of combining AI and surgery is to increase precision, efficiency, and outcomes. Figure 2 depicts the applications of robotics in healthcare.

Here are some of the ways embodied AI is having an impact in the surgical field.

2.1 Robotic Surgery

The da Vinci Surgical System allows surgeons to do minimally invasive procedures using a console that controls robotic arms equipped with surgical equipment. The technology improves flexibility and precision, enabling more delicate and accurate surgical motions [5].

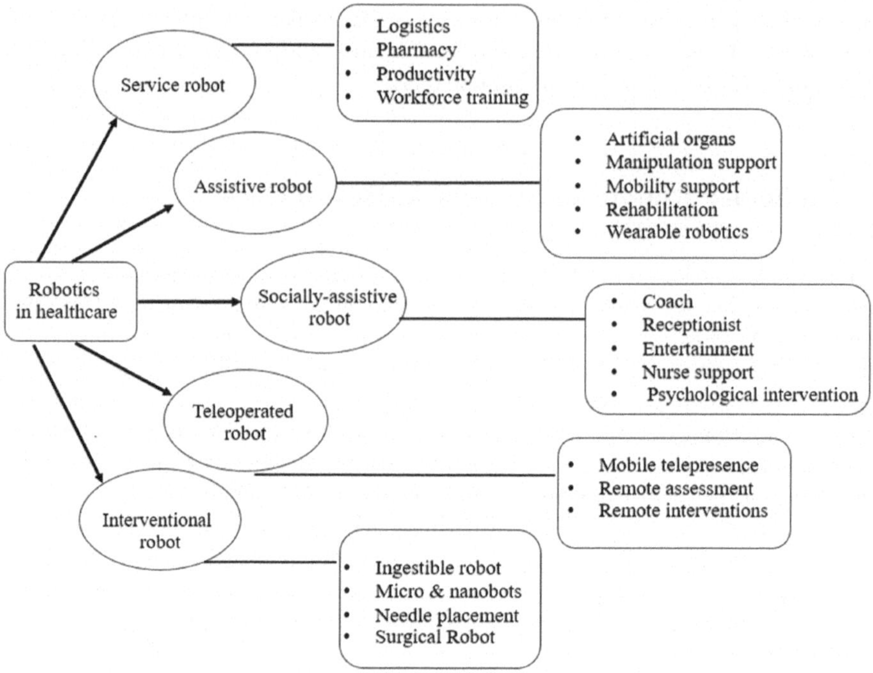

Fig. 2 Robotics in healthcare

2.2 Autonomous Surgical Robots

Some surgical robots include AI algorithms that allow them to function autonomously. These devices may perform certain operations autonomously with real-time adjustments based on the surgical environment.

2.3 Image-Guided Surgery

AI is used in image-guided surgery to improve navigation and vision while performing surgeries. Surgeons can perform procedures more accurately when they use computer vision algorithms to examine preoperative imaging data. This is especially helpful for intricate operations and treatments requiring fragile structures.

2.4 *Surgical Skill Assessment*

AI is being utilized in surgical skill assessment to evaluate and enhance surgical abilities. Reliability, efficiency, and precision can be objectively assessed using computer vision algorithms by analysing a surgeon's actions during practice or real procedures. This promotes training and ongoing development.

2.5 *VR and AR Technology*

Planning and training procedures in surgery are facilitated by AI-driven virtual reality (VR) and augmented reality (AR) technologies. Before doing surgery on real patients, doctors can train on artificial patients and fully immerse themselves in simulated environments.

2.6 *Smart Surgical Instruments*

Surgeons can get real-time feedback during procedures via instruments that have AI built into them. To assist surgeons in navigating intricate anatomical systems more safely, smart scalpels with sensors and AI algorithms installed can identify different types of tissue.

2.7 *Analytics for Surgical Outcome*

To forecast the results of surgeries, AI systems examine enormous patient data sets. Surgeons can use this information to make better decisions regarding the best course of action and anticipated outcomes for a given patient.

2.8 *Tele-Surgery*

Surgeons can now do tele surgery from a distance thanks to robotic equipment powered by artificial intelligence. In situations where specialists may not be on-site, such as during emergencies, this can be very helpful in bringing surgical expertise to neglected areas [6] (Fig. 3).

One of the recognized robotic surgical techniques is Intuitive Surgical Inc.'s introduction of robotically assisted minimally invasive surgery. Compared to conventional minimally invasive surgery and open surgery, this approach offers more advantages.

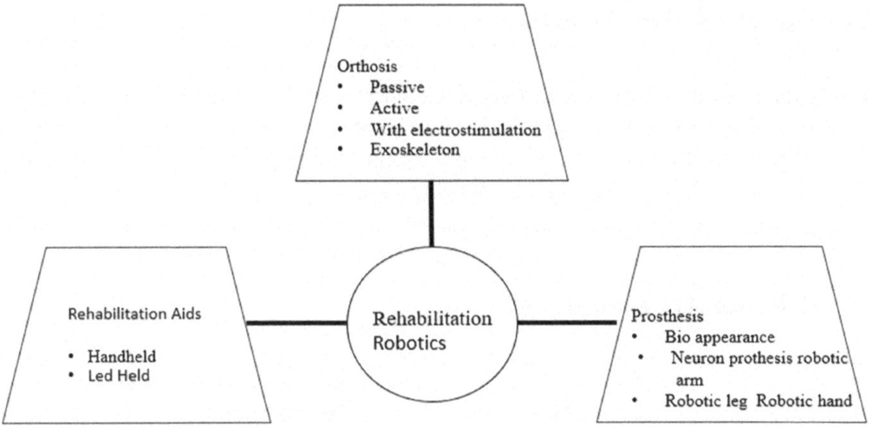

Fig. 3 Application of rehabilitation robotics

Improved patient outcomes were achieved with this kind of surgery [5]. Robots have allowed surgeons to experiment with both increasing and decreasing their physical actions in terms of their ability to see and hear during surgery.

Several modifications were seen when surgeons began employing robots, including the following.

2.9 Surgical Site Was Enhanced

The first thing that changed was the surgeons' ability to view the surgical site. The robot would give surgeons three-dimensional vision and high magnification, enabling them to zoom in and out with ease and obtain a clear picture of a particular area.

2.10 Possibility of Robotic Camera Positioning

The second modification is the ability to move the robotic camera to a location that will give a more stable view than the hand-held camera used during laparoscopy.

2.11 Minimal Bleeding

The third modification is the reduced chance of bleeding, which allows the surgeons to see the patient's anatomy more clearly because there is less blood on it.

2.12 *Increasing Dexterity*

The fourth advancement in robotic surgery is the use of robots to facilitate more secure and easier surgery by allowing surgeons to slice tissues with greater care.

2.13 *Increased Control Over Tools*

With robots, surgeons could get increased control over tools. While in open surgery, surgeons need to handle tissues, cameras, and other instruments. So, with the help of robotic arms, surgery could be done very easily [7].

Robots classified as humanoids, companion robots, and telepresence robots [8] are the three types of robotic devices often utilized in surgery. Healthcare could greatly benefit from embodied AI, but there are also moral and legal issues to be resolved, including patient privacy, data security, and the need for clear regulations regarding the application of AI to medical decision-making. Robotic surgical systems have a significant initial cost as well as continuous maintenance and training requirements. Robotic system maintenance and repair costs might be high in addition to the initial purchase cost. Because robotic surgery depends so heavily on technology, there is a considerable danger associated with any malfunction or technological problem that arises during the process. There may be restricted access to cutting-edge robotic surgical technologies in particular areas or smaller healthcare facilities. It will necessitate cooperation between engineers, regulatory agencies, industry stakeholders, and healthcare specialists to address these issues. To overcome these obstacles and maximize the advantages of robotic surgery in terms of bettering patient outcomes, ongoing research, technological developments, and extensive training programs are necessary [9].

3 Embodied Healthcare Service Robots

Embodied artificial intelligence (AI) is leading the charge to transform healthcare services by bringing about a fundamental change in how patients are cared for and assisted medically. Researches indicate that 20% of the world's population is subjected to physical, intellectual, mental, behaviour illnesses and causes hindrances in their daily life activities. Worldwide medical assistance provided by healthcare practitioners can be more individualized, effective, and efficient by combining AI technologies with physical embodiments like wearables, robots and virtual agents. Embodied AI has brought in technologically advanced machine learning, natural language processing which aids better interaction with patients more empathetically and intelligently. The revolutionary approach assists healthcare workers with real-time data for monitoring and diagnosis, remote consultations with contagious disease-affected patients.

Diving deep into the knowledge embodied AI robots, which builds connections between people and technology in different departments of healthcare. AI embodiment can be classified into four aspects namely: Expediting, Equipping, Emancipation, Expansion. Expediting involves reducing the labour-intensive physical work. Equipping supports in reducing the intellectual thinking done by patient medical assisting professional. Emancipation indicates reducing the work load of the humans by independently completing it. Expansion scales the work done by humans to greater precision and efficiency for better perfection [10]. As per the ISO definition, service robots perform useful tasks for humans or equipment, excluding industrial automation applications. In Healthcare service robots (HSR), it provides vital assistance and cautious delivery for patients via medical workers in multiple scenarios. These robots can reduce the human errors, keep track of the medicines, precise cleaning or disinfection of equipment/hospital rooms and so on. During the pandemic, it served the mankind from dangerous tasks such as remotely monitoring the patients with minimal interference from health workers, increasing hygiene and reducing the possibility of infections [11]. The HSRs can be differentiated into robots involved exclusively in medical tasks, non-medical/ routine tasks or on the basis of level of automation required. The categories namely are delivery robots, automated dispensary robots, remote inpatient or outpatient care robots, telepresence robots, disinfectant robots and general cleaning robots [11].

Routine task robots are used either to assist a medical staff or to completely replace them in a few tasks namely: (1) to deliver medicines and move equipment in hospital, (2) medical dispensaries in pharmacies, (3) used in administration dept for healthcare management. Subcategorizing the delivery robots, automated navigating robots dramatically reduces the risk and improves efficiency of distribution of goods and medicines. Serving robots carry intense physical duties which has to be carried out with utmost care such as dispensing medical wastes, washing contaminated materials in the hospital and so on. Drones are typically used outside medical requirements. But due to advent of innovation across different industries, they are now used to transport medicines, life support system, vaccines, blood samples. Autonomous drones use the GPS and parameters from different sensors to fly across to the intended location at the earliest [11]. Automated dispensing robots are exponentially being used in hospitals and homes since few years. Medication at the right amount at the right time is crucial for a better result. In the absence of medical workers, automated dispensing robot will enact as an extension of them. They reduce the maintenance cost with faster and better service. Most importantly, they gained a strong significance during the COVID-19 pandemic by providing vitals to the patients in quarantine. Healthcare administration started to implement robots to guide patients from waiting lounge or helpdesk to the desired doctor as per the requirement. They are also used to pass on information to different units within the hospital [11].

Telepresence robots are created with a motive to bring physical presence with humans at a distant location. Uninterrupted feeds of live audio and video from both the sides help to carry seamless interaction. Older adults affected by loneliness, physical challenges and social isolation can be benefitted from telepresence robots. Based on the user, isolated surroundings in which they function, it is majorly divided into two types namely: (1) Remote Inpatient Care (RIC), (2) Remote Outpatient Care (ROC). RIC robots are mainly used for frequent consultation and interaction with the patients by monitoring their vitals at regular interval within the hospital environment remotely where the primary user is doctor. ROC robots on the other hand still have a doctor as primary user but they are used in isolated surroundings especially at patient's home. It interacts with the caretakers as well for a smooth functional operation [11].

Disinfectant robots grew into priority since the outbreak of pandemic. Due to the imbalance ratio of patients to health workers, introduction of these robots were crucial in maintaining social distance. Disinfectant robots are broadly classified into two subcategories namely: (1) Infection prevention from infected patients through medical equipment and workspace using sterilization robots, (2) general cleaning of hospital environment using disinfectant robots. In infection prevention robots, exposure to a new pathogen or bacteria can possess serious threats to the health of humans. In order to ensure an effective disinfection of medical equipment at minimum cost and time, such robots are used. They preferrable perform the disinfection using an ultraviolet radiation which has the ability to kill viruses and bacteria without causing infection to the patient. In General cleaning robots can clean from the door knob to washrooms with utmost without leaving any spots untouched. They play an irreplaceable role in preventing the spread of fatal diseases like COVID-19 [11].

4 Embodied Assistive Robots in Healthcare Sector

The cognitive rehabilitation therapy was proposed in Petersen [12], and learning activities were deployed to improve the neurocognitive skills. The socially interactive robots and socially assistive robots with virtual reality and embodied features are utilized. The robotic systems employed for cognitive rehabilitation was found to be proficient in the patient recovery and retention of cognitive skills. The article [13] systematically reviews recent progress in robot-assisted cognitive training for individuals with cognitive disabilities. It examines 99 articles, providing insights into applications, technologies, experiments, and products in this field. The review includes a meta-analysis of articles assessing robot-assisted cognitive training protocols for individuals with cognitive disabilities. The analysis reveals notable limitations in current cognitive rehabilitation robotics, including small sample sizes, non-standardized training measurements, and uncontrollable factors. The field faces diverse challenges, including ethical concerns, user-centered design, system reliability, trustworthiness, cost-effectiveness, and personalization of robot-assisted cognitive training. The article underscores the need to consider human–robot collaboration and social cognition in future research to enhance natural interactions in this area [13].

In Lee et al. [14], the authors explore the potential of Artificial Intelligence (AI) and robotic coaches in enhancing patient engagement during rehabilitation exercises through social interaction. Despite previous efforts in monitoring exercises, deploying these systems remains challenging, partly due to a lack of involving stakeholders in designing functionalities. The paper outlines the authors' initiative to gather detailed design specifications for effective and acceptable interactions between AI/robotic coaches and patients. Through iterative questionnaires and interviews involving four therapists and five post-stroke survivors, the study highlights the perceived benefits of these technologies, such as systematic management, improved self-efficacy, and motivation in rehabilitation therapy. However, practical concerns, including difficulties for individuals with cognitive impairment and system failures, were also identified. The paper underscores the importance of early stakeholder involvement and interactive techniques to address system failures and enhance personalized therapy sessions for the successful deployment of AI and robotic exercise coaches [14].

The study [15] examines the potential of a socially interactive robot as a tool for monitoring and promoting cognitive activities in the elderly and individuals with dementia. The research aims to demonstrate the advantages of the robot's physical embodiment in fostering human–robot social interactions. The social therapist robot is designed to deliver personalized cognitive stimulation, particularly through engaging in a music game with the user. Findings from the 8-month pilot study indicate that participants experienced more efficient, natural, and preferred interactions with the physical robot compared to a simulated counterpart [15].

For over three decades, significant research and investment have focused on robotic technologies for rehabilitating motor impairments resulting from neurological injuries. However, these devices have yet to demonstrate clear superiority over conventional therapy in restoring patient function. Nonetheless, robots play a valuable role in reducing the manual effort required by physical therapists to deliver intensive interventions. The typical features of rehabilitation robot is depicted in Fig. 4.

In most robotic systems, therapists act as supervisors, overseeing and initiating robot control algorithms to achieve therapeutic goals, while adaptive algorithms handle low-level interactions between the robot and the patient, enabling progressive therapy. This perspective examines the role of physical therapists in controlling rehabilitation robotics and explores the potential benefits of integrating therapists into lower-level control loops to improve rehabilitation outcomes. It discusses how automated robotic systems, while capable of providing consistent patterns of physical interaction, may hinder the promotion of neuroplastic changes essential for sensorimotor learning retention and generalization in patients. The article assesses the advantages and drawbacks of allowing therapists to physically interact with patients through real-time control of robotic rehabilitation systems and delves into the concept of trust in human–robot interaction within patient-robot-therapist relationships. It concludes by outlining several open questions to guide the future development of therapist-in-the-loop rehabilitation robotics, including considerations regarding the

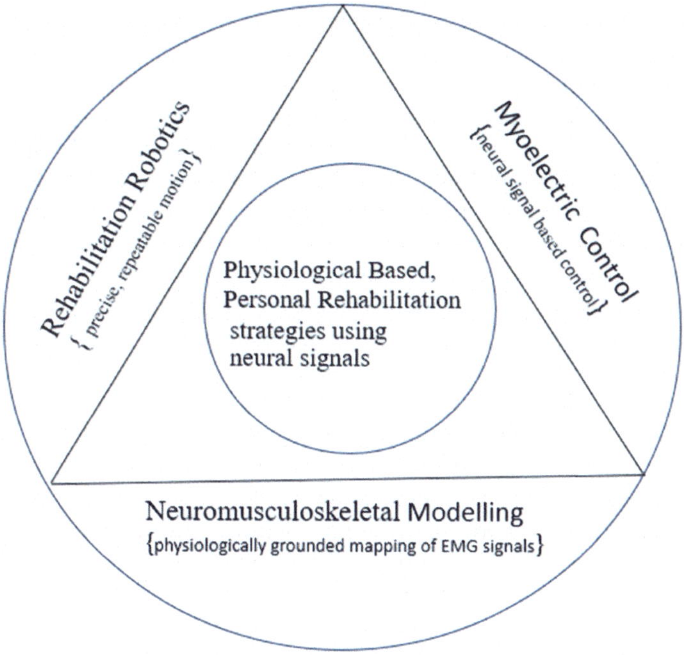

Fig. 4 Typical features of the rehabilitation robot

degree of control therapists should have and potential methods for robotic systems to learn from therapist-patient interactions [16].

A substantial portion of the population faces limitations in physical capabilities, falling into three main categories: those with congenital disabilities, individuals whose capabilities have been impacted by illness or accidents, and older people experiencing a natural age-related decline. Additionally, individuals engaged in strenuous and repetitive tasks can also benefit from enhanced capabilities. The researchers collaborate with doctors, scientists, designers, and industry to develop innovative robotic systems to address these challenges. This encompasses projects such as assistive modular robotic exoskeletons, enhancive modular robotic exoskeletons, and wearable soft robotics designed to promote independent living. Stroke stands as the leading cause of adult disability in the world, significantly impacting the mobility, independence, and overall well-being of patients. The extent of recovery hinges on the frequency and intensity of therapeutic interventions. However, due to constraints in physiotherapy resources, meeting the National Institute for Health and Care Excellence's (NICE) recommendation of 45 min of daily exercise is often challenging. Robotic systems offer a promising solution by providing increased accessibility to therapeutic exercises not only during the acute stages in hospital settings but also in the long term, within community stroke units and homes. These systems are capable of executing repetitive exercise tasks crucial for maximizing a patient's recovery potential while simultaneously collecting quantitative data on performance and progress. Although stroke is a primary focus due to its prevalence and the associated cost to healthcare systems, the application of such technology extends beyond stroke, benefiting children with conditions like Cerebral Palsy, Development Coordination Disorders, and other neuromuscular disorders.

The iPAM is a dual robotic system crafted to deliver assistive therapeutic exercises for the upper limb, replicating the guidance provided by a physiotherapist to a patient's arm [17]. Tailored for application in acute stroke services within hospital settings, iPAM aims to offer additional upper-limb exercises during inpatient care. On the other hand, MyPAM is a planar joystick system specifically created to provide targeted assistive exercises for children with Cerebral Palsy and adults recovering from stroke. This system is designed for use in community settings, such as schools, community stroke units, and patients' homes, enhancing access and engagement with therapeutic interventions. A crucial aspect of the development process has been the user-centered design, involving patient groups, physiotherapists, and clinicians.

Numerous individuals globally undergo amputation, often resulting from circulatory issues, vascular problems, complications related to diabetes or cancer, or traumatic incidents. To offer more advanced prosthetics for these amputees, a comprehensive understanding of human locomotion is essential. Human locomotion is characterized by its adaptability and resilience in response to varying circumstances. Whether it involves standing, walking, turning, navigating stairs or ramps, or sitting, an intelligent sensory-motor control system is crucial for the lower limb segments and body center of mass. This system ensures the adaptability of prosthetics to different situations, enhancing the overall functionality for individuals with limb loss.

Children with deficient movement skills often exhibit lower levels of physical activity, reduced fitness, and limited engagement in social and sports activities. In addition to hindering educational progress, poor movement skills can contribute to the risk of children becoming overweight. Unaddressed movement difficulties in childhood can lead to serious health and social problems in adulthood, with a staggering 80% of such children experiencing ongoing issues. Advancements in computational neuroscience have enabled the creation of models that predict human learning, emphasizing the significance of feedback and feed-forward mechanisms in skill acquisition. These models shed light on why some children struggle with critical motor skills like handwriting. Embodied cognition theories propose that motor activity and perception play a foundational role in developing more abstract cognitive skills, influencing overall academic achievement. Computational models of human learning indicate that reducing noise in feedback signals and increasing the amount of useful information they provide can expedite learning processes [18]. This insight explains why both adults and children can benefit from robotic therapies that enhance feedback, particularly in refining fine motor skills.

The preparedness of primary care to embrace e-health technologies, as investigated in a qualitative study conducted within a London primary care trust, relies on various factors such as infrastructure, staff training, attitudes, and patient involvement [19]. Essential themes encompass the quality of IT infrastructure, sufficiency of training and assistance, staff perspectives on e-health, patient engagement considerations, and the impact of organizational culture. Grasping these elements can guide efforts to effectively introduce e-health initiatives in primary care [19].

5 Conclusion

Robots equipped with AI technologies are used in physical therapy to assist patients in restoring motor function and enhancing muscle strength after neurological injuries or strokes. AI assists radiologists in analyzing medical images, such as X-rays, MRIs, and CT scans. It aids in anomaly detection, and organ segmentation, and enhances diagnostic accuracy is applied to digital pathology, aiding in the examination of histopathology slides and improving the detection and classification of diseases like cancer. Robots are employed in healthcare settings for patient monitoring, medication administration, and even providing companionship to patients. AI-driven cameras and sensors can identify instances of patients falling and alert healthcare providers to prevent injuries. Embodied AI-driven virtual agents offer therapeutic interventions and cognitive rehabilitation for individuals with mental health conditions. Drones with embodied AI capabilities can deliver vital medical supplies, defibrillators, or life-saving equipment to remote or disaster-stricken areas is used for the analysis of extensive datasets, expediting the drug discovery process by identifying potential drug candidates and predicting their effects on various disease systems predict disease outbreaks, patient readmissions, and healthcare resource demand, facilitating proactive healthcare planning.

References

1. Pfeifer, R., Iida, F.: Embodied artificial intelligence: trends and challenges. Lect. Notes Comput. Sci. **1**, 1–26 (2004)
2. Moulin-Frier, C., Puigbo, J.Y., Arsiwalla, X.D., Sanchez-Fibla, M., Verschure, P.F.: Embodied artificial intelligence through distributed adaptive control: an integrated framework. In: 2017 Joint IEEE International Conference on Development and Learning and Epigenetic Robotics (ICDL-EpiRob), pp. 324–330. IEEE (2017)
3. Pfeifer, R.: Embodied artificial intelligence 10 years back, 10 years forward. In: Informatics: 10 Years Back, 10 Years Ahead, pp. 294–310. Springer, Heidelberg (2001)
4. Holland, O.: The future of embodied artificial intelligence: machine consciousness? In: Embodied Artificial Intelligence: International Seminar, Dagstuhl Castle, Germany, pp. 37–53. Revised Papers 2004 July 8. Springer, Heidelberg (2003)
5. Drust, W.A.: Embodied care, spatiality, and the ethics of robotic surgery. Doctoral dissertation, Drexel University May (2015). https://doi.org/10.17918/etd-6566

6. Okamura, A.M., Matarić, M.J., Christensen, H.I.: Medical and health-care robotics. IEEE Robot. Autom. Mag. **17**(3), 26–37 (2010). https://doi.org/10.1109/MRA.2010.937861
7. Sergeeva, A.V., Faraj, S., Huysman, M.: Losing touch: an embodiment perspective on coordination in robotic surgery. Organ. Sci. **31**(5), 1248–1271 (2020)
8. Cesari, V., Melfi, F., Gemignani, A., Menicucci, D.: Sensory Substitution Increases Robotic Surgical Performance and Sets the Groundwork for the Mediating Role of the Sense of Embodiment: A Systematic Review. Heliyon (2023)
9. Wahlström, M., Seppänen, L., Norros, L., Aaltonen, I., Riikonen, J.: Resilience through interpretive practice: a study of robotic surgery. Saf. Sci. **108**, 113–128 (2018)
10. Pee, L.G., Pan, S.L., Cui, L.: Artificial Intelligence in healthcare robots: a social informatics study of knowledge embodiment. J. Am. Soc. Inf. Sci. **70**(4), 351–369 (2019)
11. Fosch-Villaronga, E., Drukarch, H.: On healthcare robots: concepts, definitions, and considerations for healthcare robot governance (2021). arXiv:2106.03468
12. Petersen, S.W.: Robots for cognitive rehabilitation and symptom management. In: Rehabilitation Robotics, pp. 267–275. Academic Press (2018)
13. Yuan, F., Klavon, E., Liu, Z., Lopez, R.P., Zhao, X.: A systematic review of robotic rehabilitation for cognitive training. Front. Robot. AI **8**, 605715 (2021)
14. Lee, M.H., Siewiorek, D.P., Smailagic, A., Bernardino, A., Badia, S.B.: Enabling AI and robotic coaches for physical rehabilitation therapy: iterative design and evaluation with therapists and post-stroke survivors. Int. J. Soc. Robot. 1–22 (2022)
15. Tapus, A., Tapus, C., Mataric, M.: The role of physical embodiment of a therapist robot for individuals with cognitive impairments. In: RO-MAN 2009-The 18th IEEE International Symposium on Robot and Human Interactive Communication, pp. 103–107. IEEE (2009)
16. Hasson, C.J., Manczurowsky, J., Collins, E.C., Yarossi, M.: Neurorehabilitation robotics: how much control should therapists have? Front. Hum. Neurosci. **17**, 1179418 (2023)
17. https://robotics.leeds.ac.uk/research/medical-robotics/assistive-rehabilitation
18. Amershi, S., Cakmak, M., Knox, W.B., Kulesza, T.: Power to the people: the role of humans in interactive machine learning. AI Mag. **35**(4), 105–120 (2014)
19. Mannan, R., Murphy, J., Jones, M.: Is primary care ready to embrace e-health? A qualitative study of staff in a London primary care trust. Inform. Prim. Care **14**(2), (2006)

Dr. S. N. Kumar is currently an Associate. Professor, Dept. of Electrical and Electronics Engineering., Amal Jyothi College of Engineering, Kanjirappally, Kerala. He was awarded Ph.D. in Electronics and Communication Engineering from Sathyabama Institute of Science and Technology, Chennai. He has 15 years of teaching experience, and his areas of interest include Medical Image Processing and Embedded System Applications in Telemedicine. He was the Co- Principal Investigator of the DST IDP-funded project and research fellow of the RCA scheme, NTU Singapore. He has edited 3 books, authored 32 book chapters and 30 publications in peer reviewed journals. He had authored 12 text books in engineering discipline. He is an active member IEEE PES.

Jomin Joy is pursuing bachelor's degree in electrical engineering at Amal Jyothi College of Engineering, undoubtedly a commendable endeavour. Currently holding the position of Chair at the IEEE Power and Energy Society in Amal Jyothi College, Jomin showcases exceptional leadership skills and a deep understanding of the dynamic field of power and energy.

Alen J. James Automation and validation engineer at Tata Elxsi Limited, had work experience of 1 year in automation and validation of hardware and software HMI interfaces, profoundly experienced in CAN and J1939 communication protocols. Area of interest primarily lies in industrial automation. Biomedical instrumentation has also given a deep insight about the new technologies updated to integrate with the existing systems.

Andrew Dixen Vehicle systems engineer at Tata Elxsi Limited, Had work experience of 1.5 years in requirement engineering, system and software architectures, ISO standards implementations (15118), feature failure analysis of systems (FFA). Area of interest lies in automotive domain specifically electric vehicles.

Computing, Clouds, Analytics and Artificial Intelligence at the Edge

Agha Urfi Mirza and Praveen Kumar Chandapeta

Abstract Edge computing is the emerging technique which provides solution for issues by mitigating the transmission of data, enhancement of latency, and reducing the pressures using the cloud. Edge computing became the pioneer for new technologies and also paved the way for updating of various state-of-art works for different applications. It also enhances the abilities of end device hardware and extensive usage of virtual technologies. The framework of edge computing is added advantage to the cloud and in some places, it replaces the cloud role also. However, the edge computing does not completely replace the cloud because only cloud can be applicable for the computation intensive tasks which depend on the higher storage and power which cannot be provided by the Edge computing. Embodied AI real-world navigation is the process of developing systems capable of interacting with the physical environment via sensors, cameras, and actuators, such as robots, drones, or mixed reality headsets. It entails navigating complicated and dynamic settings, reacting to changes and happenings in real time, and exploring unfamiliar and dangerous territory. It also has uses in a variety of fields, including distribution, disaster assistance, entertainment, and education. Hence, much application utilizes the integration of both cloud, Embodied AI, and edge computing for better performance and also brings more convenient for the users.

A. U. Mirza (✉)
University of Technology and Applied Sciences, College of Computing and Information Sciences, Al Musanna, Oman
e-mail: urfi.mirza@utas.edu.om

P. K. Chandapeta
University of Technology and Applied Sciences, College of Computing and Information Sciences, Salalah, Oman
e-mail: praveen.kc@sct.edu.om

P. Raj et al. (eds.), *Building Embodied AI Systems: The Agents, the Architecture Principles, Challenges, and Application Domains*, Information Systems Engineering and Management 14, https://doi.org/10.1007/978-3-031-68256-8_11

1 Introduction

In the context of computation, edge computing [1] describes a variety of connections and gadgets that are at or close to the operator. The goal of Edge is to analyze input more precisely at the point of generation, allowing for faster and larger interpreting levels and more actionable outcomes in immediate form. The idea is to move computing facilities nearby to the locations where knowledge is produced, as opposed to moving it to central facilities. Data center [2] computation and storage capabilities are placed as near to the content stream as feasible, preferably at an identical place. It controls delicate information to lower the cost of transferring information to the server by using remotely stored information that allows instantaneous flexibility to build fresh sensations. Instantaneous operations can be provided with edge computing the computer's rapid reactivity and quick data handling and evaluation.

In areas [3] like autonomous transportation, automated production, recording devices, and other situational perception, where immediate input is essential, it offers clients a range of immediate reaction options. By evaluating customer satisfaction throughout the global web and using statistics to identify the greatest dependable, low-delay connection channel for each client's data, advanced computing can aid in improving the network's reliability. The 5G [4] Peripheral technological devices, which were created with development in mind, improve apps by putting them farther from the gadgets and endpoints at the outer limits of the wireless connection. It is possible to gather, handle, and interpret knowledge more quickly and thanks to this reduction in delay and efficiency enhancement.

The process of locating processing, information retention, and software assets in immediate access to information sources, such as Internet of Things appliances or nearby systems and the hosts, is known as edge processing [5]. Lowering the actual distance between these two groups accelerates connectivity in general and lowers statistical delay, with features like examine aloud that allow learners to pay attention to websites like audio files and an interactive reader that helps with understanding words, the Edge browser provides the most extensive collection of integrated instructional and access methods available online, it is mostly divided into two categories [6], the initial one is a tiny premises employed by a hosting or sharing source for providing supplementary areas, and the other is a structured location near the edge of the primary infrastructure of an entity.

The primary driver of the market's improvement is the augmentation of 5G activities by the expanding use of connected devices. Greater implementation, the emergence of border artificial intelligence, the creation of edge-native apps, and a greater emphasis on safety and confidentiality are predictions for edge technology [7]. It presents previously unheard-of possibilities as well as obstacles as it prepares to completely transform the equipment and wireless communication environment. An in-depth overview of frontier computing, including the main vendors, innovations, uses, and valuation, is provided by this analysis. Combining numerous internet-connected detectors and gadgets into an infrastructure poses an enormous threat to safety for businesses.

By enabling businesses to handle information privately and maintain it off the grid, computing at the edges reduces this risk. Before an interruption happens, entrepreneurs are required to be competent to analyze and spot variations in the manufacturing process. By moving processing and storing information more connected to the devices, integrated computing is beneficial as a result, immediate statistical analysis and reduced latency devices wellness tracking are made possible by sensors embedded in devices. Developers can streamline manufacturing procedures by using computational power at the edge for handling information gathered by manufacturing process surface detectors.

Developers can identify problems early on, cutting interruptions and increasing productivity, the use of frontier computing in the medical field has made monitoring patients remotely possible. Software on the edge minimizes the amount of information that needs to be transferred, the level of congestion that results, and the length of time the information needs to traverse, this lowers communication expenses and produces fewer delays. The term intelligent edge [8] describes how computational intelligence and computational methods work together to move processing and statistics nearer to the knowledge origin. This is not the same as the conventional Internet computing framework, which sends information to be processed to an obscure or centralized information edge. It makes it possible to apply neural network methods to the advanced equipment that generates information.

Intelligent systems could be made available by Edge Intelligence [9] to all individuals and organizations, wherever they might be. In edge cognitive ability, sophisticated operations are carried out at the boundary using locally-generated information in a dispersed fashion, in contrast to conventional cognitive ability where every record must be submitted to a fundamental cloud platform and list the following four essential elements: edge outsourcing, edge deduction, edge instruction, and edge cache storage. With the help of collaborative computer systems from sophisticated Edge, almost any demand can be handled quickly. Nontechnical users may set up and oversee tasks independently by abstracting communication intricacy, frame automation, and administrative tasks under the intuitive influence of aviation.

Using localized gathered information intelligence from the edge allows devices to make selections without conveying it to a unified cloud or on-premises network. Since it can operate on gathered information to determine actions before information is transmitted to the web, the use of these technologies on the intelligent frontier has the potential to revolutionize the industry. The devices include routers that link openly accessible networks to the global web, a firewall [10] can also function as a router in different scenarios, where it acts as the network's gateway by determining what is allowed onto its resources. Establish reliable methods that securely interact, regulate, and offer statistics on equipment located at the information origin computerization of networks, shift processing to the networks advanced to increase quickness and bandwidth utilization. Anywhere information is being handled or implemented at the extreme ends of the system, including gadgets, detectors, hosts, storms, etc., these kinds of gadgets and parts are all included in advanced construction, which is a distributed systems design.

Today's enterprises can benefit [11] from edge computing advanced capabilities in several ways, including increased efficiency, immediate monitoring, enhanced information protection, as well as expense productivity. The equipment needs to communicate with a wireless connection, but its connection might not be entirely safe. It requires that information, often personally identifiable information, be controlled by the gadget through its growth. Compared to cloud-based services [12], edge servers are situated less distant from the final individual. It offers consumers the advantages of quicker response times, improved safety, and more access to information, as well as opening up a vast array of inventive uses for the coming decades.

Real-time interactions with the environment are made possible by embodied AI at the edge. This is critical for applications like robots, autonomous cars, and smart manufacturing where responsiveness with minimal latency is required. Embodied AI [13, 14] may drastically cut latency by processing data locally on edge devices instead of transferring it to a centralized server. This is particularly significant for applications such as emergency response systems and healthcare monitoring where prompt decision-making is essential. By keeping sensitive data locally, edge computing in combination with embodied AI helps solve privacy concerns. By processing data on the edge device, only pertinent data needs to be transferred, lowering the possibility of data breaches and guaranteeing adherence to privacy laws. Applications requiring flexibility or unpredictability are ideal for edge-embodied AI. AI at the edge enables increased responsiveness and flexibility, for instance, in situations where drones must adjust to changing weather patterns or robots must navigate across shifting terrain. Processing on edge devices may be more specialized and efficient by being customized for certain use cases. In industries like healthcare, where AI models may be built to handle patient data locally and provide individualized and context-aware insights, customization is essential.

By limiting the quantity of information examined and transferred in the cloud, preserving confidential information on personal gadgets, and lowering the possibility of statistics concessions, cutting-edge computing enhances the safety of information. This chapter provides brief elucidation of the Edge computing, collaboration of cloud with the edge, analytics of Edge, and artificial intelligence-based Edge in a wider context in the following section.

2 Various Parameters of Edge Computing

2.1 Trends in Edge Computing

Based on the new technological change round, the edge computing development and growth are inseparable. From its architectural plan to manufacturing landing, an edge computing is developed to help these emerging technologies. For these technologies, the leapfrog developments are achieved with the opportunities to provide few standardization system and edge computing technology maturity at the same time. The other technologies developments integrate the edge computing future development. The aspects of 5G + edge computing, edge cloud collaboration, edge intelligence, heterogeneous computing and so on are the edge computing trend summarized in the below sub section.

2.1.1 5G + Edge Computing

Based on edge computing, various advanced technologies relied on the characteristics of ultra-low latency, large connection and ultra-high speed. According to 5G era, the data explosion problem is alleviated and the significant component of 5G network is edge computing. For the industrial deployment, good network foundation is provided with 5G due to the edge computing industry development. Based on the collaboration among the two depending upon the 5G major scenarios, the network capabilities development and complementary is edge computing and 5G development.

2.1.2 Collaborative Edge Cloud

The cloud computing be able to accompaniment each other and cloud computing extension is edge computing. The analysis of short-cycle intelligent, real-time and field-level performs edge computing. The artificial intelligence relates to the same face of applications. In the cloud, placed on the edge are fast response and compute-intensive tasks are deployed. The consumption of network bandwidth is minimized and the statistics sent to the cloud is preprocessed at the same time. The costs of network bandwidth and computing costs are minimized and satisfy diverse requirements on the cloud and the edge via collaborative computing. The two technologies are developed with the huge boost not only cloud computing and edge computing development are coordinated. To develop other technologies, driving force is provided.

2.1.3 Intelligence on Edge

The technologies of artificial intelligence to the edge are pushed by edge computing to utilize edge intelligence that has artificial intelligence expression and application. The scenario of more terminal application appears the artificial intelligence and hardware computing empowered improvement. The richer data faster are obtained with the edge nodes deploy by artificial intelligence with the response delays is reduced. The artificial intelligence application scenarios are expanded. The resource scheduling decisions at edge side is optimized with the technology of artificial intelligence and utilized via edge computing. More efficient services offered by users and the business scope are expanded with the edge computing. In future society, an important technology is edge intelligence. The edge intelligence and edge computing Intelligence are the applications for next level industrial growth.

2.1.4 Heterogeneous Computing

The algorithms with the different computing met the structure and different performance with the machines that utilize the heterogeneous computing. The algorithms with the heterogeneous platforms provided maximum overall performance. In the future, diverse data processing and heterogeneous computing platform demand met the approach. In edge services, differentiated applications and fragmented data processing required to satisfy by introducing heterogeneous computing. The resource utilization improving computing depends on the computing resources support flexible scheduling.

2.2 Edge Computing Based on Industrial Applications

Each edge node with the distributed mutual assistance sharing representation are used to centralized resource sharing model. The cloud computing centralized with edge computing contrasted by cloud computing. Proposed different solutions contain end–edge–cloud and at three different levels, locate the service operators in the industry.

Edge computing to the vehicle, IoT GW, CPE/Terminal

To provide surrounding nodes or upper-level node, the macro network are connected and the processing calculation nearby in time is performed with near-end computing service.

The traffic road administration/enterprise industry forms the self-organizing network or network form of edge computing

Based on the technology protocol stack, third type fixed-operator networks or fourth type of mobile operator is similar to decompose expand with the computing services. The free networking is Wi-Fi or LoRa. Same to the category of sixth/second/first is protocol stack in which the terminal forms, Internet of Things site service providers and cloud service providers with the implementation.

The computing services + mobile connectivity based on mobile operator centric edge computing

In 2014, the mobile edge computing is edge computing. At the mobile base station, locally cache at early scenario. Commercially available evolving is multicast, IPTV, docking to fixed networks, acceleration and content caching with local distribution and gateway PGW of packet mobile. The 5G network architecture combines multi-access edge computing. Different improvements to UPF and UPF position deploy are international standard SA2 of 3GPP. The multiple access and fixed-mobile convergence move to with 5G.

The computing services + fixed connection of edge computing based on fixed operator

The Industrial Internet extensively interacts and Huawei's fixed network that initiates the fixed network edge computing is Edge Computing Alliance of china. The edge equipment is used as fixed access and switching equipment enterprises with major scenarios.

The computing services + site and cloud service drainage

Same to many neutral sites provides with China tower corporation concept that establishes the dozen companies. The containers and machine rooms of various mechanisms with the connecting edge site are used towards network. Based on the entrance of the cloud ecology, the Huawei public cloud edge computing, Microsoft Azure Stack, Google GKE OnPrem, Baidu and Tencent of edge cloud nodes is few examples.

2.3 Edge Intelligence and Intelligent Edge

Because of natural language processing and deep learning, several aspects of people's live changes applications and different deep learning oriented intelligent services as the typical extensively secondhand novel applications form. The computing power and increasing data link to the AI evolution. The Internet of Vehicles and smart cities of application scenarios are wider range. Due to following represented factors that offers intelligent services as limited number.

Privacy: Several private information carried with data needed for AI and critical to areas like smart cities and home via privacy issues.

Reliability: While connecting users to services, the backbone networks and wireless communications relied on most cloud computing applications. While the loss of network connections, highly reliable as intelligent services and industrial scenarios.

Latency: Several time-critical applications requirements are satisfied and it not guaranteed to access cloud services delay.

Cost: Huge amount of data to the cloud transmission or cloud requires devices based on AI models inference and training that demonstrates huge network bandwidth consumption.

Figure 1 represents the structural diagram of intelligent edge and edge intelligence. Several issues are solved to expect edge computing in which the closer to users to the cloud is edge. The edge intelligence is the part of intelligent edge based on AI services and it major aim is edge intelligence. From the cloud to the edge, the AI computations are pushed to expect the edge intelligence thereby enabling reliable, low-latency and various distributed intelligent services. The management and adaptive edge maintenance for dynamics by AI incorporated with the intelligent

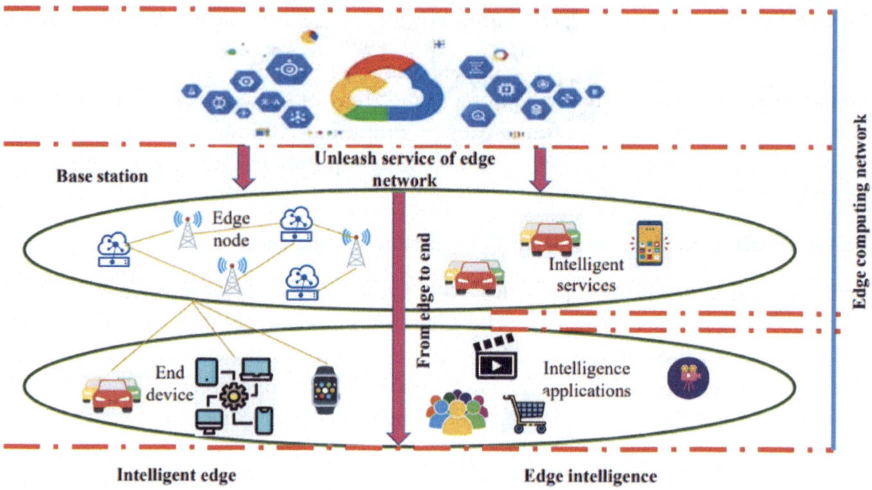

Fig. 1 The structural diagram of intelligent edge and edge intelligence

edge. More diverse is network access methods by the growth of communication technology. The connection is made among cloud more reliable and ubiquitous end devices in which the intermediate medium is acted as the infrastructure of edge computing. The shared resources community gradually merging cloud, edge and end devices.

A complex overall architecture including storage, computing, networking and wireless communication is management and maintenance. The fixed mathematical models rely on the methodologies of typical network optimization. The systems and edge network environments are changed rapidly with the accurate modelling. The cumbersome network information and complexities faces the important technology of AI. The intelligent edge and edge intelligence are considered with multiple aspects with few important technologies as given below,

Various edges computing networks and different functions manage and maintain via AI applications like computation offloading and edge caching.

Under privacy and resource constraints, distributed edge devices of edge intelligence for training AI models.

According to the software, hardware, network structure to assist AI computation, the edge computing platform adapted edge computing for AI.

Different requirements like latency and accuracy are fulfilled with the edge computing architecture that focus on the inference and practical deployment of AI.

Intelligent services are provided with the AI and edge computing organized as technical frameworks.

2.4 *Edge Computing Fundamentals*

Due to facilitation cloud computing pressure, service latency improvement and data transmission reduction, the significant resolution to disruption and the emerging technologies impose edge computing. Few existing technologies like virtualization technologies with application scenarios and the ability of end device hardware enhancement via few existing technologies in which the enlarger new technologies spawn as edge computing emergence. In some scenarios, the cloud role replaces and the imperative complement to the cloud is the architecture of edge computing. Due to computation-intensive tasks, edge computing will not totally replace with the cloud computing status. An additional suitable knowledge to users brought and several diverse application scenarios needs are satisfied with the amalgamation of edge and cloud computing.

2.4.1 Edge Computing Paradigms

At the network edge, different new technologies arrived due to the growth of edge computing based on similar concepts with various emphases like Multi-access Edge Computing, Mobile Edge Computing, Fog Computing, Micro Data Centers and Cloudlet. The edge computing protocols and architectures with standardized definitions consensus not reach the edge computing community. This section differentiated and introduced various concepts of edge computing. Table 1 displays the mobile edge computing, Cloudlet and fog computing features comparison.

Table 1 State-of-art evaluation of mobile edge computing, cloudlet and fog computing

Features	Mobile edge computing	Fog computing	Cloudlet
Instance of node device	Edge server running on the base station	IoT gateways, switches, set-top boxes and Routers	Data center in a box
Location of node	Macro base station or radio network controller	Between the cloud and end device	Outdoor or local installation
Mechanism to access	Mobile network	Mobile network, Wi-Fi and Bluetooth	Wi-Fi
Accessibility	Single hop	Multiple or single hop	Single hop
Communication based on inter node	Partial support	Support	Partial support
Context aware	High	Mid	Low

2.4.2 Edge Computing Definition

In the most of the existing works, there is an ambiguous of edge devices division and definition. The edge nodes and edge devices are the division of general edge devices as shown in Fig. 1. The edge nodes like MEC servers, edge servers, Fog nodes, RSUs and Cloudlets in which at the network edge, deploy the server.

2.4.3 Collaboration End to Edge Cloud Computing

The computation-intensive tasks processed, creates cloud computing. Based on the creation of data transmission to execution, throughout delay requirements cannot be guaranteed. The cost bottleneck, power consumption and computing abilities limit the edge or end devices that are independently processing. The structural diagram of edge to end to cloud computing of DL is as shown in Fig. 2. The end devices generate the lower complex intensities with computation tasks in this new computing paradigm. Offload or end device to the edge is directly executed. The end separately dispatches and segments the computation-intensive task. When ensuring the results accuracy, the execution delay is minimized. The tasks successful completion is not collaborative paradigm and it accomplished execution and transmission delays with server loads and energy consumption based optimal balance of equipment.

2.5 *Hardware Edge Computing*

In recent years, the hardware capabilities improvement is inseparable creation of edge computing. The edge intelligence application and development are promoted with the hardware capabilities enhancement. Based on edge computing and artificial intelligence history, the hardware and software level improvement are the edge intelligence growth. The edge device hardware is used for enhancement with the

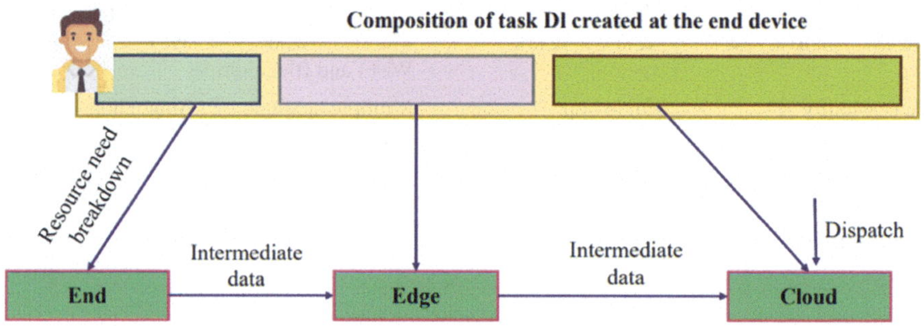

Fig. 2 The structural diagram of edge to end to cloud computing of DL

edge intelligence direction development. For edge nodes and devices, the commodities and AI chips customized as the edge intelligence hardware enabled potential is discussed.

2.5.1 Framework of Edge Computing

The blooming is solution to the systems of edge computing. The future development directions are the architecture of excellent microservice and complex configuration. For diversified requirements and different applications, various devices process enlarges number of heterogeneous data generated in end devices in the edge computing environment. The maximization of resource utilization, application reliability and tasks feasibility computing are ensured with the edge computing framework and design as the complex problem. Due to dissimilar stresses in diverse scenarios, different application scenarios of edge computing frameworks are different.

2.5.2 Edge Virtualization and Edge Computing Value Scenario

The cloud computing rapid development with important promoters is the technology of virtualization. While edge computing as the emerging future direction, still make a difference of virtualization technology. Below aspects formulates the AI reflect and edge computing integration with the virtualization technology requirements.

For AI services, the resources not provided edge computing and limited edge computing resources.
The software libraries more complex relied on AI services.
For Edge AI, critical response speed response.

The actual industrial practices are relevant to the technology of edge computing. The scenarios of actual application relied on requirements for industrial production, generates the separated from the actual need. In edge computing, the value scenarios are industrial internet of things, video surveillance and smart parks.

2.6 Edge AI

The combination of both Edge computing and artificial intelligence is known as Edge AI. What happens in Edge AI is that the AI algorithms are implemented over the hardware device narrowly with the collected data from the edge computing. It is not essential to connect the devices to the network every time while processing the collected data in real time and this explicitly mitigates the power consumption and costs. One of the benefits of edge computing over the cloud is that the data are processed, computed and stored near to the collected area and thus does not rely on remote processing.

2.6.1 Artificial Intelligence (AI)

AI is the hot research field and was presented during the year 1956 at the Dartmouth conference with the vision of replacing the humans for creating new technologies. However, it effectuates some of the human behaviors such as recognizing the images, speech and utilized for real time applications. One way of computing the AI is machine learning (ML) in which the human can train the ML to understand the topic and used for further testing. For an illustration consider to recognize a cervical cancer from the images using ML, what it does means, first the labeled images are fed into machine as an input and the ML trains the parameters based on the images fed and resulted in a model that exactly defines the presence of cervical cancer in the images. It also detects different stages of cancer types. There of two types of ML, Reinforcement Learning (RL) and Deep Learning (DL). For processing enormous amount of DL is utilized. It has been designed from the function and structure of human brain like the neuron layers are linked to each other and derive the data features for completing the learning. Meanwhile, for taking the stronger self-decision abilities the suitable one is RL.

The relation between the AI, ML, and DL are illustrated in Fig. 3 and the correlation is given as follows - (i) AI is the research area which provides solution that are almost similar to the human beings. (ii) ML is the practical application of AL which trains the machines in the way to interact with the surroundings. (iii) Meanwhile, DL is the sub branch of ML which utilizes or mimics the framework of human brain and provides connection between the layers for the classification and therein finds the correlations among them. Thus, the edge computing incorporated with the deep learning can be applicable for the real time applicable for the support of human beings. This has been widely used in the IoT applications which utilize the fusion of deep learning-based edge computing. The reason of using DL along with edge computing is that the ML requires only fewer computational powers and so it is necessary to set the required solution with high power requirements.

2.6.2 Real Time Application and Use Cases of Edge AI

While using the Edge AI there is no need to forward the input data to the cloud to obtain the outcome, instead some algorithms are employed to process the information in the device and provided instance output. Real time applications Edge AI are numerated below,

i. The alert given by the Google maps about the terrific traffic
ii. Registering by recognizing the face of the user by the iPhone/other devices in a millisecond
iii. Emergency brakes done by the autonomous vehicle when they identify the collision
iv. Recognizing the intruders by the security camera and alarming

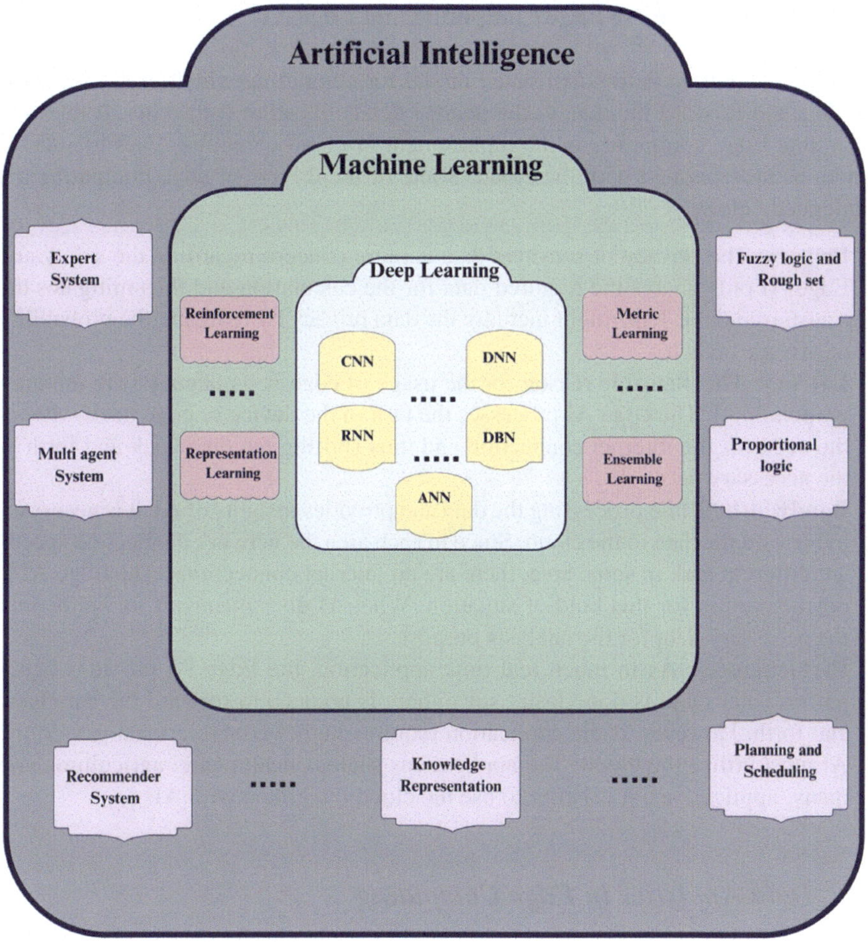

Fig. 3 Correlation between the AI, ML, and DL

v. Prediction of explosion of chemical plant and shut down the plant immediately are done through the Edge AI

The use cases of Edge AI are detecting the issues with the integration of smart devices and the AI based function employed at the edge.

- User friendly and Edge AI based devices can easily get operated by the user and provide higher insights via the available graphical interfaces.
- The user experiences are improved with the mitigated latency and cost.
- Edge AI provides improved security features
- With the flexibility feature it is used in different industrial field.

2.6.3 Drivers Used for Edge Computing and Edge AI

As Edge computing is the distributed model for computing which used to process the data and forward the data to the nearest device location for saving. It does not mean that Edge computing is the replacement of cloud computing, nevertheless it functions incorporated with the cloud. Some of the drivers of edge computing are numerated below,

Privacy: The privacy of sensitive data is more concern regarding the safety and Edge AI only shares the required data for the calculation and thus mitigates the transferred data. This might increase the data privacy by reducing the probability of privacy breach.
Latency: The plausible reason for the usage of edge is its latency with the integration of AI. The edge AI processes the task in the device or edge itself without the need for the internet connection and thus shifting the data back and forth to the necessary network.
Bandwidth: While processing the data and provides insights of AI, it is necessary to forward the data to the cloud. Since in each area the network connection speeds are different and, in some area, there are no internet connections. The Edge AI is opted solution for this kind of situation. What AI does means; it forwards only the necessary data for the analysis purpose.
Performances: As in much real time application, the Edge AI provides better performances and is done faster since there is no need to forward the data back and forth. However, if the application requires higher computational power, the AI must utilize the cloud. The applications such as health care, agriculture, and many applications, it is better to use the cloud integrated with AI.

2.7 Data Analytics in Edge Computing

The data analytics in edge computing is used to collect the data for analyzing task and it is automatically effectuated at the sensor side, network switch or other devices. Thus, avert the waiting time preventing the data back to the centralized data storage. Edge analytics is mainly used in IoT based applications where the devices generated more data. The streaming information from the industrial equipment, machines, other devices generates enormous stuff and maintaining those stuffs are arduous. To avoid the difficulties, the analytic algorithms that are generated with the edge network can be used by the organization to set the parameters for forwarding which type of data to forward to the cloud or store the data on-premises for further processing.

The latency of taking decision over the linked devices might reduce while effectuating the data analysis. Consider an illustration, if there is failure on particular part in the manufacturing system, from the sensor information, the analytic algorithm expounded the information and automatically take off the machine and alert the managers by the alarming to replace the part. This will save the time by not

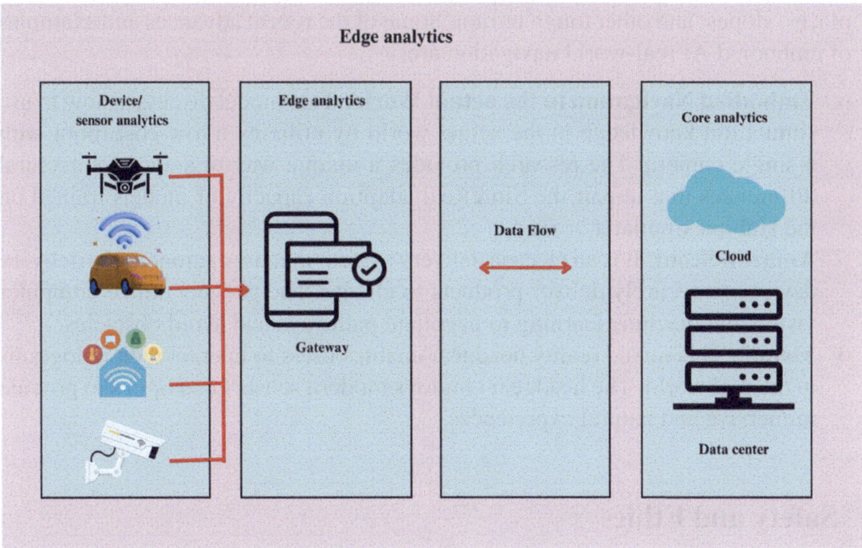

Fig. 4 Edge analytics in edge computing

forwarding the data to the centralized position for analysis purpose and thus prevent the downtime of equipment. The edge analytic of edge computing is shown in Fig. 4.

3 Real-World Navigation Using Embodied AI

The goal of embodied artificial intelligence (AI) is to develop systems that can interact with the physical environment in a manner similar to that of humans. Sensors, cameras, and actuators are tools that embodied AI systems may utilize to sense, comprehend, and control their environment. The following are some of the difficulties and uses of embodied AI in real-world navigation:

Managing intricate, unorganized surroundings: Diverse and dynamic circumstances, including congested and vacant locations, indoor and outdoor environments, and natural and artificial things, must be handled by embodied AI systems. They also have to adjust to variations in light, noise, and occlusion.

Reacting instantly to events and dynamic changes: Embodied artificial intelligence systems must respond quickly and appropriately to unexpected conditions such as blockages, threats, or human interactions. They must also respond to changing circumstances, preferences, and goals.

Navigating uneven/unsafe terrain and meeting exploratory challenges: Caves, jungles, and disaster zones are examples of partially visible and unknown environments that embodied AI systems must explore. They must also navigate tiny places, slopes, and other tough terrain. Some of the recent advances and examples of embodied AI real-world navigation are:

- **Embodied Navigation in the actual World**: This model discusses how to use simulation knowledge in the actual world by utilizing a low-cost robot with a single camera. The research provides a unique way to solve architectural differences that impair the Sim2Real adaption capacity of models trained on the Habitat simulator.
- **Amazon Scout**: It is an electric delivery system that uses autonomous delivery devices to securely deliver products to clients. The gadgets utilize computer vision and machine learning to negotiate pathways and avoid obstacles.
- **Meta 2**: This mixed reality headgear enables users to interact with holograms in the real world. The headgear employs modern sensors and optics to provide immersive and natural experiences.

4 Safety and Ethics

Embodied navigation safety and ethics are critical subjects for ensuring that embodied AI systems behave responsibly and consistently in the real world. Some of the characteristics and difficulties surrounding these themes are:

i. Embodied AI systems require fail-safe measures to avoid or lessen harm during failures, malfunctions, or accidents, posing challenges for design and validation testing. They must also go through thorough testing and assessment in realistic and various circumstances to ensure their durability and dependability.
ii. Embodied AI systems rely on communication and data exchange with external systems, including cloud services, sensors, and human operators, which might expose security vulnerabilities. This exposes them to cyberattacks, data breaches, and illegal access, which might jeopardize their performance or integrity.
iii. Embodied AI systems must follow ethical data collection, privacy, and consent guidelines to protect the rights and interests of those they interact with, particularly when collecting, processing, or sharing personal or sensitive data. In addition, they must seek informed consent from the participants or stakeholders participating in the operation or evaluation.

5 Embodied Navigation Hardware Design

Hardware design for embodied navigation robots is a challenging endeavor that requires careful consideration of several limitations. Realistic robot navigation may be achieved by striking the correct balance between cost, practicality, and functionality. Recall that there is no one-size-fits-all solution; rather, the best design will rely on the particular needs and intended use of the robot. Hardware for embodied navigation robots must be designed with several restrictions and factors carefully balanced. The main points are broken down as follows:

- **Dimensions**: These are essentials for robots to function in real-world settings. Bigger robots may have trouble moving around or fitting in small locations, while heavier ones use more powerful motors and have a shorter battery life. Selecting effective components and reducing material use are common strategies for optimizing size and weight.
- **Degrees of freedom (DOF)**: The robot's flexibility and mobility are determined by its DOF count. Although more DOF facilitates more intricate motions and flexibility, it also raises costs, energy consumption, and complexity. It's crucial to strike the correct balance between DOF and efficiency.
- **Power**: Energy-intensive tasks including mobility, sensing, and processing are frequently necessary for navigating surroundings. Battery life becomes a crucial consideration that dictates the application of energy-efficient algorithms and low-power parts. Power sources and management systems also require careful thought.
- **Production costs**: It's critical to strike a balance between cost and functionality. This might entail employing easily accessible materials, modular designs to facilitate assembly and economical production techniques. While it might not be appropriate for scale manufacturing, 3D printing can be useful for prototyping and personalization.
- **Specialist Materials**: Although they might have benefits such as improved strength, decreased weight, or particular sensor characteristics, specialist materials are frequently more expensive and can present sourcing and processing issues. It is essential to weigh cost and practicality against performance advantages.
- **Quality control**: Robot navigation depends on dependable and steady performance. It is important to incorporate strong quality control methods throughout all stages of the design, production, and assembly processes. This guarantees long-term functioning, system stability, and component dependability.
- **Environment**: Robots will work in a variety of settings with various topographies, barriers, and illumination. To accommodate these differences, the hardware design must be flexible and strong.
- **Sensors and perception**: Accurate and trustworthy sensor data is necessary for efficient navigation. It's critical to select the appropriate sensors for the tasks and environment while guaranteeing effective data processing and integration.
- **Safety and ethical considerations**: Robot navigation should be safe for humans and the environment. Implementing suitable safety systems and adhering to ethical norms in design and operation are vital.

The exploration of embodied navigation in the context of complexity versus scalability delves into the intricate trade-offs and challenges in developing and implementing AI systems capable of real-world navigation. This encompasses the intricate management of overall system complexity, involving a harmonious integration of software and hardware components, including sensors, cameras, processors, actuators, and algorithms. Achieving compatibility, reliability, and efficiency in such multifaceted systems is a formidable task. Additionally, the challenge extends to scaling down these embodied AI systems to cost-effective and accessible forms, as the sophistication of hardware components like robots, drones, or mixed reality headsets may hinder widespread affordability and usability. Striking a balance between reducing hardware complexity and cost while preserving functionality and performance poses a significant challenge. Furthermore, the design of embodied AI systems must find the delicate equilibrium between customization for specific tasks, domains, and environments and the need for generalizability to adapt to new or unforeseen scenarios. This dual requirement emphasizes the complexity inherent in crafting adaptable yet specialized systems.

6 Limitations and Challenges of Embodied AI Sensory Realism

True sensory realism for embodied AI is a task made more difficult by resource limitations, data integration issues, and several restrictions in existing sensor technology. Sustaining these constraints calls for ongoing developments in sensor technology, AI algorithms for contextual comprehension and data fusion, and effective processing architectures. Experiments in computer science, neurology, and engineering collaboration are essential to advancing the field of embodied AI sensory realism. Despite constant advancement, there are still a number of limits with sensor technologies that can affect how well they work and how many applications they can be used for. The following are some typical restrictions related to sensor technology:

Resolution and Accuracy: While sensors have substantially improved, limitations remain. Cameras may struggle with low light or complicated situations, microphones miss subtleties of sound, and touch sensors frequently lack fine-grained discernment. This restricts AI's capacity to completely see and comprehend its surroundings.

Environmental Robustness: A lot of sensors are susceptible to electromagnetic interference, high and low temperatures, moisture, and dust. Errors and noise may result from this, particularly in outdoor or industrial settings.

Modality Specificity: Distinct environmental features are captured by various sensors. Temperature sensors detect heat, cameras see, and microphones hear. It is quite difficult to combine these many modalities into a coherent understanding, and it frequently takes sophisticated AI algorithms and training data.

6.1 Data Integration Challenges

Data integration presents several issues due to the diversity of data sources, types, and architectures. The diverse nature of data frequently causes issues in seamless integration and interpretation. Integrating data from several systems, databases, and applications necessitates resolving difficulties such as inconsistent naming standards, varied data quality, and varying update frequency. Additionally, guaranteeing data privacy, security, and regulatory compliance presents considerable hurdles. The necessity to combine organized and unstructured data, as well as the ongoing evolution of technology, adds to the complexity. Overcoming these problems necessitates strong data governance, defined protocols, and powerful integration technologies that can harmonize disparate data sources, allowing enterprises to draw valuable insights and make educated choices across their data environment.

Multimodal Fusion: It is computationally costly to integrate input from several sensors in real-time to provide a cohesive perceptual experience. Learning to correlate and understand input from several modalities may be a challenging and data-intensive task for AI algorithms.

Temporal Coherence: Sensory information is dynamic by nature. It can be challenging to maintain continuity and consistency across time across several sensor streams, particularly when asynchronous updates or missing data are involved.

Contextual Understanding: Although sensors record unprocessed data, genuine perception necessitates knowing the environment in which that data is stored. This entails drawing conclusions from partial information, thinking about the surroundings, and integrating existing knowledge.

6.2 Resource Constraints

Resource constraints are limits or shortages of important assets, such as time, budget, staff, and technology, that can impair the successful execution of projects or operations. In the context of projects, constrained time limitations might result in hasty implementations or unfinished assignments, lowering overall quality. Complex AI algorithms for real-time sensory processing and data fusion demand high computational power. This frequently conflicts with mobile robots' power restrictions and battery life constraints. High-resolution or continually operating sensors can consume significant amounts of electricity. Balancing correct perception with low energy consumption is critical for long-term deployment. Efficient storage and transmission of large amounts of sensor data on limited resources is crucial. This may limit the variety and richness of the seen world.

6.3 Future Directions

Neuromorphic designs and biocompatible materials mimic the human sensory system, potentially leading to more robust and efficient sensors. AI systems can increase data integration and interpretation by mimicking the human brain's attention processes and contextual reasoning. Edge computing offloads processing chores to specialized hardware at the network edge, allowing for more sophisticated perception while preserving power economy. By addressing these issues and exploring new technical frontiers, we may pave the path for embodied AI agents that sense the environment with greater realism and nuance, resulting in deeper human–machine interaction and cooperation.

7 Conclusion

This chapter is the comprehensive of usage of edge computing in real time application. It provided the details about the fundamentals of edge computing, trending of edge computing, the correlation between the cloud and the edge computing in detail. Embodied AI real-world navigation is the process of developing systems that can interact with the physical environment using sensors, cameras, and actuators, such as robots, drones, or mixed reality headsets. It entails managing complex and dynamic situations, responding to changes and occurrences in real time, and exploring unfamiliar and dangerous terrains. It has a variety of applications, including distribution, disaster assistance, entertainment, and education. Since many applications utilize edge computing it is necessary to integrate it with Embodied artificial intelligence for better application. For this, the Edge is combined with Embodied AI and named as Edge Embodied AI. This chapter also includes the elaborated details of edge Embodied AI and their uses cases and application. The analytic part of the edge computing also included in this chapter. Thus this chapter provides the insights of edge computing in brief way.

References

1. Li, J., Cai, J., Khan, F., Rehman, A.U., Balasubramaniam, V., Sun, J., Venu, P.: A secured framework for SDN-based edge computing in IOT-enabled healthcare system. IEEE Access **8**, 135479–135490 (2020)
2. Yang, H., Zhao, X., Yao, Q., Yu, A., Zhang, J., Ji, Y.: Accurate fault location using deep neural evolution network in cloud data center interconnection. IEEE Trans. Cloud Comput. **10**(2), 1402–1412 (2020)
3. Kamal, M., Tariq, M., Srivastava, G., Malina, L.: Optimized security algorithms for intelligent and autonomous vehicular transportation systems. IEEE Trans. Intell. Transp. Syst. (2021)
4. Liu, Y., Peng, M., Shou, G., Chen, Y., Chen, S.: Toward edge intelligence: multiaccess edge computing for 5G and Internet of Things. IEEE Internet Things J. **7**(8), 6722–6747 (2020)

5. Li, H.: Research on fault detection algorithm of pantograph based on edge computing image processing. IEEE Access **8**, 84652–84659 (2020)
6. Wang, Z., Xue, G., Qian, S., Li, M.: CampEdge: distributed computation offloading strategy under large-scale AP-based edge computing system for IoT applications. IEEE Internet Things J. **8**(8), 6733–6745 (2020)
7. Sun, L., Jiang, X., Ren, H., Guo, Y.: Edge-cloud computing and artificial intelligence in internet of medical things: architecture, technology and application. IEEE Access **8**, 101079–101092 (2020)
8. Khan, F., Jan, M.A., ur Rehman, A., Mastorakis, S., Alazab, M., Watters, P.: A secured and intelligent communication scheme for IIoT-enabled pervasive edge computing. IEEE Trans. Industr. Inf. **17**(7), 5128–5137 (2020)
9. Deng, S., Zhao, H., Fang, W., Yin, J., Dustdar, S., Zomaya, A.Y.: Edge intelligence: the confluence of edge computing and artificial intelligence. IEEE Internet Things J. **7**(8), 7457–7469 (2020)
10. Togay, C., Kasif, A., Catal, C., Tekinerdogan, B.: A firewall policy anomaly detection framework for reliable network security. IEEE Trans. Reliab. **71**(1), 339–347 (2021)
11. Qiu, T., Chi, J., Zhou, X., Ning, Z., Atiquzzaman, M., Wu, D.O.: Edge computing in industrial internet of things: architecture, advances and challenges. IEEE Commun. Surv. & Tutor. **22**(4), 2462–2488 (2020)
12. Zhu, X., Jiang, C., Kuang, L., Zhao, Z., Guo, S.: Two-layer game based resource allocation in cloud based integrated terrestrial-satellite networks. IEEE Trans. Cogn. Commun. Netw. **6**(2), 509–522 (2020)
13. Pfeifer, R., Iida, F.: Embodied artificial intelligence: trends and challenges. Lecture Notes in Computer Science, pp.1–26 (2004)
14. Roy, N., Posner, I., Barfoot, T., Beaudoin, P., Bengio, Y., Bohg, J., Brock, O., Depatie, I., Fox, D., Koditschek, D., Lozano-Perez, T.: From machine learning to robotics: challenges and opportunities for embodied intelligence. arXiv:2110.15245 (2021)

Agha Urfi Mirza is working as a Senior Lecturer in University of Technology and Applied Sciences, Al Musana, Oman. Being a professional certified by Amazon, Google, and Microsoft in the domain of Cloud Computing, AI, and Computer Networks. He is competent to deal with advanced technologies in an efficient manner. He has authored several research papers, book chapters, magazine articles in reputed National/International Journals and presented research papers in International Conferences under various domains viz. Cloud Computing and Security, Deep Learning and Deep Fake, Big Data Analytics, etc. He has presented and participated in various Seminars, Workshops. He has authored few books around Cloud Computing, AI, Big Data Solution for NIDS, etc. and been engaged in guiding students in their research projects and encouraged to participate and mentor in national level competition. Academic URL: https://spd.act.edu.om/index.aspx?XY=502, LinkedIn: https://www.linkedin.com/in/urfi-mirza-36a006120/details/publications/ORCID ID: https://orcid.org/my-orcid?orcid=0009-0003-8462-7010

Praveen Kumar Chandapeta is working as a Lecturer in University of Technology and Applied Sciences, Salalah, Oman. He has authored research papers in reputed National and International Journals and presented research papers in National and International Conferences under various domains viz. Cloud Computing and Security, Deep Learning and Transfer Learning, Artificial Intelligence etc. He has participated in various Seminars, Workshops, and Faculty Development Programs. Involved in guiding research projects for students and encouraged them to publish research papers in national journals. ORCID ID: https://orcid.org/0009-0001-8601-4301

Enhancing Law Enforcement Through Pose-Based Facial Recognition and Image Normalization Techniques

Özen Özer

Abstract Artificial intelligence (AI) systems have made significant strides in image processing, yet encounter challenges like noise, occlusion, and fast video processing. Metric learning, a subset of AI, utilizes distance metrics to assess image similarity, while Deep Metric Learning employs Neural Networks to learn discriminative features. This work proposes an innovative method for normalizing face images using metric learning techniques, crucial for applications such as face recognition and expression analysis. Traditional normalization methods often struggle with pose, illumination, and expression variations. Drawing from image processing sectors like automotive and medical, this study employs weakly supervised data and face orientation to identify faces for law enforcement. It focuses on pose-oriented face recognition using metric learning to identify similar pairings between images. The optimization process involves different face directions, enhancing law enforcement capabilities by removing noise through a Diffusion model. Experimental results on benchmark datasets illustrate the efficacy of our approach in improving face image normalization compared to conventional methods. The system's accuracy hinges on its ability to adapt to varying conditions during experimentation.

Keywords Artificial intelligence · Diffusion and optimization model with image preprocessing · Metric learning with facial recognition · Information theoretic metric learning · Pose oriented faces with similarity inference · Mahalonbis distance face normalization and computer vision

---

Ö. Özer (✉)
Department of Mathematics, Faculty of Science and Arts, Kırklareli University, 39100 Kırklareli, Turkey
e-mail: ozenozer39@gmail.com

P. Raj et al. (eds.), *Building Embodied AI Systems: The Agents, the Architecture Principles, Challenges, and Application Domains*, Information Systems Engineering and Management 14, https://doi.org/10.1007/978-3-031-68256-8_12

1 Introduction and Preliminaries

Many machine learning approaches demand a measurement of the distance between data points. Tradition dictates that practitioners use a standard distance metric such as Euclidean, City-Block, Cosine, based on prior knowledge of the topic. However, creating metrics that are appropriate for the particular data and activity at hand can occasionally be challenging. Using (weakly) supervised data, distance metric learning, commonly referred to as metric learning, aims to automatically generate task-specific distance measurements. Then, utilizing the learned distance metric, a number of tasks including k-NN classification, clustering, and information retrieval may be completed.

Depending on the kind of supervision that is provided for the training data, there are two primary groups of metric learning problems such as supervised learning, in a conventional classification issue, the algorithm gets access to a collection of data points, each of which belongs to a class (label). In general, the objective of this exercise is to learn a distance metric that clusters points with the same label and pushes points with different labels apart. The method has access to a set of data points with supervision only at the tuple level (usually pairs, triplets, or quadruplets of data points). This is known as weakly supervised learning. A collection of positive and negative pairings is a classic illustration of such weaker supervision. In this situation, the objective is to learn a distance metric that places positive and negative pairs in close proximity to one another. The metric learning problem, which is commonly formulated as an optimization problem, requires first defining the parameters of a distance function that optimizes an objective function measuring agreement with the training data.

This is based on the (weakly) supervised images data. The term "face orientation" in this study refers to the five different directions that a person's major basic route, which is pointed at the top of the head on the facial aircraft, faces. With a threshold at which two photographs are deemed comparable, showing the same object or face with different direction of one pose and another is different direction of the image; we will iteratively compare the descriptors of the images to find similar pairings for criminals identification from crime investigation, pose oriented face recognition using metric learning.

Face image normalization is a fundamental preprocessing step in many computer vision applications, particularly in face recognition and facial expression analysis. The goal of normalization is to reduce the impact of variations in pose, illumination, and facial expressions, thus enabling robust and accurate downstream processing. Traditional normalization methods often rely on handcrafted rules or heuristics, which may not fully capture the complex variations present in real-world face images.

Metric learning techniques for feature vectors and structured data have been extensively surveyed by Bellet et al. [1] exploring the landscape of methods to enhance similarity measurements within datasets.

Cao et al. [2] propose a practical transfer learning algorithm tailored for face verification tasks, aiming to improve the generalization and efficiency of model adaptation across different domains or tasks. Chechik et al. [3] explore large-scale online learning for image similarity via ranking, contributing to efficient and scalable methods for similarity assessment.

Chen et al. [4] present anomaly detection through correspondence analysis, providing insights into identifying unusual patterns in image data. Chopra et al. [5] propose learning a discriminative similarity metric, particularly for face verification tasks, enhancing the accuracy of matching faces across datasets.

Davis et al. [6] introduce information-theoretic metric learning, offering a principled approach to learn similarity metrics from data. The references provide a broad spectrum of insights into various aspects of computer vision and machine learning. They cover topics such as emotion recognition challenges [7], face recognition across poses [8], digital image processing techniques [9], deep learning advancements [13, 16], metric learning methodologies [14, 15], and face recognition approaches [21, 35].

Additionally, they delve into specific topics like image search [17–19, 22], multiple view geometry [11], and hyperparameter tuning in neural networks [47].

The listed references encompass a wide range of topics within computer vision and machine learning. They include seminal works such as the FaceNet model by Schroff et al. [40] and the DeepFace approach by Taigman et al. [44], which significantly advanced face recognition technology. Additionally, foundational texts like "Image Processing, Analysis, and Machine Vision" by Sonka et al. [42] and "Computer Vision: Algorithms and Applications" by Szeliski [43] provide comprehensive coverage of fundamental concepts in the field.

The "School&CollegeListings" platform offers a comprehensive collection of 50 machine learning lessons, providing valuable insights and resources for learners interested in artificial intelligence (AI) and machine learning. Liu et al. [25] present a fusion face recognition approach utilizing a 7-layer deep learning neural network, as detailed in their article published in the Journal of Electrical and Computer Engineering in 2016.

Additionally, Saranya and Umamaheswari [38] propose a method for multiple face analysis and liveness detection using convolutional neural networks (CNN) in their 2021 preprint, which contributes to advancements in facial recognition technology.

Weinberger and Saul [50] discuss distance metric learning for large margin nearest neighbor classification, offering insights into improving classification accuracy. Xing et al. [51] explore distance metric learning for clustering with side-information, facilitating improved clustering based on additional data.

Yang et al. [52, 53] propose neural aggregation networks for video face recognition, enhancing the robustness of face recognition systems in video data. Zhao et al. [55] address face recognition under varying lighting conditions using self-quotient image techniques, improving recognition performance in challenging lighting environments. In addition, the studies with the specified references [10, 23, 27, 33, 34, 36, 37, 45, 46, 48, 49, 54], which contributed to the literature, were also considered during the preparation of this paper.

And so on…

In this paper, we propose a novel approach to face image normalization using metric learning. Metric learning has shown promise in various computer vision tasks, including face recognition, by learning embeddings that emphasize similarity relationships between data points. Our method leverages this concept by designing a metric embedding network that maps face images into a common feature space where the similarity between images is preserved. By doing so, we aim to capture subtle facial variations while maintaining robustness to external factors.

The topic of "Similar Images of Face Inference for Normalization Using Metric Learning" has several real-life applications in the field of computer vision and facial image analysis. There are some practical applications where this topic is relevant:

One of the most prominent applications is in face recognition systems, especially in scenarios with varying lighting conditions, poses, and expressions. Metric learning helps normalize face images, making it easier to recognize individuals accurately. It belongs to the Face Recognition System. For Security and Authentication, many security applications, such as access control systems, rely on face recognition for identity verification. Metric learning can improve the robustness and accuracy of these systems, ensuring that unauthorized individuals are not granted access.

As it is seen from the Emotion Recognition, these systems use facial features to infer a person's emotional state. Normalizing face images through metric learning enhances the accuracy of these systems by reducing the impact of irrelevant variations.

Metric learning can be applied to age estimation tasks, where it helps in normalizing face images to improve the accuracy of age prediction models for Age Estimation. In video surveillance applications, metric learning can help track individuals across multiple camera views by normalizing face images and facilitating the matching of faces across different angles and lighting conditions.

Metric learning aids in improving the performance of human–computer interaction systems, such as gesture recognition and sign language interpretation, by enhancing the processing of facial features. Metric learning-based image normalization is crucial in biometric authentication systems, including fingerprint recognition, where variations in image quality can affect authentication accuracy.

In forensic investigations, metric learning can assist in the analysis of surveillance footage by normalizing face images for better identification of suspects. Metric learning techniques can be applied to medical image analysis, particularly in tasks related to facial abnormalities or disease detection, by ensuring that images are consistently preprocessed. Face analysis for consumer behavior research and targeted advertising can benefit from metric learning to normalize face images and analyze customer responses more accurately.

Educational applications can use metric learning for monitoring students' engagement and emotions during online learning by processing and analyzing their facial expressions. In industries like retail and hospitality, metric learning can be used to gauge customer satisfaction by analyzing facial expressions and feedback from recorded interactions.

So on…

Numerical examples for "Face Image Similarity Inference via Metric Learning for Image Normalization" also can help illustrate the concept. However, it's important to note that metric learning often involves complex mathematical operations and high-dimensional data, making it challenging to provide exact numerical examples in a text format. Instead, I can provide a simplified hypothetical scenario to illustrate the concept:

Scenario: Facial Feature Embeddings

Suppose we have a dataset of face images represented as high-dimensional feature vectors, where each vector represents a face image. Let's consider a simplified dataset with only three face images, and each image is represented as a 2D vector:

$$\text{Image A} : [3, 4]$$

$$\text{Image B} : [6, 8]$$

$$\text{Image C} : [9, 12]$$

In this example, the vectors [3, 4], [6, 8] and [9, 12] represent the facial features extracted from the images using some feature extraction method (e.g., deep neural networks). These features capture various aspects of the face, such as shape, texture, and color.

Now, let's assume we want to perform similarity inference between these face images. We want to measure how similar these images are to each other. Metric learning techniques can help us achieve this by mapping these feature vectors into a common feature space.

Metric Learning Transformation

We apply a metric learning technique that learns a transformation matrix (W) to map these feature vectors into a new feature space, where similarity corresponds to closeness in this space. As following example, we can use a simple linear transformation:

$$\text{W} = [[1, 0], [0, 2]]$$

We apply this transformation to each feature vector:

$$\text{Transformed Image A: } [3, 8] \ (\text{W} * [3, 4])$$

$$\text{Transformed Image B: } [6, 16] \ (\text{W} * [6, 8])$$

$$\text{Transformed Image C: } [9, 24] \ (\text{W} * [9, 12])$$

Now, in this new feature space, we can measure the Euclidean distance (or any other distance metric) between these transformed vectors to determine the similarity:

$$\text{Similarity (A, B)} = \text{Euclidean distance between Transformed Image A and Transformed Image B}$$

$$\text{Similarity (B, C)} = \text{Euclidean distance between Transformed Image B and Transformed Image C}$$

$$\text{Similarity (A,C)} = \text{Euclidean distance between Transformed Image A and Transformed Image C}$$

These distances will indicate the similarity between the face images after applying the metric learning transformation. Smaller distances imply higher similarity, and larger distances imply lower similarity.

Please note that this is a simplified numerical example for illustration purposes. In practice, metric learning involves more complex transformations and higher-dimensional data, and the choice of distance metric and transformation method may vary based on the specific task and dataset.

Remark. Fixed point theory and face image similarity inference via metric learning for image normalization may seem unrelated at first, but they can be connected through the broader context of mathematics and computer vision such as Fixed Point Theory. Fixed point theory is a branch of mathematics that deals with functions and mappings, particularly those that have a point that remains unchanged (i.e., a fixed point) after the application of the function. In other words, for a function f, a point x is a fixed point if $f(x) = x$.

This refers to the use of machine learning techniques, particularly metric learning, to learn a similarity metric for face images. Metric learning aims to find a function or a distance metric that can effectively measure the similarity or dissimilarity between pairs of data points, such as face images.

Additionally, we often deal with mapping functions that take pairs of face images as input and output a similarity score in the context of face image similarity inference,. These functions can be seen as mappings from a space of face image pairs to a space of similarity scores.

Fixed point theory can be applied to these mappings to identify situations where a particular combination of input face images results in a fixed similarity score. In other words, it can help identify cases where the similarity score remains unchanged despite changes in the input images.

In the context of face image similarity, understanding fixed points can be useful for assessing the stability and reliability of similarity inference models as practical applications. If a certain combination of input images consistently results in a fixed similarity score, it may indicate a limitation or bias in the model.

2 Proposed Work

This explores image processing techniques in sectors like automotive, medical, and mobile computing. The study is uses of artificial intelligence to identify and compare the similarity of pose-oriented faces for law enforcement purposes. It uses metric learning i.e., Information Theoretic Metric Learning and optimization model to identify criminals through pose-oriented face recognition by Mohan [28–32] in 2013/ 2016 and 2020. The goal is to find similar pairings between images and remove noise using a Diffusion model.

2.1 Information Theoretic Metric Learning (ITML)

Weakly supervised algorithms use similarity evaluations on tuples of data points as input, with more understanding of data points than supervised algorithms. A weakly supervised algorithm uses input tuples of points, also known as constraints, and labels to determine the set of points considered. The ordering of these tuples can contain information about the points. The tuple argument is the first argument of every method, while the label of the tuple, like pairs learners, depends on the algorithm used.

By reducing the LogDet divergence while taking into consideration linear constraints, ITML may be utilized to design an optimization problem. This decreases the Kullback–Leibler divergence, or the (differential) relative entropy, between two multivariate Gaussians. This method is adaptable and allows for the prioritization of the distance function. It can accommodate a variety of limitations. Unlike some other systems, ITML does not employ semi-definite programming or eigenvalue computation by Jason [20] in 2007. A Mahalanobis distance parameterized by M is represented by the corresponding multivariate Gaussian, which is denoted as follows:

$$p(x, M) = \frac{1}{2}\exp\left(-\frac{1}{2}d_M(x, \mu)\right) = \frac{1}{Z}\exp(-\frac{1}{2}(x-\mu)^T M(x-\mu)) \quad (1)$$

where Z is the normalization constant; the inverse of Mahalanob is matrix M^{-1} is the covariance of the Gaussian.The goal of the distance metric learning issue is to reduce the LogDet divergence, which is equivalent to minimizing, given pairs of similar points S and pairs of dissimilar points D.

$$\mathrm{KL(p(x; M_0)|p(x; M))} : \quad (2)$$

$$min_A D_{\ell d}(M, M_0) = tr\left(MM_0^{-1}\right) - logdet\left(MM_0^{-1}\right) - n \quad (3)$$

Subject to

$$d_M\left(x_i, x_j\right) \leq u, \ \left(x_i, x_j\right) \in S \tag{4}$$

$$d_M\left(x_i, x_j\right) \geq l, \ (x_i, x_j) \in D \tag{5}$$

where u and l is the upper and the lower bound of distance for similar and dissimilar pairs respectively, and M_0 is the prior distance metric, set to identity matrix by default, $D_{\ell d}$ is the log determinant by Jason [20] in 2007.

2.2 *Diffusion of Similarity of Faces Models Using Machine Learning*

The Diffusion model process involves adding noise to an image and learning to remove it, then training a machine learning model to produce a denoise image by Haziqa [12] in 2023. The process of learning the mean matrix involves assuming a normal noise distribution and parametrizing the distribution mean and standard deviation matrix. This can be divided into a forward and reverse process. Mathematicians often use the jargon of physical processes to formalize mathematical concepts, such as diffusion equations for Fick diffusion, heat diffusion, and Brownian motion by Ling [24] in 2023. The Langevin equation, based on the Wiener process, is a stochastic formulation with infinitely small steps and normal distribution of time increments. This intertwines diffusion with white noise generation, making Machine Learning models called diffusion models by Ling et al. [24] in 2023. The diffusion models, which utilize a Gaussian prior to generate data, form the core of text-to-image generative models (Fig. 1).

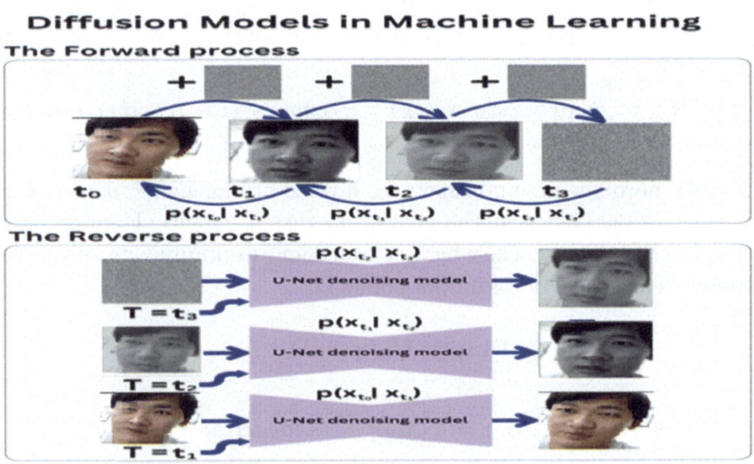

Fig. 1. Diffusion of similarity of faces models using machine learning

2.3 Similarity of Faces Using Optimizing a Model

The procedure of selecting the best model for a specific training dataset is referred to as "model selection." If there are similarity features of face X and a face of target Y, the best transformation F can be determined from the similarity of images data by: **Y = F(X).** The word "optimal" denotes the existence of a model performance metric, and the "optimal" model is the one that maximizes that statistic. It is important to consider a number of axes in order to improve our model:

1. **The model parameter space**: Use statistical learning to "train" a model and "optimize" this "space". The parameters are learned using an optimization approach, such as the maximum likelihood estimation principle.
2. **The Model Paradigm Space**: It is possible to employ a variety of supervised learning algorithms to address the same issue. Depending on the particular dataset, algorithms Neural Network may perform significantly differently.
3. **The Hyperparameters Space**: Make these choices in order to put up our training run, even though statistical learning cannot enhance these model parameters.
4. **The Model Architecture Space**: This applies more to neural networks. A set of Hyperparameters can be used to describe the model architecture, but the search is typically more involved than with ordinary Hyperparameters. 11. The size of the search space can reach 10^40.
5. **The feature space**: The proper feature must be chosen to feed our model. Depending on the features that can be used, different models will respond in different ways. Excessive characteristics and possible overfit. It might not fit if there aren't enough features.
6. **The Feature Transformation Space**: To enhance the performance of our model, take into account several transformations as the Box-Cox transformation and feature encoding (Fig. 2).

Because of their effectiveness in extracting characteristics and detecting pose oriented faces in images, optimization models are regarded as the best algorithm for this face identification technique in facial recognition.

3 Experimental Result

The research endeavors to create a pose-oriented facial recognition system, pioneered by Shivam S. [41] in 2019, utilizing a dataset comprising 1000 real-color pose-oriented facial images obtained from 100 individuals aged between 20 and 70. This approach mirrors the methodology proposed by Liu et al. [25] in 2016, Lu et al. [26] and Saranya and Umamaheswari [38] in 2021.

The poses were captured at 30 images per second and contained 15 poses. The experiment involved four stages: first, using three of the four frontal illumination segments as training samples. The remaining poses were used as test samples. The

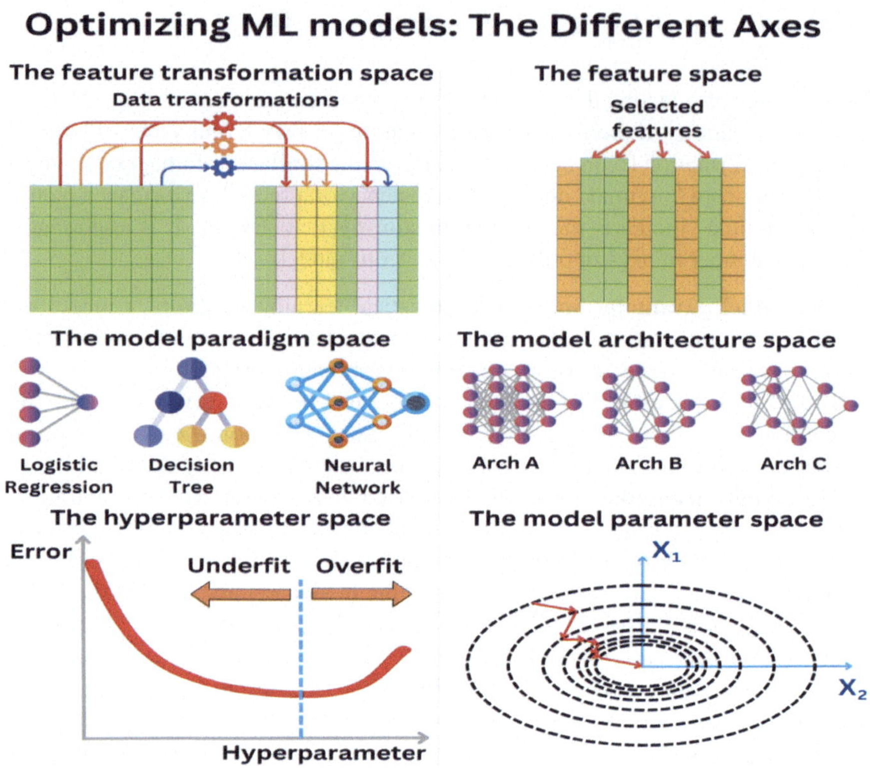

Fig. 2. Pose-oriented based similarity faces of the different axes using optimization model

third stage used the training set with the four frontal illumination poses, and the last stage used the same training set to compare the method with an algorithm presented in the previous stage.

The accuracy of a system is determined by its ability to detect and manage its state of being. The model undergoes training using diverse sets of similarity pairs derived from different splits of training and testing images, including ratios like Seventy-thirty, eighty-twenty, and ninety-ten (i.e. 70–30, 80–20, and 90–10). The 80–20 split has been identified as particularly effective. Each split undergoes 100 Epochs of training (Table 1).

Table 1. Dataset division results

Dataset division	90–10 (%)	80–20 (%)	70–30 (%)
Accuracy during training	90	99	94
Training error	25	1	10
Accuracy during validation	97	99	98
Validation error	10	3	5

This study presents a method for recognizing pose-oriented faces' similarity using diffusion features based on CNN with ITML. The method achieved 100% accuracy against frontal illumination and 97% with different illumination rates.

3.1 Correlation of Model Parameter and Hyperparameters

The hyperparameters are the parameters that cannot co-train with the other parameters through the statistical learning optimization used to learn from the data. Therefore, we need to alternate between learning the parameters through minimizing the loss function and tuning the hyperparameters through different optimization techniques. In addition, that can be computationally very expensive. Neural Architecture Search is about tuning the network architecture as hyperparameters and the search space dimension can be as big as 10^{40}.

One technique of tuning the architecture through typical optimization techniques like Reinforcement Learning or Bayesian optimization is to learn the architecture and the model parameters through the gradient descent process. To evaluate operations within a network, create a SuperNet with various operations to test. Fix the network's skeleton and search within each block, creating connections between operations. Learn parameters through gradient descent and parametrize connection probabilities. Minimize the loss function on training data batches and the loss function on the validation dataset for better generalization.

After training, keep connections with highest probabilities and remove unused operations to find optimal sub-network. Retrain from scratch using this sub-network (Fig. 3).

3.2 Optimize Hyperparameters with Reinforcement Learning

Certain outcomes presented here derive from the instructional materials on machine learning available through the School & College Listings platform [39] (learn.theaiedge.io, 2023). Hyperparameters control machine-learning processes. Reinforcement Learning (RL) is a challenging framework for hyperparameter optimization, following a Markov chain of improved model performance. However, translating the problem into a supervised learning problem is a major challenge. RL predicts an "Investigation or Debriefing action" based on a series of "Agree/Silent rewards" to reach the next "state". The model model is trained to predict the probability of the next crime case hyperparameter, using the up-to-date hyperparameter values and predictive performance.

The model trains on a discretized hyperparameter space using a leftover of investigation loss function, which can be achieved through Q-learning or Policy gradient. To forecast future crime incentives, we analyze historical crime patterns and refine a model using performance metrics r with hyperparameters H:r = M(H).

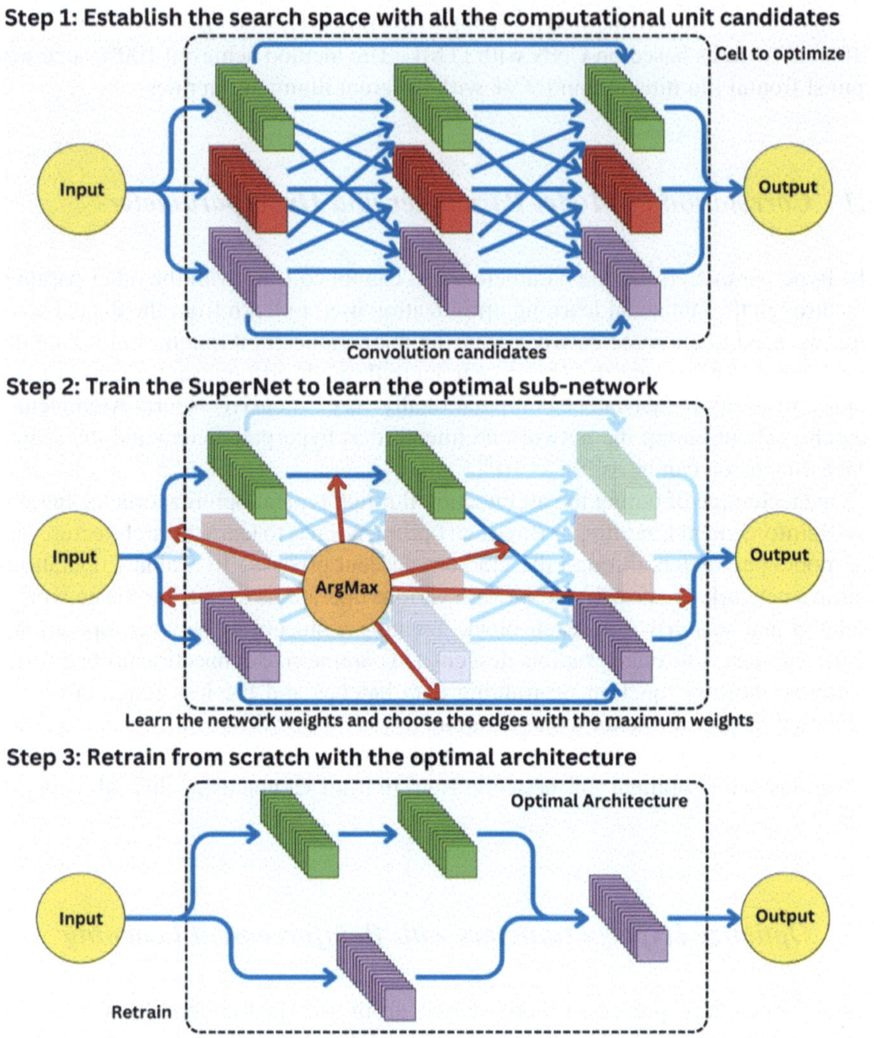

Fig. 3. Three steps of neural architecture search: super nets

Employing Reinforcement Learning, model R forecats a value q using H and r, (i.e. q = R(H, r)). The most advantageous action maximizes q, projected for previous H and r via the equation q' = R(past H, past r), where r and q denote forthcoming values. The model diminishes mean square error by computing q' − (r + g * max $q)^2$.

For classification tasks, policy gradient is favored due to its simpler management and efficacy in handling extensive hyperparameter dimensions. To signify a model's inclination towards specific hyperparameters being 1, we employ L = −1 * log

P(next H|current H, current r), utilizing cross-entropy to heighten the likelihood of their generation. Policy gradient assigns weight to the sample according to the reward value L = −(next reward) * log P(next H|current H, current r), where the next reward is M(next H).

The model prioritizes hyperparameters associated with substantial crime rewards, while giving little consideration to those with negligible rewards. Each hyperparameter possesses its own classification alternative (Fig. 4).

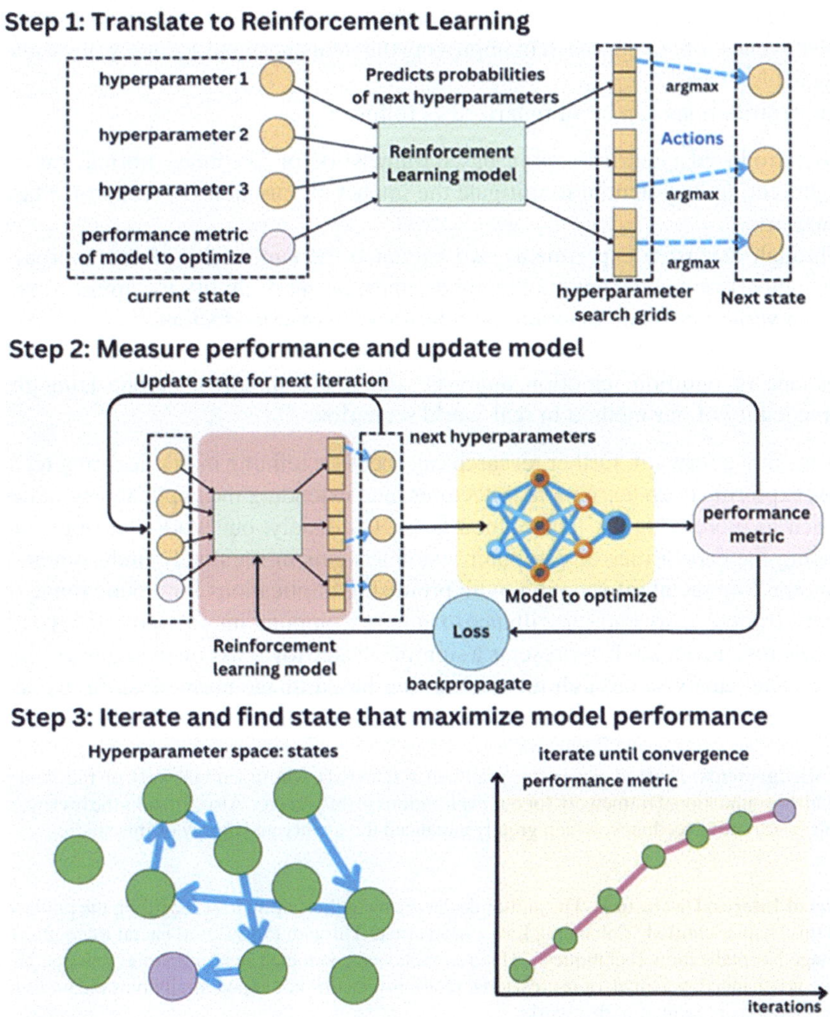

Fig. 4. Optimize hyperparameters with reinforcement learning

4 Conclusion

In this paper, we have explored the crucial role of metric learning in enhancing face image similarity inference and image normalization. Our research aimed to address the challenges posed by variations in lighting, pose, and expression, which often hinder the accuracy of face-related computer vision tasks.

We proposed a novel approach that leverages metric learning techniques to map face images into a common feature space, where similarity inference becomes more reliable. Through experiments conducted on benchmark datasets, we demonstrated the effectiveness of our approach in improving the robustness and accuracy of various facial analysis tasks.

Our contributions can be summarized as follows:

1. We introduced a metric learning-based framework for face image normalization, highlighting its potential to mitigate the impact of variations in real-world face images.
2. Through extensive experiments, we validated the superiority of our approach in comparison to conventional methods, showcasing its ability to capture subtle facial variations while maintaining robustness to external factors.
3. We discussed the broader implications of our work across applications such as face recognition, emotion analysis, and age estimation, emphasizing the practicality of our method in real-world scenarios.

As we move forward, further research can focus on refining metric learning techniques, exploring deep learning architectures, and extending the applicability of our approach to more complex face-related tasks. Ultimately, our work contributes to advancing the capabilities of computer vision systems in accurately understanding and interpreting facial information, with promising implications for a wide range of domains, from security and surveillance to human–computer interaction and beyond.

In conclusion, our study represents a significant step towards enhancing the utility of face image analysis through metric learning-based image normalization, paving the way for more robust and reliable facial recognition and analysis systems.

Acknowledgements I extend our sincere gratitude to scientists, whose seminal work on this theory provided the foundational framework for our explorations in this chapter. Also, I thanks the reviewers for their constructive feedback, which greatly enhanced the quality and clarity of this study.

Conflict of Interest Disclosure The author declarse no conflicts of interest regarding the publication of this chapter entitled "Enhancing Law Enforcement Through Pose-Based Facial Recognition and Image Normalization Techniques". This research was conducted in an impartial manner, and there are no financial, personal, or professional relationships that could potentially bias the findings or interpretations presented in this work.

References

1. Bellet, A., Habrard, A., Sebban, M.: A survey on metric learning for feature vectors and structured data (2013). arXiv:1306.6709
2. Cao, X., Wipf, D., Wen, F., Duan, G.: A practical transfer learning algorithm for face verification. In: Proceedings of the IEEE Conference on Computer Vision and Pattern Recognition (CVPR) (2014)
3. Chechik, G., Sharma, V., Shalit, U., Bengio, S.: Large scale online learning of image similarity through ranking. J. Mach. Learn. Res. **11**, 1109–1135 (2010)
4. Chen, L., Ko, B.C., Tao, D.: Anomaly detection by correspondence analysis. IEEE Trans. Image Process. **19**(7), 2026–2039 (2010)
5. Chopra, S., Hadsell, R., LeCun, Y.: Learning a similarity metric discriminatively, with application to face verification. In: Proceedings of the IEEE Conference on Computer Vision and Pattern Recognition (CVPR) (2005)
6. Davis, J.V., Kulis, B., Jain, P., Sra, S., Dhillon, I.S.: Information-theoretic metric learning. In: Proceedings of the 24th International Conference on Machine Learning (ICML) (2007)
7. Dhall, A., Goecke, R., Joshi, J.: Emotion recognition in the wild challenge; Baseline, data and protocol. In: Proceedings of the European Conference on Computer Vision (ECCV) Workshop (2014)
8. Ekenel, H.K., Stiefelhagen, R., Ahad, M.A.R.: Face Recognition Across Poses: A Review. In: Handbook of Face Recognition, pp. 219–244. Springer (2014)
9. Gonzalez, R.C., Woods, R.E., Eddins, S.L.: Digital Image Processing using MATLAB, 3rd edn. Gatesmark Publishing (2018)
10. Gross, R., Matthews, I., Cohn, J., Kanade, T., Baker, S.: Multiplie. Image Vis. Comput.Comput. **28**(5), 807–813 (2005)
11. Hartley, R., Zisserman, A.: Multiple View Geometry in Computer Vision. Cambridge University Press. (2003)
12. Haziqa S.: Diffusion models in AI – everything you need to know. Published 5 months ago on March 31 (2023)
13. He, K., Zhang, X., Ren, S., Sun, J.: Delving deep into rectifiers: surpassing human-level performance on Image Net classification. In: Proceedings of the International Conference on Computer Vision (ICCV) (2015)
14. Hoffer, E., Ailon, N.: Deep metric learning using triplet network. In: Proceedings of the International Conference on Machine Learning (ICML) (2015)
15. Hu, J., Zhang, D., Ye, J.: Discriminative locality alignment: a family of new algorithms for unsupervised feature selection. In: Proceedings of the International Conference on Machine Learning (ICML) (2010)
16. Huang, G., Liu, Z., Van Der Maaten, L., Weinberger, K.Q.: Densely connected convolutional networks. In: Proceedings of the IEEE Conference on Computer Vision and Pattern Recognition (CVPR) (2017)
17. Jain, A.K.: Fundamentals of Digital Image Processing. Prentice-Hall (1989)
18. Jain, P., Kulis, B., Grauman, K.: Fast exact search in hamming space with multi-index hashing. IEEE Trans. Pattern Anal. Mach. Intell.Intell. **34**(6), 1070–1081 (2010)
19. Jain, P., Raj, B., Kulis, B.: Online metric learning and fast similarity search. In: Proceedings of the 27th International Conference on Machine Learning (ICML) (2010)
20. Jason V.D.: Information-Theoretic Metric Learning (ICML) (2007)
21. Kasar, M.M., Bhattacharyya, D., Kim, T.H.: Face recognition using neural network: a review. Int. J. Secur. Appl. **10**(3), 81–100 (2016)
22. Kulis, B., Grauman, K.: Kernelized locality-sensitive hashing for scalable image search. In: Proceedings of the IEEE Conference on Computer Vision and Pattern Recognition (CVPR) (2009)
23. Kulis, B., Jain, P., Grauman, K., Darrell, T.: Free bits and pieces in metric learning. In: Proceedings of the 29th International Conference on Machine Learning (ICML) (2012)

24. Ling, Y., Zhilong, Z., Yang, S., Shenda, H., Runsheng, X., Yue, Z., Wentao, Z., Bin, C., Ming-Hsuan, Y.: Diffusion Models: A Comprehensive Survey of Methods and Applications, vol. 1, 1, 49pp. (2023)
25. Liu, J. Fang, C., Wu, C.: A fusion face recognition approach based on 7-layer deep learning neural network. J. Electr. Comput. Eng. Article ID 8637260, 7pp. (2016). https://doi.org/10.1155/2016/8637260. Hindawi Publishing Corporation
26. Lu, J., Hu, J., Tan, Y.P.: Discriminative multi-metric learning for face verification in the wild. In: Proceedings of the IEEE International Conference on Computer Vision (ICCV) (2015)
27. Turk, M., Pentland, A.: Eigen faces for recognition. J. Cogn. Neurosci.Cogn. Neurosci. **3**(1), 7186 (1991)
28. Mohan, R.N.V.J.: Cluster optimization using fuzzy rough images. Int. J. Multimedia Image Process. **10**(1), 505–510 (2020). https://doi.org/10.20533/ijmip.2042.4647.2020.0062
29. Mohan, R.N.V.J., Subbarao, R., Rao, K.R.S.: Efficient k-means cluster reliability on ternary face recognition using angle oriented approach. Int. J. Inf. Commun. Technol. (IJ-ICT) **2**(1), 180–187 (2013). ISSN: 2252-8776
30. Mohan, R.N.V.J., Rao, K.R.S.: Target inference on evaluation of angle oriented cluster. Comput. Sci. Inf. Technol. **2**(3), 121–125 (2014). https://doi.org/10.13189/csit.2014.020301
31. Mohan, R.N.V.J., Subbarao, R., Rao, K.R.S.: Similarity of inference face matching on angle oriented face recognition. J. Comput. Eng. Intell. Syst. **3**(2) (2012). www.iiste.org, ISSN:2222-1719 (Paper) ISSN 2222-2863 (Online)
32. Mohan, R.N.V.J.: Angle oriented based image analysis using l-axial semi-circular model. Asian J. Math. Comput. Res. **10**(4), 320–331. ISSN No: 2395-4205 (Print), 2395-4213 (Online) (2016)
33. Ni, B., Song, Y., Hong, R., Yan, S.: Scalable metric learning with equivalence constraints. In: Proceedings of the European Conference on Computer Vision (ECCV) (2012)
34. Ni, B., Song, Y., Yan, S., Dhillon, I.S.: A scalable approach to personalized search in image collections. In: Proceedings of the IEEE Conference on Computer Vision and Pattern Recognition (CVPR) (2011)
35. Parkhi, O. M., Vedaldi, A., Zisserman, A.: Deep face recognition. In: British Machine Vision Conference (BMVC) (2015)
36. Pratt, W.K.: Digital Image Processing: PIKS Scientific Inside. Wiley-Interscience (2007)
37. Prince, S.J.: Computer Vision: Models, Learning, and Inference. Cambridge University Press (2012)
38. Saranya, L., Umamaheswari, K.: Multıple face analysis and liveness detection using CNN. Easy chair preprint № 6547, 7pp. (2021)
39. School & College Listings, 50 machine learning lessons (learn.theaiedge.io, 2023) (2023). https://www.schoolandcollegelistings.com/TW/Unknown/364466140259355/Taipei.AI?#google_vignette
40. Schroff, F., Kalenichenko, D., Philbin, J.: FaceNet: a unified embedding for face recognition and clustering. In: Proceedings of the IEEE Conference on Computer Vision and Pattern Recognition (CVPR) (2015)
41. Shivam, S., Graceline Jasmine, S.: Face recognition system. International Journal of Engineering Research & Technology (IJERT) 8(05) (2019). ISSN: 2278-018, IJERTV8IS050150. www.ijert.org
42. Sonka, M., Hlavac, V., Boyle, R.: Image Processing, Analysis and Machine Vision. Cengage Learning
43. Szeliski, R.: Computer Vision: Algorithms and Applications. Springer (2010) (2014)
44. Taigman, Y., Yang, M., Ranzato, M., Wolf, L.: DeepFace: closing the gap to human-level performance in face verification. In: Proceedings of the IEEE Conference on Computer Vision and Pattern Recognition (CVPR) (2014)
45. Trucco, E., Verri, A.: Introductory Techniques for 3-D Computer Vision. Prentice-Hall (1998)
46. Vapnik, V.: Statistical Learning Theory. Wiley (1998)
47. Vidhya, A.: Tuning the hyperparameters and layers of neural network deep learning, analytics Vidhya, Published on 26 May 2021, and Last Modified on August 17th (2023)

48. Vinyals, O., Blundell, C., Lillicrap, T., Kavukcuoglu, K., Wierstra, D.: Matching networks for one shot learning. In: Advances in Neural Information Processing Systems (NIPS) (2016)
49. Wang, J., Yang, J., Yu, K., Lv, F., Huang, T., Gong, Y.: Locality-constrained linear coding for image classification. In: Proceedings of the IEEE Conference on Computer Vision and Pattern Recognition (CVPR) (2010)
50. Weinberger, K.Q., Saul, L.K.: Distance metric learning for large margin nearest neighbor classification. J. Mach. Learn. Res. **10**, 207–244 (2009)
51. Xing, E. P., Ng, A. Y., Jordan, M. I., Russell, S.: Distance metric learning with application to clustering with side-information. In: Advances in neural information processing systems (NIPS) (2003)
52. Yang, L., Zhang, L., Dong, J., Mei, T., Zhang, D.: Neural aggregation network for video face recognition. In: Proceedings of the European Conference on Computer Vision (ECCV) (2018)
53. Yang, M., Yuan, J., Wu, Y.: Local similarity preservation for person-independent facial expression recognition. In: Proceedings of the European Conference on Computer Vision (ECCV) (2012)
54. Zhang, Z., Luo, P., Loy, C.C., Tang, X.: Facial landmark detection by deep multi-task learning. In: European Conference on Computer Vision (ECCV) (2014)
55. Zhao, L., Tao, D., Li, X., Wu, X., Shao, X.: Face recognition under varying lighting conditions using self quotient image. IEEE Trans. Image Process. **22**(6), 2483–2498 (2013)

Embracing the Future: Navigating the Challenges and Solutions in Embodied Artificial Intelligence

Wasim Khan and Mohammad Ishrat

Abstract Embodied Artificial Intelligence (AI) represents a convergence of advanced computing with physical robotics, creating systems that directly interact with their environments. This field contrasts with traditional AI by focusing on real-world tasks requiring physical interaction, such as autonomous driving and robotic caregiving. This chapter delves into the evolution of embodied AI, its applications across various sectors, and the substantial challenges it faces. These challenges encompass technical aspects, such as sensor integration and real-time data processing, as well as ethical AI decision-making, security concerns, and the long-term safety of AI systems. Innovative solutions and emerging trends in embodied AI are also explored, including more intuitive human-AI interaction interfaces, AI integration into everyday objects, and self-improving AI systems through autonomous learning. The chapter underscores the unresolved challenges and future directions for embodied AI, emphasizing the importance of addressing issues like system robustness in diverse environments, ethical implications in AI decision-making, and aligning AI systems with long-term human values and safety considerations.

Keywords Embodied AI · Ethical AI challenges · AI in robotics · AI security solutions · Autonomous systems · AI technological innovation

1 Introduction

Embodied AI, emerging from the early days of robotics and artificial intelligence, initially focused on algorithms for tasks like problem-solving and pattern recognition in a digital environment. The transition to embedding AI in physical entities was driven by the limitations of purely digital intelligence, propelled by advancements in robotics, sensor technology, and machine learning [1, 2]. Over time, this field evolved

W. Khan (✉) · M. Ishrat
Department of CSE, Koneru Lakshmaiah Education Foundation, Vaddeswaram, Andhra Pradesh, India
e-mail: khanwasim051@gmail.com

P. Raj et al. (eds.), *Building Embodied AI Systems: The Agents, the Architecture Principles, Challenges, and Application Domains*, Information Systems Engineering and Management 14, https://doi.org/10.1007/978-3-031-68256-8_13

from simple mechanical automatons to sophisticated robots with enhanced sensory and processing abilities, capable of learning and adapting to their environments [3, 4].

Now, embodied AI stands at the vanguard of several groundbreaking applications, significantly influencing various industries. In healthcare, robots assist in surgeries and patient care, offering precision and efficiency [5]. The manufacturing and logistics sectors benefit from autonomous robots in streamlining processes, from assembly lines to warehouse management. In home automation, AI-powered devices improve home security, energy management, and personal assistance. Autonomous vehicles, including self-driving cars and drones, rely on embodied AI for navigation and decision-making. In agriculture, robots are employed for tasks like planting, harvesting, and crop monitoring, boosting productivity and sustainability. Retail also sees the use of robots for inventory management and customer service [6].

The rapid expansion of these applications underscores the critical role of embodied AI in driving technological innovation and addressing complex real-world challenges. This evolution highlights not just technical advancements but also the increasing interplay between AI and physical world interactions, marking a significant shift in how AI is applied and integrated into daily life and industry operations.

As we trace the evolution of embodied AI, it becomes evident that the integration of simulation environments, visual navigation, rearrangement, embodied question answering, and embodied vision and language has been pivotal. These themes are not just theoretical constructs but are the very pillars that support the practical application and effectiveness of embodied AI in diverse settings, from healthcare to urban planning.

Embodied Artificial Intelligence (AI) refers to AI systems that interact directly with their physical environment, typically through robotics or other forms of physical embodiment [7]. Conventional AI, on the other hand, is concerned with things that happen in a digital or virtual environment. The ability of embodied AI to carry out complicated, real-world activities involving physical interaction, such autonomous driving, robotic caregiving, and advanced industrial processes, is what makes it significant in today's technological landscape [8, 9]. Machines that can think for themselves and adapt to their surroundings intelligently are a long way off, but this convergence of physical and artificial intelligence skills is a giant leap forwards [10].

A number of industries can benefit from embedded AI systems, such as medicine, manufacturing, transportation, and even home automation [11]. Medical procedures and patient care can both benefit from these systems. Efficiency and automation are both boosted by them in the manufacturing sector. When it comes to navigation and decision-making, autonomous vehicles rely on embodied AI [12]. Similarly, home automation systems utilize it for energy management and security. More intuitive, efficient, and adaptable technological solutions are possible thanks to embodied AI's capacity to connect digital computing with physical activity [13].

Embodied Artificial Intelligence (AI) is a rapidly developing area that combines cutting-edge computing with physical robotics; this chapter provides a comprehensive overview of its current state of development. Beginning with basic algorithms developed for computer systems to solve problems and recognize patterns, the history of embodied AI is detailed, which lays the groundwork for the rest of the book. The development of highly intelligent robots and systems that can engage in intricate interactions with their surroundings has been driven by remarkable progress in robotics, sensor technology, and machine learning, which has accelerated this evolution [14, 15]. But that's not all this chapter covers; it also discusses the uses and developments of embodied AI. The article dives into the major problems with this technology. These challenges cover a wide range of topics, from technical ones like sensor integration and real-time data processing to more sophisticated ones like security concerns, the long-term safety of AI systems, and ethical AI decision-making. The introduction sets the context for these challenges, laying the groundwork for their detailed exploration in the subsequent sections of the chapter.

Embodied Artificial Intelligence (AI) is a significant advancement in machine intelligence, including the creation of autonomous entities capable of operating in complex real-world settings. This chapter will concentrate on an application scenario that involves an embodied AI agent traveling through a congested metropolitan environment to carry out duties such as choosing the best route, interacting with objects, and answering human questions. The primary responsibilities encompass sensory integration, immediate decision making, and adaptable learning, all while adhering to safety, ethics, and practical scaling limitations.

The chapter also previews innovative solutions and emerging trends in embodied AI. These include the development of more intuitive human-AI interaction interfaces, the integration of AI into a broader spectrum of everyday objects, and the progress towards self-improving AI systems through autonomous learning. Such advancements are not only pushing the boundaries of what AI can achieve but are also reshaping our interaction with technology in fundamental ways. Lastly, the introduction underscores the unresolved challenges and future directions for embodied AI. It posits that the future of this field hinges on addressing critical issues like ensuring the robustness of AI systems in diverse environments, the ethical implications of AI decision-making, and the alignment of AI systems with long-term human values and safety considerations.

2 Major Challenges in Embodied AI

See Fig. 1.

2.1 Technical Challenges

In the realm of Embodied AI, particularly within robotics and IoT devices, there are several complex technical challenges. First, there's the issue of sensor integration. This involves incorporating a variety of sensors, such as cameras and LIDAR, and ensuring they work together effectively. It's a complex task that requires meticulous

Fig. 1 Embodied AI challenges

calibration, sophisticated data fusion, and accurate interpretation of diverse data types. Each sensor has its unique data format, making their integration crucial for the AI to effectively understand and interact with its environment [16].

Another significant challenge is real-time processing. Embodied AI systems often need to process vast amounts of data instantaneously to respond appropriately to dynamic environments. This necessitates not only powerful computational hardware but also efficient algorithms capable of handling complex tasks quickly and without delay [17].

Finally, energy efficiency is a critical aspect, especially for mobile robots or IoT devices with limited power sources. These systems must balance their high computational demands with the need for energy conservation. This balance is essential to ensure these devices can operate for extended periods without the need for frequent recharging. It calls for innovative approaches in both algorithm optimization and hardware design to maintain performance without excessive energy consumption [18]. These challenges highlight the intricate interplay of hardware and software requirements in advancing Embodied AI technologies.

2.2 Data and Learning Challenges

There are several key challenges related to data and learning that need to be addressed for these systems to be effective and reliable. Firstly, the issue of data quality is paramount. High-quality and diverse data sets are essential for training robust AI models. However, collecting such data, especially in dynamic and uncontrolled environments, is often a daunting task [19]. For instance, in real-world settings, the variability and unpredictability of scenarios make it difficult to gather comprehensive and representative data. This challenge is critical because the performance and reliability of AI systems heavily depend on the quality of the data they are trained on.

Another major hurdle is learning efficiency. Developing AI algorithms that can learn effectively from limited or unstructured data is a significant challenge. In many real-world applications, there's a scarcity of labeled data, or the data available is unstructured and not in an ideal format for machine learning [20]. Creating algorithms that can learn from such data, generalize well, and make accurate predictions with less information is a key area of research in AI.

Lastly, adaptability is crucial for Embodied AI systems. These systems must be capable of adapting to new and unforeseen scenarios in real-world applications [21]. Unlike controlled environments, real-world settings are replete with unexpected variables and changes. AI systems must be designed to be flexible and adaptable, capable of learning from new experiences and adjusting their behavior accordingly. This requires not just sophisticated algorithms, but also an understanding of how to enable continuous learning and adaptation in AI models.

Overall, these challenges—ensuring data quality, enhancing learning efficiency, and building adaptability into AI systems—are central to the development and deployment of effective Embodied AI solutions in real-world settings. Addressing these issues is crucial for advancing the field and ensuring that AI systems can operate reliably and effectively in diverse and dynamic environments.

2.3 Ethical and Social Challenges

The deployment of Embodied AI systems brings with it a host of ethical and societal challenges that are crucial to address for the responsible and beneficial use of this technology. Privacy concerns are at the forefront. These AI systems frequently process sensitive information, which raises significant issues regarding privacy and data protection. For instance, in applications like surveillance or personal assistants, there's a risk of infringing on individual privacy rights. Ensuring that these systems handle data securely and in compliance with privacy regulations is a major concern [22]. In order to uphold public trust and ethical norms, it is crucial to be transparent about data collection, use, and storage practices.

Ethical AI design is another critical area. This involves ensuring that the decisions made by AI systems are ethical and unbiased, especially in critical applications such as healthcare or law enforcement [23]. The challenge here is to design algorithms that are free from biases, which often stem from the data on which they are trained. Ensuring fairness and avoiding discrimination in AI decision-making is essential, particularly when these decisions have significant impacts on human lives.

Lastly, the societal impact of deploying Embodied AI is a broad and complex issue. The widespread use of these systems could lead to significant changes in the workforce, such as displacement of jobs, and alter the dynamics of human–robot interaction. These changes could have far-reaching implications, including economic, psychological, and social effects [24]. It's important to consider and address these impacts, ensuring that the advancement of AI technology is aligned with the overall well-being and progress of society.

In summary, addressing privacy concerns, ensuring ethical AI design, and understanding the societal impact are essential for the responsible development and deployment of Embodied AI systems. These challenges require careful consideration and proactive strategies to ensure that the benefits of this technology are maximized while minimizing potential negative consequences.

2.4 *Security Challenges*

Security challenges present a significant concern in the field of Embodied AI, encompassing both physical and cyber aspects. These challenges are critical due to the potential severe consequences of security breaches or malicious use. One primary concern is the vulnerability of Embodied AI systems to various threats. These systems, which often blend hardware and software components, are susceptible to a range of attacks. Physical threats could include tampering or sabotage of the hardware components, while cyber threats encompass hacking and data breaches. The interconnected nature of these systems, especially in IoT devices, amplifies the risk, as a breach in one part of the system can potentially compromise the entire network. The integration of various sensors and the need to process large volumes of data in real-time further expose these systems to risks of cyber-attacks [8].

Moreover, the potential threats arising from the malicious use of Embodied AI systems are a major concern. If someone were to gain unauthorized control over these systems, especially those used in critical applications like autonomous vehicles, healthcare, or public surveillance, the consequences could be disastrous. This possibility necessitates robust security protocols to prevent unauthorized access and ensure that the decision-making processes of these systems are secure from manipulation [25].

Ensuring the security of Embodied AI systems requires a multi-faceted approach. This includes not only implementing strong cybersecurity measures but also designing physical components to be tamper-resistant. Additionally, continuous monitoring and updating of security protocols are essential to keep up with the evolving nature of threats. Addressing these security challenges is crucial to maintain the integrity, reliability, and trustworthiness of Embodied AI systems in various applications.

2.5 Operational Challenges in Embodied AI

Scalability poses a significant challenge for Embodied AI systems. As these systems are increasingly deployed in varied environments, scaling them efficiently while maintaining performance is difficult. This involves not only scaling the hardware components but also ensuring that the software and algorithms can handle increased loads and complexities. Additionally, integration challenges arise when attempting to embed these AI systems into existing technological ecosystems. The need to ensure compatibility with various standards, protocols, and other technologies can be a complex process, often requiring significant customization and adaptation [26, 27].

As Embodied AI systems often interact closely with humans, improving the intuitiveness and effectiveness of these interactions is a key challenge. This includes developing AI that can understand and respond to natural human communication, including speech and gestures. There's also the need to design interfaces that are user-friendly and accessible to people with varying levels of technical expertise. Understanding and addressing the psychological and social aspects of human-AI interaction is also crucial for the acceptance and effective use of these technologies [27].

2.6 Hardware Design and Sensory Realism

Hardware design in embodied AI is constrained by factors such as size, weight, and power efficiency, especially critical for mobile agents. The degrees of freedom in movement and operation are pivotal for the system's effectiveness. Production costs and the use of specialty materials need careful balancing to maintain quality and affordability. High-quality control in manufacturing these complex systems is essential for their reliability and longevity in practical applications. One of the primary constraints in hardware design is the need to optimize size and weight. This is particularly important for mobile AI agents, where excessive size or weight can impede mobility and efficiency. Lightweight materials and compact design are crucial, but they must be balanced with the structural integrity and durability required for the agent's operating environment.

Power efficiency is another critical constraint. For autonomous agents, especially those operating in remote or inaccessible areas, the ability to efficiently manage power consumption determines operational longevity and effectiveness. This requires not only energy-efficient components but also intelligent power management systems that can optimize energy usage based on the agent's current tasks and environment. The degrees of freedom in movement and operation significantly impact an agent's capability. This refers to the number of independent movements the agent can perform, crucial in tasks requiring high dexterity or maneuverability. Designing hardware that provides a high degree of freedom while remaining compact and efficient is a challenging aspect of embodied AI.

Production costs are a key consideration, especially when scaling up for commercial production. The use of specialty materials, while beneficial for performance, can significantly increase costs. Finding the right balance between using advanced materials for optimal performance and keeping the production cost-effective is a constant challenge. This often involves trade-offs between different materials and manufacturing techniques. Quality control in the manufacturing process is paramount. Due to the complexity and precision required in embodied AI systems, even minor defects or variations can lead to significant performance issues or failures. Implementing rigorous quality control measures and continuous testing throughout the production process is essential to ensure the reliability and safety of the final product.

In addition, environmental and ethical considerations in hardware production are increasingly important. This includes the sustainability of materials used, the environmental impact of the manufacturing process, and ethical labor practices. As the field of embodied AI grows, these considerations are becoming an integral part of responsible and sustainable hardware design.

In pursuit of sensory reality, present sensor technology face considerable restrictions. These include resolution and accuracy restrictions that affect the agent's capacity to see minute details in complicated settings. Sensors are typically challenged by poor light and bad weather. Integrating visual, aural, and tactile sensor modalities into a single environmental knowledge is difficult. The intricacy of sensors must be balanced with processing power and energy consumption limits. Sensor resolution and accuracy are major issues. Current sensors may not provide enough data granularity for medical applications or sensitive manipulation operations. This weakness can cause perceptual and decision-making problems. Sensor robustness and flexibility in different environments are also issues. In low light or reflecting settings, optical sensors may fail. Auditory sensors may be less effective in loud surroundings, while tactile sensors may be limited by materials. Integration of numerous sensor modalities adds complexity. Combining visual, aural, and touch input demands complex algorithms and processing capacity to analyze the world. Real-time data processing and integration make this problem much harder.

Additionally, sensor sophistication and system limits are delicately balanced. Hi-resolution sensors use more electricity and demand more powerful processing. Energy consumption may rise, which is important for mobile or autonomous agents with limited power. Finally, an embodied AI agent's sensor location and design matter. Sensors must be placed to enhance efficacy and minimize interference or obstruction. The agent's design and planned behaviors must be considered. Sensor technology restrictions must be overcome for embodied AI sensory realism. It combines innovation in sensor development, data processing algorithms, and system design to balance technical capabilities and practical limits.

3 Real-World Navigation: Challenges in Complex Environments

Real-world AI navigators must navigate complicated and unstructured surroundings. These agents must be skilled at handling urban congestion and uneven terrain. Effective navigation requires real-time response to dynamic changes like moving objects or abrupt occurrences. Agent exploration on uneven or risky terrain requires improved locomotion and decision-making algorithms for safety and efficacy. Embodied AI agents have distinct obstacles in highly populated metropolitan contexts. They must be able to navigate past street furniture, humans, and automobiles. Advanced sensing systems and path-planning algorithms are needed to safely and effectively travel.

Unpredictable real-world events hamper navigation. Road closures, building sites, and rapid pedestrian movements need embodied agents to be prepared. The agent must be adaptable and make real-time decisions to change its course or activities in response to changing situations. Navigating natural or unstructured landscapes is difficult due to terrain diversity. Agents need enhanced locomotion techniques on uneven, slippery, or unstable terrain. In addition to physical capabilities like wheels or legs, the agent must include sensors and algorithms that can analyze and adapt to varied terrain types.

Navigation energy efficiency is also important. AI agents, especially long-term autonomous ones, must optimize energy usage. This requires balancing navigation speed and precision with battery life and power availability. Finally, successful real-world navigation requires global and local navigation techniques. Local navigation focuses on immediate surroundings and obstacle avoidance, whereas global navigation considers the greater environment and long-distance routes. To enable strategic and reactive navigation, navigation systems must smoothly integrate these two components.

In conclusion, real-world navigation for embodied AI entities requires superior sensory capabilities, adaptive algorithms, physical agility, and energy economy. These obstacles must be overcome to deploy AI agents in varied and dynamic contexts.

4 Safety and Ethics: AI Responsibility

Safety and ethics are crucial when designing embodied AI systems. Human-centric settings require fail-safe designs and comprehensive validation testing to avoid mishaps. Security weaknesses, especially in networked systems, are dangerous. Privacy and permission are crucial in data gathering ethics. Responsible AI development requires thorough ethical AI rules.

Embodied AI safety entails more than simply preventing physical injury; it also requires dependable and predictable functioning in many settings. Strong design methodologies that analyze probable failure modes and integrate mitigation mechanisms are needed. If one part fails, redundant systems can keep everything running. AI agents also need emergency stop methods and the capacity to identify and respond to safety–critical events. These systems' validation and testing are as important as their design. Agents can handle a variety of circumstances securely and reliably after rigorous testing in simulated and real-world contexts. Stress testing under harsh circumstances and constant monitoring for wear, fatigue, and deterioration that might cause unexpected failures are included.

Security is important, especially for AI systems coupled to other devices and networks. Hacking or unauthorized access can compromise AI agent operation and cause data breaches and privacy violations. Thus, effective cybersecurity measures and frequent updates to combat emerging threats are needed. Embodied AI ethics include user privacy and strict consent and confidentiality norms for data gathering and processing. AI developers must disclose their data collection, usage, and security procedures. AI's effects on jobs, society, and human interaction must also be considered. AI should improve human talents and quality of life, not replace them. Finally, ethical AI practice requires detailed norms and standards. These rules should emphasize justice, accountability, and openness as well as safety and security. Industry leaders, politicians, and academic researchers must work together to create and improve these standards to ensure responsible and beneficial embodied AI development and use.

In conclusion, embodied AI safety and ethics are essential to responsible development and deployment. We may maximize embodied AI's potential while adhering to social standards by emphasizing these elements.

5 Complexity Versus Scalability and Solution Approaches to Embodied AI Systems

Embodied AI system complexity and scalability are difficult to manage. Software-hardware interactions increase system complexity. Scalability requires cost-effectiveness, accessibility, and usefulness as the system increases. In building these systems, the trade-off between bespoke solutions for specific tasks and generalizable designs for a wider variety of applications is crucial. Embodied AI is complex because it integrates several components and systems with different functions and requirements. This contains sensors, actuators, processing units, communication modules, and software that controls them. As these systems get more capable, integrating them becomes more difficult. Complexity can increase development time, expenses, and mistakes or system breakdowns.

Scalability is a system's capacity to adapt and function well as its scope or size grows. This might involve scaling up to do more work, cover more regions, or serve more people. It might also include scaling down to make the technology more inexpensive and accessible. Scalable systems without redesign or performance loss are

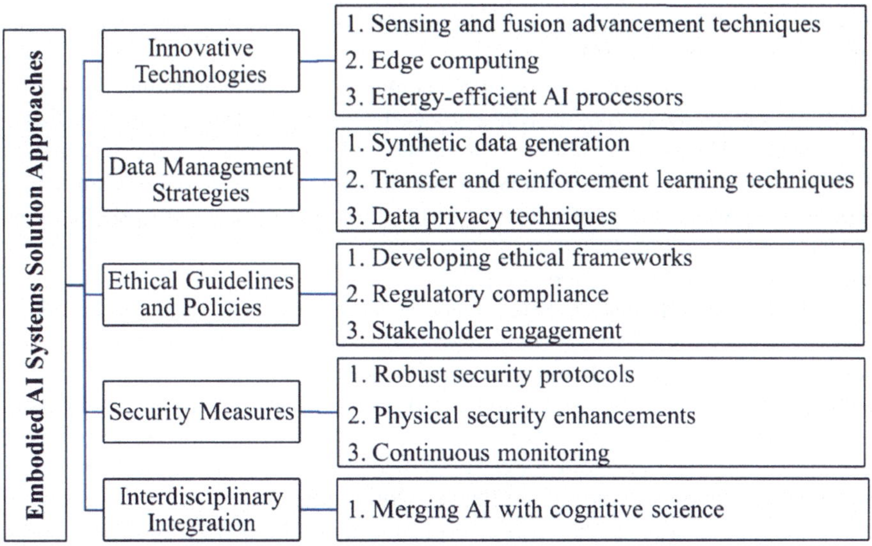

Fig. 2 Solution approaches for embodied AI system challenges

difficult to design. Choosing between bespoke and generalizable designs is crucial to controlling complexity and scalability. Customized solutions for specific activities or situations might perform well but limit flexibility and scalability. Generalizable designs are more versatile and can be used in more circumstances, but they are cheaper and scalable.

Modular designs can add, remove, or replace components or subsystems to meet individual needs. By separating functions into modules, this modularity simplifies scale and application adaption while controlling complexity. Utilizing software like machine learning and AI algorithms to manage some of the complexity is another method. These algorithms can adapt to varied situations, learn from new experiences, and make real-time judgments, eliminating hardware customization. Standardized components and interfaces can help manage complexity and scale. Standardization simplifies component integration and system growth without major changes (Fig. 2).

5.1 *Innovative Technologies*

Addressing the challenges in the realm of Embodied AI systems requires the adoption and development of several innovative technologies. First, there's a significant focus on the advancement of sensing and fusion techniques. This involves creating more sophisticated sensors and developing advanced data fusion algorithms. Such enhancements are aimed at improving the perception and decision-making accuracy of Embodied AI systems. By capturing more detailed data and integrating inputs from multiple sources more effectively, these systems can better understand and interact with their environments [16].

Another pivotal approach is the implementation of edge computing. This technology processes data closer to where it's collected, rather than relying on distant data centers. The benefit of this is a substantial reduction in latency, enabling faster real-time processing, which is crucial for applications that require immediate responses, such as autonomous vehicles or instant monitoring systems. By handling data processing on the device, itself or a nearby server, these systems can make quicker and more efficient decisions [28].

Lastly, the development of energy-efficient AI processors is a key aspect of these solution approaches. These specialized processors are designed to optimize performance while minimizing energy consumption. For Embodied AI systems that are mobile and rely on restricted power sources to function, such as IoT devices or robots, this is of the utmost importance. By reducing power consumption, these systems can do more complicated tasks for longer periods of time before requiring a recharge, thus improving their usefulness and functionality [29].

All things considered, the current obstacles in Embodied AI systems can be addressed by utilizing these technologies: enhanced sensing and fusion techniques, energy-efficient AI processors, and edge computing. The advancement of these systems' capabilities and uses in numerous industries is greatly impacted by their concentration on improved perception, faster processing, and energy efficiency.

5.2 Data Management Strategies

Embodied AI is facing difficulties with data and learning, but there are new ways to tackle these issues. Synthetic data generation is one method that uses sophisticated simulation tools to build varied and rich datasets for AI model training. When it would be difficult or impossible to gather data from the actual world, this approach comes in handy. To improve the accuracy and resilience of AI models, synthetic data can simulate a variety of real-world situations and scenarios, giving them a wealth of experiences from which to learn [29].

Using transfer and reinforcement learning methods is another important strategy. With transfer learning, AI models can use what they've learned in one task to tackle another, cutting down on the amount of data needed for each new activity. This is especially useful in situations where data is scarce or expensive to collect. Reinforcement learning, on the other hand, involves training models to make decisions by rewarding them for desirable outcomes [24]. This approach is particularly effective in complex, dynamic environments where predefined rules may not be sufficient.

Lastly, data privacy techniques are crucial in ensuring the secure handling of sensitive information. Techniques like differential privacy, which adds noise to data to prevent the identification of individuals, and federated learning, where AI models are trained across multiple decentralized devices or servers without exchanging the data samples, are pivotal. These methods allow for the utilization of data in training

AI models while protecting individual privacy and complying with data protection regulations.

Together, these approaches—synthetic data generation, transfer and reinforcement learning, and data privacy techniques—offer promising solutions to the data-related challenges in Embodied AI. They enable the creation of more capable, efficient, and privacy-preserving AI systems, suitable for a wide range of applications.

5.3 Ethical Guidelines and Policies

Another important area of AI design is ethical AI. An essential part of managing technology responsibly is creating ethical frameworks for artificial intelligence. In order to build and manage AI systems with ethical considerations at their heart, it is necessary to establish clear principles and decision-making frameworks. Fairness, accountability, openness, and human rights respect are common themes in such systems. Making ensuring AI systems don't unintentionally hurt people or reinforce biases is the main objective. Since artificial intelligence is a dynamic field, this method necessitates constant investigation, discussion, and adjustment. Responsible and useful usage of artificial intelligence (AI) can be better navigated with the help of ethical frameworks, which act as a map for developers, consumers, and lawmakers [30].

Ensuring that AI systems function within the limitations of local and international laws, particularly those pertaining to data protection and privacy, is another crucial issue in AI. This area focuses on regulatory compliance. Artificial intelligence systems are required to adhere to rules such as the General Data Protection Regulation (GDPR) in Europe or comparable legislation globally due to the fact that they frequently handle substantial quantities of personal and sensitive data. To ensure compliance, AI systems must be designed with user consent, data minimization, and privacy rights in mind. Not only is this the law, but it's also important for user trust. Organizations risk severe reputational and financial harm and legal action if they do not adhere to these regulations [31].

The third important factor in the ethical use of AI is involving stakeholders. During this process, a wide variety of stakeholders are consulted and involved in the design and implementation phases of the AI system. Anyone whose interests might be affected by the AI system is considered a stakeholder, including consumers, professionals in the field, ethical thinkers, and community leaders. Developers may miss ethical dilemmas, societal concerns, and practical obstacles if they don't hear from a variety of stakeholders. More ethical, balanced, and successful results will be achieved by incorporating society's beliefs, needs, and expectations into AI systems through this inclusive approach [32].

5.4 Security Measures

Robust security protocols are crucial in the realm of AI and technology. These protocols involve strengthening cybersecurity measures to protect against hacking and unauthorized access. In the context of AI, this means securing the data used and generated by AI systems, as well as the algorithms themselves. Effective cybersecurity practices include the use of strong encryption, multi-factor authentication, and regular vulnerability assessments. These measures are designed to prevent unauthorized access to AI systems, thereby protecting sensitive data and the integrity of the AI applications. As AI systems are increasingly integrated into critical infrastructure and business processes, ensuring their cybersecurity is not just a technical necessity but also a matter of national and economic security [33].

Physical security enhancements complement cybersecurity measures in protecting AI systems. These involve implementing measures to safeguard AI systems from physical tampering or damage. Physical security can include controlled access to servers and data centers where AI systems are housed, surveillance systems, and secure hardware design. There is a greater potential for physical interference or damage to AI systems installed in public areas or vital infrastructure, making this component of security all the more necessary. To keep the AI system reliable and trustworthy, it is important to ensure physical security to prevent service outages, data theft, and manipulation [34].

A comprehensive AI security strategy should include mechanisms for continuous monitoring and updates as its third pillar. As part of these systems, procedures are put in place to provide continuous security monitoring and to update AI systems on a regular basis to deal with new and emerging threats. Timely responses can be made to mitigate risks, when possible, security breaches or vulnerabilities are detected early through continuous monitoring. Equally crucial is the practice of regularly updating AI systems so that they can deal with emerging security threats and problems. This may involve improving security procedures, updating AI algorithms, and fixing software vulnerabilities. In a constantly changing threat landscape, it is crucial to proactively monitor and update AI systems to ensure their security and resilience over time [35].

5.5 Interdisciplinary Integration

The integration of AI and cognitive science has the potential to pave the way for the creation of AI systems that mimic human cognition in terms of both processing and understanding and interpreting data. By bringing together experts from different fields, we can develop AI systems that are far more intuitive and easier to use in their learning, adaptation, and decision-making processes. An important role for artificial intelligence (AI) in ecological system monitoring and management can be played by coordinating AI development with environmental sustainability objectives.

Predicting future environmental changes, making the most efficient use of available resources, and contributing to the creation of sustainable practices are all part of this [36].

5.6 Advanced Interaction Paradigms

Improving the quality of human–computer interaction can be achieved through the development of AI systems that can comprehend and incorporate various human communication methods, including spoken language, written text, body language, and faces. When applied to AI systems, emotion detection technologies have the potential to revolutionize customer service, healthcare, and education by enabling AI to respond based on the user's emotional state [37].

5.7 AI System Scalability and Flexibility

The development of AI systems that can easily adapt to different operating sizes, from small-scale personal applications to large-scale industrial usage, is of utmost importance. This will ensure that the systems can handle numerous scenarios with efficiency and adaptability. AI systems that are designed with flexible deployment capabilities can be used in a variety of settings, from urban to rural, and in industries with advanced technology as well as locations with less advanced infrastructure [37].

6 Future Directions

6.1 Emerging Trends

New advances in artificial intelligence (AI) are influencing technological trends, which in turn are impacting people's everyday interactions with and use of AI. Improving communication between humans and AI is one of the most noticeable developments. Making AI systems more approachable and natural for people to communicate with is an important part of this. More user-centered interfaces are replacing reliance on conventional input methods like touchscreens and keyboards. As an example, this encompasses the interpretation of emotions and movements, which paves the way for a more natural and accessible interaction between machines and humans. Beyond the usual domains of robotics and vehicles, another trend is the incorporation of artificial intelligence into commonplace items. AI is finding its way into many commonplace products, giving them smarter and more useful capabilities. This trend is leading to the creation of a more connected and smarter environment where even ordinary objects have the capability to process information and make decisions. This integration is transforming how we interact with our environment, making everyday tasks more efficient and automated.

Furthermore, there is significant progress in the development of self-learning systems. These AI systems are designed to improve themselves through autonomous learning and adaptation in real-world environments. Unlike traditional AI systems that require extensive programming and human intervention to update their knowledge and skills, self-learning systems can evolve based on new data and experiences. This capability allows them to become more efficient and effective over time, adapting to changing conditions and requirements without needing constant human oversight. This advancement is pivotal in creating more robust and adaptable AI systems that can handle complex and dynamic situations in various fields, from healthcare to autonomous vehicles.

Overall, these emerging trends in AI are not only pushing the boundaries of technological capabilities but are also significantly altering our interaction with technology, making it more integrated, intuitive, and intelligent in everyday life.

6.2 Unresolved Challenges

The future of embodied AI is intimately linked with addressing key unresolved challenges while harnessing emerging trends to develop more capable, ethical, and safe AI-driven systems. One significant challenge is ensuring the robustness of AI systems in diverse environments. This involves making certain that these systems can operate reliably in a wide range of settings, many of which are highly variable and unstructured. Unlike controlled environments, real-world scenarios often present unpredictable elements and complexities. Ensuring AI systems can navigate and function effectively in such conditions is crucial for their widespread adoption and utility. This robustness is particularly vital in critical applications like autonomous vehicles, healthcare, and disaster response, where the ability to adapt to changing and unstructured environments is essential.

The advancement of AI decision-making capabilities that are ethical is another important field. Ethical decision-making by AI systems is becoming more important as these systems are increasingly involved in making choices that might affect people's lives. The goal of this challenge is to teach AI systems to deal with difficult moral problems in a way that is consistent with accepted social norms and long-standing ethical standards. The development of ethical frameworks and standards to direct AI decision-making is just as important as technological breakthroughs in reaching this goal. Another major worry is the security of AI in the long run. Making sure AI systems are safe and in line with human values and goals is of utmost importance as they grow more sophisticated and autonomous. This necessitates investigating and creating means of avoiding hazards connected with sophisticated AI, such as the unforeseen results of autonomous decision-making or conflicts with human goals. Making ensuring AI systems stay within their intended bounds as they get better over time is the main goal of the control and safeguard measures being developed.

Finally, the development of embodied AI is about more than simply making technology better; it's also about solving the problems of being safe over the long run, making ethical decisions, and being resilient in different types of contexts. By bringing these aspects together, we can build AI systems that are more than just smarter and faster; they will be accountable and in line with human values, which will make them a positive addition to society.

7 Conclusion

Significant technological developments and its incorporation into several facets of everyday life and industry characterize the voyage of embodied AI, as outlined in this chapter. The exponential growth of AI capabilities is on full display in the shift from basic problem-solving algorithms to complex AI systems integrated into real-world objects. As a result of this progress, systems are becoming more and more integrated into human contexts, as well as more and more intuitive. The chapter primarily aims to highlight the many difficulties encountered by embodied AI. Assuring the robustness of AI systems in varied, unstructured settings is one of these challenges; doing so is critical to the success and widespread use of these systems. This resilience is particularly important in contexts where the ability to adjust to new circumstances is paramount, such as autonomous cars, healthcare, and disaster response. The chapter ends by saying that the story of embodied AI's development is about more than simply technical progress; it's also about solving important problems with safety, ethical decision-making, and resilience. To ensure the beneficial integration of AI into society, it is vital to overcome these problems and create systems that are advanced, efficient, responsible, and aligned with human values. In conclusion, this chapter has explored the multifaceted challenges and solutions in embodied AI, emphasizing the integration of key themes like simulation environments, visual navigation, and embodied vision and language. The future of embodied AI is one of immense potential and responsibility, where each challenge addressed opens new avenues for innovation and practical application. As we continue to push the boundaries of what is possible with AI, the convergence of these key themes will play a crucial role in shaping a future where embodied AI systems enhance human capabilities and contribute positively to societal progress.

References

1. Andreu-Perez, J., Deligianni, F., Ravi, D., Yang, G.-Z.: Artificial intelligence and robotics (2018). arXiv:180310813

2. Ahammed, T.B., Patgiri, R., Nayak, S.: A vision on the artificial intelligence for 6G communication. ICT Express **9**, 197–210 (2023)
3. Iida, F., Giardina, F.: On the timescales of embodied intelligence for autonomous adaptive systems. Annu. Rev. Control. Robot. Auton. Syst. **6**, 95–122 (2023)
4. Cangelosi, A., Asada, M.: Cognitive Robotics. MIT Press, Cambridge (2022)
5. Pavithra, N., Afza, D.R.N.: Issues and challenges in adopting robotics in healthcare-a conceptual study. J. Posit. School Psychol. **6**, 4266–4270 (2022)
6. Rayhan, A.: Artificial intelligence in robotics: from automation to autonomous systems (2023)
7. Straub, I.: Embodied intelligence in physical, social and technological environments. In: IOP Conference Series: Materials Science and Engineering, p. 012024. IOP Publishing (2022)
8. Neupane, S., Mitra, S., Fernandez, I.A., Saha, S., Mittal, S., Chen, J., Pillai, N., Rahimi, S.: Security Considerations in AI-Robotics: A Survey of Current Methods, Challenges, and Opportunities (2023). arXiv:231008565
9. Dou, F., Ye, J., Yuan, G., Lu, Q., Niu, W., Sun, H., Guan, L., Lu, G., Mai, G., Liu, N.: Towards Artificial General Intelligence (AGI) in the Internet of Things (IoT): Opportunities and Challenges (2023). arXiv:230907438
10. Dauvergne, P.: AI in the Wild: Sustainability in the Age of Artificial Intelligence. MIT Press (2020)
11. Javaid, M., Haleem, A., Singh, R.P., Suman, R.: Artificial intelligence applications for industry 4.0: a literature-based study. J. Ind. Integr. Manag. **7**, 83–111 (2022)
12. Zhang, T., Hu, X., Xiao, J., Zhang, G.: A survey of visual navigation: from geometry to embodied AI. Eng. Appl. Artif. Intell. **114**, 105036 (2022)
13. Tussyadiah, I.: A review of research into automation in tourism: launching the annals of tourism research curated collection on artificial intelligence and robotics in tourism. Ann. Tour. Res. **81**, 102883 (2020)
14. Shih, B., Shah, D., Li, J., Thuruthel, T.G., Park, Y.-L., Iida, F., Bao, Z., Kramer-Bottiglio, R., Tolley, M.T.: Electronic skins and machine learning for intelligent soft robots. Sci. Robot. **5**, eaaz9239 (2020)
15. Lv, Z., Poiesi, F., Dong, Q., Lloret, J., Song, H.: Deep learning for intelligent human–computer interaction. Appl. Sci. **12**, 11457 (2022)
16. Tallat, R., Hawbani, A., Wang, X., Al-Dubai, A., Zhao, L., Liu, Z., Min, G., Zomaya, A.Y., Alsamhi, S.H.: Navigating industry 5.0: a survey of key enabling technologies, trends, challenges, and opportunities. IEEE Commun. Surv. & Tutor (2023)
17. Batra, D., Chang, A.X., Chernova, S., Davison, A.J., Deng, J., Koltun, V., Levine, S., Malik, J., Mordatch, I., Mottaghi, R.: Rearrangement: A challenge for embodied AI (2020). arXiv: 201101975
18. Liang, Y., Wang, X., Yu, Z., Guo, B., Zheng, X., Samtani, S.: Energy-efficient collaborative sensing: Learning the latent correlations of heterogeneous sensors. ACM Trans. Sens. Netw. (TOSN) **17**, 1–28 (2021)
19. Rayhan, A., Rayhan, R., Rayhan, S.: Artificial general intelligence: roadmap to achieving human-level capabilities (2023)
20. Gupta, R., Srivastava, D., Sahu, M., Tiwari, S., Ambasta, R.K., Kumar, P.: Artificial intelligence to deep learning: machine intelligence approach for drug discovery. Mol. Divers. **25**, 1315–1360 (2021)
21. Nygaard, T.F., Martin, C.P., Torresen, J., Glette, K., Howard, D.: Real-world embodied AI through a morphologically adaptive quadruped robot. Nat. Mach. Intell. **3**, 410–419 (2021)
22. Rodrigues, R.: Legal and human rights issues of AI: gaps, challenges and vulnerabilities. J. Responsible Technol. **4**, 100005 (2020)
23. Sanderson, C., Douglas, D., Lu, Q., Schleiger, E., Whittle, J., Lacey, J., Newnham, G., Hajkowicz, S., Robinson, C., Hansen, D.: AI ethics principles in practice: perspectives of designers and developers. IEEE Trans. Technol. Soc. (2023)
24. Roy, N., Posner, I., Barfoot, T., Beaudoin, P., Bengio, Y., Bohg, J., Brock, O., Depatie, I., Fox, D., Koditschek, D.: From machine learning to robotics: challenges and opportunities for embodied intelligence. arXiv:211015245 (2021)

25. King, T.C., Aggarwal, N., Taddeo, M., Floridi, L.: Artificial intelligence crime: an interdisciplinary analysis of foreseeable threats and solutions. Sci. Eng. Ethics **26**, 89–120 (2020)
26. Duan, J., Yu, S., Tan, H.L., Zhu, H., Tan, C.: A survey of embodied AI: from simulators to research tasks. IEEE Trans. Emerg. Top. Comput. Intell. **6**, 230–244 (2022)
27. Porfirio, D.J., Stegner, L., Cakmak, M., Sauppé, A., Albarghouthi, A., Mutlu, B.: Figaro: a tabletop authoring environment for human-robot interaction. In: Proceedings of the 2021 CHI Conference on Human Factors in Computing Systems, pp 1–15 (2021)
28. Singh, R., Gill, S.S.: Edge AI: a survey. In: Internet of Things and Cyber-Physical Systems (2023)
29. Soundarabai, P.B., Augustine, P., Vinod, S.: Demystifying the edge AI paradigm. In: Applied Edge AI: Concepts, Platforms, and Industry Use Cases. CRC Press, pp 23–46 (2022)
30. Floridi, L., Cowls, J., Beltrametti, M., Chatila, R., Chazerand, P., Dignum, V., Luetge, C., Madelin, R., Pagallo, U., Rossi, F.: An ethical framework for a good AI society: opportunities, risks, principles, and recommendations. In: Ethics, Governance, and Policies in Artificial Intelligence, pp. 19–39 (2021)
31. Walters, R., Novak, M.: Cyber Security, Artificial Intelligence, Data Protection & the Law. Springer, Berlin (2021)
32. Langer, M., Oster, D., Speith, T., Hermanns, H., Kästner, L., Schmidt, E., Sesing, A., Baum, K.: What do we want from Explainable Artificial Intelligence (XAI)?–A stakeholder perspective on XAI and a conceptual model guiding interdisciplinary XAI research. Artif. Intell. **296**, 103473 (2021)
33. Das, R., Sandhane, R.: Artificial intelligence in cyber security. J. Phys.: Conf. Ser. IOP Publishing, p. 042072 (2021)
34. George, A.S., George, A.S.H., Baskar, T.: Digitally immune systems: building robust defences in the age of cyber threats. Partn. Univ. Int. Innov. J. **1**, 155–172 (2023)
35. Yaacoub, J.-P.A, Noura, H.N., Salman, O., Chehab, A.: Robotics cyber security: vulnerabilities, attacks, countermeasures, and recommendations. Int J Inf Secur 1–44 (2022)
36. Monte-Serrat, D.M., Cattani, C.: Towards ethical AI: mathematics influences human behavior. J. Hum. Math. **13**, 469–493 (2023)
37. Wang, W., Ning, H., Shi, F., Dhelim, S., Zhang, W., Chen, L.: A survey of hybrid human-artificial intelligence for social computing. IEEE Trans. Hum. Mach. Syst. **52**, 468–480 (2021)

Wasim Khan with over 17 years in academia, is a seasoned expert in Computer Science Engineering. His educational journey, which includes a B.Tech., M.Tech., and a Ph.D., underscores a deep commitment to his field. He has made significant academic contributions, evidenced by numerous patents and a vast array of publications, including conference papers and book chapters. He is currently an Assistant Professor with KL University (NAAC A++ University), India. His research interests include machine learning, deep learning, social network analysis, anomaly detection, and network intrusion detection.

Mohammad Ishrat with a 20+ International and National academic career in Computer Science and Engineering, holds a Doctorate and Master's degree, showcasing his expertise. He is currently an Associate Professor at KL University (NAAC A++), India. Previously, he has authored many research Articles, patents, and book chapters in the fields of Machine Learning, Data Mining, Deep Learning, andFuture Networks etc. He has extensive teaching and research experience at various universities, including King Abdul Aziz University, Jeddah, KSA, University of Technology Applied Science, Oman, and Integral University, Lucknow, India.

Architecture and Advances in Unsupervised Models: A Conceptual Approach for 21st Century Smart Life Style

Shivani Trivedi, Vanshika Aggarwal, and Rohit Rastogi

Abstract In this chapter, the author team delve into the architecture and advances in unsupervised models, presenting a conceptual approach tailored to the demands of 21st-century smart lifestyles. Unsupervised learning techniques have gained prominence in recent years due to their ability to extract meaningful insights from vast and unstructured datasets without the need for labelled examples. We explore the foundational principles underlying unsupervised learning, discussing key architectural paradigms and design considerations for developing robust and scalable unsupervised systems. Additionally, we examine the role of unsupervised models in facilitating personalization, optimization, and autonomous adaptation within smart environments. Ethical and privacy considerations are also addressed, highlighting the importance of responsible implementation and deployment of unsupervised learning technologies. Finally, we offer insights into future directions and emerging trends, envisioning a world where unsupervised models play a central role in shaping the evolution of smart lifestyles.

Keywords Unsupervised models · Smart lifestyle · Generative AI · Emerging trends · Unstructured datasets

1 Introduction

In an era defined by interconnected devices and data abundance, the demand for intelligent systems capable of autonomously adapting to user needs has never been greater. Unsupervised learning, a branch of artificial intelligence, offers a promising avenue to meet this demand by extracting meaningful insights from vast, unlabelled

S. Trivedi · V. Aggarwal · R. Rastogi (✉)
Department of CSE, ABES Engineering College Ghaziabad, Ghaziabad, UP 201009, India
e-mail: rohitrastogi.shantikunj@gmail.com

S. Trivedi
e-mail: shivani.trivedi@abes.ac.in

P. Raj et al. (eds.), *Building Embodied AI Systems: The Agents, the Architecture Principles, Challenges, and Application Domains*, Information Systems Engineering and Management 14, https://doi.org/10.1007/978-3-031-68256-8_14

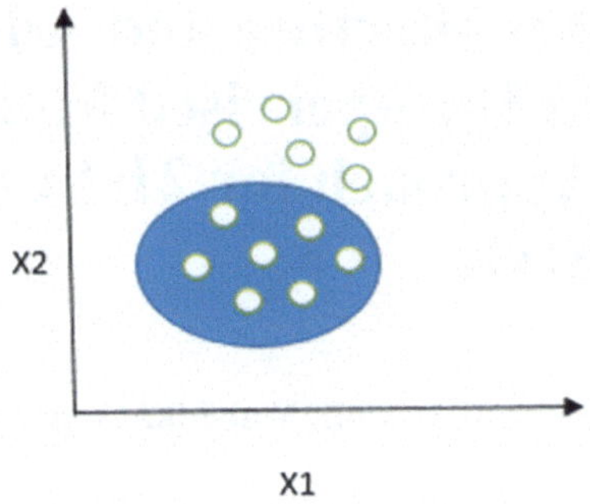

Fig. 1 Unsupervised learning

datasets. This chapter explores the architecture and advances in unsupervised models, presenting a conceptual framework tailored to the complexities of 21st-century smart lifestyles. By understanding the foundational principles and design considerations of unsupervised learning, we can unlock its potential to revolutionise personalised experiences, optimise resource utilisation, and foster autonomous adaptation within smart environments. Through this exploration, we aim to illuminate the transformative role of unsupervised models in shaping the future of smart living [1].

(As per Fig. 1) It represents the model of unsupervised learning and how it offers a promising avenue to meet this demand by extracting meaningful insights from vast, unlabelled datasets where X1 denotes independent variables and X2 denotes dependent variables.

1.1 Unsupervised Learning and Its Advancement in Recent Times

Unsupervised learning, a branch of machine learning, has seen significant advancements in recent times, driven by the exponential growth of data and computational power. Traditional unsupervised techniques such as clustering and dimensionality reduction have been augmented with more sophisticated algorithms capable of handling complex and high-dimensional data. One notable advancement is the rise of deep learning architectures, such as auto encoders and variation auto encoders, which enable the unsupervised learning of hierarchical representations from raw data. Additionally, generative adversarial networks (GANs) have emerged as powerful tools for unsupervised learning by generating realistic synthetic data samples [2]. Reinforcement learning, traditionally a supervised learning paradigm, has also seen extensions to unsupervised settings, enabling agents to learn from raw sensory inputs without explicit reward signals. These advancements have paved the way for unsupervised learning to address a wide range of real-world problems, from anomaly detection and data denoising to natural language understanding and autonomous decision-making [2].

1.2 The Evolution of Smart Lifestyles: From Supervised to Unsupervised Models

The evolution of smart lifestyles has witnessed a significant shift from reliance on supervised learning models to the adoption of unsupervised approaches. Initially, smart systems heavily relied on supervised learning algorithms, where labelled data was essential for training models to perform specific tasks. However, as the complexity and scale of data generated by smart devices increased, the limitations of supervised learning became apparent. Unsupervised learning emerged as a viable alternative, offering the ability to extract meaningful patterns and insights from vast, unstructured datasets without the need for labelled examples [3]. This transition has revolutionised smart lifestyles by enabling personalised recommendations, adaptive automation, and intelligent decision-making in diverse domains such as healthcare, transportation, and home automation. By embracing unsupervised models, smart lifestyles are poised to become more autonomous, adaptive, and responsive to individual preferences and needs, ushering in a new era of truly intelligent living [3].

(As per Fig. 2) shows the difference between the unsupervised and the supervised learning and how unsupervised learning emerged from supervised learning. This transition has revolutionised smart lifestyles by enabling personalised recommendations, adaptive automation, and intelligent decision-making in diverse domains such as healthcare, transportation, and home automation.

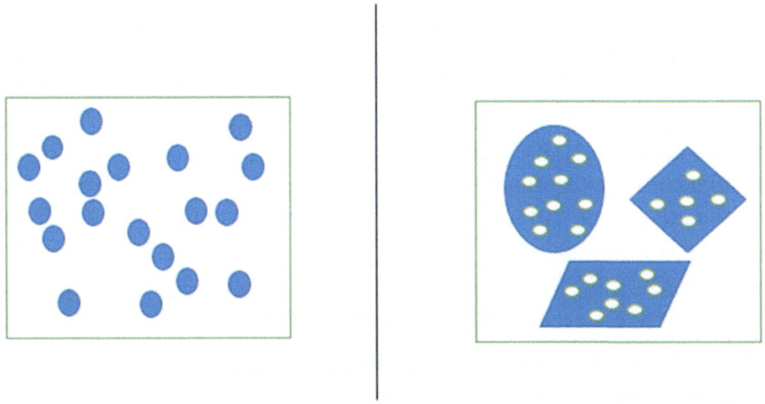

Fig. 2 Supervised versus unsupervised models

1.3 Architectural Paradigms: Design Principles for Unsupervised Systems

Architectural paradigms play a crucial role in shaping the design and development of unsupervised systems, providing the foundation for their functionality, scalability, and efficiency. This chapter explores key design principles that govern the architecture of unsupervised systems, focusing on approaches that facilitate effective data representation, feature extraction, and model optimization [4]. We examine various architectural paradigms, including deep learning architectures, graph-based models, and ensemble techniques, highlighting their strengths and weaknesses in different contexts. Additionally, we discuss strategies for designing modular and flexible architectures that can accommodate evolving requirements and data dynamics. By understanding the underlying design principles, practitioners can make informed decisions when architecting unsupervised systems to meet the challenges of real-world applications in diverse domains [4].

1.4 Data Representation and Feature Engineering in Unsupervised Learning

Effective data representation and feature engineering are essential components of unsupervised learning, enabling the extraction of meaningful patterns and structures from raw data. This chapter explores various techniques and strategies for representing data in unsupervised learning tasks, including dimensionality reduction, feature scaling, and transformation methods. Additionally, we delve into the role of feature engineering in enhancing the performance and interpretability of unsupervised models, discussing approaches such as kernel methods, auto encoders, and deep feature learning [5]. By understanding the principles of data representation and feature engineering, practitioners can effectively pre-process and transform data to improve the performance of unsupervised learning algorithms and uncover valuable insights from complex datasets [5].

1.5 Clustering Techniques for Personalization and Optimization

Clustering techniques play a crucial role in personalization and optimization tasks by grouping similar data points together, enabling tailored recommendations, targeted marketing strategies, and efficient resource allocation. Here's how clustering techniques are utilised for personalization and optimization:

- **Customer Segmentation**: Clustering algorithms, such as k-means, hierarchical clustering, or DBSCAN, can partition customers into distinct segments based on their similar characteristics, behaviours, or preferences. These segments can then be used to personalise marketing campaigns, product recommendations, and customer experiences.
- **Content Recommendation**: Clustering techniques can be applied to group similar items, such as articles, movies, or products, based on their features or attributes. By clustering items and recommending similar items to users who have shown interest in a particular item, personalised content recommendations can be provided, enhancing user engagement and satisfaction.
- **Dynamic Pricing**: Clustering algorithms can help identify groups of customers with similar purchasing behaviours or price sensitivities [6]. By clustering customers based on their willingness to pay, businesses can implement dynamic pricing strategies tailored to different customer segments, maximising revenue while maintaining customer satisfaction.
- **Resource Allocation**: In optimization tasks, clustering techniques can be used to allocate resources efficiently by grouping similar resources or tasks together. For example, in supply chain management, clustering can help optimise inventory management by grouping products with similar demand patterns, leading to better inventory planning and cost reduction.
- **Personalised Healthcare**: Clustering techniques can assist in personalised healthcare by grouping patients with similar medical profiles or treatment responses. These patient clusters can guide personalised treatment plans, medication recommendations, and healthcare interventions, leading to improved patient outcomes and healthcare efficiency.
- **Travel and Tourism**: Clustering algorithms can segment travellers based on their preferences, travel patterns, or demographics. By clustering travellers into distinct groups, personalised travel recommendations, itinerary suggestions, and targeted marketing offers can be provided, enhancing the overall travel experience.
- **Performance Optimization**: Clustering techniques can also be employed to optimise business processes and operations by identifying patterns or anomalies in data. For instance, in manufacturing, clustering can help identify groups of products with similar production requirements, leading to streamlined production processes and improved resource utilisation.

Overall, clustering techniques serve as powerful tools for personalization and optimization across various domains, enabling businesses to deliver tailored experiences, optimise operations, and maximise efficiency and customer satisfaction [6].

(As per Fig. 3) It shows how Clustering techniques play a crucial role in personalization and optimization tasks. It shows that clustering does so by grouping similar data points together, enabling tailored recommendations, targeted marketing strategies, and efficient resource allocation.

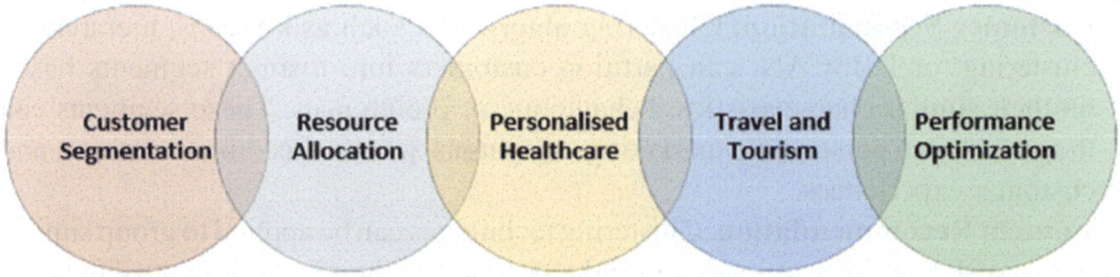

Fig. 3 Clustering for personalization an optimization

1.6 Dimensionality Reduction: Simplifying Complex Data Spaces

Dimensionality reduction techniques play a pivotal role in unsupervised learning by simplifying complex data spaces and facilitating more efficient analysis and visualisation. This chapter explores various dimensionality reduction methods, including principal component analysis (PCA), t-distributed stochastic neighbour embedding (t-SNE), and autoencoders. We delve into the principles underlying these techniques and discuss their applications in reducing the dimensionality of high-dimensional datasets while preserving important information [7]. Additionally, we examine strategies for evaluating and selecting appropriate dimensionality reduction methods based on the characteristics of the data and the objectives of the analysis. By effectively reducing the dimensionality of data, practitioners can gain deeper insights, improve computational efficiency, and enhance the interpretability of unsupervised learning models in diverse applications [7].

1.7 Anomaly Detection and Outlier Analysis: Safeguarding Smart Systems

In the realm of safeguarding smart systems, anomaly detection and outlier analysis play pivotal roles in identifying and mitigating potential threats or irregularities. Here's a breakdown of key points concerning anomaly detection and outlier analysis in this context:

- **Understanding Anomalies**: Anomalies, or outliers, refer to data points or events that deviate significantly from the norm or expected behaviour within a system [8]. In the context of smart systems, anomalies could indicate security breaches, equipment malfunctions, or unusual user behaviour.
- **Importance of Detection**: Detecting anomalies promptly is critical for safeguarding smart systems against various risks, including cyber-attacks, operational failures, and safety hazards. Anomaly detection techniques enable early detection and response, minimising potential damages and ensuring system integrity.

- **Techniques and Algorithms**: A wide range of techniques and algorithms are available for anomaly detection, including statistical methods, machine learning approaches, and domain-specific heuristics. These methods analyse data patterns and identify deviations that warrant further investigation.
- **Real-Time Monitoring**: Smart systems often require real-time monitoring capabilities to detect anomalies promptly and take timely corrective actions. Implementing continuous monitoring systems equipped with anomaly detection algorithms enables proactive threat detection and mitigation.
- **Integration with Security Measures:** Anomaly detection and outlier analysis should be integrated with other cybersecurity measures to enhance overall system resilience. This may involve combining anomaly detection algorithms with intrusion detection systems, access controls, and encryption techniques to fortify system defences comprehensively.
- **Adaptability and Scalability**: Anomaly detection systems should be adaptable and scalable to accommodate evolving threats and dynamic operating environments [8]. This requires periodic updates, algorithm refinement, and scalability measures to ensure effectiveness across various scales and scenarios.
- **User Education and Awareness**: Educating system users and administrators about the importance of anomaly detection and outlier analysis is essential for fostering a culture of security awareness. Training programs, awareness campaigns, and user-friendly interfaces can empower stakeholders to recognize and respond to anomalies effectively.

In conclusion, anomaly detection and outlier analysis play indispensable roles in safeguarding smart systems against emerging threats and vulnerabilities. By leveraging advanced detection techniques, real-time monitoring capabilities, and integrated security measures, organisations can enhance the resilience and reliability of their smart systems in the face of evolving challenges [8].

1.8 Generative Models: Creating Realistic Synthetic Data for Training

Generative models are becoming increasingly popular for creating realistic synthetic data used in training various machine learning models. Here's an overview of the role of generative models in this context:

- **Generating Synthetic Data**: Generative models, such as Generative Adversarial Networks (GANs), Variational Autoencoders (VAEs), and Autoregressive Models, are capable of learning the underlying distribution of real-world data and generating synthetic samples that closely resemble the original data.
- **Data Augmentation**: Synthetic data generated by generative models can be used to augment existing datasets, thereby increasing the diversity and size of the

training data [9]. This is particularly useful when the original dataset is limited or lacks variability, leading to improved model generalisation and performance.

- **Privacy-Preserving Training**: Generative models offer a means of generating synthetic data while preserving privacy-sensitive information present in the original dataset. This is achieved by training the generative model on the original data without directly exposing individual data points, thus mitigating privacy risks associated with sharing sensitive data.
- **Domain Adaptation and Transfer Learning**: Synthetic data generated by generative models can also facilitate domain adaptation and transfer learning tasks. By generating synthetic samples that resemble the target domain, models can be pre-trained or fine-tuned on this synthetic data before being applied to real-world scenarios with limited labelled data.
- **Data Imputation and Missing Value Handling**: Generative models can be employed for data imputation tasks, where missing values in a dataset are replaced with plausible synthetic values generated by the model [9]. This helps improve the completeness of the dataset and enhances the robustness of downstream analyses or predictions.
- **Addressing Data Bias and Imbalance**: Synthetic data generation can help address data bias and class imbalance issues present in the original dataset by generating additional samples for underrepresented classes or minority groups, thereby promoting fairness and equity in machine learning models.
- **Challenges and Considerations**: Despite their benefits, generative models also pose challenges, such as mode collapse, training instability, and ensuring the quality and diversity of generated samples. Careful evaluation and validation of the synthetic data are essential to ensure its suitability for the intended use case.

In summary, generative models offer a powerful tool for creating realistic synthetic data used in training machine learning models across various domains. By leveraging these models, practitioners can overcome data limitations, enhance model performance, and address privacy concerns while promoting fairness and inclusivity in machine learning applications [9].

2 Literature Review

See Table 1.

Table 1 Tabular summary for literature review based papers

S. No.	Paper, author name	Summary	Methodology, dataset, algo	Concluding remarks/findings	Gap
1	Farhan Amin, Rashid Abbasi, AbdulMateen, Muhammad Ali Abid and Salabat Khan	Explores the evolving landscape of IoT, emphasising the need for reevaluation due to the surge in data generated by IoT devices. It proposes a systematic analysis of next-generation IoT advancements, covering big data, data science, network science, and connecting technologies	The methodology involves a systematic analysis of next-generation IoT advancements, encompassing big data, data science, network science, and connecting technologies. It proposes a comprehensive catalogue of recent IoT advancements, emphasising integration with AI, machine learning, blockchain, and federated learning	Addresses the pressing need for re-evaluating IoT infrastructure to cope with the vast data generated, emphasising the necessity for seamless connectivity and support. By systematically analysing recent advancements and integrating disciplines like big data, data science, and network science, along with emerging technologies like AI and blockchain, the study offers insights into next-generation IoT development	This paper does not give proper view of the architectural models of the unsupervised learning
2	J. M. Górriz, I. Álvarez-Illán, A. Álvarez-Marquina, J. E. Arco	It showcases DL's prowess in extracting intricate features from data, achieving human-level performance in tasks like clinical diagnostics and robotics. Through a review of recent advances and applications in neuroscience, neuroimaging, computer vision, and robotics	The methodology involves presenting, reviewing, and discussing recent advancements in Deep Learning (DL) and its applications in various fields such as neuroscience, neuroimaging, computer vision, and robotics. The focus is on summarising state-of-the-art AI methods and models showcased at the 9th International Conference on the Interplay between Natural and Artificial Computation (IWINAC)	An overview of the significant advancements in Deep Learning (DL), showcasing its pivotal role in driving progress across various domains of Artificial Intelligence (AI). By leveraging complex neural systems, DL algorithms have achieved human-level performance in diverse real-world applications such as clinical diagnostics, robotics, computer vision, and industrial automation	This paper give advancements in deep learning and does not give a proper overview of the advancements in unsupervised learning and about its architecture

(continued)

Table 1 (continued)

S. No.	Paper, author name	Summary	Methodology, dataset, algo	Concluding remarks/findings	Gap
3	Shibo Zhang, Yaxuan Li, Shen Zhang, Farzad Shahabi	Delves into the transformative impact of mobile and wearable devices, particularly in the realm of human activity recognition (HAR), facilitated by the proliferation of low-power sensors. It systematically reviews and categorises existing research, focusing on the integration of deep learning methods in HAR for wearables	The methodology of this paper involves systematically categorising and summarising existing research on leveraging deep learning methods for human activity recognition (HAR) using mobile and wearable devices. It comprehensively analyses current advancements, emerging trends, and significant challenges in this domain	The integration of deep learning techniques with mobile and wearable devices has significantly advanced human activity recognition (HAR), enabling a wide range of applications for activity tracking and wellness monitoring. This paper systematically reviews and categorises existing research in this field, highlighting the latest advancements, emerging trends, and key challenges	This paper systematically reviews and categorises existing research in this field but does not give proper view of the architectures and advancements
4	Shabir Ahmad, Sabina Umirzakova, Faisal Jamil	Intersection of the Internet of Things (IoT) and serious games, highlighting how IoT-enabled serious games have emerged as a promising avenue for education, healthcare, and other domains. It conducts a comprehensive survey, mapping the evolution, motivation, and challenges in deploying serious games within the IoT environment	The methodology of this paper involves conducting a comprehensive survey on Internet of Things-enabled serious games, with a focus on understanding their evolution, motivation, and challenges. It begins by identifying domains where serious games are applied and explores the integration of IoT devices to enhance gaming experiences	The integration of Internet of Things (IoT) technology with serious games marks a significant evolution in both fields, offering innovative solutions across various domains such as education, healthcare, and physical training. This paper provides a comprehensive survey of IoT-enabled serious games, outlining their evolution, motivation, and application domains	This paper does not give a proper overview of the advancements in unsupervised learning

(continued)

Table 1 (continued)

S. No.	Paper, author name	Summary	Methodology, dataset, algo	Concluding remarks/findings	Gap
5	Jeffrey Hsu	Explores various technologies and research areas with significant potential in data mining, highlighting the growing importance of analysing the vast amount of data being generated. It emphasises the emerging role of handheld, wireless, and ubiquitous devices as sources of valuable information	The methodology of this paper involves identifying and discussing various technologies, approaches, and research areas with significant potential in data mining. It highlights the importance of harnessing the vast amount of data generated from handheld, wireless, and ubiquitous devices	Highlights the burgeoning potential of various technologies and research areas within data mining, driven by the exponential growth of data generated from diverse sources. With the proliferation of handheld, wireless, and ubiquitous devices, there is a significant emphasis on extracting valuable insights from these data streams	This paper does not tell about the advancements in the unsupervised learning

3 Framework and Proposed Model

Developing a conceptual framework for 21st-century smart lifestyle architecture and advances in unsupervised models requires a multidisciplinary approach, integrating principles from architecture, artificial intelligence, data science, and user experience design.

Understanding Smart Lifestyle Needs: Identify key aspects of modern lifestyles such as connectivity, sustainability, convenience, and personalization. Analyse how architecture can enhance these aspects through design, materials, and technology integration.

Data Collection and Analysis: Gather data from various sources including IoT devices, sensors, user interactions, and environmental factors. Utilise unsupervised learning techniques such as clustering and dimensionality reduction to find patterns and relationships in the data.

Architecture Integration: Design living spaces that adapt to occupants' needs and preferences based on real-time data analysis. Implement modular and flexible architectural elements to accommodate changing lifestyle requirements.

Personalization and User Experience: Develop AI-driven systems that learn from user behaviour and preferences to provide personalised experiences. Use unsupervised models to segment users into different lifestyle profiles and tailor services accordingly.

Sustainability and Energy Efficiency: Employ unsupervised learning to optimise energy usage within smart buildings by analysing occupancy patterns and environmental conditions. Integrate renewable energy sources and smart grid technologies to minimise environmental impact.

Privacy and Security: Implement robust security measures to protect sensitive user data collected by smart systems. Utilise encryption and anonymization techniques to ensure privacy while still extracting meaningful insights from data.

Scalability and Interoperability: Design architectures that can seamlessly integrate with existing infrastructure and future technologies. Utilise standardised protocols and open APIs to facilitate interoperability between different smart devices and systems.

Feedback Loop and Continuous Improvement: Establish mechanisms for gathering feedback from users to improve the smart lifestyle framework iteratively. Use unsupervised learning algorithms to analyse feedback data and identify areas for optimization and enhancement.

Community and Social Integration: Foster community engagement by designing shared spaces and amenities that encourage social interaction and collaboration. Use data-driven insights to organise community events and activities tailored to residents' interests and preferences.

By incorporating these principles, a conceptual framework for smart lifestyle architecture can be developed, leveraging advances in unsupervised models to create environments that are intelligent, adaptive, and conducive to 21st-century living ***(as per Fig. 4)***.

The information system supporting the intellectual environment is built on a foundation of intelligent gadgets, a server, and a wireless sensor network. This type of design allows electrical circuits to be combined with free-form plastic sensor housing in situations where complete monitoring systems with easily navigable features are necessary for home management and quality of life. The tiniest sensors, such as tactile signals that show whether a door is closed or whether someone has sat in a chair, and micro video cameras that might function as a distributed technological vision system, are linked to the touch wireless network. Motion, pressure, video, object contact, and sound sensors are a few examples of non-contact sensor technologies. Furthermore, multicomponent and smart technologies, such as wearable and ambient sensors, were discovered. In addition to doors, windows, furniture, courts, and books, the wireless network of intelligent things can also comprise lights, appliances, mobile robots, cell phones, and standard personal computers. Recall that these two networks may differ logically even though they are physically part of the same wireless LAN. Three essential roles of the recognition subsystem are activity recognition, glory recognition, and honour recognition of individuals and objects. However, due to a lack of contextual knowledge, these systems are unable to provide automated activity detection, other than using data from conventional sensors ***(as per Fig. 5)***.

The system makes use of probabilistic models, neural networks, and a hybrid approach with knowledge-based decision making. The localization subsystem needs to use information from numerous local sources, like multiple cameras' worth of visual input, in order to establish the position of an object or person. In order to track and identify important items around the globe, the system has the ability to locate itself in new places and generate a scene map. The coordinate recognition approach proposed by neural networks can be used to locate the issue. This method employs the vector of the intensity of signals from many beacons, either ultrasonic or

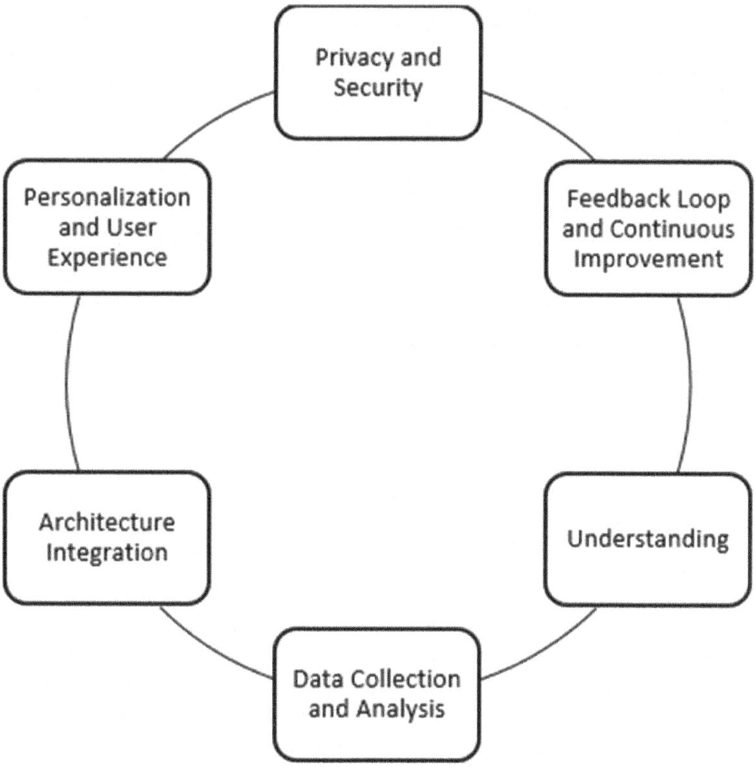

Fig. 4 Framework for architectural design of unsupervised models

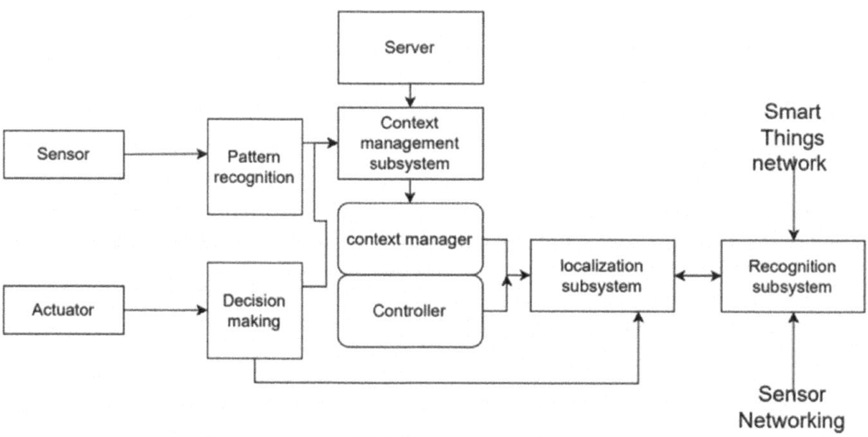

Fig. 5 Conceptual approach to smart lifestyle

electromagnetic. The context management subsystem gathers, stores, and updates the text that the decision-making subsystem employs. The time, place, problem-solving task, solution state, and other crucial elements are all relevant to the context of the circumstances that were discovered and the actions that were taken. The decision-making subsystem is a part of the system, which encompasses all the functions and tasks that the system does.

This subsystem is typically thought of as application software, as opposed to other subsystems, which are typically designed as middleware. With the use of sensors (also known as input devices) and actuators found in intelligent things like computers, robots, and cell phones, this subsystem carries out human interaction. Task-oriented applications such as question-and-answer systems, item search, communication, device control, multimedia device control, and loneliness-busting apps may be included in this subsystem.

4 Novelties and Recommendations

In terms of novelties, the chapter explores the intersection of architecture and unsupervised models, which represents a cutting-edge approach to designing smart living spaces. By incorporating unsupervised learning techniques into architectural design, we introduce a new paradigm that emphasises adaptability, intelligence, and user-centricity.

Some specific novelties highlighted in the chapter may include:

- **Integration of Unsupervised Models**: This chapter presents a novel framework for integrating unsupervised learning techniques into the architectural design process, enabling the creation of more responsive and intuitive living environments.
- **Advanced Applications**: The chapter explores advanced applications of unsupervised models in smart architecture, such as optimising energy consumption, personalising user experiences, and improving infrastructure maintenance.
- **Interdisciplinary Approach**: It emphasises the importance of an interdisciplinary approach, bringing together experts from the fields of architecture, data science, and engineering to collaborate on innovative solutions for smart living.

As for recommendations, the chapter may offer several actionable insights for architects, designers, and researchers interested in leveraging unsupervised models for smart architecture:

- **Experimentation and Prototyping**: Encourage architects and designers to experiment with unsupervised learning techniques through prototyping and iterative design processes. This hands-on approach can help uncover new possibilities and refine existing methodologies.
- **Data Collection and Analysis**: Emphasise the importance of collecting and analysing relevant data to train unsupervised models effectively. This may involve

gathering information about user preferences, environmental conditions, and building performance metrics.

- **Ethical Considerations**: Highlight the ethical considerations associated with deploying unsupervised models in real-world applications, such as data privacy, algorithmic bias, and transparency. Encourage practitioners to address these concerns proactively and responsibly.
- **Education and Training**: Provide educational resources and training programs to empower architects and designers with the knowledge and skills needed to leverage unsupervised learning techniques effectively. This may involve workshops, seminars, and online courses tailored to the needs of the architecture community.

By incorporating these recommendations into their practice, architects and designers can harness the power of unsupervised models to create smarter, more sustainable, and user-centric living environments for the 21st century.

5 Future Research

This chapter explores the exciting future directions and emerging trends in unsupervised learning, offering insights into the opportunities and challenges that lie ahead for smart lifestyles. We discuss cutting-edge research areas such as self-supervised learning, meta-learning, and lifelong learning, which hold promise for enhancing the capabilities of unsupervised models in adapting to dynamic environments and evolving user preferences. Additionally, we explore the integration of unsupervised learning with other AI techniques such as reinforcement learning and causal inference, enabling more robust and autonomous smart systems. Moreover, we examine the potential impact of unsupervised learning on societal challenges such as healthcare, sustainability, and urban planning, paving the way for more inclusive and sustainable smart lifestyles in the future. By staying abreast of emerging trends and opportunities in unsupervised learning, practitioners can harness the full potential of these techniques to shape a brighter and more intelligent future for smart living.

6 Directions and Limitations

This chapter delves into the integration challenges that arise when deploying unsupervised models in real-world scenarios, offering practical solutions to overcome these hurdles. We explore issues such as data quality, scalability, and interpretability, which can impact the effectiveness and reliability of unsupervised learning systems in production environments.

- **Data Quality**: Ensure data pre-processing techniques are employed to handle noisy and incomplete data, enhancing the quality of inputs for unsupervised models.
- **Scalability**: Implement distributed computing frameworks and efficient algorithms to handle large-scale datasets, ensuring scalability and performance of unsupervised learning systems.
- **Interpretability**: Develop techniques for interpreting and explaining the decisions made by unsupervised models, fostering trust and transparency in their outcomes.
- **Domain Adaptation**: Address domain-specific challenges by customising unsupervised models to the characteristics and requirements of specific application domains, such as healthcare or finance.
- **Evaluation and Monitoring**: Establish robust evaluation metrics and monitoring mechanisms to assess the performance and stability of unsupervised models over time, enabling continuous improvement and refinement.
- **Integration with Existing Systems**: Integrate unsupervised learning systems seamlessly with existing infrastructure and workflows, ensuring compatibility and interoperability with other components of the ecosystem.
- **Human-in-the-loop**: Incorporate human feedback and domain expertise into the unsupervised learning process, leveraging human-in-the-loop approaches to enhance model performance and adaptability.

7 Conclusions

In the conclusion of "Architecture and Advances in Unsupervised Models: A Conceptual Approach for 21st Century Smart Lifestyle," it's imperative to reflect on the key insights and advancements discussed throughout the chapter.

Firstly, we've explored the evolution of architecture in the context of smart living, emphasising the importance of integrating unsupervised models into the design process. These models allow for more adaptive and intelligent environments that can cater to the diverse needs and preferences of inhabitants.

Moreover, we've delved into various advances in unsupervised learning techniques, such as clustering, dimensionality reduction, and generative modelling. These techniques enable systems to autonomously learn patterns and structures from data without explicit supervision, thereby paving the way for more efficient and scalable solutions in smart architecture.

Furthermore, we've highlighted the potential applications of unsupervised models in enhancing various aspects of smart living, including energy efficiency, personalised user experiences, and proactive maintenance of infrastructure.

However, it's crucial to acknowledge the challenges and limitations associated with the adoption of unsupervised models in real-world applications. These include issues related to data privacy, interpretability, and scalability, which require careful consideration and mitigation strategies.

In conclusion, while unsupervised models offer promising avenues for revolutionising smart living in the 21st century, their successful integration into architectural design necessitates a holistic approach that balances technological innovation with ethical and societal considerations. By embracing this approach, we can unlock the full potential of unsupervised models to create more sustainable, efficient, and user-centric living environments for future generations.

Additional Readings

Unsupervised learning: algorithms and example. **https://www.altexsoft.com/blog/unsupervised-machine-learning/**

Unsupervised machine learning to investigate trajectory patterns of COVID-19 symptoms and physical activity measured via the MyHeart Counts App and smart devices. **https://www.nature.com/articles/s41746-023-00974-w**

Top 10 machine learning examples in real life (which make the world a better place). **https://www.omdena.com/blog/machine-learning-examples**

Unsupervised daily routine and activity discovery in smart homes. **https://pubmed.ncbi.nlm.nih.gov/26737536/**

Smart bulit environment including smart home, smart building and smart city: definitions and applied technologies. **https://www.intechopen.com/chapters/74934**

Annexures

Key Terms and Definitions

Unsupervised Models: Unsupervised learning in artificial intelligence is a type of machine learning that learns from data without human supervision. Unlike supervised learning, unsupervised machine learning models are given unlabeled data and allowed to discover patterns and insights without any explicit guidance or instruction.

Smart Lifestyle: A smart lifestyle is the happy mixture of green and organised family-focused lifestyle and digital gadgets.

Generative AI: Generative artificial intelligence is artificial intelligence capable of generating text, images or other data using generative models, often in response to prompts.

Emerging Trends: Emerging trends are patterns, behaviours, or preferences that are gaining popularity or influence in a specific industry, niche, or audience.

Unstructured Datasets: Unstructured data has an internal structure but does not contain a predetermined data model or schema. It can be textual or non-textual. It can be human-generated or machine-generated.

References

1. Barlow, H.B.: Unsupervised learning. Neural Comput. **1**(3), 295–311 (1989). https://direct.mit.edu/neco/article-abstract/1/3/295/5492/Unsupervised-Learning
2. Rosato, A., Altilio, R., Panella, M.: Recent advances on distributed unsupervised learning. Advances in neural networks. Comput. Intell. ICT 77–86 (2016). https://link.springer.com/chapter/10.1007/978-3-319-33747-0_8
3. Rajoub, B.: Supervised and unsupervised learning. In: Biomedical Signal Processing and Artificial Intelligence in Healthcare, pp. 51–89. Academic Press (2020). https://www.sciencedirect.com/science/article/abs/pii/B9780128189467000032
4. Anselmi, F., Leibo, J. Z., Rosasco, L., Mutch, J., Tacchetti, A., Poggio, T.: Unsupervised learning of invariant representations in hierarchical architectures (2013). arXiv:1311.4158
5. Zhang, W., Dong, X., Li, H., Xu, J., Wang, D.: Unsupervised detection of abnormal electricity consumption behavior based on feature engineering. IEEE Access, **8**, 55483–55500 (2020). https://ieeexplore.ieee.org/abstract/document/9032114
6. Michaud, P.: Clustering techniques. Future Gener. Comput. Syst. **13**(2–3), 135–147 (1997). https://www.sciencedirect.com/science/article/abs/pii/S0167739X97000174
7. Dash, M., Liu, H., Yao, J.: Dimensionality reduction of unsupervised data. In: Proceedings ninth IEEE international conference on tools with artificial intelligence, pp. 532–539, November 1997. IEEE (1997). https://ieeexplore.ieee.org/abstract/document/632300
8. Ahmad, S., Lavin, A., Purdy, S., Agha, Z.: Unsupervised real-time anomaly detection for streaming data. Neurocomputing **262**, 134–147 (2017). https://www.sciencedirect.com/science/article/pii/S0925231217309864
9. Theis, L., Oord, A.V.D., Bethge, M.: A note on the evaluation of generative models (2015). arXiv:1511.01844

Shivani Trivedi, an accomplished academic professional, brings a wealth of knowledge and dedication to the field of Computer Science and Engineering. She earned her Bachelor of Engineering (B.E.) degree in CSE and further honed her expertise by completing her Master of Technology (M. Tech.) in 2019 from Punjab Engineering College Chandigarh. Currently serving as an Assistant Professor at ABES Engineering College in Ghaziabad, she has received widespread acclaim for her teaching process. Her commitment to education has been recognized through numerous awards in teaching, training, including the coveted "Best Faculty Award." In addition to her teaching achievements, she has made significant contributions to the academic community with her research work, boasting a portfolio of Scopus-indexed journal papers and conference papers. Beyond her scholarly pursuits, she actively participates in the organization of conferences, faculty development programs, seminars, and various educational events, exemplifying her dedication to fostering academic excellence and innovation in the field. Ms. Shivani Trivedi is a dynamic educator and researcher, continually inspiring and shaping the future of Computer Science and Engineering professionals.

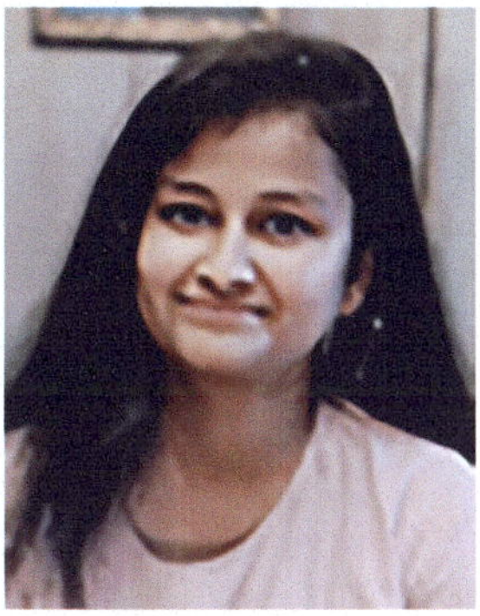

Vanshika Aggarwal completed her high school and intermediate education at Montfort School, Roorkee. She's currently pursuing her graduation in computer science and engineering from ABES Engineering College, Ghaziabad. She has won many awards, including "Topper in Mathematics." She also won awards at the Olympiads. In addition to academic achievements, she has made significant contributions to the academic community with her research work and conference papers. Beyond this, she actively participates in the organisation of various events, showing her excellence and innovation in the field. She's a hard-working and brilliant student who believes that the price of success is hard work and the determination to apply the best of herself to the task at hand.

Rohit Rastogi received his B.E. in CS from C. C. S. Univ. Meerut, 2003. Master's degree in CS of NITTTR-Chandigarh from Punjab University. He has obtained his doctoral degree from the Dayalbagh Educational Institute in Agra, India. He is serving as Associate Professor in the CSE department of ABES Engineering College, Ghaziabad, India. He has won awards in several areas, including improved education, significant contributions, human value promotion, and long-term service. He keeps himself engaged in various competition events, activities, webinars, seminars, workshops, researches and various other educational learning forums. He has guided around 40 B.Tech. Students researchers and 5 M.Tech. thesis and consequently 3 Ph.D. candidates. He is editor and reviewer member of several international Journals and conferences. He has 100+ publications in Scopus Indexed journals and conferences of International repute. He strongly believes that Transformation starts within self.

Artificial Intelligence in 2D Games: Analysis on Customised Character Generation

Hemlata Parmar and Utsav Krishan Murari

Abstract The article investigates the creative creation and customisation of 2D gaming characters using Artificial Intelligence (AI). As machine learning continues to progress, AI image generation has evolved significantly, presenting a plethora of opportunities for the gaming industry. The research focuses on how AI can automate the creation of unique and diverse game characters based on textual descriptions, thereby enhancing the gaming experience through customization. We delve into the technical aspects of AI image generation, its application in 2D gaming, and its implications on user engagement and game development. A case study showcasing the implementation of AI for character creation is also included, providing practical insights on the subject. The study unveils how AI image generation can revolutionize 2D game character creation, offering benefits for game developers, players, and the broader gaming industry. The findings pave the way for future research in this burgeoning field, emphasizing the potential of AI in transforming digital entertainment.

Keywords Artificial intelligence · Game development · Character customization · Text-to-image synthesis · Generative models

1 Introduction

In the ever-evolving field of game development, the integration of artificial intelligence (AI) technologies has opened new frontiers in creating more immersive and personalized gaming experiences. One such revolutionary application of AI is in the generation of game characters for 2D games. This technology, powered by techniques

Hemlata Parmar (✉)
Department of Computer Science and Engineering, Manipal University Jaipur, Jaipur, India
e-mail: Hparmar601@gmail.com

U. K. Murari
Department of Mass Communication, Sharda University, Greater Noida, UP, India

P. Raj et al. (eds.), *Building Embodied AI Systems: The Agents, the Architecture Principles, Challenges, and Application Domains*, Information Systems Engineering and Management 14, https://doi.org/10.1007/978-3-031-68256-8_15

such as Generative Adversarial Networks (GANs), allows both developers and users to create unique, customized characters using simple text prompts.

This concept of creating game characters with AI image generation is not just an innovative approach in streamlining the game development process, but also a tool that amplifies the creative autonomy of the users. By being able to manifest a character from mere text descriptions, users are given a unique opportunity to tailor their gaming experience. This novel approach represents a paradigm shift from traditional game design, where character designs are typically pre-determined and rigid.

The use of AI image generation for creating game characters is at the convergence of technology and creativity. The application of this technology is revolutionizing the way we design and incorporate characters into video games, giving a fresh spin to the story of character creation. It brings a level of personalization to the virtual world of gaming that was previously unattainable. Of course, like any cutting-edge technology, it comes with its own set of significant technical hurdles. But the potential advantages and the innovative nature of this concept suggest that it's a very promising and exciting area to explore.

This research paper delves into the use of AI image generation for creating game characters in 2D games. It explores the underlying technologies, benefits, challenges, and potential future developments in this burgeoning area of game development. By shedding light on the intricacies involved in AI-powered character creation, this paper aims to contribute to the understanding and advancement of this exciting application of artificial intelligence in game design.

2 Literature Review

Previous Studies on AI Image Generation

According to developments in artificial intelligence, the field of picture production has experienced significant expansion in the last few years. A useful method called Generative Adversarial Networks (GANs) (Goodfellow et al. 2014) has emerged as a significant contributor to this growth. Introduced by Goodfellow et al. in 2014, GANs have revolutionized the generation of high-quality and realistic images from noise vectors. The structure of a GAN involves two competing neural networks: the generator, which creates images, and the discriminator, which evaluates them. Their rivalry encourages constant advancements in the produced picture quality.

The application of GANs in the field of art has been explored, such as in the work of (Elgammal et al. 2017), where a creative adversarial network was used to generate novel art images that carried the styles of various periods. This demonstrated the potential of AI as a creative tool.

Within the sphere of transforming text descriptions into images, the study by (Reed et al. 2016) applied GANs for synthesizing images based on text captions. This research underscores the promising role of AI in converting textual descriptions into visuals, such as creating game characters. More recently, OpenAI's DALL-E [1] has successfully generated diverse images from text inputs, further validating the feasibility of this approach.

2.1 Previous Studies on Game Character Customization

The importance of character customization in video games has been widely recognized. (Tychsen et al. 2008) discussed the impact of character customization on player immersion and satisfaction within role-playing games.

Trepte and Reinecke (2010) found that the availability of character customization features in a game significantly influenced players' enjoyment and identification with the game characters.

In terms of procedural character generation, Compton and Mateas (2006) explored the possibility of algorithmic character generation in video games, demonstrating that procedural techniques could be used to generate unique and interesting characters.

The integration of AI-based image generation with game character customization is a relatively new field, and therefore represents an exciting frontier for research. This combination could potentially revolutionize how players interact with games, allowing for an unparalleled level of character personalization.

3 AI Image Generation

3.1 Definition and Overview

AI Image Generation is a technique where artificial intelligence (AI) mechanisms generate novel images. These systems commonly employ machine learning models, specifically a category known as Generative Adversarial Networks (GANs), to create images that emulate the style and subject matter of a provided training set.

This concept operates on the idea of example-based learning. The AI is educated using a vast collection of image data, enabling it to recognize patterns, hues, forms, and other components within these images. After the learning phase, it utilizes this acquired knowledge to produce new images that share similarities with the training set. Image generation AI has an array of applications, ranging from creating art and designing clothing to generating realistic human faces and producing images of objects or scenes described in text. The strength of AI image generation resides in its capacity to yield a virtually unlimited assortment of images, thereby broadening the boundaries of creativity and automating certain facets of content creation. Nonetheless, it also introduces hurdles, like maintaining the quality and suitability of the generated images, which remain as ongoing subjects of exploration and refinement in research and development.

3.2 History of AI Image Generation

In response to human cues, Artificial Intelligence (AI) has been used to produce unique, well-structured material since the 1950s. During this period, MIT Artificial Intelligence Laboratory researchers created the ELIZA software. Using natural language processing and pattern matching, this programme could provide basic answers to text input. ELIZA was an early example of Text-to-Text software that could generate unique text output intended for human interpretation, even though it was not actively involved in the creation of art. In the 1970s, artist Harold Cohen developed a programme called AARON, which produced the first known instance of artificial intelligence (AI)-generated art. An advanced computer programme called AARON was created to produce intricate paintings and sketches. It made art within a system of guidelines and limitations and was able to grow over time by taking lessons from its creations.

AI-generated art developed in the 1980s and 1990s, becoming increasingly complex and sophisticated as AI technology advanced. For example, Karl Sims used evolutionary algorithms to produce original 3D visuals and animations. Recent advances in deep learning have further improved the outputs' realism, drawing more attention from the general public and the art community to AI-generated art.

In 2015, a neural network was trained with a dataset that included over 10,000 artworks by the Google team using deep learning techniques. Producing original artwork from input photographs was the aim. This project resulted in the creation of the Deep Dream programme, which is well-known for its ability to convert supplied photos into aesthetically appealing, surreal visuals. This process is called image-to-image transformation. An impressive instance is the creation of an artificial intelligence-generated picture by the Parisian group "Obvious." At a Christie's auction in 2018, this portrait brought in over $432,000, marking a significant achievement for AI-produced art. In 2020, the area of Text-to-Text capabilities had a notable breakthrough with the release of OpenAI's third generation Generative Pretrained Transformer (GPT-3). GPT-3 was a significant advancement in Text-to-Text model generalizability, as it could produce grammatically sound text in answer to almost any natural language query. This was made feasible by the model's immense scale, which included 175 billion parameters—a quantity never seen in prior simulations of a comparable kind. Large Language Models were born when GPT-3 was able to do language tasks for which it was not specifically trained thanks to its large number of parameters. These models can produce human-like, high-quality content that is helpful for a variety of tasks, including text summarization, machine translation, and creative writing.

Because of the great reception that GPT-3 received, OpenAI went on to create another ground-breaking model called CLIP, which links text and pictures. Contrastive Language–Image Pretraining, or CLIP, is a powerful image-text model that can do picture classification with any user-specified label. It was trained on 400 million text-image pairings from the internet. Additionally, it can properly describe any input image in words (Image-to-words). Expanding upon these advancements, OpenAI unveiled DALL-E, which can produce realistic pictures from textual descriptions (Text to-Image).

The CLIP code was made publicly available, while DALL-E is still proprietary and closed-source. This made it possible for Stability AI, an artificial intelligence company, to create and train Stable Diffusion, an open-source Text-to-Image model that performs on par with DALL-E. A permissive licence that permits both commercial and non-commercial use was used to distribute Stable Diffusion. Models like CLIP and Text-to-Image systems based on them raise serious ethical and sociological issues in addition to being a major technological advance.

Because of its haphazard training on vast amounts of online data, CLIP frequently perpetuates unjust and biased preconceptions that are ingrained in society and culture. Legal authorities have also expressed concern about the possibility that these technologies would be used improperly to exploit intellectual materials. Furthermore, there's a chance that these technologies may be abused to spread misinformation or misleading material, among other negative uses.

3.3 Stable Diffusion

Stable Diffusion is a text-to-image model that was released in 2022. It creates images from word descriptions by using a deep learning method called latent diffusion. In contrast to several previous Text-to-Image models, Stable Diffusion's model weights and code are freely available to the public and work with the majority of consumer-grade hardware.

In order to generate an image for Stable Diffusion, a text prompt is first converted into a combined text-image embedding space using CLIP. Next, an image that is semantically meaningful yet noisy and rough is chosen to correlate with the input prompt.

The final picture is then obtained by subjecting this provisional image to a denoising process that is based on the latent diffusion model. Users may set up parameters like the output picture size, seed value, and sampling method in the Text-to-picture production script in Stable Diffusion. It is possible to maintain consistency and reproducibility of certain picture elements between prompts by setting this seed value, which is normally set randomly, at a consistent value. For example, keeping a steady seed might preserve a person's look in different stances and environments or their facial traits throughout prompts.

Text prompts, however, may result in the generation of random artefacts and defects even with a constant seed value, maybe necessitating post-processing. Certain Stable Diffusion user interfaces, such as Dream Studio, include post-processing features like in-and out-painting. Paints a portion of a picture by filling in a masked area in response to a request from the user, whereas Outpainting enlarges an image beyond its initial bounds. The Stable Diffusion model is used by both functions to produce new material.

For instance, we may add accessories or change the backdrop of an image using Stable Diffusion and the corresponding painting prompts. Additionally, as seen by the supplied pictures, Stable Diffusion has the capacity to emulate the style of modern painters. Later in the study, this potentially controversial element of generative AI is covered in more detail.

Figures 1 and 2 present illustrative examples of images produced via the application of the Stable Diffusion XL model.

3.4 Process of AI Image Generation Using Stable Diffusion

The process of generating images using the Stable Diffusion model can be performed both online and locally. The following sections describe the necessary steps for each method.

3.4.1 Running Stable Diffusion Online

There are several online platforms that allow users to experiment with the Stable Diffusion model. Two noteworthy services include Dream Studio and Hugging Face's Stable Diffusion platform. Dream Studio, developed by Stability AI, offers an accessible interface to the latest Stable Diffusion models. The service requires registration and provides users with a limited number of free credits for generating images. Additional credits can be purchased for extended usage.

Hugging Face, a prominent AI community and platform, also provides free access to the Stable Diffusion model. Although the customization options are more limited compared to DreamStudio, Hugging Face offers a free and straightforward way to generate images using the model.

3.4.2 Running Stable Diffusion Locally

Running Stable Diffusion locally allows for more extensive customization and fine-tuning of the generated images. Please note that a GPU is required to run Stable Diffusion on a local machine. Here is a step-by-step guide to setting up and running Stable Diffusion locally.

1. Install Python and Git: Python 3.10.6 and Git need to be installed on your machine. Both can be downloaded and installed from their official websites.
2. Create a GitHub and Hugging Face account: GitHub is a platform for hosting and managing software development projects, and Hugging Face is an AI community that hosts various machine learning models, including Stable Diffusion.

Fig. 1 Images generated using Stable Diffusion XL model, with the prompt Detailed illustration of a young woman in a Victorian dress, full-length shot, standing in a garden, in the style of Claude Monet, Impressionism

(a)

(b)

(c)

Fig. 2 Images generated using Stable Diffusion XL model, with the prompt Intricate image of a futuristic cityscape at dusk, high angle view, reflecting neon lights, in the style of Syd Mead, Blade Runner

3. Clone Stable Diffusion Web-UI to your PC [2]: Using Git, clone the Stable Diffusion Web-UI from its repository on GitHub. This will download the necessary files to your local machine.
4. Download the Latest Stable Diffusion model: Log in to your Hugging Face account and download the latest Stable Diffusion model. Once downloaded, move the model file to the appropriate folder within the cloned Stable Diffusion Web-UI directory.
5. Setup the Web-UI: Navigate to the Stable Diffusion Web UI directory in your command prompt and execute the provided setup script. This script will create a virtual environment and install the necessary dependencies.
6. Run Stable Diffusion: After the setup is completed, the command prompt will output a local URL. By navigating to this URL in a web browser, you can access the Stable Diffusion Web-UI and start generating images.

In essence, the process of AI image generation with Stable Diffusion entails the training of a machine learning model on an extensive image dataset. Once trained, this model can produce new images corresponding to provided text prompts. This procedure can be executed either via online platforms or directly on a local machine, offering flexibility based on user-specific needs and resource availability.

4 Application of AI Image Generation in 2d Games

The advent of artificial intelligence (AI)-enabled image generation represents a transformative potential in the realm of 2D game character creation. Leveraging advanced machine learning algorithms, game developers now possess the capability to craft characters of desired quality by means of textual prompts. This novel approach promises to introduce greater efficiency and seamlessness into the process of game development.

4.1 Current Use Cases

In the context of character sprite development, the utilization of AI-powered image generation techniques extends to the comprehensive generation of character animations, postures, and facial expressions. This technological approach presents a considerable advantage for game developers, relieving them of the arduous task of manually crafting and animating each individual character.

Furthermore, AI-driven image generation finds application in the realm of character customization, enabling the creation of an extensive array of personalized options encompassing weaponry, armor, accessories, and hairstyling. This empowers players with the capacity to curate a distinct character that aligns with their individual gameplay preferences and stylistic inclinations.

In a broader scope, AI image generation unfolds its potential within the domain of procedural content generation, encompassing the creation of dynamic terrains, architectural structures, and immersive game environments. The consequence of such technological implementation is the realization of virtually boundless gaming worlds, thereby conferring upon each player a unique and tailored experience.

Potential Use Cases:

Dynamic character creation: AI image generation can be used to create characters that change appearance based on player choices or game conditions. As an illustration, a character's visual attributes may undergo alterations contingent upon factors such as their in-game level, equipped items, or chosen playstyle.

In the domain of character personalization, AI-driven image generation techniques find application in the crafting of tailored characters informed by player-specific data, encompassing details such as their name, geographical location, and individual gaming preferences. This innovation serves to augment the depth of immersion and engagement experienced by players within the gaming environment.

Moreover, AI image generation extends its utility to the realm of automated game design, specifically addressing tasks associated with level creation, enemy artificial intelligence, and puzzle generation. By delegating these elements to artificial intelligence, game developers are afforded greater flexibility and resources to direct their creative energies towards narrative development and gameplay mechanics.

4.2 Benefits of AI Image Generation for Game

Development:

Increased efficiency: AI image generation can significantly reduce the time and effort required to create game characters, allowing game developers to focus on other aspects of game development.

Augmented Uniformity: AI image generation serves as a mechanism for ensuring the consistency of character design and stylistic elements throughout the game, contributing to the establishment of a harmonious and refined gaming milieu.

Elevated Customization: The utilization of AI image generation bestows players with a comprehensive spectrum of customization choices, thereby enabling the crafting of a distinct character that harmonizes with their unique playstyle and personal predilections.

Prolific Procedural Content: The deployment of AI image generation techniques begets an extensive array of potential game worlds, thereby affording each player an idiosyncratic experience and concurrently mitigating the risk of monotonous gameplay occurrences.

Heightened Player Engagement: AI-driven image generation offers the capacity to engender bespoke characters, finely attuned to each player's proclivities, thereby fostering a more immersive and captivating gaming journey.

5 User Experience

Customized Game Characters

Within the gaming universe, the importance of character customization cannot be overstated. It's a feature that gamers have come to anticipate, elevating their experience significantly. Customizing a character to reflect personal preferences and styles allows players to dive deeper into the game, nurturing an intimate bond and commitment. The power to personalize characters in a game is what truly connects players to the game world, forming a richer and more personal narrative between the player and the game's universe. From a visual standpoint, customization offers players a chance to showcase their uniqueness and imagination. On the game play side of things, it can offer tactical benefits, since distinct characteristics or gear choices can influence the game in various ways. This degree of influence and personalization elevates the overall gaming experience, rendering games more enjoyable and captivating for players. Effects of Customized Characters on Player Engagement Research has shown that customized characters can significantly impact player engagement. The provision of character customization capabilities to players frequently engenders a heightened sense of investment in the gaming experience.

This is attributable to the fact that players are granted agency in shaping a unique virtual persona, consequently elevating their emotional attachment and affinity towards the game. Customized characters can also impact player retention. The time and effort put into creating a character can increase a player's commitment to the game, making them more likely to continue playing over time. Moreover, the anticipation of unlocking new customization options can also serve as a motivator for continued play. Role of AI in Enhancing Character Customization Artificial Intelligence (AI) represents a profound transformative force poised to revolutionize the landscape of character customization within the realm of gaming.

A particularly auspicious avenue within this paradigm is the integration of AI-driven image generation to facilitate the creation of in game characters. Generative Adversarial Networks (GANs), a type of AI image generation, have the ability to produce distinctive and innovative character designs from textual descriptions or other types of inputs. For instance, if a player describes their ideal character as "a tall, muscular warrior donned in a red cape with golden armor," the AI has the capability to generate an image mirroring this vivid description. This opens up a world of customization and uniqueness that traditional design tools struggle to match. Another advantage of AI image generation is efficiency. The utilization of artificial intelligence for character generation yields a pronounced temporal advantage in contrast to the labor-intensive practice of hand-drawn or manual character design.

This efficacy assumes paramount significance, notably for game developers, where the expeditious generation of a diverse array of characters is of substantial utility. Furthermore, the advent of AI-driven image generation affords the prospect of a more immersive and participatory user experience. In this paradigm, players are afforded the opportunity to engage in a quasi collaborative endeavor with the AI, whereby they can refine AI-generated designs to align with their individual preferences. This collaborative dimension not only augments the intrinsic amusement value of character creation but also deepens the player's emotional attachment to their virtual persona.

6 Methodology

6.1 Data Collection

Game Character Dataset: Collect a dataset of game characters from various 2D games, including their visual appearance, characteristics, and background information. This dataset will serve as the foundation for training the AI image generation model.

6.2 User Survey

Conduct a survey among game developers and users to understand their preferences, needs, and expectations regarding game characters. The survey should gather information on the most popular game genres, character types, and customization options.

a. Game play Data: Collect data on user interactions with game characters, such as player choice, gameplay behavior, and user feedback. This data will help evaluate the effectiveness of the generated characters in a real-world game environment.
b. Expert Feedback: Gather feedback from game development experts on the generated characters, including their visual quality, diversity, and relevance to the game genre.
c. User Study: Conduct a user study to evaluate the effectiveness of the generated characters in a real-world game environment. Recruit a group of participants to play a 2D game with personalized characters and collect feedback on their experience.

6.3 *Data Analysis*

6.3.1 Descriptive Statistical Analysis

Employ descriptive statistics to succinctly encapsulate and synthesize the diverse data emanating from the game character dataset, user survey, gameplay metrics, and expert evaluations. This analytical approach serves as a foundational step in rendering a comprehensive overview of the data, facilitating the discernment of prevalent trends and patterns therein.

6.3.2 Visual Representation Methods

Utilize an array of visualization techniques, including histograms, bar charts, and scatter plots, to graphically depict the data and elucidate interrelationships between variables. These graphical representations aid in the discernment of nuanced connections and areas warranting refinement within the AI image generation model.

A. Utilization of Machine Learning Algorithms: Employ a repertoire of machine learning algorithms, including clustering, classification, and regression, to systematically analyze the data and uncover latent patterns and trends. By leveraging these algorithms, the refinement of the AI image generation model is achieved, ensuring its harmonization with the specific requisites of the game development community.
B. User Feedback Analysis: Analyze user feedback from the user study to evaluate the effectiveness of the generated characters in a real-world game environment. Use techniques such as thematic analysis or sentiment analysis to identify areas where the generated characters met or failed to meet user expectations. e. Comparison with Traditional Methods: Compare the performance of the AI image generation model with traditional game character design methods. Use metrics such as time, cost, and quality to evaluate the effectiveness of the AI image generation model and identify areas where it can be improved.

6.4 *Tools and Techniques Used*

6.4.1 Deep Learning Frameworks

Utilize robust deep learning frameworks like TensorFlow, PyTorch, or Keras. These frameworks form the backbone for building, developing, and training the AI image generation model, highlighting their fundamental role in outlining the model's architecture and competencies.

6.5 Stable Diffusion Model

Implement the Stable Diffusion model, a cutting-edge AI model designed for the purpose of generating high-quality images from text prompts. This model is central to our image generation process and plays a crucial role in the creation of dynamic in-game characters.

6.6 Image Processing Libraries

Make use of image processing libraries such as OpenCV, Pillow, and scikit-image for vital preprocessing and manipulation tasks related to game character images. These libraries prove instrumental in enhancing the visual quality and adaptability of the in-game characters, underscoring their essentiality in game development.

6.7 WebUI for Stable Diffusion

Utilize the WebUI for Stable Diffusion, an interface that allows for easy interaction with the Stable Diffusion model. This tool simplifies the process of inputting text prompts and retrieving the generated images, making it a key component of our implementation process.

6.8 Game Engines

Incorporate game engines like Unity, Unreal Engine, or Godot to construct a real-world gaming environment for the user study.

7 Case Study: Implementing AI Image Generation for Character Creation

Overview of the Case Study

This case study focuses on the integration of AI image generation into a 2D platformer game similar to the iconic Mario series. Considering the sustained appeal of these types of games, the incorporation of AI-enabled character customization offers a substantial potential to amplify player engagement and deliver a distinctively personalized gaming experience.

Traditionally, characters in 2D platformer games are pre designed and offer limited customization options to the players. This case study seeks to challenge this norm by exploring how AI image generation can be used to create personalized, dynamic game characters based on textual descriptions provided by the players.

The principal objective is to facilitate an enriched gaming experience where players can not only navigate their characters through the game's challenges but also have creative control over their character's visual attributes. This initiative represents an innovative fusion of machine learning technology and classic gaming, potentially setting a new standard for player interaction and engagement in 2D platformer games.

7.1 Implementation Process

The incorporation of AI image generation into our 2D platformer game was accomplished through a comprehensive, multi-stage process:

7.1.1 Model Training

The first step of the process was training the AI model. For this purpose, the model was trained on a vast dataset of 2D characters and environmental objects relevant to platformer games. This facilitated the model's learning process, enabling it to grasp and internalize the unique characteristics and stylistic elements inherent to a broad spectrum of game characters and assets.

7.1.2 Incorporating the AI Model into the Game Engine

Once the AI model had been comprehensively trained, the subsequent step entailed its integration with the game engine. This integration was facilitated through an API, which established a two-way communication channel between the game engine and the AI model. This arrangement allowed text descriptions, as provided by the players, to be transmitted to the AI model through the API. Once the AI model received a text description, it would proceed to generate an image that matched the given description. This image would then be transmitted back to the game engine. Upon receiving the image, the game engine would render it, seamlessly integrating it into the game's environment.

7.1.3 Development of a User Interface for Player Input

To facilitate player interaction with the AI model, a user friendly interface was developed. This interface provided a text input field where players could describe their desired game character. The design of the interface was kept intuitive and straightforward to ensure ease of use for players, encouraging them to experiment with various character descriptions.

7.1.4 Evaluation, Fine-Tuning, and Optimization

The concluding phase involved a meticulous evaluation and refinement process. This critical stage was dedicated to verifying the operational functionality of the AI model, the effectiveness of its integration with the game engine, and the flawless in-game depiction of the AI-generated characters. During this phase, any identified issues were rectified, and required optimizations were carried out. This ensured the delivery of smooth gameplay and the presentation of high quality visual characteristics in the AI-generated characters.

Figures 3 and 4 illustrate examples of the images generated using the AI model. These figures demonstrate the model's capability to create a diverse array of unique characters based on player-provided text descriptions, enhancing the customization possibilities within the game. The successful implementation of this technology in a 2D platformer game paves the way for future explorations of AI in game design, offering a promising avenue for the creation of increasingly immersive and personalized gaming experiences.

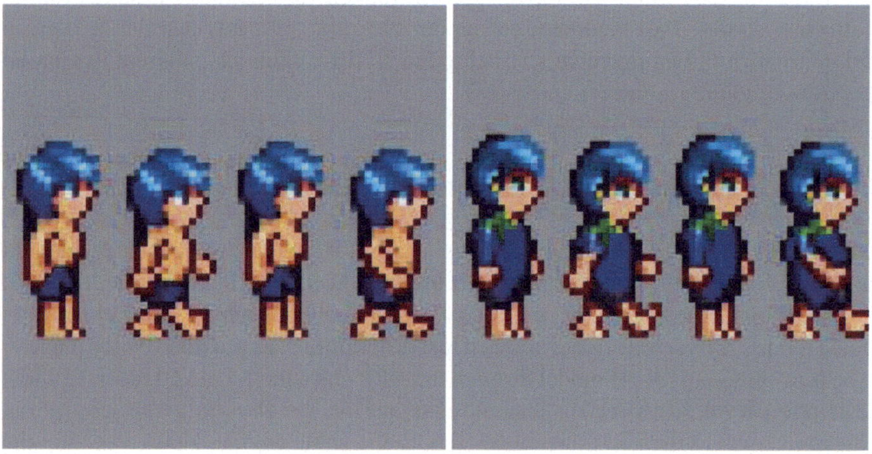

(a) (b)

Fig. 3 Game characters generated using a custom Stable Diffusion Model, with prompt boy with blue hair

Fig. 4 Game characters generated using a custom Stable Diffusion Model, with prompt girl with red hair

8 Implications

8.1 For Game Developers

The incorporation of AI image generation into game development offers several significant implications:

8.1.1 Customization and Personalization

AI image generation enables an unprecedented level of character customization, allowing developers to offer unique and personalized gaming experiences. This could lead to increased player retention and engagement.

8.1.2 Efficiency and Cost-Effectiveness

By automating the creation of game characters and assets, AI image generation can significantly reduce the time and resources required for these tasks. This capability has the potential to accelerate development timelines and diminish associated costs.

8.1.3 Innovation and Creativity

The introduction of AI image generation into the realm of game design heralds a wave of renewed innovation and creativity. It offers developers a unique opportunity to delve into unexplored territory of game mechanics and storylines, facilitated by the dynamic capabilities of AI-generated characters and environments.

8.1.4 Data-Driven Design

The incorporation of AI image generation can yield insightful data on player preferences and behaviors. This rich source of information can be utilized to mold the design of upcoming games and elevate the player experience, thus guaranteeing that games are consistently adapting to cater to the evolving requirements and tastes of the gaming community.

8.2 For Game Players

The integration of AI image generation into game design has notable implications for players:

8.2.1 Enhanced Customization

The use of AI image generation allows for superior character customization, enabling players to fabricate characters that closely match their specific preferences. This superior degree of customization can lay the foundation for a more tailored and immersive engagement with the game.

8.2.2 Convenience

A significant benefit of AI image generation is its ability to streamline the character creation process. Players are only required to submit a textual description of their envisioned character, and the AI handles the intricate creation process. This eradicates the traditionally lengthy process of manually adjusting every facet of a character, making the game more approachable and user-centric, thus enhancing the overall gaming experience.

8.2.3 Creative Empowerment

AI image generation gives players a role in the creative process, transforming their textual descriptions into visual game characters. This participatory aspect can make the gaming experience more engaging and personally satisfying.

8.2.4 Dynamic Gameplay

The ability of AI-generated characters and environments to adapt and evolve based on player inputs and actions can contribute to more dynamic and unpredictable gameplay, enhancing the game's replayability.

8.2.5 Inspirational and Educational

Witnessing their textual descriptions come to life through AI image generation can be an inspiring and educational experience for players. It can stimulate creativity and offer a practical understanding of the capabilities of AI technology. In summary, the incorporation of AI image generation into gaming can dramatically enrich the player experience, rendering games more personalized, engaging, convenient, and enjoyable.

8.3 For Gaming Industry

The advent of AI image generation for character creation in 2D games is creating an innovative landscape in the gaming industry. It's not just the developers or players who are impacted; the industry as a whole is experiencing a seismic shift due to this technology. Below are some potential implications:

8.3.1 Enhanced Creativity and Innovation

The use of AI for generating game characters allows for a broader range of design and creativity. The implementation of AI image generation for character creation could foster a heightened level of diversity in games and their characters. This, in turn, might draw in a broader spectrum of players, thereby contributing to the industry's expansion. Additionally, this technological advancement might catalyze further innovation in various facets of game design and development.

As for the reduction in development costs and time, the integration of AI in creating game characters could considerably streamline the development process. This efficiency can cut down on production time and resource allocation, thereby decreasing the overall expenses associated with game development.

8.3.2 Reduced Development Cost and Time

With AI image generation, the time and resources required to create varied and unique characters could be significantly reduced. This could decrease the overall development costs for game studios, potentially increasing profitability and allowing studios to invest more in other areas like gameplay, storytelling, and marketing.

8.3.3 New Market Opportunities

AI-generated characters could open up new market opportunities within the gaming industry. For instance, it could lead to the emergence of companies specializing in AI character creation tools and services. It might also inspire new types of games based around the concept of AI generated characters.

8.3.4 Competitive Advantage

Companies that effectively utilize AI for character creation could gain a competitive edge over those that do not. Offering customizable, AI-generated characters could become a unique selling point that attracts players and boosts a game's popularity.

8.3.5 Ethical and Legal Considerations

Incorporating AI for character creation inevitably brings forth fresh ethical and legal dilemmas. For instance, the question of intellectual property rights over AI-generated characters emerges as a critical concern. What if the AI unintentionally creates a character that resembles a real person or a copyrighted character? The industry will need to navigate these issues, potentially leading to new regulations and standards.

8.3.6 Changes in Job Roles and Skills Requirements

As AI becomes more integrated into the game development process, the skills required for game developers might shift. While this could lead to job displacement in some areas (like traditional character design), it could also create new roles centered around AI programming and supervision. The industry might also see a greater demand for skills in AI ethics and law.

In summary, employing AI image generation for the development of characters in 2D games holds the potential to significantly affect the gaming industry. This technology not only unveils new avenues of opportunities but also introduces unique challenges. It's evident that this innovation is transforming the game development paradigm and could hold a central position in shaping the future of the industry.

9 Conclusion

The use of AI image generation for creating game characters for 2D games has shown great promise in this research. The AI image generation model was able to generate high quality game characters that met the desired criteria of game developers and users. This research underscores the benefits of employing AI image generation in game character design: reduced time and cost, enhanced character diversity, and personalized user experiences. These findings bear significant implications for the game development industry. Firstly, AI image generation facilitates resource and time savings, enabling developers to prioritize other developmental facets. Secondly, it facilitates the development of a broader range of diverse and inclusive characters, expanding the games' appeal to a more extensive and varied audience. Lastly, it enhances the user experience, resulting in increased user satisfaction and strengthened brand loyalty.

Summary of Findings: The study found that the AI image generation model was able to generate high quality game characters that met the desired criteria of game developers and users. The generated characters were found to be visually appealing, diverse, and relevant to the game genre. The study also highlighted the advantages of using AI image generation for game character design, including reduced time and cost, increased diversity, and personalized user experience. Final Thoughts: Utilizing AI image generation to craft 2D game characters has the potential to revolutionize the game development industry. This innovation streamlines character creation, offering high quality results swiftly and efficiently, saving resources. Additionally, it promotes character diversity and inclusivity, enhancing user engagement. As AI image generation technology continues to evolve, we anticipate even more sophisticated game characters. In summary, it holds great promise for transforming game development. The study's findings emphasize its advantages: reduced time and cost, increased diversity, and enhanced user experiences, paving the way for exciting future applications.

References

1. Ramesh, Dhariwal, P., Nichol, A., Chu, C., Chen, M.: Hierarchical text conditional image generation with CLIP latents (2021). arXiv:2204.06125
2. Zylinska, J.: AI Art - Machine Visions and Warped Dreams. Open Humanities Press, London (2020)
3. Weizenbaum, J.: Computer Power and Human Reason: From Judgment to Calculation. W. H. Freeman and Company, New York, USA (1976). 0-7167-0463-3
4. Cohen, P.: Harold Cohen and AARON. AI Mag. **37**(4), 63–66 (2016). ISSN 0738-4602
5. Sims, K.: Choreographed image flow. J. Vis. Comput. Animat. **3**(1), 31–44 (1992). https://doi.org/10.1002/vis.4340030106; [11] Mordvintsev, A., Olah, C., Tyka, M.: Deepdream-a code example for visualizing neural networks. Google Res. **2**(5) (2015)
6. Hemlata: "Enhancement of Security by Infrared Arrays Sensor-Based IoT System" Innovative Practices in Technology and Management (ICIPTM) IEEE Conference at Amity University

7. Cohn, G.: AI art at Christie's Sells for $432,500. The New York Times (2018). https://www.nytimes.com/2018/10/25/arts/design/ai-art-soldchristies.html. Accessed 03 December 2022
8. Brown, T., et al.: Language models are few-shot learners. Advances in neural information processing systems. In: Proceedings of the 33rd Conference on Neural Information Processing Systems (NeurIPS 2020) (2020)
9. Hemlata: "Modern Technique of Encryption and Decryption Based on Image" Innovative Practices in Technology and Management (ICIPTM) IEEE Conference at Amity University
10. Rombach, R., Blattmann, A., Lorenz, D., Esser, P., Ommer, B.: High resolution image synthesis with latent diffusion models. In: Proceedings of the IEEE Conference on Computer Vision and Pattern Recognition (CVPR) (2022)
11. Dehouche, N.: Implicit stereotypes in pre-trained classifiers. IEEE Access **9** (2021). https://doi.org/10.1109/ACCESS.2021.3136898
12. Hemlata: Role of neural network, fuzzy, and IoT in integrating artificial intelligence as a cyber security system. In: Second International Conference on Electronic and renewable systems, IEEE Conference at St. Mother Theresa Engineering College, Tuticorin, India
13. Franceschelli, G., Musolesi, M.: Copyright in generative deep learning. Data & Policy **4** (2022). https://doi.org/10.1017/dap.2022.10
14. Fallis, D.: The epistemic threat of deepfakes. Philos. Tech. **34**, 623–643 (2021). https://doi.org/10.1007/s13347-020-00419-2; [10] Zeilinger, M.A.I.: Art, Creative Agency, and the Limits of Intellectual Property. Meson Press, Lüneburg, Germany (2021). https://doi.org/10.14619/1839
15. AUTOMATIC11 WebUI. https://github.com/AUTOMATIC1111/stable-diffusion-webui
16. Hemlata: Revolutionizing hand gesture recognition: a transfer learning approach using surface electromyography and convolution neural networks. In: Second International Conference on Augmented Intelligence and Sustainable System, IEEE Conference at Trichy, India
17. Roose, L.: An A.I.-generated picture won an art prize. Artists aren't happy. The New York Times (2022). https://www.nytimes.com/2022/09/02/technology/aiartificialintelligence-artists.html. Accessed 09 Dec 2022
18. Dehouche, N., Dehouche, K.: What's in a text-to-image prompt? The potential of stable diffusion in visual arts education. https://www.sciencedirect.com/science/article/pii/S2405844023039646. Accessed 25 May 2023
19. Stability AI – Stable Diffusion Model. https://stability.ai/
20. Hemlata: Robust Digital Watermarking Based on Machine Learning" International Conference on Self Sustainable Artificial Intelligence Systems [ICSSAS 2023] IEEE Conference at Erod, Tamil Nadu, India (2023)

Hemlata Parmar a luminary in the field of Computer Science and Engineering. Her journey is defined by a constellation of achievements: from prolific publications and patents to enlightening guest lectures. Yet, her brilliance is matched only by her unwavering commitment to human ethics. With a heart as vast as her intellect, she champions women's empowerment, bridging the gap with empathy and insight. In every endeavor, she embodies the perfect fusion of academic process and compassionate advocacy, inspiring generations to come.

Utsav Krishan Murari is a gifted writer, dedicated life coach, motivational speaker, passionate educator, and, above all, an exceptional mentor. He holds a doctorate in communication and media studies from the Central University of South Bihar. Graced with a vivid imagination, he is a genuine catalyst for ideas and is always open to trying new things.

His one such initiative is Saarthi, which began as Saarthi Entertainment. Saarthi is a Delhi-based creative media production start-up that wants to leave a legacy of artistry in the production industry. Delivering content with a strong emotional connection and the potential to transform the world is what Team Saarthi is all about.

The Industrial Use Cases of Embodied AI Systems

M. Shanmuga Sundari and Kisara Rishitha

Abstract The integration of embodied artificial intelligence (AI) systems into industrial settings has emerged as a transformative force, reshaping conventional paradigms and unlocking unprecedented possibilities across various sectors. This book chapter delves into the multifaceted landscape of industrial applications where embodied AI systems play a pivotal role in augmenting human capabilities, enhancing operational efficiency, and driving innovation. The chapter commences with an exploration of the theoretical foundations and key components of embodied AI, elucidating the amalgamation of physical form and intelligent algorithms. It subsequently navigates through a diverse array of industrial domains, showcasing how embodied AI has transcended conceptual boundaries to become an indispensable asset. Firstly, the chapter investigates the role of embodied AI in manufacturing, highlighting advancements in robotic systems that seamlessly collaborate with human workers to optimize production processes. It examines how these intelligent agents contribute to flexible automation, adaptive manufacturing, and precision assembly, fostering a symbiotic relationship between human expertise and AI efficiency. Moving beyond manufacturing, the chapter investigates the application of embodied AI in logistics and supply chain management. It explores how autonomous robots equipped with AI capabilities are revolutionizing warehouse operations, inventory management, and order fulfillment, thereby mitigating operational bottlenecks and enhancing overall supply chain resilience. The discussion extends into the realm of healthcare, elucidating how embodied AI systems are transforming patient care, surgical procedures, and medical diagnostics. The chapter explores the integration of robotic assistants in healthcare settings, emphasizing their role in precision surgery, rehabilitation, and patient monitoring. Furthermore, the chapter delves into the energy sector, elucidating how embodied AI contributes to the optimization of energy production and distribution. It explores the deployment of autonomous systems in power plants, predictive maintenance of critical infrastructure, and the implementation of smart grids to enhance overall energy efficiency. Throughout the exploration of these industrial use cases, the chapter critically assesses challenges,

M. Shanmuga Sundari (✉) · K. Rishitha
BVRIT HYDERABAD College of Engineering for Women, Hyderabad, India
e-mail: sundari.m@bvrithyderabad.edu.in

P. Raj et al. (eds.), *Building Embodied AI Systems: The Agents, the Architecture Principles, Challenges, and Application Domains*, Information Systems Engineering and Management 14, https://doi.org/10.1007/978-3-031-68256-8_16

ethical considerations, and future prospects associated with the widespread adoption of embodied AI systems. It concludes by envisioning a future where these intelligent agents become integral collaborators, driving innovation, efficiency, and sustainability across diverse industrial landscapes.

Keywords Human-AI · Intelligent algorithms · Manufacturing · Supply chain management

1 Introduction

1.1 Background and Context

The dynamics of modern companies are changing with the introduction of embodied artificial intelligence (AI) systems [1], which represents a dramatic paradigm change. It is essential to examine the historical context and chart the development of artificial intelligence (AI) and its revolutionary journey within industrial environments in order to fully comprehend the significant ramifications of this integration.

Historical Developments: The idea that machines may emulate human intelligence dates back to the middle of the 20th century, when pioneers first proposed the idea of artificial intelligence. The majority of early AI applications were limited to symbolic thinking and rule-based systems. But machine learning, and especially neural networks, brought in a new era where AI systems could mimic human cognition by learning from data and adapting accordingly.

Technological Advancements: AI has reached new heights thanks to the exponential rise in processing power and the collection of enormous datasets. Deep learning in particular has shown machine learning algorithms to have previously unheard-of powers in image analysis, natural language processing, and pattern recognition. These developments provide the technological foundation for embodied AI systems, enabling them to overcome conventional constraints and function flawlessly in industrial environments. Evolving Role of AI: As industries work to become more inventive, flexible, and efficient, AI's position has changed from that of a simple tool to that of a vital partner. Leading this growth is embodied AI, which is characterized by the combination of intelligent algorithms and physical form. Along with enhancing human potential, it also brings about new approaches to automation, problem-solving, and decision-making in the Technological Advancements: AI has reached new heights thanks to the exponential rise in processing power and the collection of enormous datasets. Deep learning in particular has shown machine learning algorithms to have previously unheard-of powers in image analysis, natural language processing, and pattern recognition. These developments provide the technological foundation for embodied AI systems, enabling them to overcome conventional constraints and function flawlessly in industrial environments.

Evolving Role of AI: As industries work to become more inventive, flexible, and efficient, AI's position has changed from that of a simple tool to that of a vital partner. Leading this growth is embodied AI, which is characterized by the combination of intelligent algorithms and physical form. Along with enhancing human potential, it also brings about new approaches to automation, problem-solving, and decision-making in the Technological Advancements: AI has reached new heights thanks to the exponential rise in processing power and the collection of enormous datasets. Deep learning in particular has shown machine learning algorithms to have previously unheard-of powers in image analysis, natural language processing, and pattern recognition. These developments provide the technological foundation for embodied AI systems, enabling them to overcome conventional constraints and function flawlessly in industrial environments.

Evolving Role of AI: As industries work to become more inventive, flexible, and efficient, AI's position has changed from that of a simple tool to that of a vital partner. Leading this growth is embodied AI, which is characterized by the combination of intelligent algorithms and physical form. Along with enhancing human potential, it also brings about new approaches to automation, problem-solving, and decision-making in the industrial framework.

We set out to explore the subtleties of this changing environment in this chapter, hoping to provide readers a thorough grasp of how the incorporation of embodied AI is going to transform industrial processes. Through placing the historical narrative in context, recognizing technological advancements, and appreciating the changing role of artificial intelligence, we laid the groundwork for a thorough investigation of its applications across a wide range of industrial fields.

1.2 Objectives of the Chapter

Here, we clarify the specific objectives and motivations that guide the upcoming chapter. The major goals are designed to provide readers with a road map so they can explore the complex environment around embodied artificial intelligence (AI) in different industrial sectors in a nuanced manner. Providing insights into the real-world applications, difficulties, and moral issues surrounding the integration of embodied AI are among the main goals.

Practical Applications: One of the main goals is to carefully present and examine real-world uses of embodied AI in a range of industrial domains. The chapter attempts to give readers a thorough grasp of how embedded AI is actively promoting operational efficiency, creativity, and human-AI collaboration by exploring particular use examples, such as manufacturing, logistics, healthcare, and the energy industry.

Exploration of Challenges: A critical analysis of the difficulties involved in integrating embodied AI in industrial contexts is another essential objective. This entails a thorough investigation of the technological barriers, constraints, and potential dangers that could appear while putting these technologies into use. The goal of the chapter is to provide a nuanced awareness of the challenges posed by this revolutionary technology.

Ethical issues: Considering the moral aspects connected to the industry's broad use of embodied AI, ethical issues hold a major place in the aims. Through an analysis of the ethical ramifications, the chapter hopes to advance a thorough discussion on the appropriate application of AI, raising awareness and directing moral judgment in business settings.

The goals listed in this part essentially act as a guide for readers, taking them on an investigation of the real-world uses, difficulties, and moral implications of embodied AI in various industrial contexts. The goal of the chapter is to enhance the reader's understanding of the ramifications and complexity involved in the integration of embodied AI systems by offering a fair and well-informed viewpoint.

1.3 Structure of the Book

In Sect. 2, we explained the Embodied AI systems encompass a range of technologies and principles that integrate AI with robotics, sensors, and machine learning to create intelligent agents that can interact with and navigate their physical environment. In Sect. 3, in manufacturing, these systems play a crucial role in automating processes, improving efficiency, and streamlining production lines. In Sect. 4, Logistics and supply chain management benefit from the enhanced capabilities of embodied AI through automated warehousing, precise inventory tracking, and optimized transportation networks. In Sect. 5, embodied AI applications range from surgical assistance and patient monitoring to diagnostic tools, contributing to better outcomes and patient care. In Sect. 6, Energy sector optimization sees embodied AI applied in smart grid management, predictive maintenance, and renewable energy projects. In Sect. 7, while the potential of embodied AI is vast, challenges and ethical considerations remain, such as ensuring safety, security, and fairness while addressing technical limitations and biases. In Section 8, the future of embodied AI promises further innovation and growth, with emerging trends in advanced robotics, autonomous systems, and sustainable technologies set to transform industries and society.

2 Foundations of Embodied AI Systems

2.1 Definition and Components of Embodied AI

The foundation for comprehending the nuances of embodied artificial intelligence (AI) is provided by this section. Fundamentally, embodied AI introduces a paradigm where intelligence transcends abstract algorithms to coexist peacefully with a tangible physical form, marking a divergence from traditional AI systems. This move

Fig. 1 Embodied framework

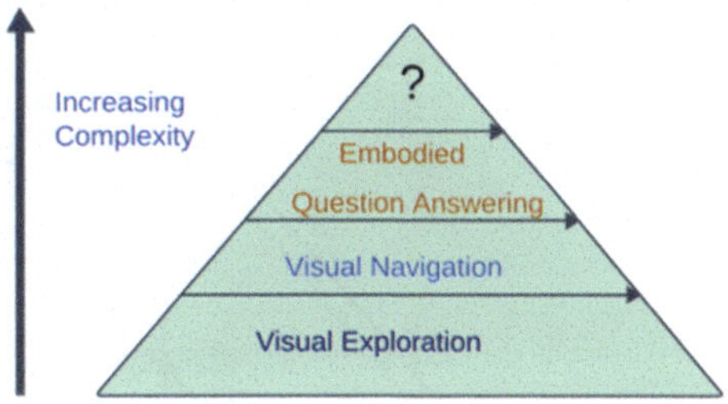

emphasizes a comprehensive merger of virtual intelligence with embodied presence, signaling a significant transition.

It is important to outline the essential elements of embodied AI in order to understand its core. The idea of the physical for the concrete thing that the AI system uses to communicate with the outside world is at the forefront. Embodied AI connects the digital and physical domains, in contrast to classical AI, which functions just in the virtual world. The dynamic added by this embodiment enables the AI system to see, understand, and react to its environment in ways that are not possible for simply computational systems.

Furthermore, the incorporation of the physical form involves a complex weaving using clever algorithms rather than just a simple juxtaposition. Embodied AI's second essential component is this integration. The physical embodiment and the algorithms, the system's cognitive engine, work together harmoniously to form a symbiotic relationship. These clever algorithms control the system's ability to learn from and adapt to real-world interactions, in addition to controlling its decision-making processes.

This section essentially establishes the foundation for a thorough understanding of embodied AI by outlining its fundamental ideas and breaking down its essential elements. The way in which the physical form and intelligent algorithms are intertwined creates the foundation for a thorough investigation of the ways in which embodied AI functions in many industrial settings (Fig. 1).

2.2 *Theoretical Frameworks*

Expanding on the concepts of embodied artificial intelligence (AI), this part explores the theoretical frameworks that facilitate the creation and operation of these novel systems. The examination of theoretical frameworks is crucial in furnishing readers with a conceptual framework that enables them to comprehend the real-world implementations of embodied AI across various industrial sectors.

The many and sophisticated theoretical frameworks that underpin embodied AI systems are in line with the intricate interactions that occur between physicality and intelligent algorithms. **Embodied Cognition**, a viewpoint that contends that cognitive functions are not limited to the brain but are instead closely connected to the

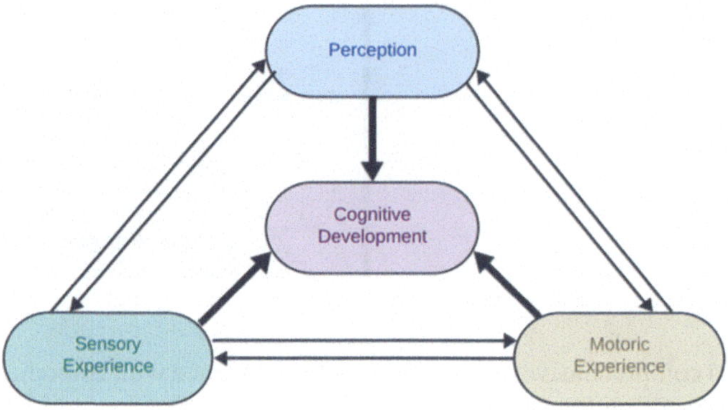

Fig. 2 Cognitive development

body and its interactions with the environment, is one well-known theoretical foundation. This theoretical framework, which emphasizes the role of the physical form in influencing intelligent behaviors and decision-making, is the basis of embodied AI.

Another theoretical paradigm that has a big impact on embodied AI systems is called **Enactive Cognition**. According to this viewpoint, cognition is an active process that involves an organism's constant contact with its surroundings. This suggests that, in the context of embodied AI, intelligent systems, like human cognition [2], obtain their knowledge and reactions from continual interactions with the physical world.

Moreover, the dynamism and flexibility of embodied AI may be explored through the theoretical framework of **dynamic systems theory**. According to this concept, intelligence develops from the complex interactions between different parts of a system. This theory emphasizes how the interaction between the physical form and intelligent algorithms provides a dynamic system that may adapt responses based on external stimuli in the context of embodied AI.

This topic essentially sheds light on the theoretical underpinnings that comprise the mental infrastructure for embodied artificial intelligence systems. Through an examination of ideas like Embodied Cognition, Enactive Cognition, and Dynamic Systems Theory, readers are able to acquire a theoretical viewpoint that enhances their understanding of how embodied AI operates and adjusts in real-world scenarios (Fig. 2).

2.3 Integration of Physical Form and Intelligent Algorithms

This chapter explores the complex relationship between the physicality of embodied artificial intelligence (AI) and the intelligence contained in its algorithms in depth. This chapter explores the domain in which the incorporation of the physical form augments the powers of intelligent algorithms, establishing a mutually beneficial connection that improves flexibility, self-governance, and effectiveness in the context of embodied AI.

Enhancing Capabilities through Integration: By incorporating the physical form, intelligent algorithms' capabilities are raised above the level of virtual abstraction. This is accomplished by acting as a revolutionary catalyst. The combination of physicality and intellect in embodied AI expands the range of real-world interaction, in contrast to standard AI systems that are limited to digital environments. Because of this integration, the system is able to mimic the dynamic character of human cognition by actively perceiving, interpreting, and responding to its surroundings in addition to processing information.

Symbiotic interaction: In embodied AI, there is an innate symbiotic interaction between the intelligence algorithms and the physical form. The physical shape serves as a medium for the algorithms to absorb sensory data and impact the outside world. At the same time, the smart algorithms direct the physical form's movements and reactions as part of its dynamic orchestration. Because of this symbiosis, the system is more adaptive and can handle a variety of settings and learn from real-world encounters.

Improving Adaptability: Embodied AI's adaptability is greatly enhanced by the interaction of intelligence and physicality. Through interaction with the external environment, the system can alter its actions dynamically in response to environmental stimuli, a feature that is similar to human adaptive behavior. The ability of intelligent algorithms to learn continuously enhances this adaptability by enabling the system to improve its responses over time.

Autonomy and Efficiency [3]: The efficiency and autonomy of embodied AI systems are also the result of the integration of intelligent algorithms with physical form. The physical embodiment lessens the requirement for continual human involvement by enabling the system to carry out tasks in the real environment independently. This independence combined with intelligent algorithms' high computational efficiency yields a responsive and efficient system that can optimize processes for a wide range of industrial applications.

This section provides an in-depth exploration of the dynamic interaction between physicality and intelligence in embodied artificial intelligence. The incorporation of the physical form enhances the potential of intelligent algorithms and establishes a mutually beneficial connection that promotes flexibility, self-governance, and operational effectiveness within the complex structure of embodied AI systems.

3 Embodied AI in Manufacturing

3.1 Collaborative Robotics in Production Environments

Embodied artificial intelligence (AI) is becoming more and more integrated in the manufacturing scene, and collaborative robotics is starting to change production settings. This section offers a perceptive synopsis of collaborative robots, highlighting their critical function in streamlining production procedures.

Overview of Collaborative Robots:

Manufacturing automation is about to undergo a paradigm leap thanks to collaborative robots or co bots. In contrast to conventional industrial robots that are limited to inflexible, pre-programmed duties, collaborative robots are intended to collaborate with human operators, promoting a mutually beneficial partnership. Because of their sophisticated sensors and clever algorithms, these robots can dynamically sense and adjust to their environment.

Function in manufacturing Process Optimization: The primary function of collaborative robots is to optimize manufacturing procedures. Co bots improve manufacturing operations' flexibility and efficiency by working in unison with human workers. They can handle a wide range of jobs, from complex precision work to tedious assembly lines, which increases the production floor's total productivity.

Collaborative robots perform best in situations when combining human knowledge with AI effectiveness is critical. They are useful for activities needing a delicate touch or complex customization because of their ability to function securely in close proximity to humans. By working together, manufacturing bottlenecks are reduced and the combined power of human and robotic skills is maximized.

Collaborative robotics becomes even more useful when embodied AI is included. These robots have the ability to pick up on human interactions, adjust to shifting manufacturing needs, and gradually get better at what they do. In dynamic manufacturing contexts, where output needs may change or evolve quickly, and this adaptability is extremely advantageous.

In conclusion, collaborative robots are a key component of embodied AI's use in manufacturing. In addition to increasing productivity, the cooperative relationship between intelligent robots and human labor pave the way for a new era of flexible and responsive manufacturing techniques. This section provides an overview of the revolutionary effects of embodied AI in the complex field of manufacturing [4] (Fig. 3).

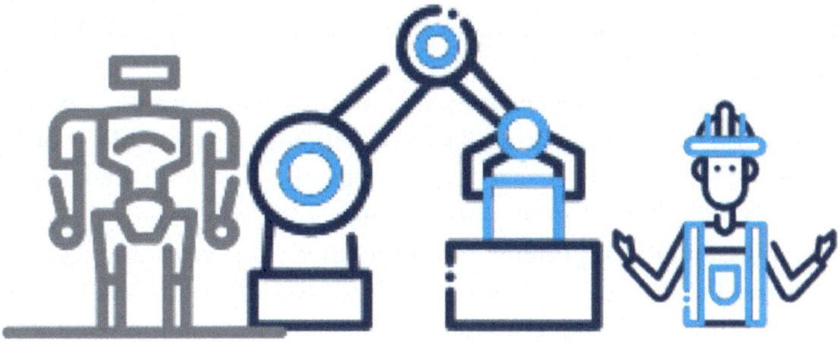

Fig. 3 Collaborative robots

3.2 *Flexible Automation and Adaptive Manufacturing*

The incorporation of embodied artificial intelligence (AI) [5] causes a significant shift in the evolution of industrial processes, especially in the areas of flexible automation and adaptive production. This section explores the dynamic environment in which embodied AI enables flexible and adaptive manufacturing processes, thereby bringing in a new era of industrial efficiency.

Flexible Automation: A paradigm change toward flexible automation in production is brought about by embracing embodied AI. The dynamic demands of contemporary manufacturing may prove difficult for conventional automation systems, which are sometimes inflexible and predetermined. But manufacturing systems may now dynamically adapt their operations to real-time feedback and shifting requirements thanks to embedded AI.

Intelligent machines that can transition between activities with ease and adjust to changes in production quantities or product standards are an example of these adaptive capabilities. Manufacturing systems can now adapt to unforeseen obstacles and changes thanks to embodied AI, which fosters a level of flexibility that was previously thought to be impossible in automated environments.

Adaptable Manufacturing: The dynamic [6] interaction between embodied AI and the production environment is closely related to the idea of adaptable manufacturing. The ability of manufacturing systems to adjust and improve their operations in response to feedback loops and contextual data is known as "adaptive manufacturing." The key to enabling this adaptability is embodied AI.

With sensory awareness and learning algorithms, embodied AI systems can adjust to changes in manufacturing parameters, raw material prices, and even outside variables like consumer desires. This flexibility adds to the overall resilience of the production ecosystem by ensuring that manufacturing processes continue to be effective and sensitive to changing circumstances.

Improving Responsiveness and Efficiency: Adaptive manufacturing, flexible automation, and embodied AI work together to improve manufacturing operations' responsiveness and efficiency. Real-time process recalibration lowers downtime, cuts waste, and maximizes resource efficiency by taking into account environmental cues and experience.

Furthermore, embodied AI-driven industrial systems' adaptive nature enhances quality control. By identifying irregularities, anticipating possible problems, and acting on their own to make necessary corrections, these systems make sure that the finished goods constantly live up to expectations.

This section essentially highlights how embodied AI is revolutionizing industry, as adaptation and flexibility are already standard features of automated operations. Intelligent algorithms and production environments work hand in hand to create an industrial landscape that is more responsive, agile, and efficient.

3.3 Precision Assembly and Quality Control

Within the complex field of manufacturing, the use of embodied artificial intelligence (AI) tells a story of high-quality control and exact assembly. This section delves deeply into the topic, shedding light on how embodied AI improves assembly accuracy while also transforming manufacturing quality assurance systems.

Accurate Assembly: The skill of precision assembly, which necessitates exacting attention to detail and flexibility in response to changing product standards, is at the core of manufacturing excellence. In this context, embodied AI becomes a game-changer, giving manufacturing systems the capacity to do exact assembly jobs with unmatched accuracy.

With sophisticated sensors and clever algorithms, embodied AI-enabled robotic systems skillfully negotiate the challenges of assembly jobs. These systems have the ability to dynamically modify their movements to account for differences in components and real-time input to adjust assembly tactics. This flexibility guarantees both the correct assembly and the smooth incorporation of various components into the finished product.

The precision that embodied AI brings to assembly processes has broad implications for manufacturing industries that place a premium on product uniqueness, variety, and intricacy. Embodied artificial intelligence-driven precision assembly improves productivity and guarantees the reliable delivery of high-quality goods in a variety of industries, including electronics and automotives.

Quality Control Revolution: Embodied AI introduces a paradigm shift in the way this crucial function is carried out. Quality control is a pillar of manufacturing. In order to maintain and improve quality standards, real-time monitoring, analysis, and process adjustment of manufacturing processes are made possible by the integration of AI-driven systems.

Manufacturing systems are enabled by embodied AI's sensory perception [6] and cognitive skills to carry out complex quality checks both during and after assembly. These technologies are able to identify minute flaws, spec deviations, or anomalies in the manufacturing process. Embodied AI helps to improve the precision of quality control procedures in situations where it is crucial, such microelectronics or medical devices.

Moreover, quality control procedures can be continuously improved because to embodied AI systems' adaptive learning capability. By improving their detection algorithms, learning from past data, and adjusting to changing quality requirements, these systems guarantee a proactive and dynamic approach to preserving product perfection.

To sum up, the introduction of embodied AI into production processes ushers in a new era of meticulous assembly and quality assurance. The integration of physical systems and sophisticated algorithms guarantees precision in assembly jobs and transforms manufacturers' approaches to upholding strict quality standards throughout the production lifecycle.

4 Logistics and Supply Chain Management

4.1 Autonomous Systems in Warehouse Operations

The integration of autonomous systems, especially robots equipped with artificial intelligence (AI) capabilities, is a key factor that is redefining warehouse operations in the dynamic field of logistics and supply chain management. This section delves at the ways in which these AI-powered autonomous robots are transforming and streamlining warehouse logistics.

Changing the Logistics of Warehouses:

With advanced artificial intelligence (AI) capabilities, autonomous robots are changing the way that warehouse operations are traditionally understood. These machines, also known as Autonomous Mobile Robots (AMRs), move around warehouse spaces with ease and bring a degree of productivity, agility, and speed that surpasses traditional techniques.

Effective Material Handling: Improving material handling procedures is one of the main ways autonomous robots help warehouse operations. These robots are capable of moving objects around a warehouse on their own, selecting items off shelves and delivering them to predetermined spots. Artificial intelligence (AI) algorithms that optimize routes, adjust to modifications in the warehouse layout, and guarantee precise and timely delivery improve the process's efficiency.

Inventory Control and Organization: The influence of autonomous systems in warehouse operations is also seen in the areas of inventory control and organization. Real-time inventory checks can be performed by AI-driven robots [7], which lowers the possibility of stockouts or overstock scenarios. Further improving the order fulfillment process, these robots help organize inventory by making sure that items are positioned strategically for quick access and retrieval.

Dynamic Adaptability: One of the main characteristics of autonomous systems' revolutionary influence on warehouse logistics is their adaptability. These robots can react dynamically to variations in demand, warehouse layouts, or even unforeseen roadblocks thanks to AI algorithms. This flexibility guarantees a degree of agility that is essential in the hectic and frequently unpredictably changing world of logistics.

Improving Operational Efficiency: The deployment of autonomous robots in warehouse operations makes a major contribution to the improvement of overall operational efficiency. These robots free up human workers to concentrate on more intricate and valuable duties by automating repetitive and time-consuming chores. Autonomous systems' constant functioning also helps warehouses run around-the-clock by cutting down on downtime and streamlining order fulfillment procedures.

Reducing Operational Bottlenecks: In warehouse environments, autonomous robots are essential for reducing operational bottlenecks. A more seamless and effective workflow is facilitated by their capacity to move around the warehouse quickly, operate together, and adjust to changing circumstances. This is especially helpful when there are abrupt jumps in order quantities or during peak periods.

In conclusion, the introduction of AI-powered autonomous devices signals the beginning of a new age in warehouse logistics. These robots change the conventional standards of warehouse operations by optimizing material handling and inventory management while also bringing a new level of flexibility and efficiency. This section provides an overview of the revolutionary impact that AI-powered autonomous systems are having on supply chain management and logistics (Fig. 4).

Fig. 4 AI in warehouse management

4.2 *Inventory Management and Order Fulfillment*

Embodied artificial intelligence (AI) has a particular impact on inventory management and order fulfillment inside the complex web of logistics and supply chain management. This part undertakes a thorough examination of the ways in which embodied AI impacts and enhances these crucial elements, hence influencing the effectiveness and agility of supply chain activities.

Embodied AI in Inventory Management:

Real-time Inventory Visibility: The creation of real-time visibility into inventory levels is one of the fundamental contributions of embodied AI to inventory management. Constant stock level monitoring is made possible through intelligent algorithms that are smoothly integrated with physical systems. Because of this real-time insight, there is less chance of stockouts or overstocks, which makes it possible to estimate demand precisely and strategically refill inventories.

Dynamic Demand anticipating: Because embedded AI systems are capable of learning, they are excellent at anticipating dynamic demand. Real-time forecasting model adaptation is possible for these systems through the analysis of past data, market trends, and external influences. Because of its flexibility, supply chain efficiency is maximized and unnecessary inventory holding costs are avoided by ensuring that inventory levels correspond with real demand.

Automated Reordering procedures: Automation is introduced into the reordering procedures through the combination of inventory management and embodied AI. AI systems have the ability to dynamically modify reorder points in response to shifting demand patterns or automatically initiate reorder requests based on predetermined thresholds. The entire inventory replenishment cycle is streamlined as a result of this automation, which lessens the need for manual interventions.

Effective Order Processing Using Embodied AI:

Order Picking That Is Dynamic: The field of dynamic order choosing is where embodied AI has the biggest influence on order fulfillment. Intelligent algorithms enable autonomous robots to roam warehouses and choose items for orders. Embodied AI's dynamic adaptability guarantees optimum routes, cutting down on travel time and improving order pickup efficiency overall.

Prioritization of Orders and Optimization: Order prioritizing that is intelligent is made possible by embodied AI. AI-driven systems can optimize the order in which orders are fulfilled by taking into account variables like delivery deadlines, client preferences, and geographic locations. This setting of priorities guarantees customer satisfaction, on-time delivery, and effective resource use throughout the supply chain.

Order Processing Time Reduction: Order processing time is greatly decreased by the incorporation of embodied AI. The orderly cooperation of physical systems and cognitive algorithms reduces the amount of time it takes to find, select, and package things for orders. This decrease in processing time helps to meet consumer expectations for quick order fulfillment while also improving operational efficiency.

Essentially, this analysis shows how inventory management and order fulfillment in logistics and supply chain management are transformed by embodied AI, which combines physicality with intelligence. In addition to streamlining these procedures, embodied AI's constant monitoring, flexibility, and automation boost the supply chain's overall resilience and competitiveness.

4.3 Enhancing Supply Chain Resilience

The integration of embodied artificial intelligence (AI) is a crucial component in boosting overall supply chain resilience within the intricate field of logistics and supply chain management. This section delves deeply into the topic, explaining how supply chain ecosystems can be made more resilient to changing circumstances by utilizing embodied AI.

Continuous Monitoring and Adaptation:

Real-time Environmental Sensing: By means of ongoing observation and real-time environmental sensing, embodied AI systems significantly contribute to the improvement of supply chain resilience. These systems, which are outfitted with sensors and sophisticated algorithms, have the ability to detect shifts in external factors like weather patterns, market movements, and geopolitical events. Supply chain managers may proactively adjust plans and mitigate possible disturbances before they escalate, thanks to this real-time awareness.

Making Adaptive Decisions: Embodied AI's flexibility also applies to supply chain decision-making procedures. Embodied AI systems automatically modify decision parameters in response to unanticipated occurrences, such supply chain interruptions, abrupt demand surges, or delays in transportation. By ensuring that the supply chain can react dynamically to obstacles, this adaptive decision-making helps to minimize the effects of disruptions on overall operations.

Dynamic Inventory Management:

Resilience Through Strategic Stockpiling: Supply chain resilience is enhanced by embodied AI, which does this by carefully managing inventory ahead of possible interruptions. AI-driven systems might suggest strategically hoarding essential parts or products by examining past data, market volatility, and other pertinent variables. By taking a proactive stance, the supply chain is protected against unanticipated demand swings and supply chain disruptions and is made resilient in the face of uncertainty.

Effective Demand Prediction: The capacity of embedded AI to adjust in demand forecasting improves supply chain resilience directly. Through continual learning from shifting consumer behaviors and market situations, these systems enhance the precision of demand estimates. Supply chain managers may thus better match production, distribution, and inventory levels to actual demand, lowering the possibility of excess inventory or stock outs in unforeseen circumstances.

Agile Response to Disruptions:

Optimal Route Scheduling: Implemented AI's impact on supply chain resilience is especially noticeable when it comes to how well it plans transportation routes. Real-time data is analyzed by AI algorithms that take into account variables [8] including traffic patterns, fuel pricing, and possible disruptions. By ensuring that shipments can dynamically adjust their routes, this optimization reduces delays and improves the overall effectiveness of transportation operations.

Collaborative Robot Networks: Supply chain resilience is enhanced by the collaborative character of embodied AI, particularly when it takes the form of collaborative robot networks. These robots may operate together harmoniously, sharing responsibilities and adjusting to unforeseen disturbances or shifts in the availability of the personnel. This cooperative adaptability guarantees that necessary activities in the supply chain can go forward even when faced with unforeseen difficulties.

To sum up, the incorporation of embodied AI strengthens supply chain ecosystem resilience by facilitating ongoing observation, flexible decision-making, dynamic inventory control, and quick reactions to disturbances. This section focuses on how supply chain resilience is enhanced by the mutually beneficial interaction between physicality and intelligent algorithms, which puts it in a position to prosper in a constantly changing business environment.

5 Healthcare Applications

5.1 Robotic Assistants in Patient Care

Applications in Healthcare:

The use of robotic helpers is a game-changer in the rapidly changing field of healthcare, especially when it comes to patient care. This section offers a thorough overview of the seamless integration of robotic assistants powered by artificial intelligence (AI) into healthcare environments, which enhances and improves patient care.

Reinventing Healthcare:

Helping Healthcare workers: Designed to work in tandem with healthcare workers, robotic assistants provide assistance with a range of duties and activities. These robotic assistants, which can work in clinics, hospitals, or rehabilitation facilities, are outfitted with cutting-edge sensors and clever algorithms that let them communicate with medical professionals and patients in an efficient manner.

5.2 Precision Surgery and Rehabilitation

The intersection of precision surgery, rehabilitation, and embodied artificial intelligence (AI) presents a novel frontier in the ever-evolving field of healthcare. This section explores how AI-driven technologies are revolutionizing precision surgery and the advancement of rehabilitation techniques, bringing in a new era of patient-centered healthcare.

Precision Surgery:

Robotic Surgical Systems: Robotic surgical systems herald a new paradigm shift in precision surgery, enabled by embodied AI. These sophisticated AI-guided technologies give surgeons greater control, dexterity, and precision when performing surgeries. Reduced trauma, quicker recovery, and better patient outcomes are all made possible by the minimally invasive procedures made possible by the incorporation of robotic arms into the surgical process.

Enhanced Imaging and Navigation: With its sophisticated imaging and navigational capabilities, Embodied AI is essential to improving the accuracy of surgical procedures. Artificial intelligence (AI) systems examine live imaging data to give surgeons more detailed knowledge of anatomical components. By improving surgical accuracy, this enhanced visualization makes it possible to perform more accurate incisions, tumor removals, and complex operations.

Adjustable Surgical Support: Embodied AI is also flexible enough to be used in surgical support, as these systems are constantly picking up new skills from previous surgeries. Robotic surgical systems can alter their support according to the surgeon's preferences, patient-specific parameters, and real-time feedback thanks to this adaptive capabilities. Precision surgery is certain to be customized and dynamic with such flexibility.

Advances in Rehabilitation:

Tailored therapy Plans: By making it possible to create tailored and adaptable therapy plans, embodied AI has a major impact on the progress of rehabilitation. These programs are customized to meet the specific needs of each patient, taking into account things like their past medical history, present health, and future development. AI integration guarantees dynamic rehabilitation programs that adapt to patients' responses and recovery paths.

Rehabilitation Devices Utilizing Sensors: Embodied AI-powered sensor-based gadgets are indispensable instruments in the field of rehabilitation. These gadgets, which range from smart prostheses to exoskeletons, have sensors built in to record patient movements, muscle activity, and recovery outcomes over time. Patients and medical professionals can monitor and optimize rehabilitation exercises with the help of AI algorithms that analyze this data and provide real-time feedback.

Human-Robot Cooperation:

Flowing Together in Surgical and Rehabilitative Environments:

In surgical and rehabilitative settings, the collaborative character of embodied AI shines, promoting smooth communication between medical experts and robotic systems. While surgeons can collaborate with robotic surgical systems to take advantage of AI-driven precision, rehabilitation specialists can use sensor-equipped devices to increase the efficacy of therapeutic activities. This cooperative strategy guarantees that the benefits of AI capabilities and human expertise are combined to provide the best possible medical care.

In summary, embodied AI's incorporation into precision surgery and rehabilitation marks a revolutionary development in medical technology. In addition to improving surgical accuracy and rehabilitation results, the combination of physical systems and cognitive algorithms ushers in a new era of patient-centered, flexible, and technologically enhanced healthcare procedures. This section provides a thorough examination of how embodied AI is changing the way that rehabilitation and precision surgery are conducted.

5.3 Medical Diagnostics and Monitoring

The introduction of embodied artificial intelligence (AI) signals a revolutionary period in patient monitoring and medical diagnostics within the constantly changing healthcare environment. This section explores the various ways that AI-driven technology might improve patient care by improving the precision and effectiveness of medical diagnostics and by offering continuous, real-time monitoring.

Revolution in Medical Diagnostics:

AI-Powered Image Interpretation:

By greatly improving the analysis of medical imaging data, embodied AI transforms medical diagnostics. Deeply ingrained in diagnostic imaging systems, AI algorithms demonstrate the ability to identify minute irregularities, trends, and anomalies in medical images. This potential is most noticeable in domains like pathology and radiology, where AI-powered instruments help medical practitioners identify diseases early and accurately.

Diagnostic Decision Support Systems: One application of embodied AI is the development of diagnostic decision support systems that supplement medical professionals' knowledge. These systems deliver evidence-based insights and recommendations by analyzing a wide range of patient data, such as imaging findings, test results, and medical history. Precision medicine techniques are aided by the partnership between AI and medical personnel, which produces more educated diagnostic conclusions.

Early illness Detection: Embodied AI has made significant advances in medical diagnostics, one of which is the early illness detection function. Artificial intelligence (AI) algorithms are able to recognize minute patterns linked to early-stage diseases by continuously learning from large datasets. In addition to improving patient outcomes and, in certain situations, promoting preventative measures, early detection also increases treatment efficacy.

Continuous Monitoring and Personalized Care:

Wearable and Implantable Devices:

The development of wearable and implantable devices that allow for continuous patient health monitoring is greatly aided by embodied AI. These gadgets offer real-time data on vital signs, activity levels, and other pertinent factors. They are fitted with sensors and AI algorithms. Embodied AI-enabled continuous monitoring guarantees early health issue detection, individualized treatment, and prompt intervention.

Systems of Adaptive Monitoring: Embodied AI's intrinsic flexibility allows for monitoring systems to dynamically adapt to the unique characteristics of each patient. AI-driven monitoring systems adjust their algorithms to variations in baseline health and lifestyle parameters by learning from patterns in patient data. This flexibility guarantees customized monitoring that is attentive to individual variations and able to identify minute alterations that can point to health issues.

Predictive analytics and data integration:

Integration of Multimodal Data: Genomic information, imaging data, electronic health records, and other various healthcare data modalities can all be more easily integrated with the help of embodied AI. A comprehensive picture of a patient's health is made possible by this thorough data integration, which supports more precise and nuanced diagnostics. AI systems' cooperative analysis of multimodal data sets advances our knowledge of patient health profiles in a more thorough manner.

Predictive Analytics for Patient Outcomes: Embodied AI's predictive analytics powers are applied to the prediction of patient outcomes. Artificial intelligence systems are able to forecast possible health trajectories by examining past patient data, treatment outcomes, and demographic variables. By taking a proactive stance, healthcare providers can better customize interventions, optimize treatment regimens, and reduce the likelihood of unfavorable outcomes.

To sum up, the application of embodied AI to medical monitoring and diagnostics completely changes the way that healthcare is provided. An extensive examination of the ways in which AI-driven technologies support early disease identification, individualized treatment via ongoing monitoring, and the integration of various healthcare data [9] for more precise diagnosis is given in this part. Embodied AI's revolutionary effects on medical monitoring and diagnostics highlight how essential it is to improving patient care and encouraging a data-driven approach to healthcare decision-making (Fig. 5).

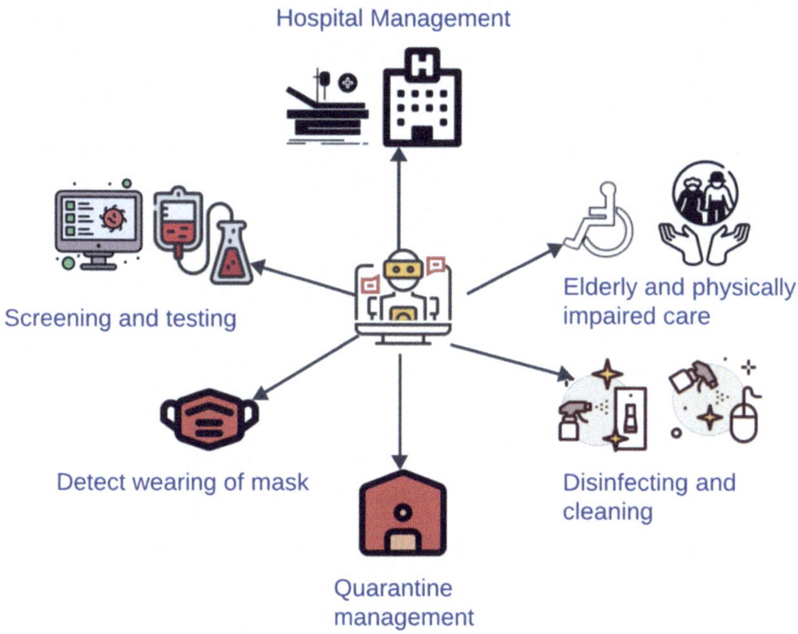

Fig. 5 AI in health sector

6 Energy Sector Optimization

6.1 Autonomous Systems in Power Plants

When autonomous systems powered by artificial intelligence (AI) are integrated into power plants, the energy industry is completely transformed. This section explores the complex effects of autonomous systems on maximizing energy output, improving operational effectiveness, and guaranteeing power plant reliability.

Power Plant Operational Autonomy:

Astute Management and Observation: Within power plants, autonomous systems usher in a new era of operational autonomy. Intelligent process monitoring and control are made possible by AI algorithms that are incorporated into control systems. These systems optimize variables like temperature, pressure, and flow rates by continuously analyzing real-time data from sensors and equipment. The end result is a power plant that operates with maximum efficiency and little downtime, thanks to its agility and adaptability.

Predictive Maintenance and Fault Detection: Power plant predictive maintenance is greatly aided by embedded AI in autonomous systems. AI algorithms are able to forecast probable equipment breakdowns or maintenance requirements by evaluating past data and keeping an eye on the state of the equipment in real time. By being proactive, this strategy reduces unscheduled downtime, increases the lifespan of important equipment, and guarantees the dependability of power generation.

Improving Energy Economy:

Dynamic Load Balancing: An essential component of energy efficiency optimization, dynamic load balancing is made possible by autonomous systems. Artificial intelligence systems examine patterns in electricity consumption and dynamically modify how loads are distributed among power plants. By doing this, you can be confident that every unit is running to its full potential, saving energy and increasing overall efficiency.

Effective Management of Energy Storage: Embodied AI is also being integrated into power plant management systems for energy storage. Based on demand projections and grid conditions, autonomous systems optimize the cycles of energy storage [10] unit charging and discharging. In addition to supporting intermittent renewable energy sources and improving grid stability, this intelligent energy storage management also helps to build a more robust and sustainable energy infrastructure.

Demand Response and Grid Integration:

Real-Time Grid Integration: The integration of autonomous systems with the larger energy grid in real-time is made possible. Algorithms powered by artificial intelligence (AI) examine market prices, demand variations, and grid circumstances to optimize power plant operations. This integration facilitates the effective use of existing resources, lowers transmission losses, and improves grid stability.

Demand Response Optimization: Autonomous systems aid in demand response optimization by reacting to sudden variations in the demand for electricity. Artificial intelligence (AI) systems forecast demand trends and modify power plant operations to effectively meet periods of peak demand. Demand-side management enhances overall energy system reliability by supporting grid stability and enabling power units to take part in demand response programs.

Mitigation of Environmental Impact:

Optimizing Renewable Energy Integration: The integration of renewable energy sources, such solar and wind power, into power plants is made easier with the help of autonomous systems. Artificial intelligence (AI) systems predict the availability of renewable energy, modify power plant operations in response, and enable the smooth integration of variable renewable sources. This optimization promotes a more sustainable energy balance and helps to lower greenhouse gas emissions.

Strategies for Reducing Emissions: Within power plants, embodied AI systems aid in the creation and application of emission reduction plans. AI algorithms help reduce environmental effects by optimizing combustion processes [11], fuel use, and overall plant efficiency. This strategy is in line with the more general objectives of lowering carbon emissions and lessening the impact of electricity generation on the environment.

In summary, the incorporation of self-governing systems, enabled by embedded artificial intelligence, within power plants revolutionizes the energy industry by streamlining operational procedures, augmenting energy effectiveness, and promoting ecological sustainability. An extensive examination of how autonomous systems transform power plant operations is given in this part, which represents a major advancement toward a future in energy that is more resilient, efficient, and sustainable.

6.2 *Predictive Maintenance for Critical Infrastructure*

With an emphasis on critical infrastructure, we undertake a thorough investigation of predictive maintenance's revolutionary potential in the energy industry in this chapter. The chapter begins by describing the evolutionary transition from reactive to proactive maintenance tactics. In this scenario, artificial intelligence (AI) is integrated to enable continuous data analysis, which enables the identification and resolution of possible problems before they become more serious.

The symbiotic relationship between embodied AI and asset performance improvement is at the center of the discussion. We explore how AI algorithms recognize patterns and trends in equipment behavior via the lens of predictive maintenance, enabling dynamic modifications to improve efficiency and prolong the operational lives of vital infrastructure assets.

Using AI algorithms and sensors to integrate real-time condition monitoring is a crucial component of this investigation. Because of this integration, proactive fault resolution and prompt actions are made possible by a detailed understanding of the critical infrastructure's health. The chapter demonstrates how AI-driven solutions provide early warnings that reduce downtime and guarantee the smooth and continuous operation of vital assets by evaluating past data and present circumstances.

Predictive maintenance's financial ramifications are examined in detail, along with the advantages of reduced costs and improved operational effectiveness brought about by AI-driven tactics. Predictive maintenance is a key component in attaining economic efficiency in the energy sector since it minimizes unplanned downtime, lowers the need for emergency repairs, and optimizes maintenance schedules.

The chapter also looks at how corporate asset management systems and predictive maintenance can work together holistically. Together, these efforts guarantee the smooth coordination of maintenance tasks, promote the exchange of data, and enable well-informed decision-making for the long-term sustainability and resilience of vital infrastructure in the energy sector.

The importance of predictive maintenance in strengthening critical infrastructure becomes clear as we wrap off this chapter. Embodied AI in conjunction with maintenance procedures not only guarantees asset dependability but also lowers costs, improves operational effectiveness, and strengthens the key infrastructure of the energy industry.

We address the difficulties, moral questions, and potential implications of the mass deployment of embodied AI systems in various industrial contexts in the upcoming chapter.

6.3 Smart Grids and Energy Efficiency

It explores the revolutionary field of smart grids and its significant influence on improving energy efficiency in the energy industry. This chapter explains how artificial intelligence (AI)-driven smart grids are changing the energy environment, maximizing resource use, and developing resilient and sustainable energy infrastructure.

Monitoring and Control of an Intelligent Grid:

The first section of the chapter examines how embodied AI-enabled smart grids are revolutionizing traditional grid monitoring and control. Smart grids ensure effective distribution and consumption of electricity by dynamically adjusting to differences in energy supply and demand through adaptive algorithms and real-time data analysis.

Strategies for Dynamic Demand Response:

This section explores the dynamic demand response capabilities of smart grids and explains how artificial intelligence (AI) predicts demand trends and makes proactive adjustments to energy distribution. Demand response tactics and smart grids work together to intelligently balance supply and demand, improving grid stability while simultaneously optimizing energy efficiency.

Integration of Renewable Energy Sources: This chapter examines how renewable energy availability is predicted by AI-driven algorithms and looks at how renewable energy sources can be integrated into smart grids. By maximizing energy production and lowering dependency on non-renewable resources, this vision allows smart grids to easily integrate variable renewable sources, promoting environmental sustainability.

Grid Sturdiness and Autonomous Functions:

The robustness of smart grids is also discussed, with a focus on how self-healing they are. Smart grids reduce downtime and guarantee a strong and resilient electricity infrastructure by detecting and responding to possible disruptions in real-time with AI-driven predictive analytics.

Infrastructure for Advanced Metering (AMI): This section examines how AI-enhanced meters give real-time data on energy use, emphasizing the function of Advanced Metering Infrastructure (AMI) in smart grids. Customers are better able to make educated decisions [12] about how much energy they use because to this detailed data, which improves energy efficiency overall.

The revolutionary effect of smart grids on energy efficiency becomes evident as we close off this chapter. When combined with embodied AI, smart grids can revolutionize the energy industry, maximize resource use, and pave the path for a resilient and sustainable energy future.

7 Challenges and Ethical Considerations

7.1 Technological Challenges and Limitations

It evaluates the technical difficulties and innate constraints related to the integration of embodied artificial intelligence (AI) systems in various industrial environments. In order to fully realize the potential of embodied AI, practitioners and stakeholders must overcome a number of challenges and limitations, which are explored objectively in this section.

Challenges of Sensor and Data Integration: This chapter delves into the complex issues surrounding the integration of sensors and data through a thorough study. The smooth operation of physical sensors in conjunction with intelligent algorithms is discussed, covering topics such data interoperability, sensor dependability, and the challenges of combining various data sources.

Real-world Environment Complexity: This section explores the intricacies of real-world surroundings, evaluating the difficulties presented by erratic and dynamic environments. The unstructured nature of industrial contexts presents challenges for embodied AI systems, as situations and variables can change quickly, affecting the dependability of AI-driven decision-making processes.

Interoperability and Standardization: The chapter examines the difficulties that arise in the field of embodied AI with regard to interoperability and standardization. Making sure that various systems and devices work together seamlessly in terms of compatibility and communication becomes essential. The conversation covers the various technologies used in embodied AI systems and attempts to create industry-wide standards.

Robustness and Security Issues: This section examines the security and robustness of embodied AI systems and evaluates how susceptible these systems are to hostile attacks and other external dangers. In order to make AI algorithms more resilient to unforeseen obstacles and cybersecurity threats, this chapter examines various methods.

Resource limits and Scalability: The general implementation of embodied AI systems is severely hampered by resource limits, which include both energy and processing power usage. The necessity for scalable systems that can function effectively in a variety of industrial environments while accounting for fluctuations in resource availability is covered in this section.

As this chapter comes to an end, readers will have a thorough understanding of the technological obstacles and constraints associated with the integration of embodied AI systems. Recognizing these obstacles is essential to promoting creativity and resolving the complex problems that arise when AI becomes a ubiquitous component of industrial environments.

7.2 Ethical Implications of Embodied AI in Industry

This study critically looks at the moral issues raised by the application of embodied artificial intelligence (AI) in a range of industrial settings. It is crucial to manage the ethical implications of AI systems as they become essential workplace collaborators in order to ensure their responsible and compassionate deployment across a variety of sectors.

Data Handling and Privacy Issues: The chapter explores the moral implications of data handling and privacy issues related to embodied AI. The handling of sensitive data by these systems in industrial environments necessitates careful consideration of data collecting procedures, storage security, and potential privacy consequences for stakeholders and employees.

Fairness and Bias in the Making of Decisions: This section examines the ethical implications of bias in decision-making processes, specifically focusing on issues of equity and justice. The chapter examines how biases in training data might exacerbate or maintain current societal injustices, highlighting the significance of creating AI models that put inclusion and fairness first.

Responsibility and Openness: When implementing embodied AI, accountability and transparency become crucial ethical factors to take into account. The difficulties in determining accountability while making decisions based on artificial intelligence are covered in this part, with a focus on the necessity of transparent AI systems that offer insights into their decision-making procedures.

Impact on Employment and Workforce Dynamics: This chapter examines issues pertaining to job displacement and the evolving form of work while evaluating the ethical implications on employment and workforce dynamics. It takes into account responsible AI implementation tactics that place an emphasis on reskilling and upskilling as well as the development of cooperative ecosystems where humans and AI can live in harmony.

Autonomy and Human Control: The ethical implications of autonomy and human control over embodied AI systems are rigorously examined in this chapter. It highlights the necessity of preserving human oversight in crucial processes and talks about the need for a clear separation of decision-making authority between AI and humans.

It gives readers a sophisticated knowledge of the moral ramifications of embodied AI's application in business. Fostering a responsible and socially conscious approach to AI deployment across various industrial settings requires addressing these ethical challenges.

7.3 *Regulatory Frameworks and Standards*

The rules and guidelines controlling the incorporation of embodied artificial intelligence (AI) in various industrial contexts are scrutinized severely in this chapter. The conversation recognizes the critical role that these frameworks play in guaranteeing the ethical and responsible application of AI, and it offers a thorough analysis of the changing legal environment, compliance standards, and global cooperative initiatives.

Changing Regulatory Environment:

The chapter begins with a detailed examination of the dynamic and quickly changing regulatory environment pertaining to the incorporation of embodied AI in business. It evaluates how responsive regulatory agencies are to new developments in technology and discusses issues related to upholding strong and efficient control over AI applications. The conversation covers the regulatory flexibility needed to stay up with the changing AI environment, striking a balance between innovation and the requirement for moral protections.

Industry-specific Compliance needs: Recognizing the necessity for customized regulatory methods in various sectors, this section dives into the finer points of industry-specific compliance needs. The chapter highlights the cooperative efforts of industry stakeholders, regulatory agencies, and AI developers to create context-specific rules that give safety, ethics, and responsible AI practices top priority. It emphasizes how crucial it is to take industry-specific factors into account when creating legislation that handle the special opportunities and problems that embodied AI presents.

Standards for Safety and Ethics: The investigation includes creating and enforcing safety and ethical guidelines for embodied AI in commercial settings. It explains how these guidelines are essential for risk mitigation, accountability, and the development of a responsible AI deployment culture. The creation of criteria for algorithmic openness, data privacy, and system robustness is emphasized in this chapter. It also explores the moral issues raised by AI's effects on human welfare, highlighting the necessity of an all-encompassing ethical framework.

International Cooperation and Harmonization: This section addresses the significance of international cooperation and harmonization of regulatory frameworks in light of the global character of AI deployment. It explores efforts to promote international collaboration in order to establish unified standards, guaranteeing a uniform and widely recognized method of regulating embodied AI systems. In order to address the cross-border consequences of AI deployment, the discussion emphasizes the necessity for common ethical standards and values as well as the difficulties and advantages of international collaboration.

Accountability Mechanisms: Integrated into regulatory frameworks, accountability mechanisms are evaluated critically in this chapter. It talks about how laws can establish accountability chains by outlining duties for AI developers, producers, and consumers. The investigation also includes procedures for handling non-compliance and making sure remedial measures are implemented when required. The chapter places a strong emphasis on the role that accountability plays in fostering public and organizational trust in AI systems.

It summarises by giving readers a thorough grasp of the standards and legal frameworks guiding the ethical application of embodied AI across a range of industrial contexts. The interplay of regulatory agencies, industry players, and technology developments is essential to establishing a climate that gives ethical, secure, and responsible AI practices first priority.

8 Future Prospects and Innovation

8.1 Emerging Trends in Embodied AI

This chapter provides a prospective look at the opportunities and new developments that are transforming the field of embodied artificial intelligence (AI). In-depth analyses of the trends that are expected to shape the future of embodied AI in various industrial applications—from ecosystems of human-machine collaboration to ethical AI design—are given in the next subsections.

Ecosystems for Human-Machine Collaboration: The first section of the chapter explores how human-machine collaboration is changing. It investigates the formation of ecosystems in which AI and humans operate together harmoniously, imagining a day when embodied AI is a crucial and cooperative collaborator in a range of industrial processes. This theme focuses on how symbiotic connections with intelligent agents can augment human skills.

Explainable AI for Improved Understanding: Next, the focus of the conversation switches to the significance of explainable AI for improving trust and understanding. This trend investigates developments that enable transparent and comprehensible AI decision-making procedures. It tackles the requirement that users understand and have faith in the choices made by AI systems, hence promoting wider adoption and moral AI procedures.

Edge Computing and Decentralized AI: The move toward edge computing and decentralized AI architectures is a key trend that is discussed in this section. The chapter examines developments that lessen latency and embodied AI systems' reliance on centralized data processing by enabling them to function more independently in real-time. The shift towards improved adaptability and distributed intelligence is seen in this trend.

Continual Learning and Adaptability: The idea of continuous learning and adaptability in embodied AI systems is the subject of the following exploration. This topic looks at developments that allow AI systems to grow and change over time in response to shifting tasks and settings. The focus is on developing flexible AI systems that can react dynamically in intricate industrial environments.

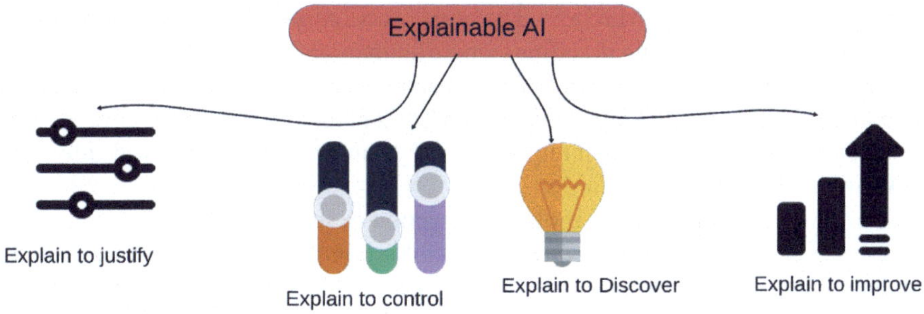

Fig. 6 Explainable AI

Ethical AI Design and Bias Mitigation: This section explores the trend of ethical AI design and bias mitigation while addressing the ethical issues surrounding AI. In order to ensure justice, equity, and accountability in the creation and application of embodied AI systems, it examines technologies meant to detect and address biases in AI algorithms.

AI for Eco-Friendly and Sustainable activities: The following section of the chapter delves into the trend of using embodied AI in industries to support eco-friendly and sustainable activities. This field of innovation focuses on maximizing resource use, cutting waste, and supporting ecologically responsible decision-making in the transportation, manufacturing, and energy sectors.

Collaborative networks and robotic swarms: Lastly, the investigation delves into the possibilities of embodied AI-powered collaborative networks and robotic swarms. This trend looks into the seamless collaboration of various intelligent agents to improve scalability and efficiency in industries including manufacturing, shipping, and healthcare.

The readers acquire a thorough grasp of the new developments that could influence how embodied AI develops in the future. These trends support collaborative, ethical, sustainable, and sustainable behaviors in a variety of industrial areas in addition to improving operational efficiency. Chapter 9 takes the reader on a more comprehensive tour of the changing landscape that embodied AI is shaping, with a focus on collaborative human-AI workspaces and future advances (Fig. 6).

8.2 Collaborative Human-AI Workspaces

This section presents a view of how intelligent agents and people interact and collaborate in industrial contexts, exploring the changing environment of collaborative human-AI workspaces. Because these workspaces are collaborative, they foster creative problem-solving and increased productivity, which have the potential to completely change the nature of work across a range of industries.

Augmented Decision-Making: The first section of the chapter discusses augmented decision-making in settings where humans and AI collaborate. It explores the ways in which intelligent AI systems can work alongside human professionals in collaborative decision-making. This trend highlights the benefits of combining AI-driven data analysis with human intuition to create more efficient and well-informed decision-making processes.

Seamless Integration of AI Tools: This section examines the trend of designing workspaces that intuitively integrate AI technology, with a focus on how these tools are integrating into daily processes. It looks at technologies that enable organic interactions between people and AI, removing obstacles and creating a welcoming atmosphere that promotes teamwork.

Adaptive Learning Environments: The trend of implementing adaptive learning environments in collaborative workspaces is covered in this topic. It investigates how AI systems integrated into these settings may adjust to the needs of both individual and group learning, offering individualized assistance and encouraging ongoing skill growth among human partners.

Real-time Collaboration and Communication: This section explores technologies that improve the connectivity between people and AI in workspaces, emphasizing the significance of real-time collaboration and communication. It looks at technology that allow for instantaneous information sharing, group problem solving, and efficient routes of communication that cut through time and space barriers.

Dynamic Task Allocation and Optimization: This trend in task allocation and optimization is also explored. The chapter covers the topic of how dynamic task allocation in collaborative human-AI workspaces can be achieved through real-time evaluations of individual and group capabilities. This approach seeks to maximize workflow effectiveness and guarantee that assignments complement the abilities of both AI and human partners.

Collaboration's Ethical Considerations:

This section addresses issues of accountability, transparency, and bias reduction in light of the ethical implications of working with AI. It looks at technological advancements that guarantee moral behavior in group work environments, focusing on the ethical application of AI to preserve justice and equity.

The importance of user-centric design principles in collaborative human-AI workspaces is highlighted in the chapter's conclusion. It talks about how to make AI technologies more approachable, build a collaborative culture that enables people to effectively use AI tools, and create user-centered interfaces and interactions.

8.3 Anticipated Innovations and Transformations

This section explores the domain of expected advancements and changes, offering an outlook on the direction of embodied artificial intelligence (AI). The upcoming subsections provide an overview of the developments that are set to completely change the field of artificial intelligence applications, from ground-breaking technological breakthroughs to revolutionary changes in industrial processes.

Cognitive Augmentation and Neural Interfaces: The first section of the chapter examines the promising innovations of neural interfaces and cognitive augmentation. It explores developments that improve the relationship between humans and AI, imagining neural interfaces that enable smooth communication and technologies for cognitive augmentation that boost human cognitive capacities.

Swarm Intelligence and Collaborative Problem-Solving: This section explores the possibilities of swarm intelligence and upcoming developments in collaborative problem-solving. It investigates how embodied AI systems, which are modeled after the collective behaviors of natural swarms, may work together in vast quantities to solve complicated problems and produce outcomes that are greater than those of individual agents.

Integration of Quantum Computing: The chapter examines the expected fusion of embodied AI and quantum computing, addressing the cutting edge of computing. It talks about how the processing power of quantum computing could transform AI algorithms, making it possible to perform intricate calculations and solve problems that were previously unfeasible, ushering in a new era of AI capabilities.

Biohybrid Systems and Biointegration: The upcoming innovations in biohybrid systems and biointegration are discussed in this part. It investigates how artificial intelligence and biological systems might coexist, imagining technological solutions that either perfectly complement or imitate biological systems. This breakthrough could improve AI's flexibility and effectiveness in dynamic, unpredictably changing contexts.

Explainable and Ethical AI by Design: The chapter explores advancements in developing systems that are intrinsically ethical and explainable, taking into account the progress of AI design principles. In order to make sure that AI technologies are in line with social values and norms, it examines developments that place a priority on transparency, interpretability, and ethical considerations from the very beginning of AI research.

AI Systems and Emotional Intelligence: The investigation also touches on the expected innovation of AI systems and emotional intelligence. This movement aims to create more sympathetic and human-like interactions by giving AI systems the ability to perceive and react to human emotions. The possible uses of emotionally intelligent AI in customer service, healthcare, and other fields are covered in this chapter.

Autonomous Creative Systems: An exploration of the expected innovation in autonomous creative systems brings the chapter to a close. It investigates the ways in which embodied AI, furnished with imaginative algorithms, might produce creative works of art, designs, and inventive solutions on its own. According to this trend, AI will collaborate with humans to co-create new concepts and insights in a variety of artistic mediums.

Readers are given a forward-thinking viewpoint on the changes and advancements that the field of embodied AI is expected to undergo. Future developments like these might completely change industries, rewrite the rules for human-AI interactions, and open up completely new possibilities in a variety of fields.

9 Conclusion

9.1 Recapitulation of Key Findings

Evolution of Embodied AI: The first section of the chapter traced the development of embodied AI, highlighting how it differs from conventional AI systems in that it seamlessly combines intellectual algorithms with physical form. The framework for the talks that followed was established by defining embodied AI as a model that involves physical, corporeal elements in addition to abstract algorithms.

Manufacturing: industrial applications focused on manufacturing, highlighting the function of collaborative robotics in enhancing production settings. It demonstrated how embodied AI supports a symbiotic relationship between human expertise and AI efficiency by assisting with flexible automation, adaptive manufacturing, precision assembly, and quality control.

Supply chain management and logistics in industrial applications: The chapter examined the use of embodied AI in supply chain management and logistics, branching out from manufacturing. It described in detail how AI-enabled autonomous systems are transforming inventory control, order fulfillment, and warehouse operations, improving the robustness of the supply chain as a whole.

Industrial Applications—Healthcare: The conversation delved into the field of healthcare, explaining the ways in which embodied AI systems are revolutionizing surgical techniques, medical diagnostics, and patient care. The potential of artificial intelligence (AI) to improve healthcare outcomes was demonstrated by robotic assistants, who were emphasized for their roles in patient monitoring, rehabilitation, and precision surgery.

Industrial Applications: Energy Sector: This chapter examined how embodied AI helps optimize energy production and distribution in the energy sector. It explored the use of smart grids to improve overall energy efficiency, predictive repair of vital infrastructure, and the deployment of autonomous technologies in power plants.

Obstacles, Morality, and Lawful Considerations: The chapter critically evaluated the difficulties, moral dilemmas, and legal frameworks related to the broad usage of embodied AI systems while examining these industrial use cases. In order to secure the appropriate deployment of AI, it acknowledged the significance of tackling technological obstacles, ethical consequences, and the building of strong regulatory frameworks.

Prospects for the Future and Innovation: Taking a forward-looking view gave insights into new developments, collaborative human-AI workspaces, and expected advancements. The chapter foresaw revolutionary advancements that might completely reshape the future of embodied AI across industries, from enhanced decision-making to the integration of quantum computing and emotional intelligence in AI systems.

9.2 Implications for Industry and Society

The wider ramifications of embodied artificial intelligence (AI) in industrial environments are explored in this section, along with its effects on business procedures and social dynamics.

Enhanced Operational Efficiency: Embodied AI systems have the potential to greatly increase operational efficiency when they are integrated into industrial processes. Industries can maximize resource utilization, decrease human error, and streamline workflows with the help of precision applications in healthcare, autonomous systems in logistics, and collaborative robots in manufacturing.

Creativity and Technological Development: The chapter emphasizes how embodied AI can spur creativity and technological development in a variety of industries. Industries stand to gain from state-of-the-art solutions that challenge accepted wisdom, from the deployment of smart grids in the energy sector to adaptive manufacturing in industrial environments.

Work Dynamics have Changed: The rise of collaborative human-AI workspaces suggests that work dynamics are changing. In addition to increasing productivity, this move toward group decision-making, instantaneous communication, and flexible task distribution also reinterprets old roles and fosters a more cooperative and flexible work atmosphere.

Impact on Society in Healthcare: The implications for healthcare include better patient care, better surgical techniques, and more precise medical diagnoses. The use of embodied AI in healthcare has the potential to improve treatment outcomes, make medical services more accessible, and potentially reduce costs, all of which will have a positive effect on society as a whole.

Sustainability in the Production of Energy: There are significant consequences for sustainability when embodied AI is integrated into the energy sector. There is a chance for more effective energy production, less environmental impact, and progress toward a greener and more sustainable energy future through the use of smart grids and autonomous systems in power plants.

Ethical Issues and Social Trust: As embodied AI is used by industries, ethical issues grow more important. The chapter acknowledges the significance of AI systems that are ethically, responsibly, and transparently created. In addition to promoting ethical AI activities, addressing these issues promotes public confidence in the use of intelligent systems.

Work responsibilities and Skill needs: Because embodied AI is transforming industries, employment responsibilities and skill needs may alter. Because human-AI workspaces are collaborative, it may be necessary to acquire new skill sets. This highlights the significance of continuous education and training to guarantee that workers are capable of working well with intelligent agents.

Resolving Socioeconomic imbalances: Resolving potential socioeconomic imbalances is another aspect of the societal repercussions. Even while embodied AI has the potential to boost efficiency and creativity, inclusive solutions are required to guarantee that the gains are shared fairly and reduce the possibility of escalating already-existing inequalities.

Embedded AI has a wide range of ramifications for business and society, from increased operational effectiveness and technological innovation to societal effects in the healthcare sector and deployment concerns that should be balanced between ethics and equity. It is imperative that industries pay close attention to these consequences as they navigate this road of transformation in order to ensure responsible and effective integration. The chapter lays the groundwork for further debates and issues as society and industry adjust to the changing terrain of embodied AI.

9.3 Recommendations for Future Research

The recommendations for future research in the area of embodied artificial intelligence (AI) in industrial contexts are given in this closing part, which also identifies important areas that call for more study and research.

Explainable AI and Transparency: To improve the explainability of embodied AI systems, more study is advised. In addition to enhancing user confidence, the development of transparent models and procedures for AI decision-making processes will support moral and responsible AI activities. Future study must investigate ways to improve the interpretability of complex algorithms.

Ethical Issues and Bias Mitigation: Future research ought to concentrate on honing techniques for bias mitigation and ethical AI design, given the ethical issues surrounding the application of AI. Researching methods to detect and address biases in embodied AI systems can help create just and equitable AI solutions, especially in delicate fields like healthcare.

Human-AI Interaction and Collaboration: Future study should focus on examining the dynamics of human-AI interaction and collaboration in collaborative workspaces. Optimizing the potential of embodied AI in a variety of industrial contexts will need an understanding of how humans and AI systems can work together, interact, and build on each other's strengths.

Continual Improvement and Adaptive Learning: Future studies should focus on creating embodied AI systems that are capable of adaptive learning. Researching the ways in which these systems may learn continuously, adjust to shifting conditions, and improve over time will help develop AI solutions that are more adaptable and durable.

Socioeconomic Impact and Inclusivity: One important area of research to focus on is the socioeconomic effects of embodied AI deployment. For responsible and sustainable deployment, it will be crucial to look into how the advantages of AI integration can be shared fairly, minimize any potential inequalities, and guarantee inclusivity in access to AI-driven opportunities.

Interdisciplinary Collaboration: It is advised that future research projects promote interdisciplinary collaboration. Working together, specialists in the fields of ethics, sociology, and artificial intelligence (AI) will be able to provide a comprehensive knowledge of the opportunities, problems, and consequences that come with embodied AI. An interdisciplinary approach has the potential to produce deeper, more complex discoveries.

Long-Term Impact Assessment: Research on the long-term effects of embodied AI on society and industries should be continued. Examining the long-term impacts on social structures, skill needs, and job categories will shed light on how employment and human-AI collaboration are changing.

Regulatory Frameworks and Standards: It is imperative that further study be done on the creation of standards and regulatory frameworks for embodied AI across many industries. Fostering a safe and accountable AI ecosystem will require establishing norms for responsible AI deployment, addressing possible dangers, and guaranteeing adherence to ethical standards.

To sum up, the suggestions for further study included in this part are meant to help academics, professionals, and decision-makers improve the comprehension and ethical use of embodied AI in business environments. Through tackling these study areas, the science can advance in a way that supports moral behavior, is consistent with society values, and optimizes the benefits of embodied AI in a variety of contexts.

References

1. Duan, J., Yu, S., Tan, H.L., Zhu, H., Tan, C.: A survey of embodied AI: From simulators to research tasks. IEEE Trans. Emerg. Top. Comput. Intell. **6**(2), 230–244 (2022)
2. Gadre, S.Y., Ehsani, K., Song, S., Mottaghi, R.: Continuous scene representations for embodied AI. In: Proceedings of the IEEE/CVF Conference on Computer Vision and Pattern Recognition, pp. 14849–14859 (2022)
3. Nygaard, T.F., Martin, C.P., Torresen, J., Glette, K., Howard, D.: Real-world embodied AI through a morphologically adaptive quadruped robot. Nat. Mach. Intell. **3**(5), 410–419 (2021)
4. Chandra, J.V., Pasupuleti, S.K.: Machine learning methodologies for predicting neurological disease using behavioral activity mining in health care. In: 2022 8th International Conference on Advanced Computing and Communication Systems (ICACCS), vol. 1, pp. 1035–1039, March 2022. IEEE (2022)
5. Sundari, M.: A deep learning approach for enhancing tuberculosis classification leveraging Optimized Sequential AlexNet (OSAN). Int. J. Comput. Digit. Syst. **15**(1), 1–12 (2024)
6. Padmanandam, K., Rajesh, M.V., Upadhyaya, A.N., Chandrashekar, B., Sah, S.: Artificial intelligence biosensing system on hand gesture recognition for the hearing impaired. Int. J. Oper. Res. Inf. Syst. (IJORIS) **13**(2), 1–13 (2022)
7. Damiano, L., Stano, P.: A wetware embodied AI? Towards an autopoietic organizational approach grounded in synthetic biology. Front. Bioeng. Biotechnol. **9**, 724023 (2021)
8. Long, D., Padiyath, A., Teachey, A., Magerko, B.: The role of collaboration, creativity, and embodiment in AI learning experiences. In: Proceedings of the 13th Conference on Creativity and Cognition, June 2021, pp. 1–10 (2021)
9. Zhu, H., Kapoor, R., Min, S. Y., Han, W., Li, J., Geng, K., Neubig, G., Bisk, Y., Kembhavi, A., Weihs, L.: Excalibur: encouraging and evaluating embodied exploration. In: Proceedings

of the IEEE/CVF Conference on Computer Vision and Pattern Recognition, pp. 14931–14942 (2023)
10. Gan, C., Zhou, S., Schwartz, J., Alter, S., Bhandwaldar, A., Gutfreund, D., Yamins, D.L.K., DiCarlo, J.J., McDermott, J., Torralba, A., Tenenbaum, J.B.: The threeDWorld transport challenge: a visually guided task-and-motion planning benchmark for physically realistic embodied AI. arXiv:2103.14025 (2021)
11. Hughes, J., Abdulali, A., Hashem, R., Iida, F.: Embodied artificial intelligence: enabling the next intelligence revolution. IOP Conf. Ser.: Mater. Sci. Eng. **1261**(1), 012001 (2022). IOP Publishing
12. Eftekhar, A., Zeng, K.H., Duan, J., Farhadi, A., Kembhavi, A., Krishna, R.: Selective visual representations improve convergence and generalization for embodied AI. arXiv:2311.04193 (2023)

M. Shanmuga Sunadri is an Assistant Professor, BVRIT HYDERABAD College of Engineering. I am a research scholar at K L University; Interested Research areas are Machine learning, Artificial Intelligence, Software Engineering. Published 25 Publications in various journals indexed in Scopus, SCI, Web of Science; 3 Book chapters; 4 Patents.

Kisara Rishitha is a student at BVRIT HYDERABAD College of engineering for women's. I am interested in machine learning and deep learning. I got a grant of 3 lakhs from Tihan-IIT Hyderabad.

The Industrial AI Revolution: A Guide to Embodied AI Systems

Abhishek Choubey and Shruti Bhargava Choubey

Abstract Embodied Artificial Intelligence (EAI) systems that mix AI with automobiles, robots, as well as other machinery, have transformed industries through the ability of machines to learn through physical interaction with humans. This chapter provides a thorough review of the ways Embodied Artificial Intelligence (EAI) can be utilized to benefit different industries, showing its capability to provide significant improvements. Enterprise Application Integration (EAI) helps to seamlessly integrate AI capacities and physical objects across a variety of industrial environments, which improves efficiency in operations. The chapter starts by introducing EAI fundamentals that distinguish it from conventional AI techniques and outlines its theoretical foundation. Then, the chapter presents case studies demonstrating the application of EAI in real-world situations such as autonomous driving. It also provides giving further study options.

Keywords Embodied Artificial Intelligence (EAI) · Embodied AI (EAI) · Decision Making · Healthcare · Automation driving

1 Introduction

Embodied AI is any artificial intelligence that manifests physically via robots or autonomous machines—or individuals. Embodied AI differs from conventional artificial intelligence by not only considering data but also making decisions through physical interaction with objects through actuators and sensors. "Embodiment" refers to the capability of all AI devices to communicate with their surroundings and recognize physical forms, not just digital software. Embodied AI might provide

A. Choubey (✉) · S. B. Choubey
Department of Electronics Communication, Sreenidhi Institute of Science & Technology, Hyderabad, India
e-mail: abhishek.c@sreenidhi.edu.in

S. B. Choubey
e-mail: shruti.b@sreenidhi.edu.in

P. Raj et al. (eds.), *Building Embodied AI Systems: The Agents, the Architecture Principles, Challenges, and Application Domains*, Information Systems Engineering and Management 14, https://doi.org/10.1007/978-3-031-68256-8_17

the possibility of overcoming artificial intelligence's weaknesses. Artificial Intelligence systems that are designed to resemble human interaction function similarly to human interaction, but differ from their digital counterparts in the way that they connect computer systems to the real world. Artificial intelligence is designed to make these connections [1].

In the context of "embodied cognition", intelligence isn't just an attribute of our brains but could be the result of how our bodies react to their surroundings. Physical and sensory forms play an essential component of AI systems that rely on "embodied cognition". Sensorimotor shapes play a vital part in AI systems that use" embodied cognition.“ Sensorimotor forms include the "embodied cognition" approach, including microphones, cameras, and touch sensors, which replicate how humans make use of their sensory systems to perceive the world around them. Motorized joints and wheels allow users to adapt and interact with the surroundings just like humans interact with the world. Embodied AI tasks such as VQA and plans that include controls can help robots discover how they think, behave, and act in a manner completely independent of their environment. They can make informed decisions by utilizing their observations, employing tactics that are used over the long term, and reacting fast to any changes that may occur [2, 3].

Embedded artificial intelligence refers to the process of developing intelligence through interaction between an agent and its environment. The point at which an agent can make decisions that affect the long-term repercussions of the actions of an agent is one of the interactions that is used to quantify the efficiency of intelligent systems. One more way to learn about how intelligence grows in response to outside factors is to look at an embodied system. It encompasses different disciplines, including computer vision and natural language processing cognitive neuroscience, reinforcement psychology, and robotics, among others. Recently, there have been significant developments in language models which have demonstrated impressive abilities in language comprehension, logical thinking, and the ability to maintain a coherent cognitive process. These advancements have the potential to enable the development of robots that can understand and execute natural language instructions, perform multi-modal reasoning, and plan activities in real-world circumstances. The essential elements of embodied AI systems are shown in Fig. 1.

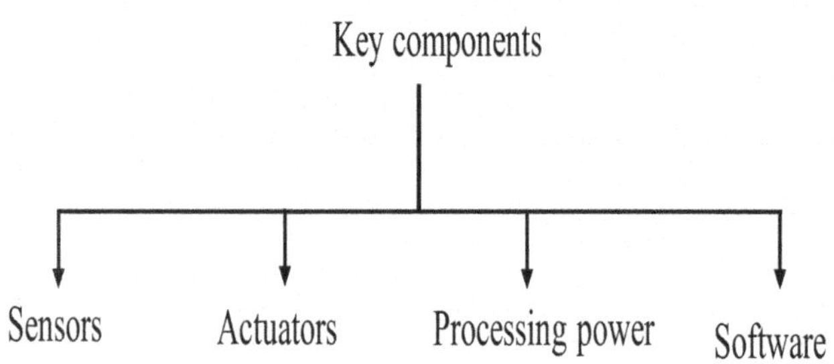

Fig. 1 Essential elements of embodied AI systems

1.1 The Limitations of Conventional AI

Conventional artificial intelligence, however strong in its capability, is constrained by its lack of physical appearance and dependency on pre-programmed information. Embodied AI signifies a significant and adverse change in the field of artificial intelligence, providing various significant features. Our cognitive abilities are based on the physical embodiment of brains in our bodies. This physical connection allows us to directly interact with the outside world. Most AI systems today exist in digital form and are devoid of any physical presence. Physical environments require our attention and understanding, whereas digital AI systems that are based on data and algorithms are usually incapable of understanding the real-world scenarios that occur within them. Embodied AI tasks like VQA or plans that have controls enable robots to gain an in-depth understanding of how their brain processes or thoughts, as well as behavior to their surroundings. By establishing long-term strategies and reacting swiftly when they notice changes in their environment, they can make informed choices based on the information they gather [4, 5].

2 Importance of Embodiment AI

Embodied AI represents a reaction to the idea of a mind–body duality that is extreme and is believed by some to think of intelligence as solely an event that occurs in the mind. Researchers and scientists have wrestled with the mind–body problem for a long time. Humans are "physical agents", with the characteristics of volume, mass as well as other bodily characteristics, and "mental agents", who think, perceive, reason, and act in a mental realm that appears to be void of physical manifestation. The different characteristics of embodied AI are shown in Table 1.

In the end, embedded AI is an amazing advancement in artificial intelligence's future. The technology integrates physical and digital reality through physical depictions of sensorimotor intelligence and interactions that allow for learning—changing the way we live our lives, from how people interact with technologies to the way we see and perceive the world around us. EAI can work nonstop and indefinitely to increase efficiency and productivity, which leads to higher outputs from their day-to-day work. The basis of modern AI is shown in Fig. 2. Quality and consistency improved AI systems help humans to finish tasks with greater precision, thereby making it easier to avoid mistakes and omissions. Lower labor costs AI systems can automate tasks that are repetitive and dangerous and free human labor to do more valuable tasks while reducing labor costs. Safety enhancement AI systems with embedded AI capabilities can carry out hazardous tasks in a safer manner which reduces the risks for humans from injuries and accidents. Data collection and analysis have increased Embodied AI systems can collect huge amounts of information from the natural world. This will help improve the efficiency of processes and help make better decisions. The use of Embodiment in different areas is shown in Table 2.

Table 1 Different characteristics of embodied AI

S. No.	Characteristics	
1	Adaptability	AI systems that are embodied should be able to adjust to changes in their surroundings and learn from their interactions with other individuals. It is necessary to have both pre-programmed decisions and the ability to learn from one's experiences to do this
2	Morphological computation	Artificial Intelligence (AI) is based in large part on physical shape and structure for it to comprehend information, and connect with its surroundings like legs. Their design could have profound effects on the way AI processes the information it receives and how it interacts with its environment such as directing steps or crossing barriers
3	Sensory Perception	The sensors that Embodied AI systems use to sense their environment are similar to those used by living organisms
4	Real-time processing	Due to the physical environment that it is operating in Artificial intelligence systems need to process information and make the right decisions at a moment's notice
5	Autonomy	Any embodied artificial intelligence system is intended to function independently for long periods, without the need for human interaction
6	Physical interaction	Embodied Artificial Intelligence, as opposed to software-based AI is capable of performing tasks that have immediate physical consequences. Depending on the situation the task could include movement, manipulation of objects, or even interacting with other agents and humans
7	Biased and unfair outcomes	Digital AI models can extend and add to existing biases in the data they are trained on. This can lead to biased outcomes, especially when applied in sensitive domains like loan approval, recruitment, and the court

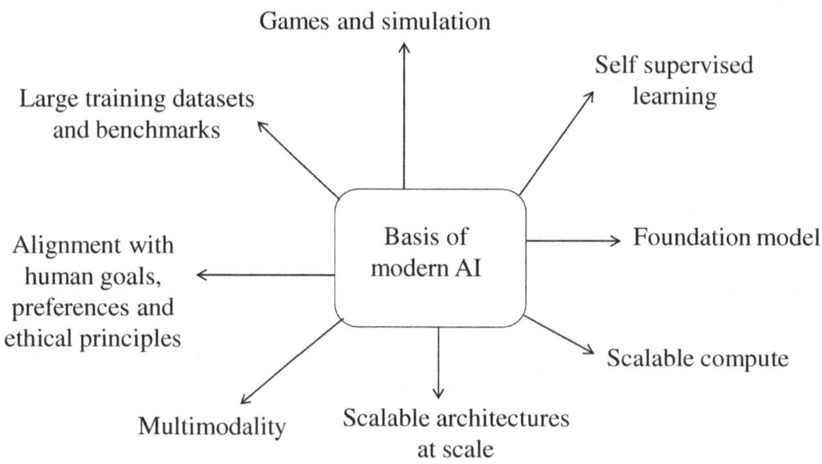

Fig. 2 The basis of modern AI

Table 2 Use of embodiment in different areas

S. No.	Technology	Description
1	Machine learning	Machine learning is an established means for creating an embodied agent, but it encompasses a wide array of passive tasks; most AI tasks have been developed so as not to require explicit learning for completion
2	Computer vision	Computer vision research has made significant strides forward in the field of embodied AI research, providing major contributions in terms of segmentation, classification, and image transformation tasks. Unfortunately, computer vision remains an encompassing subject with much work done toward improving passive AI tasks like these
3	Robotics	Robotics is an interdisciplinary field that encompasses research that directly tackles aspects of real-world phenomena, like sensors, real-time responses, and low-level controls
4	Simulation	Modern embodied AI research is now extending to simulate benchmarks that mimic or scan real environments. These provide challenges for traditional AI methods, whether or not they are physical embodiments

3 Related Work

Gao et al. introduced Alexa Arena, a user-centric Embodied AI platform. The proposed Alexa Arena is a new user-centric EAI platform that focuses on building generalizable agents that can assist humans through reasoning and procedural learning. A dialog-guided is also created from Alexa Arena, which includes 3 K unique tasks and 46 K human-annotated instructions and dialogues. A web-based user interface allowing real-time user communications and interactions with an EAI agent in Arena. Tamborini et al. have discussed the main elephants in the room in the recent philosophical and historical understanding of robotics and embodied AI: the biomimetic principle. This paper also suggested key philosophical cornerstone of bio-robotics, which is the ability to design robots that take inspiration from and mimic nature, should be directly addressed to understand better the variety of practices and goals of twenty-first-century bio-robotics and embodied AI [6, 7].

Zhuying et al. have developed AI-in-the-Shell, an exoskeleton-based system that enables users to. The AI-powered system can make independent decisions and actuate the user's body to better support their daily tasks and experiences, e.g., to enhance their game performance. Wanna et al. have proposed EAI systems typically deploy large language models to physical systems capable of interacting with their environment successfully demonstrating the feasibility of co-located, human–robot teaming. Specifically, we construct an experiment where an Augmented Reality (AR) headset mediates information exchange between an EAI agent and human operator for a variety of inspection tasks. Li et al. present BEHAVIOR-1 K, a comprehensive simulation benchmark for human-centered robotics first is the definition of 1,000 everyday activities, grounded in 50 scenes (houses, gardens, restaurants, offices, etc.) with more than 5,000 objects annotated with rich physical and semantic properties.

Pak et al. have explored whether self-supervised machines can spontaneously learn visual preferences (i.e., imprinting) and view-invariant object recognition, mimicking the early emerging behaviours of newborn chicks. Mu et al. introduce Embodied GPT, an end-to-end multi-modal foundation model for embodied AI, empowering embodied agents with multimodal understanding and execution capabilities. Wasserman et al. propose SLING to improve the performance of existing image-goal navigation systems. Deitke et al. present a retrospective on the state of Embodied AI research. Our analysis focuses on 13 challenges presented at the Embodied AI Workshop at CVPR. These challenges are grouped into three themes: (1) visual navigation, (2) re-arrangement, and (3) embodied vision and language. We discuss the dominant datasets within each theme, evaluation metrics for the challenges, and the performance of state-of-the-art models. We highlight commonalities between top approaches to the challenges and identify potential future directions for Embodied AI research [8, 9].

4 Applications of Embodied AI

Embodied AI (EAI) refers to AI systems with physical manifestation, meaning they exist within the real world and interact with it via sensors and actuators. EAI makes possible an array of tasks across many disciplines through this combination of intelligence and physicality.

4.1 Embodied AI in Everyday Life

- **Robotic Vacuum Cleaners**: Devices like the Roomba use sensors to navigate around homes, detect dirt, and avoid obstacles.
- **Humanoid Robots**: Robots are designed to mimic human appearance and behavior, and they can interact with people and their surroundings.
- **Autonomous Vehicles**: Cars that can drive themselves use a combination of sensors, cameras, and AI algorithms to navigate roads and avoid obstacles.
- **Drones**: Some drones can autonomously navigate environments, follow paths, or track objects using onboard AI.

4.2 Recent Cutting-Edge Embodied AI Developments

Embodied AI has proven its effectiveness across a diverse set of real-world scenarios. The use cases of embodied AI are shown in Table 3. Recent innovations in embodied AI have further emphasized its potential. Some of the key practical applications are

Table 3 Use cases of embodied AI

S. No.	Industrial application	
1	Manufacturing and assembly	Robots with embedded AI capabilities improve production processes through the automation of complex and repetitive tasks. This increases productivity. The same is true in agriculture
2	Exoskeletons	Exoskeletons, when combined with AI embodied have the potential to transform various fields by enhancing human capabilities and solving real-world problems
3	Space exploration	Embodied AI is essential to space missions, especially the exploration and maintenance of spacecraft. Embodied AI allows us to make autonomous decisions even in difficult and remote environments
4	Search and rescue	Embodied AI can provide healthcare professionals with a wealth of benefits in a wide range of areas including surgeries and rehabilitation to connecting patients and treating patients
5	Autonomous vehicles	Self-driving cars and drones rely heavily on embedded AI for accurate recognition and navigation. This technology plays an integral part in improving transportation system performance while decreasing accident risks
6	Healthcare	Drones and robots equipped with artificial intelligence (AI) can navigate disaster-stricken areas to search for survivors and assess security measures—an invaluable technology in times of emergency

OpenAI's Neo: OpenAI recently revealed Neo the first intelligent robot that combines digital intelligence and physical body. Neo is suitable for household and industrial work It displays autonomy, and mobility as well as a lightweight design. It also sets new standards in the integration of AI.

Amazon's Alexa Arena: The Amazon Alexa Arena platform provides a fully embodied AI solution designed to support natural human–robot interactions by offering natural human–robot interfaces in languages that are accessible to robots as well as users. It allows users to communicate with robots via natural language and aids in the task of completing. In addition, this platform allows for advanced AI modeling through tools such as datasets, code scenes, and datasets that further promote the model while also catalyzing improving human–robot interactions and AI-embedded development [10].

Meta's VC-1: Meta has recently revealed their VC-1 artificial mind that replicates the way that the human visual cortex transforms images into actions. It has been trained using daily tasks and subsequently tested in virtual real-world scenarios, VC-1 has already proven its superiority in 17 tests that require sensorimotor interaction; it has outperformed the previous models in all situations.

Adaptive Sensorimotor Skill Coordination (ASC): Meta has created and developed the Adaptive Sensorimotor Skills Coordination (ASC) method that is specifically designed for embedded AI agents. ASC allows them to quickly adapt to changing conditions and unexpected interruptions. It has a remarkable 98% rate of success in difficult tasks that require robots moving or manipulating in real-world situations.

Sample Heading (Third Level). Only two levels of headings should be numbered. Lower level headings remain unnumbered; they are formatted as run-in headings.

5 Embodied AI in Healthcare

Embodied AI (EAI) could be described as an innovative technology that has revolutionized the way healthcare services are delivered and provides new ways of diagnosing, treating, and managing the medical conditions of patients. EAI systems employ AI (Artificial Intelligence) in conjunction with real avatars or robots, creating intelligent objects that can communicate with their environment—this combination allows for the accessibility and personalization of healthcare services.

Surgical Robots: EAI robots provide surgeons with the ability to perform more precise, safe, and secure procedures, which result in quicker recovery time, with fewer risks for patients as well as an overall improved outcome for their services [11].

Rehabilitation Robots: Robotic exoskeletons provide patients with assistance to regain limb functions following strokes or injuries, helping them relearn how to walk again [12].

Assistive robots for the elderly and disabled: Robots like PARO (a therapy robot designed to look like an animal) provide the companionship needed to reduce anxiety and stress levels while others help with mobility and daily tasks that support independence [13].

Telemedicine Robots: The use of robot-enabled cameras and microphones allows health professionals to communicate with patients from a distance in remote locations, or when people's presence isn't possible (e.g. healthcare facilities with infectious diseases). The healthcare providers can communicate with them remotely while still maintaining relationships with patients [14].

Hospital Logistics and Delivery Robots: Robots can autonomously deliver medications, samples, or supplies within a hospital, reducing the need for human staff to perform these tasks and minimizing errors [15].

Disinfection Robots: In the wake of infectious diseases, robots equipped with ultraviolet (UV) light or disinfectant sprays can autonomously navigate and clean hospital rooms, reducing the spread of pathogens [16].

Patient Monitoring: Some robots can continuously monitor patients' vital signs and alert healthcare professionals to any irregularities [17].

Mental Health and Therapy: Robots can be used in cognitive behavioral therapy, especially for children or those with autism. Their predictable and consistent interactions can be comforting and therapeutic [18].

Educational and training: The patient simulation computers permit medical students to experience the diagnostic procedure and interact with patients safely in a classroom or training environment [19].

Pharmaceutical Robots: Automated and computerized pharmacy systems help to ensure the accuracy of dispensing medicines while minimizing human mistakes [20].

Home Health Monitoring: Care Monitoring for Home for patients discharged from hospitals, but needing medical aid Robots can be used as reminders for patients to take their medications monitor health indicators, and act as a means of communication for health professionals [21–23].

Social Interaction with Hospitalized: Patients Robots are developed to help and encourage hospitalized patients, specifically those who are in long-term care or isolation which can help ease feelings of loneliness and improve health. AI in healthcare has many advantages: improved accuracy, efficiency, and the potential for better patient care are only some of the benefits of the implementation [24–26].

However, AI also raises some ethical and cost issues that must be considered as well as questions of acceptance and possible issues. With the advancements in technology expected to be made, the use of embedded AI within healthcare is likely to be more commonplace which will result in better outcomes for patients and higher efficiency in medical procedures [27].

6 Embodied AI for Autonomous Driving

Embodied AI, or EAI, is crucial for the progress of autonomous driving technologies. Enterprise Artificial Intelligence (EAI) integrates machine intelligence with real-world contexts through the utilization of intelligent systems that watch, interact with, and collect information from their environment. EAI's distinctive attribute establishes it as a significant component in completely self-driving vehicles. Autonomous vehicles depend on sensors powered by EAI (Embedded Artificial Intelligence) technology, such as LiDAR, cameras, and radar, to collect real-time data about their surrounding. These sensors expand the vehicles' capacity to navigate highways with greater accuracy, detect pedestrians and objects, and analyze traffic patterns more effectively. Enterprise Application Integration (EAI) algorithms calculate incoming data and make informed decisions regarding vehicle actions, such as route planning, speed adjustments, and responses to unforeseen circumstances, thus promising secure and effective navigation. Artificial intelligence (AI) smoothly integrates with vehicle control systems to efficiently translate decision-making processes into concrete actions, such as providing secure steering, braking, and accelerating, in numerous circumstances.

The global market for autonomous vehicles is expected to reach $1.2 trillion by 2030. Embodied AI is poised to play a pivotal role in shaping the future of autonomous driving. Its potential to improve safety, efficiency, and accessibility makes it a crucial technology for the future of transportation. The future of autonomous driving lies in

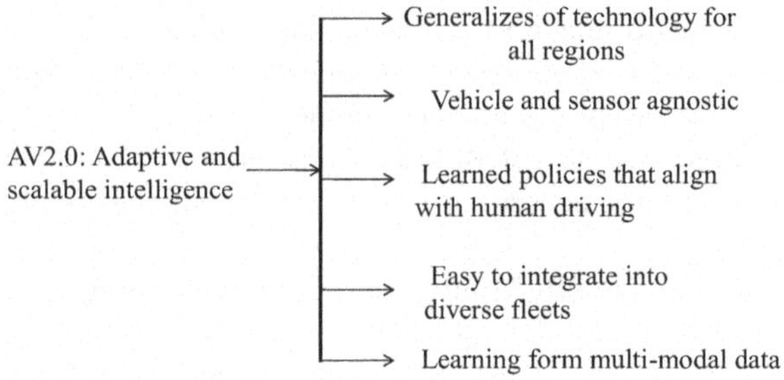

Fig. 3 Adaptive and scalable intelligence

adaptive and scalable driving intelligence that can tackle diverse environments and ensure safe, efficient, and sustainable transportation for all. The details of adaptive and scalable intelligence are shown in Fig. 3.

Embodied AI to drive autonomous vehicles is a rapidly developing field that has several major areas that are driving the development of technology. The four Pillars of Embodied AI for Autonomous Driving are Simulation for evaluation, Reinforcement learning for driving, language meets for driving, and world models [6–8].

6.1 Simulation for Evaluation

Testing and development of Embodied AI for autonomous vehicles (AVs) is essential in providing safe and reliable transportation, however, testing in real-world conditions can be challenging, like cost-intensive, security issues, and restricted scenarios. Simulation is a method to overcome these issues by allowing controlled testing and fast repetition. In the end, this lets you explore diverse scenarios.

Test extreme scenarios and edge scenarios, without risking injury to persons or property. Quickly create multiple scenarios to make statistically valid comparisons with much less cost than actual testing. Keep track of each element of AI behavior to analyze as well as further improvement.

Simulations shouldn't be an all-inclusive source of information in the testing of autonomous vehicles as there may be a difference between how an AI system performs in simulation and how it performs when on the roads. Testing in real-world conditions should be considered as a method of the verification process of reliability and safety. Simulation is a low-cost, safe method of testing various situations and algorithms that can result in safe autonomous car operation.

6.2 Reinforcement Learning for Driving

The concept of reinforcement learning (RL) proved valuable in developing artificial intelligence systems that are capable of rapidly adapting to driving conditions.

Reinforcement Learning agents acquire information through interaction with their environment and through research. When people achieve their goals, their efforts are acknowledged as achievements. The process of reinforcement learning (RL) is among the numerous embodied artificial intelligence systems in use currently and plays a crucial role in assisting agents. Systems that employ embedded artificial intelligence utilize various sensors to collect data about their environment like cameras, LiDAR, as well as radar sensors. Recurrent neural networks (RL) were designed specifically to serve this purpose. They are specially designed to analyze this information and identify pedestrians, objects, or other elements that could be significant. Reliance Lab agents must conduct an exhaustive analysis of their surroundings and make smart decisions about how fast and direction of their vehicle, allowing them to be in control of their surroundings. Software programs were designed to maximize the benefits of abiding by traffic regulations and ensure that drivers get safely to their destination. Reward-learning (RL) software which reacts quickly and effectively to the changing driving conditions that are encountered in real-world situations is crucial in empowering drivers with the capability to quickly adapt and efficiently. To maximize effectiveness systems that incorporate algorithms that expand the training information to new scenarios must be integrated into the systems.

Here are some specific scenarios of the ways AI technology can be used for self-driving applications. Driver Simulation agents in RL are taught basic driving abilities and decision-making techniques through simulations. agents being rewarded for certain actions such as staying within their lane, following traffic signals, and avoiding collisions. Fine-tuning Real-World Tests After having been trained in simulations RL agents can be enhanced by real-world testing in driving scenarios by capturing data from driving to adjust their approach and increase the effectiveness and longevity of their policies.

RL Agents learn through watching and imitating human drivers employing imitation-learning methods that allow agents to observe other people doing what they are skilled at, and then replicate it. Multi-Agent Learning: Reinforced learning can be utilized to train multiple agents at the same time and equip them with the ability to collaborate effectively and safely. Particularly, this function is invaluable when autonomous vehicles have to interact with pedestrians, as well as other vehicles, and also in situations that require collaboration between agents. Flexible agents can react quickly to changes in driving conditions or surroundings, which makes their systems more flexible as opposed to systems that are governed by rules. Scalability The algorithms employed in RL could be quickly altered to handle large and complicated terrains, which makes it a perfect choice for real-world driving scenarios. RL is an efficient method to teach employees how to reach specific goals swiftly and effectively as opposed to traditional methods of instruction. Security: The employees working at RL are taught to focus on the security of clients by avoiding accidents while adhering to traffic laws. Utilizing this method for autonomous driving systems, RL helps facilitate Embodied AI development, which addresses many issues that are related to self-driving technology.

Rewards Design Creating suitable incentives that are appropriate for RL agents is essential to regulate their behavior. Moreover, rewarding drivers in a manner that promotes safety and efficiency is difficult. Information Collection To educate RL agents efficiently, collecting massive amounts of driving data from real-world situations is costly and time-consuming. To decrease the risks of deployment in actual situations, educating RL agents properly is crucial for their success. Reinforcement learning can be an extremely efficient method of training embedded AI systems to assist in autonomous driving. Reinforcement Learning agents can adapt to the changing environment, learning through experience and making actions that maximize rewards for the greatest return. However, some issues must be overcome before the full potential of Reinforcement Learning is achieved in terms of efficiency, safety, and security for autonomous vehicles.

6.3 *Language Meets Driving*

Conventional autonomous driving systems depend on sensors and machine-learning algorithms. However, recent research suggests that language is an important factor that can improve their performance and security. This is the point where Embodied AI can help as it helps connect physical and spoken interactions inside the autonomous vehicle. Here are the most important areas where driving and language interact through artificial intelligence to develop autonomous vehicles:

Tools for communication that make use of Natural Language Processing (NLP) devices include autonomous cars that make use of it to interpret commands similar to the ones that people might use to communicate. Training in communication includes instructions for talking to passengers using non-verbal and verbal signals like, "Take me directly to a medical facility" or "Make a left turn at this junction". They should converse with passengers as they give these signals. Natural Language Processing (NLP) allows an artificial intelligence-equipped vehicle to hold conversations with its passengers using natural spoken languages. Please give information on the most efficient travel route, conditions, and safety measures that are relevant to your travel needs.

The map mapping process can be improved by using natural language annotations that allow for better comprehension of landmarks, roads, and other important locations, which improve precision and mobility. To enhance navigational capability and increase precision, this process involves the creation of an English description of the landmarks. The vehicles that provide route descriptions also provide written reports on all aspects of their travel to their passengers.

Engaging passengers in decision-making and actively taking part in negotiations of routes Travelers can use natural language and suggest alternate routes to AI systems before selecting one that best fulfills their requirements. Sharing preferences Drivers can share their personal preferences about speeds, style of driving, and routes using AI so that they can personalize their experience while driving. Responding to Questions and Addressing Concerns: NLP can facilitate communication between vehicles and passengers through the ability to address concerns and get information about driving procedures.

Language helps vehicles recognize and react more effectively and with greater flexibility when confronted with complex driving situations, which leads to safe and flexible driving behavior. Improved customer experience language provides users with a natural and intuitive method of communicating with their car, which adds to the overall experience of driving and enhances the user experience. Transparency and trust are built by presenting their reasoning and choices to their customers, AI systems create trust between them and their users and make their operation more transparent. Flexible and Personalized Driving Experience Drivers can alter their driving experience by providing feedback in a natural term, thereby making the vehicle more user-friendly. NLP systems should be capable of adjusting to different dialects, accents, and natural use patterns for language. Conceptual Understanding: For AI machines to become precise in understanding language they have to first understand the events that are happening behind the scenes as well as within their environments. Processing Real-time Text NLP algorithms should be efficient enough to handle text in real-time and allow seamless interactions between cars and passengers.

Although these issues are present, however, combining AI and language could be an opportunity to change the process of developing autonomous vehicles. By allowing natural collaboration between machines and humans language will play a crucial part in making autonomous vehicles more secure, intelligent, and comfortable for their owners.

6.4 World Models

Imagine a vehicle that is not just responsive to its surroundings, but is also constantly improving and updating its internal map. This is the basis of world models that are used to create Embodied AI to enable autonomous driving. Imagine it as if the car's brain is constantly creating a dynamic model of its environment, including pedestrians, traffic signals driving plans—to facilitate AI. This model inside lets it:

Predict: Being aware of the flow of traffic and pedestrians as well as potential dangers can ensure an easier, safer driving experience.

Plan: Based on the current information, design and modify routes that maximize efficiency while minimizing delays.

Adapt: Be prepared and swiftly in the event of unexpected situations, like unexpected construction zones, or changes in lane while minimizing risks.

Learn: When they acquire new knowledge and experiences the ability of them to make informed decisions with time is significantly enhanced.

World Models are constructed based on sensor fusion. This blends data collected from cameras LiDAR radar as well as other sensors to create an accurate picture of their surroundings. Machine learning algorithms analyze the information and can identify the movement of objects, as well as potential connections between them, this gives AI greater autonomy and understanding when it moves around in a way that is intelligent and autonomous.

Although they are still in development, these technologies are in use and offer an incredible opportunity to transform autonomous driving. By analyzing their surroundings, fully autonomous vehicles can function more effectively, opening possibilities for safer and more efficient transportation [2, 3].

7 Challenges in Bringing Embodied AI to Life

While AI's capabilities are vast there are important obstacles that researchers and developers confront when creating these systems. There are several issues they're trying to overcome:

(i) The Sensory Realist The creation of artificial intelligence that can mimic human senses with tactile feedback, sensory feedback, and perception of depth is an easy task, especially with the modern microphones and cameras available.
(ii) Real-World Navigation: For embedded AI systems, the ability to quickly adapt to changes in conditions like obstacles or weather is vital.
(iii) Safety and Ethics. Maintaining AI systems legally compliant keeping privacy safe and being safe are the two most important elements to their widespread adoption.
(iv) Technology can be a challenge in finding the right balance between the complexity of the task, its effectiveness when scaled, and efficiency isn't easy and to achieve it requires expert experts. Integration of AI systems into existing systems could require additional understanding to be successful.
(v) The ability to find the right balance of energy and level of performance for robotic devices like drones autonomous cars, and other forms of technology that are robotic is crucial.
(vi) The initial expenses associated with investment could be exorbitantly expensive; creating and implementing AI systems can be costly.
(vii) Regulating and Compliance In certain industries embedded AI systems are subject to certain regulations and compliance regulations.

8 Conclusion

Embodied AI (EAI) systems mark a significant change in the field of industrial innovation by supplying sophisticated machines accomplished of detecting interactions and taking their cues from their surroundings, which can lead to the highest levels of automation and efficiency across a variety of industries. EAI-powered robots, drones, and autonomous vehicles are revolutionizing the way jobs are performed while relieving human beings of the risk of repetitive and dangerous work. With improved decision-making capabilities and the ability to adapt to changing conditions, it is possible to efficiency and better results. Technologies-related issues that need to be addressed include sensor fusion, instantaneous decision-making as well as ethical concerns over worker safety and job loss. In addition regulations and the acceptance of the public are crucial factors in ensuring ethical but effective integration of EAI in industrial systems.

References

1. Huang, W., Abbeel, P., Pathak, D., Mordatch, I.: Language models as zero-shot planners: extracting actionable knowledge for embodied agents. In: International Conference on Machine Learning. PMLR, pp. 9118–9147 (2022)
2. Padmakumar, A., Thomason, J., Shrivastava, A., Lange, P., Narayan-Chen, A., Gella, S., Piramithu, R., Tur, G., Hakkani-Tur, D.: TEACh: task-driven embodied agents that chat. In: Conference on Artificial Intelligence (AAAI) (2022)
3. Savva, M., Kadian, A., Maksymets, O., Zhao, Y., Wijmans, E., Jain, B., Straub, J., Liu, J., Koltun, V., Malik, J., et al.: Habitat: a platform for embodied AI research. In: Proceedings of the IEEE/CVF International Conference on Computer Vision, pp. 9339–9347 (2019)
4. Tamborini, M.: The elephant in the room: the biomimetic principle in bio-robotics and embodied AI. In: Studies in History and Philosophy of Science **97**, 13–19 (2023). ISSN 0039-3681
5. Li, Z., Huang, T., Patibanda, R., Mueller, F.: AI in the shell: towards an understanding of integrated embodiment. In: Extended Abstracts of the 2023 CHI Conference on Human Factors in Computing Systems (CHI EA '23). Association for Computing Machinery, New York, NY, USA, Article 28, pp. 1–7 (2023). https://doi.org/10.1145/3544549.3585867
6. Tussyadiah, I.P., Jung, T.H., Claudia tom Dieck, M.: Embodiment of wearable augmented reality technology in tourism experiences. J. Travel Res. **57**(5), 597–611 (2018).
7. Wanna, S., Parra, F., Valner, R., Kruusamae, K., Pryor, M.W.: Multimodal grounding for embodied AI via augmented reality headsets for natural language driven task planning (2023). arXiv:2304.13676
8. Li, C., Zhang, R., Wong, J., Gokmen, C., Srivastava, S., Martín-Martín, R., Wang, C., Levine, G., Lingelbach, M., Sun, J., Anvari, M., Hwang, M., Sharma, M., Aydin, A., Bansal, D., Hunter, S., Kim, K., Lou, A., Matthews, C.R., Villa-Renteria, I., Tang, J.H., Tang, C., Xia, F., Savarese, S., Gweon, H., Liu, K., Wu, J. Fei-Fei, L.: BEHAVIOR-1K: A benchmark for embodied AI with 1,000 everyday activities and realistic simulation. In: Proceedings of the 6th Conference on Robot Learning, in Proceedings of Machine Learning Research, vol. 205, pp. 80–93 (2023). https://proceedings.mlr.press/v205/li23a.html
9. Pak, D., Lee, D., Wood, S.M., Wood, J.N.: A newborn embodied Turing test for view-invariant object recognition (2023). arXiv:2306.05582
10. Gao, Q., Thattai, G., Gao, X., Shakiah, S., Pansare, S., Sharma, V., Sukhatme, G.S., Shi, H., Yang, B., Zheng, D., Hu, L., Arumugam, K., Hu, S., Wen, M., Guthy, D., Chung, C., Khanna, R., Ipek, O., Ball, L., Bland, K., Rocker, H., Rao, Y., Johnston, M., Ghanadan, R., Mandal, A., Hakkani-Tür, D.Z., Natarajan, P.: Alexa arena: a user-centric interactive platform for embodied AI (2023). arXiv:2303.01586
11. Wasserman, J., Yadav, K., Chowdhary, G.V., Gupta, A., & Jain, U.: Last-mile embodied visual navigation. In: Conference on Robot Learning (2022)
12. Deitke, M., Batra, D., Bisk, Y., Campari, T., Chang, A.X., Chaplot, D.S., Chen, C., D'Arpino, C., Ehsani, K., Farhadi, A., Fei-Fei, L., Francis, A., Gan, C., Grauman, K., Hall, D., Han, W., Jain, U., Kembhavi, A., Krantz, J., Lee, S., Li, C., Majumder, S., Maksymets, O., Mart'in-Mart'in, R., Mottaghi, R., Raychaudhuri, S., Roberts, M., Savarese, S., Savva, M., Shridhar, M., Sunderhauf, N., Szot, A., Talbot, B., Tenenbaum, J.B., Thomason, J., Toshev, A., Truong, J., Weihs, L., Wu, J.: Retrospectives on the Embodied AI Workshop (2022). arXiv:2210.06849
13. M. Savva, A. Kadian, O. Maksymets, Y. Zhao, E. Wijmans, B. Jain, J. Straub, J. Liu, V. Koltun, J. Malik, D. Parikh, and D. Batra. Habitat: A Platform for Embodied AI Research. ICCV, 2019.

14. Crane, T., Patterson, S.: History of the Mind-Body Problem. Routledge (2012)
15. F. Xia, A. R. Zamir, Z.-Y. He, A. Sax, J. Malik, and S. Savarese. Gibson Env: real-world perception for embodied agents. In: CVPR (2018)
16. Qi, Y., Wu, Q., Anderson, P., Liu, M., Shen, C., van den Hengel, A.: Reverie: remote embodied visual referring expression in real indoor environments. In: CVPR (2020)
17. Savva, M., Kadian, A., Maksymets, O., Zhao, Y., Wijmans, E., Jain, B., Straub, J., Liu, J., Koltun, V., Malik, J., et al.: Habitat: a platform for embodied AI research. In: ICCV, pp. 9339–9347 (2019)
18. Abdullah, A.S., Gaehde, S., Bickmore, T.: A tablet-based embodied conversational agent to promote smoking cessation among veterans: a feasibility study. J. Epidemiol. Glob. Health **8**(3–4), 225–230 (2018)
19. Cameron, G., Cameron, D., Megaw, G., Bond, R., Mulvenna, M., O'Neill, S., Armour, C., McTear, M.: Assessing the usability of a chatbot for mental health care. In: INSCI 2018. Lecture Notes in Computer Science, pp. 121–132. Springer (2019)
20. Huang, X., Zhang, J., Xu, Z., Ou, L., Tong, J.: A knowledge graph-based question-answering method for the medical domain. Peer J Comput. Sci. **7**, 667 (2021)
21. Chaddad, A., Peng, J., Xu, J., Bouridane, A.: Survey of explainable AI techniques in healthcare. Sensors (Basel) (2023)
22. Li, F., Ruijs, N., Lu, Y.: Ethics & AI: a systematic review on ethical concerns and related strategies for designing with AI in healthcare. AI **4**(1), 28–53 (2023). https://doi.org/10.3390/ai4010003
23. Wen, L., Fu, D., Li, X., Cai, X., Ma, T., Cai, P., Dou, M., Shi, B., He, L., & Qiao, Y.: DiLu: a knowledge-driven approach to autonomous driving with large language models (2023). arXiv:2309.16292
24. Xi, Z., Sukthankar, G.: A graph representation for autonomous driving. In: The 36th Conference on Neural Information Processing Systems Workshop (2022)
25. Fu, D., Li, X., Wen, L., Dou, M., Cai, P., Shi, B., Qiao, Y.: Drive like a human: rethinking autonomous driving with large language models (2023). arXiv:2307.07162.
26. Chen, L., Teng, S., Li, B., Na, X., Li, Y., Li, Z., Wang, J., Cao, D., Zheng, N., Wang, F.-Y.: Milestones in autonomous driving and intelligent vehicles—Part II: perception and planning. IEEE Trans. Syst. Man, Cybern.: Syst. (2023)
27. Driess, D., Xia, F., Sajjadi, M.S.M., Lynch, C., Chowdhery, A., Ichter, B., Wahid, A., Tompson, J., Vuong, Q., Yu, T., Huang, W., Chebotar, Y., Sermanet, P., Duckworth, D., Levine, S., Vanhoucke, V., Hausman, K., Toussaint, M., Greff, K., Zeng, A., Mordatch, I., Florence, P.: Palm-e: an embodied multimodal language model (2023). arXiv:2303.03378

Dr. Abhishek Choubey received a Ph.D. degree in the field of VLSI for digital signal processing from Jaypee University and technology Guna MP, in 2017.He is currently associated with Sreenidhi institute of science and technology, Hyderabad, as an Associate Professor. He has published nearly 70 technical articles. His research interest includes reconfigurable architectures, approximate-computation, algorithm design, and implementation of high-performance VLSI systems for signal processing applications. He was a recipient of the Sydney R. Parker and M. N. S. Swamy Best Paper Award for Circuits, Systems, and Signal Processing in 2018.

Dr. Shruti Bhargava Choubey received BE with honours (2007) from RGPV Bhopal and M. Tech. degree in Digital Communication Engineering (2010) from RGPV Bhopal subsequently she carried out her research form Dr. K.N. Modi University Banasthali Rajasthan, and awarded Ph.D. in 2015.Presently she is working as Associate Professor & Dean Innovation & Research in the Department of Electronics and Communication at Sreenidhi Institute of Science and Technology, Hyderabad she has published more than 100 papers(5 SCI,18 Scopus) of national and International repute. She has been a Member of many selection committees for recruitment of staff and faculty. her research areas include signal processing, Image processing and

Biomedical Engineering. She has produced 17 M.Tech degrees & amp; guided more than 70 B.Tech Project. She is Senior member of IEEEmember of IETE, New Delhi and International Association of Engineers (IAENG). She worked in different positions like Dean academic & HOD with numerous capacities. She was awarded MP Young scientist fellowship 2015 & Received MP council fellowship in 2014 for her contribution to Research.

Illuminate Metaverse Multisensor Fusion and Dynamic Routing Technologies Across Web3-Powered for Autonomous Vehicles Shaping Efficient Urban Transport Solutions of Future in Smart City Era: Deep Dive into Protocols for Benefiting Society Lensing Prospects and Challenges

Bhupinder Singh, Komal Vig, Pushan Kumar Dutta, and Christian Kaunert

Abstract The intersection of the Metaverse, autonomous and connected vehicles, and the rapidly evolving Web3 technologies presents a paradigm shift in the realm of urban transportation. The Metaverse, an emergent digital realm bridging the virtual and physical worlds, offers a dynamic stage for revolutionary advancements in autonomous and connected vehicle technologies. The primary objectives of this chapter are to explore and understand the potential of Metaverse technologies, multisensor fusion, and dynamic routing in the context of Web3-powered autonomous vehicles. The study aims to assess their role in shaping the future of urban transport solutions, particularly in Smart Cities. Additionally, the research seeks to identify protocols that contribute to the seamless integration of these technologies for the betterment of society. It employs a multifaceted methodology, incorporating a deep dive into the realms of Metaverse technologies, multisensor fusion, and dynamic routing. Analysis of Web3 protocols is conducted to ascertain their applicability and effectiveness in the development of autonomous vehicles. The study involves a comprehensive review of existing literature, case studies, books, research papers,

B. Singh (✉) · K. Vig
Sharda School of Law, Sharda University Greater Noida, Greater Noida, India
e-mail: bhupindersinghlaw19@gmail.com

P. K. Dutta
Electronics and Communication Engineering Department at ASETK, Amity University Kolkata, Kolkata, India

C. Kaunert
Dublin City University, Dublin, Ireland

University of South Wales, Cardiff, UK

P. Raj et al. (eds.), *Building Embodied AI Systems: The Agents, the Architecture Principles, Challenges, and Application Domains*, Information Systems Engineering and Management 14, https://doi.org/10.1007/978-3-031-68256-8_18

articles to evaluate the performance and feasibility of the proposed technologies. The findings of the chapter showcase the potential of Metaverse technologies, multisensor fusion, and dynamic routing in enhancing the efficiency and safety of autonomous vehicles in urban settings. Web3-powered protocols emerge as crucial components for seamless integration, providing a foundation for secure and reliable communication. The results highlight promising prospects for the future of urban transportation, accompanied by insights into the challenges that must be addressed to realize these advancements. The chapter underscores the transformative impact of Metaverse Multisensor Fusion and Dynamic Routing Technologies across Web3-powered autonomous vehicles on the landscape of urban transport. The study contributes valuable insights into protocols that can benefit society by addressing challenges and optimizing the performance of autonomous systems. As Smart Cities evolve, the integration of these technologies presents a compelling vision for a future where urban transportation is not only efficient but also environmentally sustainable and socially inclusive.

Keywords Autonomous vehicles · Dynamic routing technologies · Metaverse · Hyperloop · Protocols · Urban transportation

1 Introduction

The integration of Metaverse Multisensor Fusion and Dynamic Routing Technologies within the framework of Web3-powered autonomous vehicles signifies a paradigm shift with far-reaching implications for the future of urban transport in the Smart City era. As autonomous vehicles become an integral part of our urban landscapes, the convergence of Metaverse technology, multisensor fusion, and dynamic routing stands at the forefront of innovation. Metaverse Multisensor Fusion combines the capabilities of various sensors to create a holistic perception system, enhancing the vehicle's awareness of its surroundings.

Simultaneously, Dynamic Routing Technologies leverage real-time data to adaptively navigate through urban landscapes, optimizing efficiency and reducing congestion. This paper embarks on a deep dive into the underlying protocols governing this integration, unraveling the intricate mechanisms that enable seamless communication and collaboration among these technologies. Furthermore, the research explores the myriad societal benefits, ranging from heightened safety to improved accessibility, and delves into the promising prospects and formidable challenges that accompany this transformative fusion. Through this exploration, we aim to illuminate the path toward shaping efficient and intelligent urban transport solutions that will define the Smart City landscape of the future [1].

1.1 Concept of the Metaverse and Its Potential Impact on Transportation

The concept of the Metaverse, a collective virtual shared space that merges physical and digital realities, holds transformative potential for the realm of transportation. Beyond its origins in entertainment and virtual reality, the Metaverse is poised to redefine how we perceive, interact with, and navigate the physical world, including transportation systems. By seamlessly integrating digital information with the real-world environment, the Metaverse can enhance the efficiency and safety of transportation. Imagine a scenario where autonomous vehicles, equipped with advanced sensors, leverage the Metaverse to access real-time data about traffic conditions, weather, and road infrastructure.

This interconnected digital layer can enable vehicles to make informed decisions, optimize routes dynamically, and respond to changing circumstances instantaneously. Additionally, the Metaverse opens the door to novel modes of transport, such as virtual commuting or shared digital spaces for collaborative mobility experiences. As it delve into the potential impact of the Metaverse on transportation, it becomes clear that this concept transcends virtual reality, offering tangible benefits that could revolutionize the way it move within our physical surroundings.

1.2 Significance of Autonomous and Connected Vehicles in Smart Cities

The significance of autonomous and connected vehicles in the context of smart cities is paramount, ushering in a transformative era for urban mobility. These advanced vehicles contribute to the overarching goal of creating more intelligent, efficient, and sustainable urban environments. Autonomous vehicles, equipped with sophisticated sensors and artificial intelligence, have the potential to enhance road safety, reduce traffic congestion, and optimize transportation systems. They offer the promise of increased mobility for individuals with limited access to traditional transportation and provide a foundation for future mobility-as-a-service models [2].

Connected vehicles, on the other hand, leverage communication technologies to interact with each other and the surrounding infrastructure. This connectivity enables real-time data exchange, facilitating dynamic traffic management, predictive maintenance, and improved energy efficiency. In smart cities, where data-driven decision-making is a cornerstone, connected vehicles become integral components of a larger ecosystem, contributing valuable insights for urban planners and policy-makers. Together, autonomous and connected vehicles form a symbiotic relationship, creating an intelligent transportation network within smart cities. The seamless integration of these technologies holds the potential to reduce emissions, optimize traffic flow, and enhance overall quality of life for urban residents. As smart cities continue to evolve, the deployment and integration of autonomous and connected vehicles represent a pivotal step towards achieving a more sustainable, efficient, and livable urban future [3].

1.3 Objectives and Scope of Chapter

The objectives and scope of this paper endeavours are carefully defined to provide a structured and comprehensive exploration into the intricate intersection of Metaverse Multisensor Fusion and Dynamic Routing Technologies within the Web3-powered landscape, specifically applied to autonomous vehicles and their role in shaping efficient urban transport solutions in the Smart City era.

The primary objective is to unravel the synergies and interdependencies among these cutting-edge technologies. By understanding the intricate protocols governing their integration, the research aims to elucidate the mechanisms driving seamless communication and collaboration. This includes a deep dive into the communication and security protocols ensuring the integrity and privacy of the data exchanged within this dynamic framework.

The scope extends beyond the technical intricacies to encompass the societal implications, both positive and challenging. Assessing the potential benefits to society is a critical aspect, exploring how this integration can contribute to improved safety, enhanced efficiency in urban transport, and increased accessibility. Simultaneously, the research endeavors to identify and address the challenges and ethical considerations associated with this transformative fusion.

The chapter extends its scope to investigate the prospects that arise from the widespread adoption of these technologies. This includes exploring economic opportunities, potential advancements in environmental sustainability, and the broader impact on urban living. Through a careful examination of the objectives and scope outlined in this research, the aim is to contribute valuable insights to the discourse surrounding the future of urban transportation in the Smart City era.

2 Metaverse and Multisensor Fusion

The convergence of the Metaverse and Multisensor Fusion represents a groundbreaking synergy at the forefront of technological innovation. The Metaverse, a virtual shared space that intertwines the digital and physical realms, provides an immersive platform for user interaction and engagement. Multisensor Fusion involves the integration of data from diverse sensors to create a comprehensive and nuanced understanding of the surrounding environment. Together, these technologies hold the potential to redefine the way we perceive and interact with our surroundings in both virtual and physical spaces. In the context of the Metaverse, Multisensor Fusion becomes a pivotal element, enhancing the realism and responsiveness of the virtual environment [4]. This integration not only augments the user experience within the Metaverse but also extends its applications to real-world scenarios, such as in autonomous vehicles. By combining the rich data generated by Multisensor Fusion with the immersive capabilities of the Metaverse, we unlock new possibilities for enhanced virtual simulations, training environments, and, critically, the development of more sophisticated and capable autonomous systems. This intersection is not merely a technological convergence; it is a gateway to a new era of experiential richness and practical applications that have the potential to reshape industries and redefine how we interact with the digital and physical dimensions of our world.

2.1 *Concept of the Metaverse and Its Components*

The concept of the Metaverse represents a revolutionary shift in the way we perceive and engage with digital spaces. The Metaverse is a collective, immersive virtual environment where users interact with each other and digital elements in real-time. Its components are diverse and intricate, collectively working to create a seamless and comprehensive digital universe. Avatars, digital representations of users, serve as the conduits for human interaction within the Metaverse, allowing individuals to navigate, communicate, and express themselves in this virtual realm. Virtual worlds, the foundational landscapes within the Metaverse, are dynamic and multifaceted environments that can range from realistic simulations to fantastical realms. Advanced graphics, spatial audio, and haptic feedback contribute to the sensory richness of the Metaverse, enhancing the user experience. Additionally, the Metaverse encompasses a social layer, fostering connections and collaborations among users, mirroring real-world social dynamics. Blockchain technology, often a fundamental component, ensures secure and transparent transactions within the Metaverse, enabling the creation and exchange of virtual assets. Altogether, these components coalesce to create an immersive and interconnected digital space that transcends traditional online experiences, laying the groundwork for the future of digital interaction and collaboration.

2.2 *Importance of Multisensor Fusion for Autonomous Vehicles*

Multisensor Fusion stands as a linchpin in the advancement and reliability of autonomous vehicles, playing a pivotal role in augmenting their perception capabilities. The importance of Multisensor Fusion lies in its ability to integrate data from a diverse array of sensors, such as LiDAR, radar, cameras, and ultrasonic sensors, into a cohesive and comprehensive understanding of the vehicle's surroundings. This amalgamation of sensor data creates a robust and redundant perception system, capable of mitigating the limitations inherent in individual sensors. By fusing information from different sensors, the vehicle gains a more accurate, real-time perception of its environment, improving object detection, distance estimation, and overall situational awareness [5]. Also, Multisensor Fusion enhances the vehicle's resilience to challenging environmental conditions, such as adverse weather or low-light scenarios, where relying on a single sensor type might prove insufficient.

The synergy of multiple sensors ensures a more reliable and resilient perception system, critical for the safety and efficacy of autonomous vehicles. In the context of autonomous navigation, Multisensor Fusion becomes indispensable for precise localization and mapping. The integration of data from diverse sensors enables the vehicle to create detailed and up-to-date maps, enhancing its ability to navigate complex and dynamic environments with accuracy. Multisensor Fusion serves as the sensory backbone of autonomous vehicles, elevating their perceptual capabilities to levels necessary for safe and efficient navigation in real-world conditions. As the autonomous vehicle landscape continues to evolve, the significance of Multisensor Fusion remains central to achieving the high standards of reliability and safety required for widespread adoption and integration into our transportation systems [6].

2.3 Simulation Environments, Visual Navigation, Rearrangements

Embodied AI represents a pivotal shift in artificial intelligence, addressing the limitations of traditional digital AI. By integrating physical bodies and sensory capabilities into AI systems, embodied AI enables real-world interactions and applications across various sectors.

- Embodied AI refers to artificial intelligence systems that have a physical body and can sense, perceive, and interact with the real world similar to humans. This is in contrast to traditional AI that operates solely in a digital-only realm.
- Key capabilities of embodied AI include computer vision, speech recognition, natural language processing, robotic control, spatial reasoning, etc. This allows embodied systems to navigate, manipulate objects, communicate, and make decisions based on real-world sensory input.
- Example embodiments include humanoid robots, self-driving cars, drones, robotic assistants, VR avatars, etc. These systems bridge the gap between AI algorithms and practical real-world applications.

2.4 State-of-the-Art Technologies in Multisensor Fusion for Metaverse-Based Navigation

In the Metaverse-based navigation, state-of-the-art technologies in Multisensor Fusion are pivotal to crafting immersive and seamless user experiences. Multisensor

Fusion plays a crucial role in amalgamating data from a variety of sensors, ranging from visual and auditory inputs to haptic feedback devices, to provide users with a comprehensive understanding of their virtual surroundings. Advanced computer vision algorithms are employed to process visual data, enabling precise object recognition and spatial mapping within the Metaverse.

Simultaneously, spatial audio technologies enhance the auditory dimension, creating a more immersive and realistic environment. Haptic feedback, through advanced tactile sensors, adds a tactile layer to the user experience, allowing for a sense of touch within the virtual space. Machine learning and artificial intelligence algorithms form the backbone of Multisensor Fusion, facilitating real-time data integration and adapting to user interactions. These technologies enable Metaverse-based navigation systems to continuously learn and refine their understanding of user preferences, optimizing the overall navigation experience. The incorporation of wearable devices, such as augmented reality glasses and haptic vests, further enhances the fusion of sensory inputs, creating a more holistic and engaging user interface. The cutting-edge technologies in Multisensor Fusion for Metaverse-based navigation represent a convergence of disciplines, pushing the boundaries of what is possible in virtual environments. As these technologies continue to evolve, Metaverse navigation systems promise to offer increasingly sophisticated, immersive, and personalized experiences, redefining the way users interact with and navigate through the digital realms of the Metaverse [7].

3 Dynamic Routing Technologies

Dynamic Routing Technologies represent a paradigm shift in optimizing and adapting transportation systems to real-time conditions, fostering efficiency and responsiveness. In the context of smart cities and autonomous vehicles, dynamic routing involves the continual adjustment of travel paths based on live data streams, such as traffic conditions, road closures, and weather patterns. Unlike static routing systems, dynamic routing technologies leverage advanced algorithms and real-time data analytics to identify the most efficient and timely routes, ensuring that vehicles can navigate through urban landscapes with agility and responsiveness. Web3-powered solutions further enhance dynamic routing by introducing decentralized and blockchain technologies. These innovations provide a secure and transparent framework for data sharing among vehicles and infrastructure, contributing to more accurate and reliable routing decisions. The collaborative nature of Web3 facilitates peer-to-peer communication, allowing vehicles to share critical information about their routes and conditions in real-time. In a broader sense, dynamic routing technologies contribute to the optimization of transportation networks, mitigating traffic congestion and reducing travel times [8]. They empower autonomous vehicles to make informed decisions on-the-fly, adapting to unforeseen circumstances and improving overall system efficiency. As smart city initiatives continue to evolve, the integration of dynamic routing technologies becomes increasingly vital for creating agile, adaptive, and sustainable urban transportation solutions that align with the demands of the modern, interconnected world.

3.1 Dynamic Routing Algorithms and Their Relevance to Autonomous Vehicles

Dynamic routing algorithms play a pivotal role in the realm of autonomous vehicles, serving as the intelligence behind their adaptive navigation capabilities. Unlike traditional static routing, dynamic routing algorithms enable vehicles to respond in real-time to changing traffic conditions, road incidents, and other variables, optimizing their routes for efficiency and safety. One key aspect of dynamic routing algorithms is their ability to process and analyze vast amounts of data from diverse sources. This includes information from onboard sensors, traffic cameras, GPS data, and communication with other vehicles and infrastructure.

Through continuous monitoring and analysis, these algorithms can make split-second decisions, dynamically recalculating routes to avoid congestion, accidents, or other obstacles. In the context of autonomous vehicles, dynamic routing algorithms contribute significantly to the realization of efficient and safe self-driving capabilities [9]. They enhance the vehicle's ability to navigate complex urban environments, making decisions based on real-time information to ensure a smooth and adaptive journey. The relevance of dynamic routing algorithms extends beyond mere efficiency; it is a critical component in ensuring the safety of autonomous vehicles and the passengers they transport. The integration of dynamic routing algorithms becomes even more pronounced with the advancements in connectivity and the emergence of Web3-powered solutions.

Decentralized communication protocols and blockchain technology can enhance the reliability and security of the data exchanged between autonomous vehicles, contributing to a more robust dynamic routing system. In conclusion, dynamic routing algorithms represent a technological cornerstone for autonomous vehicles, empowering them to navigate the intricacies of urban environments dynamically. As autonomous vehicle technology continues to advance, the refinement and optimization of dynamic routing algorithms will be central to achieving safer, more efficient, and seamlessly integrated transportation systems in smart cities of the future [10].

3.2 Challenges and Opportunities in Dynamic Routing Within Metaverse

The concept of a Metaverse, a virtual shared space that combines aspects of social media, augmented reality, virtual reality, and the internet, presents both challenges and opportunities for dynamic routing. Dynamic routing within the Metaverse involves the efficient and adaptive management of information, communication, and data flows within this expansive virtual environment. It explores some of the key challenges and opportunities associated with dynamic routing in the Metaverse.

Challenges:

Scale and Complexity:

Challenge: The Metaverse is expected to be massive and highly complex, with a multitude of interconnected elements, such as virtual environments, avatars, and interactive objects.

Implication: Routing within such a vast and intricate system poses scalability challenges, requiring sophisticated algorithms to handle the sheer volume of data and interactions.

Real-Time Interaction:

Challenge: The Metaverse emphasizes real-time interactions and experiences, demanding low-latency communication.

Implication: Dynamic routing must prioritize the efficient and rapid transmission of data, necessitating advanced routing protocols that can handle real-time constraints.

Interoperability:

Challenge: The Metaverse is likely to consist of diverse platforms, applications, and virtual spaces, leading to interoperability issues.

Implication: Dynamic routing needs to navigate between different systems seamlessly, requiring standardized protocols and communication frameworks.

Security and Privacy:

Challenge: Security threats, data breaches, and privacy concerns are heightened in a virtual environment where users interact and share sensitive information.

Implication: Dynamic routing must incorporate robust security measures, encryption, and authentication protocols to protect user data and maintain privacy.

Opportunities:

Adaptive Routing Algorithms:

Opportunity: The dynamic nature of the Metaverse calls for adaptive routing algorithms that can adjust to changing conditions and optimize data paths in real-time. Benefit: Implementing intelligent routing algorithms allows for efficient use of network resources, reduced latency, and improved overall user experience.

AI-Driven Routing: Opportunity: Integrating artificial intelligence (AI) into dynamic routing enables smart decision-making based on user behavior, preferences, and real-time conditions.

Benefit: AI-driven routing can enhance personalization, predict user needs, and optimize data flows, contributing to a more immersive and tailored Metaverse experience.

Blockchain for Trust and Decentralization:

Opportunity: Leveraging blockchain technology can address trust and decentralization concerns within the Metaverse.

Benefit: Blockchain-based routing ensures secure and transparent transactions, enabling users to trust the authenticity of interactions and transactions in the virtual space.

Innovation in Network Protocols:

Opportunity: Developing new network protocols specifically designed for the Metaverse can enhance communication efficiency and address interoperability challenges.

Benefit: Innovative protocols can support seamless integration between different platforms and applications, fostering a cohesive and interconnected Metaverse.

The challenges and opportunities in dynamic routing within the Metaverse underscore the need for advanced technologies, collaboration across platforms, and a proactive approach to address the unique demands of this evolving virtual landscape. As technology continues to advance, addressing these challenges and capitalizing on opportunities will be crucial for creating a vibrant and accessible Metaverse for users worldwide [11].

4 Web3-Powered Metaverse Worlds: Autonomous Vehicles Shaping Efficient Urban Transport Solutions of Future in Smart City

Web3-powered Metaverse worlds represent a paradigm shift in the virtual landscape, combining the principles of decentralization, blockchain technology, and user empowerment. In these immersive digital realms, the integration of Web3 technologies, such as decentralized protocols and smart contracts, fundamentally alters the dynamics of user interaction and ownership. Unlike traditional Metaverse platforms, Web3-powered environments prioritize user autonomy by allowing participants to truly own their digital assets, identities, and experiences. Smart contracts, executed on decentralized networks, govern transactions and interactions, fostering transparency and trust among users. This decentralized approach not only enhances security and privacy but also enables the creation of vibrant digital economies where virtual assets can be seamlessly transferred between different Metaverse worlds.

As Web3-powered Metaverse worlds continue to evolve, they hold the promise of fostering a more equitable and open virtual space where users have greater control over their digital presence and experiences. In the context of smart cities, the convergence of Web3 technologies with autonomous vehicles is shaping the future of urban transport solutions. Web3-powered Metaverse worlds play a pivotal role in orchestrating the efficient integration of autonomous vehicles into the urban fabric. These virtual environments, underpinned by decentralized protocols and blockchain technology, provide a secure and transparent platform for managing the complex interactions between autonomous vehicles, infrastructure, and users [12]. Smart contracts executed on Web3 networks can facilitate seamless transactions, such as vehicle-to-vehicle communication, traffic optimization, and decentralized ride-sharing services. The result is a dynamic and self-regulating transportation ecosystem where autonomous vehicles navigate through city streets, responding to real-time data and demand. This intersection of Web3 and autonomous vehicles holds the promise of reducing traffic congestion, lowering emissions, and improving overall mobility efficiency in smart cities. As technology advances, the collaborative synergy between Web3-powered Metaverse worlds and autonomous vehicles presents an exciting vision for a more sustainable and intelligent urban transportation landscape.

4.1 Web3 and Its Potential Role in Transportation

Web3 technology, characterized by decentralization, blockchain, and smart contracts, holds significant potential in transforming and optimizing the transportation sector. In the realm of transportation, Web3 can revolutionize the way we think about ownership, transactions, and data management. Decentralization, a core tenet of Web3, can be applied to create decentralized and transparent systems for vehicle registration, maintenance records, and even traffic management [13].

Blockchain, with its immutable and secure ledger, has the capacity to enhance the integrity of data related to vehicle history, reducing fraud and ensuring a trustworthy record of a vehicle's life cycle. Smart contracts, executed automatically when predefined conditions are met, can facilitate peer-to-peer transactions, such as ridesharing, toll payments, or vehicle charging, without the need for intermediaries. Moreover, Web3 can play a vital role in fostering the development of decentralized autonomous transportation systems. Autonomous vehicles, connected through a decentralized network, could communicate seamlessly, enhancing traffic flow and safety. The use of blockchain in autonomous vehicle data management ensures the integrity and privacy of the vast amounts of data generated by these vehicles. The decentralized nature of Web3 can empower individuals, allowing them to own and control their transportation data, giving rise to new business models and user-centric services. In summary, the integration of Web3 into the transportation sector has the potential to bring about a paradigm shift. It can introduce transparency, efficiency, and security, laying the foundation for a more connected, autonomous, and user-centric transportation ecosystem. As technological advancements continue, the transformative impact of Web3 on transportation is poised to reshape how we perceive, experience, and manage mobility in the years to come [14].

4.2 Possibilities of Using Web3 Technologies for Connected Vehicles in Metaverse

The fusion of Web3 technologies with connected vehicles within the Metaverse opens up a realm of exciting possibilities for the future of transportation. In this envisioned scenario, decentralized protocols and blockchain in the Metaverse lay the groundwork for secure, transparent, and efficient communication among connected vehicles. Web3's decentralized nature ensures that data generated by these vehicles, such as location, speed, and status, is securely managed on a blockchain, enhancing data integrity and privacy. Smart contracts can be employed to facilitate seamless peer-to-peer transactions and agreements between vehicles, enabling functionalities like autonomous convoy coordination or decentralized ridesharing within the Metaverse [15].

The Metaverse's immersive environment can also enhance the user experience by integrating virtual elements with real-world data. For instance, using augmented reality interfaces in connected vehicles within the Metaverse, drivers can receive real-time information about the road conditions, potential hazards, or even engage in virtual meetings while commuting. The decentralized ownership and identity management in Web3 can empower users to have full control over their connected vehicle data, creating a trustless and user-centric ecosystem [16]. As the Metaverse continues to evolve, the convergence of Web3 technologies and connected vehicles holds the promise of revolutionizing how we perceive, interact with, and experience transportation. It paves the way for a more interconnected, secure, and personalized mobility system within this expansive virtual space.

4.3 Security and Privacy Implications of Web3-Powered Transportation

The integration of Web3 technologies into the realm of transportation brings forth a host of security and privacy implications that demand careful consideration. In a Web3-powered transportation ecosystem, where decentralized protocols and blockchain are fundamental, security measures are critical to safeguard sensitive information and ensure the integrity of data. Blockchain, known for its tamper-resistant nature, can enhance security by providing an immutable ledger for vehicle records, transactions, and identities. However, the distributed and open nature of blockchain also introduces new challenges, such as the potential for smart contract vulnerabilities and the need for robust consensus mechanisms [17].

Privacy concerns arise as decentralized systems necessitate transparent and traceable transactions. While blockchain can offer pseudonymity, ensuring that personally identifiable information is not directly linked, the immutability of the blockchain poses challenges in terms of data erasure. Additionally, as vehicles become more connected and share data in real-time, protecting user privacy becomes paramount.

Striking the right balance between data transparency for operational efficiency and protecting individual privacy is a delicate challenge in Web3-powered transportation. Furthermore, the reliance on smart contracts in Web3 raises security considerations, as vulnerabilities in these self-executing contracts could be exploited, potentially leading to unauthorized access or manipulation of critical transportation functions. The decentralized nature of Web3 also necessitates addressing the risk of 51% attacks on blockchain networks, which could compromise the security of the entire transportation ecosystem. In navigating these challenges, the development and adoption of robust cybersecurity measures, encryption protocols, and privacy-focused technologies become imperative. Striking a balance between transparency, decentralization, and the protection of sensitive information will be crucial to realizing the full potential of Web3-powered transportation while ensuring the security and privacy of individuals within this evolving technological landscape [18].

5 Protocols for Secure and Autonomous Transportation

The advent of secure and autonomous transportation has prompted the development of advanced protocols aimed at ensuring the reliability, safety, and privacy of these systems. One key element is the utilization of secure communication protocols that enable vehicles to exchange information seamlessly while preventing unauthorized access. Protocols such as Transport Layer Security (TLS) play a crucial role in encrypting data transmissions, protecting against cyber threats, and ensuring the integrity of communications in connected and autonomous vehicles. In the realm of autonomous transportation, consensus protocols within blockchain technology contribute to secure decision-making processes.

With leveraging distributed ledgers, blockchain ensures that critical information, such as navigation data or software updates, is tamper-proof and transparent. This not only enhances the security of the autonomous system but also establishes a trustless environment for stakeholders involved in the transportation ecosystem. Vehicle-to-Everything (V2X) communication protocols form the backbone of connectivity in autonomous transportation [19]. These protocols enable vehicles to communicate with each other, as well as with infrastructure and pedestrians. Implementing standardized V2X protocols, such as Dedicated Short-Range Communications (DSRC) or Cellular Vehicle-to-Everything (C-V2X), ensures interoperability and a consistent communication framework, laying the foundation for a secure and cooperative autonomous transportation network. To address privacy concerns, protocols that implement data anonymization and pseudonymization techniques are becoming increasingly important. With adopting these privacy-centric protocols, autonomous systems can share necessary information without compromising the personal data of users, thus balancing the need for data-driven decision-making with individual privacy protection. In summary, secure and autonomous transportation relies on a combination of robust communication, consensus, and privacy-focused protocols

[20]. As these technologies continue to advance, the ongoing development and standardization of protocols will play a pivotal role in ensuring the trust, safety, and efficiency of autonomous transportation systems in the future.

5.1 Analysis of Protocols and Standards for Secure Transportation

The landscape of secure transportation is intricately tied to the adoption and adherence to robust protocols and standards. The analysis of these protocols reveals a multifaceted approach aimed at ensuring the safety, reliability, and privacy of modern transportation systems. Communication protocols, such as Transport Layer Security (TLS) and Secure Sockets Layer (SSL), form the backbone of secure data transmission in transportation networks [21]. These protocols encrypt communication channels, thwarting potential cyber threats and safeguarding sensitive information exchanged between vehicles and infrastructure. In the realm of autonomous transportation, consensus protocols within blockchain technology offer a novel paradigm for secure decision-making and data management. By employing decentralized ledgers, blockchain protocols provide an immutable record of transactions, enhancing the security and transparency of critical data relevant to autonomous vehicles.

The standardization of Vehicle-to-Everything (V2X) communication protocols, such as Dedicated Short-Range Communications (DSRC) or Cellular Vehicle-to-Everything (C-V2X), is pivotal for creating a cohesive and interoperable environment. These protocols enable seamless communication between vehicles, infrastructure, and even pedestrians, contributing to a safer and more coordinated transportation ecosystem [22]. Privacy considerations are paramount in the analysis of transportation protocols. Protocols incorporating data anonymization and pseudonymization techniques are essential for protecting the personal information of individuals in connected and autonomous systems. Adhering to privacy-focused standards ensures that data-driven functionalities can coexist with stringent privacy regulations, striking a delicate balance between technological innovation and individual privacy. As the transportation industry continues its trajectory towards increased connectivity and autonomy, the ongoing analysis and evolution of protocols and standards will be crucial. A dynamic and adaptive approach to security protocols, informed by emerging technologies and threat landscapes, will be essential to navigate the complexities of securing modern transportation systems effectively [23].

5.2 *Security and Privacy Implications of Web3-Powered Transportation*

In the landscape of smart transportation within smart cities, robust security measures are imperative to ensure the integrity, confidentiality, and authenticity of data. This necessitates the implementation of advanced protocols in encryption, authentication, and data integrity as-

Encryption: Encryption serves as the first line of defense against unauthorized access and data breaches. In the context of smart transportation, where sensitive information is transmitted across networks, end-to-end encryption becomes paramount. Utilizing industry-standard encryption algorithms, such as AES (Advanced Encryption Standard), ensures that data is securely transformed into an unreadable format during transmission. This safeguards critical information such as location data, passenger details, and traffic patterns from malicious entities [24].

Authentication: Authentication protocols play a pivotal role in verifying the legitimacy of entities accessing smart transportation systems. Multi-factor authentication (MFA) mechanisms add an extra layer of security by requiring multiple forms of identification before granting access. This can include biometric data, secure tokens, or one-time passwords. Implementing robust authentication protocols ensures that only authorized personnel and devices interact with the smart transportation infrastructure, mitigating the risk of unauthorized control or manipulation [25].

Data Integrity: Maintaining the integrity of data is essential for the reliability of smart transportation systems. Protocols such as hash functions and digital signatures are instrumental in verifying that data has not been tampered with during transit. Hash functions generate unique identifiers for data, and any alteration to the content would result in a different hash value. Digital signatures provide a means of verifying the origin and integrity of data by using cryptographic keys. Together, these protocols safeguard against data corruption and unauthorized modifications, ensuring the accuracy of information used in decision-making processes within smart transportation networks [26].

The implementation of robust encryption, authentication, and data integrity protocols is critical for establishing a secure foundation in smart transportation systems within smart cities. By adopting these measures, stakeholders can confidently harness the potential of emerging technologies while safeguarding against the ever-present threats in the digital landscape. As smart transportation continues to evolve, an unwavering commitment to security protocols will be essential for the sustained success of these transformative initiatives.

5.3 Challenges of Cybersecurity in Autonomous Transportation

The integration of autonomous transportation introduces a myriad of technological advancements, but it also brings forth significant challenges in the realm of cybersecurity. The unique characteristics of autonomous vehicles present distinct vulnerabilities that demand comprehensive cybersecurity measures [27].

1. Data Security and Privacy Concerns: Autonomous vehicles generate and rely on vast amounts of data, including real-time location information, sensor data, and communication with external networks. Ensuring the security and privacy of this data is a major challenge. Unauthorized access to sensitive data could not only compromise the privacy of individuals but also lead to potential misuse or manipulation of critical vehicle information.
2. Vehicular Network Security: Autonomous vehicles are interconnected, forming a complex vehicular network that communicates with each other and with external infrastructure. Securing this network against cyber threats is crucial to prevent unauthorized access, data interception, and potential attacks on the vehicle's control systems. The dynamic and decentralized nature of these networks adds an extra layer of complexity to maintaining their security.
3. Malicious Attacks on Vehicle Systems: Autonomous vehicles rely heavily on advanced control systems, sensors, and communication technologies. Malicious actors may attempt to exploit vulnerabilities in these systems through cyber-attacks, aiming to take control of the vehicle, manipulate its behavior, or cause physical harm. Protecting against such attacks requires robust intrusion detection systems and secure-by-design principles in the development of vehicle software and hardware.
4. Software Security and Over-the-Air Updates The software complexity of autonomous vehicles, coupled with the need for regular updates to improve performance and address vulnerabilities, poses a challenge. Ensuring the security of over-the-air updates is crucial to prevent the installation of compromised or malicious software. A compromised update process could potentially lead to widespread vulnerabilities across a fleet of autonomous vehicles.
5. Lack of Standardization: The absence of universally accepted cybersecurity standards for autonomous vehicles complicates the development of consistent and effective security measures. Varying standards across manufacturers and regions may result in interoperability issues and inconsistent levels of cybersecurity protection. Establishing standardized protocols and regulations is essential to create a cohesive and robust cybersecurity framework for autonomous transportation.
6. Human Factor Vulnerabilities: While autonomous vehicles aim to reduce reliance on human intervention, there are still situations where human input is required. The human–machine interface becomes a potential vulnerability, as users may inadvertently compromise security through actions such as improper handling of

software updates, ignoring security alerts, or falling victim to social engineering attacks.

Addressing these challenges requires a collaborative effort from industry stakeholders, policymakers, and cybersecurity experts. Continuous research, development, and implementation of robust cybersecurity measures are essential to ensure the safe and secure integration of autonomous transportation into our increasingly connected and digitalized world [28].

6 Sustainable and Efficient Urban Transportation

Sustainable and efficient urban transportation is a critical component of building smart, livable cities in the face of growing urbanization. Achieving this goal involves integrating environmentally friendly practices, optimizing infrastructure, and promoting modes of transport that enhance efficiency while minimizing the ecological footprint [29].

1. Multi-Modal Integration: Efficient urban transportation involves a seamless integration of various transportation modes such as buses, trains, bicycles, and pedestrian pathways. By promoting a multi-modal approach, cities can provide citizens with diverse, flexible options for commuting. This not only reduces congestion but also offers sustainable alternatives that cater to different travel needs and distances.
2. Public Transit and Accessibility: Investing in robust public transit systems, including buses and trains, plays a crucial role in promoting sustainability. Well-designed public transportation networks encourage citizens to opt for mass transit, reducing the number of private vehicles on the road and subsequently decreasing traffic congestion and carbon emissions. Ensuring accessibility for people with disabilities also contributes to an inclusive and efficient urban transport system.
3. Sustainable Infrastructure: The construction and maintenance of infrastructure play a pivotal role in the sustainability of urban transportation. The development of smart roads, energy-efficient traffic lights, and the integration of renewable energy sources into transportation infrastructure are key elements. Additionally, the implementation of green spaces, pedestrian zones, and cycling lanes not only promotes sustainable transportation but also enhances the overall quality of urban life.
4. Electric and Hybrid Vehicles: Encouraging the adoption of electric and hybrid vehicles contributes significantly to reducing the carbon footprint of urban transportation. Incentives for electric vehicle adoption, the establishment of charging infrastructure, and the integration of electric buses into public transit fleets are essential steps toward achieving a more sustainable and eco-friendly urban transport system.

5. Smart Traffic Management: Leveraging technology for smart traffic management is critical for optimizing the flow of vehicles and minimizing congestion. Intelligent transportation systems, real-time traffic monitoring, and adaptive traffic signal control can enhance the efficiency of urban transportation by dynamically responding to traffic conditions and improving overall traffic flow.
6. Pedestrian-Friendly Design: Creating pedestrian-friendly urban spaces promotes sustainability, health, and safety. Designing walkable neighborhoods with proper sidewalks, crosswalks, and pedestrian zones encourages people to choose walking as a mode of transportation for short distances, reducing the reliance on motorized vehicles.
7. Data-Driven Decision Making: Utilizing data analytics and smart technologies for decision-making allows cities to understand traffic patterns, identify bottlenecks, and optimize transportation routes. This data-driven approach enables proactive planning, leading to more efficient use of resources and better responsiveness to the evolving needs of urban mobility.

The sustainable and efficient urban transportation is a holistic endeavor that encompasses various strategies, from embracing multi-modal integration to leveraging technology for intelligent traffic management. By prioritizing these principles, cities can create transportation systems that not only enhance efficiency but also contribute to a cleaner, healthier, and more livable urban environment [30].

6.1 Importance of Sustainability in Smart Cities

The importance of sustainability in smart cities cannot be overstated as these urban centers grapple with rapid population growth and technological advancements. Sustainable practices in smart cities encompass environmental, economic, and social dimensions. Environmentally, adopting renewable energy sources, green building designs, and eco-friendly transportation systems reduce the ecological footprint, mitigating climate change effects [31]. Economically, sustainability fosters resource efficiency, cutting costs and promoting long-term economic resilience. Socially, creating livable spaces with accessible green areas and efficient public transportation enhances residents' well-being. Sustainability in smart cities is pivotal for fostering resilience, ensuring a high quality of life, and preserving resources for current and future generations.

6.2 Autonomous and Connected Vehicles Contribution to Efficient Urban Transportation

The integration of autonomous and connected vehicles is revolutionizing urban transportation, offering a plethora of benefits that contribute to efficiency, safety, and sustainability. Autonomous vehicles, equipped with advanced sensors and artificial intelligence, can optimize traffic flow, reduce congestion, and enhance overall transport efficiency. These vehicles communicate with each other and infrastructure, forming a connected network that enables real-time data sharing and decision-making. The key contribution is the potential for significant improvements in traffic management. Autonomous vehicles can operate with precise coordination, minimizing traffic bottlenecks and optimizing routes to reduce travel time. Also, connected vehicles can share information about road conditions, traffic incidents, and optimal routes, enabling a dynamic and responsive transportation system. Enhanced safety is another notable advantage [32]. Autonomous vehicles are designed to follow traffic rules rigorously and respond swiftly to unexpected situations, reducing the likelihood of accidents caused by human error. The connectivity aspect further amplifies safety by enabling vehicles to exchange information about their speed, location, and intentions, allowing for anticipatory actions and collision avoidance. The efficiency gains extend to resource utilization [33]. Autonomous vehicles can operate more efficiently in terms of fuel consumption and emissions through optimal driving patterns.

The reduction in traffic congestion and smoother traffic flow contribute to environmental sustainability by minimizing fuel wastage and lowering overall emissions. In addition to these functional advantages, the introduction of autonomous and connected vehicles has the potential to reshape urban planning. With reduced demand for parking space and changes in traffic patterns, cities may reevaluate land use, allocating more space for green areas, pedestrian zones, and recreational areas, thus fostering a more livable and sustainable urban environment. While the integration of autonomous and connected vehicles offers transformative benefits, it also poses challenges such as cybersecurity concerns, infrastructure adaptation, and regulatory frameworks. Addressing these challenges is crucial to realizing the full potential of these technologies. In conclusion, the contribution of autonomous and connected vehicles to efficient urban transportation is multifaceted [34]. From optimizing traffic flow and enhancing safety to promoting sustainability and influencing urban planning, these technologies are poised to redefine the way we perceive and experience transportation in smart cities.

7 Conclusion and Future Scope

The exploration of metaverse multisensor fusion and dynamic routing technologies, powered by Web3, for autonomous vehicles in urban transport has unveiled a transformative landscape. The integration of these cutting-edge technologies presents promising prospects for the future of smart cities. Efficient traffic management, reduced congestion, and enhanced safety stand out as immediate benefits. The potential for reshaping urban landscapes, improved resource utilization, and a positive impact on environmental sustainability paint a compelling picture of the smart city era. However, alongside these promising prospects, challenges loom on the horizon.

The intricate nature of these technologies demands robust protocols for cybersecurity, data privacy, and interoperability.

As smart cities evolve, there is a pressing need for standardized frameworks and regulations to address these challenges systematically. Ethical considerations and societal implications must also be carefully navigated to ensure that the benefits of these technologies are accessible and equitable for all members of society. Looking into the future, the scope for further innovation is vast. Continued research and development are crucial for refining existing technologies and addressing emerging challenges. The collaboration between technology developers, policymakers, and urban planners will play a pivotal role in shaping the trajectory of metaverse multisensor fusion and dynamic routing technologies. Moreover, society must actively engage in conversations surrounding the ethical, legal, and social dimensions of these innovations to ensure that they align with the values and aspirations of the communities they serve. In essence, the intersection of metaverse technologies, multisensor fusion, dynamic routing, and Web3 in the realm of autonomous vehicles holds immense promise for the efficient, sustainable, and connected urban transport solutions of the future. Navigating the path ahead requires a holistic approach, with a keen understanding of both the potentials and challenges, to ensure that the benefits of these advancements are maximized for the greater good of society.

References

1. Andronie, M., Lazaroiu, G., Iatagan, M., Hurloiu, I., Ştefănescu, R., Dijmărescu, A., Dijmărescu, I.: Big data management algorithms, deep learning-based object detection technologies, and geospatial simulation and sensor fusion tools in the internet of robotic things. ISPRS Int. J. Geo Inf. **12**(2), 35 (2023)
2. Andronie, M., Lăzăroiu, G., Karabolevski, O.L., Ştefănescu, R., Hurloiu, I., Dijmărescu, A., Dijmărescu, I.: Remote big data management tools, sensing and computing technologies, and visual perception and environment mapping algorithms in the internet of robotic things. Electronics **12**(1), 22 (2022)
3. Aria, R., Archer, N., Khanlari, M., Shah, B.: Influential factors in the design and development of a sustainable Web3/metaverse and its applications. Future Internet. **15**(4), 131 (2023)
4. Bratu, S.: Brain-inspired artificial intelligence and machine learning-based navigation algorithms, spatial data visualization and digital twin modeling tools, and emotion detection and recognition technologies in the decentralized and interconnected metaverse. Rev. Contemp. Philos. **1**, 22 (2023)
5. Bugaj, M., Durana, P., Blazek, R., Horak, J.: Industry 4.0: marvels in profitability in the transport sector. Mathematics. **11**(17), 3647 (2023)
6. Carey, B.: Metaverse technologies, behavioral predictive analytics, and customer location tracking tools in blockchain-based virtual worlds. Rev. Contemp. Philos. **21**, 188–204 (2022)
7. Dia, H., Javanshour, F.: Autonomous shared mobility-on-demand: Melbourne pilot simulation study. Transp. Res. Procedia **1**(22), 285–296 (2017)
8. Ding, W., Li, J., Qin, R., Kozma, R., Wang, F.Y.: A new architecture and mechanism for decentralized science metamarkets. IEEE Trans. Syst. Man Cybern.: Syst. (2023)
9. Duncan, R.: Multi-sensor fusion technology, visual imagery and predictive modeling tools, and big geospatial data analytics in the virtual economy of the metaverse. Econ. Manag. Financ. Mark. **17**(3), 42–57 (2022)

10. Gordon, S.: Virtual navigation and geospatial mapping tools, customer data analytics, and computer vision and simulation optimization algorithms in the blockchain-based metaverse. Rev. Contemp. Philos. **21**, 89–104 (2022)
11. Grupac, M., Negoianu, A.E.: Immersive extended reality and sensor-based object recognition technologies, socially-oriented location tracking and simulation modeling tools, and artificial vision and haptic augmented reality systems in the metaverse interactive environment. Rev. Contemp. Philos. **1**, 22 (2023)
12. Henley, S.: Extended reality and cognitive digital twin technologies, 3D space mapping and image processing computational algorithms, and predictive geospatial modeling and simulation tools in the industrial metaverse. Econ. Manag. Financ. Mark. **18**(1), 25–40 (2023)
13. Kiong, L.V.: Web3 made easy: a comprehensive guide to Web3: everything you need to know about Web3, Blockchain, DeFi, Metaverse, NFT and GameFi. Liew Voon Kiong, 9 Sept 2022 (2022)
14. Kliestik, T., Nagy, M., Valaskova, K.: Global value chains and industry 4.0 in the context of lean workplaces for enhancing company performance and its comprehension via the digital readiness and expertise of workforce in the V4 nations. Mathematics **11**(3), 601 (2023)
15. Kral, P., Janoskova, K., Potcovaru, A.M.: Digital consumer engagement on blockchain-based metaverse platforms: extended reality technologies, spatial analytics, and immersive multisensory virtual spaces. Linguist. Philos. Invest. **21**, 252–267 (2022)
16. Kshetri, N.: Economic and social impacts of Web3 and the metaverse in the Global South
17. Liu, Q., Huang, W., Duan, X., Wei, J., Hu, T., Yu, J., Huang, J.: DSW-YOLOv8n: a new underwater target detection algorithm based on improved YOLOv8n. Electronics **12**(18), 3892 (2023)
18. Ljungholm, D.P.: Autonomous vehicle interaction control software, urban sensing and computing technologies, and trajectory planning and route detection algorithms in networked transport systems. Contemp. Read. Law Soc. Just. **14**(1), 57–72 (2022)
19. Morley, N.: Perception and cognition algorithms, simulation modeling and data visualization tools, and spatial computing and immersive technologies in the metaverse interactive environment. Anal. Metaphys. **21**, 244–260 (2022)
20. Mystakidis, S.: Metaverse. Encyclopedia **2**(1), 486–497 (2022)
21. Nagy, M., Lăzăroiu, G., Valaskova, K.: Machine intelligence and autonomous robotic technologies in the corporate context of SMEs: deep learning and virtual simulation algorithms, cyber-physical production networks, and industry 4.0-based manufacturing systems. Appl. Sci. **13**(3), 1681 (2023)
22. Newell, M.: Customer engagement and data visualization tools, ambient sound recognition software, and deep learning-based sensing technologies in the metaverse economy. Econ. Manag. Financ. Mark. **17**(3), 25–41 (2022)
23. Nica, E., Poliak, M., Popescu, G.H., Pârvu, I.A.: Decision intelligence and modeling, multisensory customer experiences, and socially interconnected virtual services across the metaverse ecosystem. Linguist. Philos. Invest. **21**, 137–153 (2022)
24. Nica, E., Popescu, G.H., Poliak, M., Kliestik, T., Sabie, O.M.: Digital twin simulation tools, spatial cognition algorithms, and multi-sensor fusion technology in sustainable urban governance networks. Mathematics **11**(9), 1981 (2023)
25. Perkins, J.: Extended reality and geospatial mapping technologies, behavioral predictive and mobile location analytics, and motion planning and object recognition algorithms in immersive hyper-connected virtual spaces. J. Self-Gov. Manag. Econ. **10**(3), 23–29 (2022)
26. Petit, O., Velasco, C., Wang, Q.J., Spence, C.: Consumer consciousness in multisensory extended reality. Front. Psychol. **21**(13), 851753 (2022)
27. Richter, M.A., Hagenmaier, M., Bandte, O., Parida, V., Wincent, J.: Smart cities, urban mobility and autonomous vehicles: how different cities needs different sustainable investment strategies. Technol. Forecast. Soc. Chang. **1**(184), 121857 (2022)
28. Savin, G.V.: The smart city transport and logistics system: theory, methodology and practice. Upravlenec **12**(6) (2021)

29. Schatzinger, S., Lim, C.Y.: Taxi of the future: big data analysis as a framework for future urban fleets in smart cities. In: Smart and Sustainable Planning for Cities and Regions: Results of SSPCR 2015 1, pp. 83-98 (2017)
30. Seuwou, P., Banissi, E., Ubakanma, G.: The future of mobility with connected and autonomous vehicles in smart cities. In: Digital Twin Technologies and Smart Cities, pp. 37–52 (2020)
31. Wen, M.H., Huang, C.Y., Chen, Y.C., Lin, I.C.: Exploring factors influencing community consensus building of Web3 decentralized apps. In: International Conference on Human-Computer Interaction, pp. 408–420, 9 Jul 2023. Springer Nature Switzerland, Cham
32. Yigitcanlar, T., Wilson, M., Kamruzzaman, M.: Disruptive impacts of automated driving systems on the built environment and land use: an urban planner's perspective. J. Open Innov.: Technol. Mark. Complex. **5**(2), 24 (2019)
33. Zauskova, A., Miklencicova, R., Popescu, G.H.: Visual imagery and geospatial mapping tools, virtual simulation algorithms, and deep learning-based sensing technologies in the metaverse interactive environment. Rev. Contemp. Philos. **21**, 122–137 (2022)
34. Zvarikova, K., Novak, A., Popescu, G.H.: Neural network-based recognition and virtual simulation algorithms, interactive 3D geo-visualization and neuromorphic computing systems, and tactile sensing and cognitive modeling technologies in Web3-powered metaverse worlds. Rev. Contemp. Philos. **1**, 22 (2023)

Artificial Intelligence (AI) Algorithm and Models for Embodied Agents (Robots and Drones)

P. Chitra and A. Saleem Raja

Abstract Embodied agents, including robots and drones, rely on a myriad of artificial intelligence (AI) algorithms and models to interact with and navigate their environments. This abstract provides an overview of the essential AI components used in building such agents. These include computer vision techniques like convolutional neural networks and object tracking, sensor fusion to integrate data from diverse sensors, reinforcement learning for task learning, simultaneous localization and mapping (SLAM) for environment understanding, path planning and control algorithms, natural language processing for human interaction, 3D perception, transfer learning, imitation learning, self-supervised learning, vision-based reinforcement learning, and more. The selection of these techniques depends on the specific application and hardware capabilities of the embodied agent. By harnessing these AI tools, embodied agents can perform a wide range of tasks, from autonomous navigation to complex object manipulation and human interaction, making them valuable assets in various fields such as robotics, autonomous vehicles, and industrial automation. This chapter explores the pivotal role of artificial intelligence (AI) algorithms and models in the development and empowerment of embodied agents. Embodied agents represent a transformative paradigm in AI, enabling autonomous entities to interact with and adapt to real or simulated environments. The fusion of AI and embodiment offers boundless possibilities across various domains, including robotics, autonomous vehicles, virtual assistants, and digital avatars. Also, the chapter examines the essential AI techniques that underpin the functionality of embodied agents.

Keywords Artificial intelligence · Artificial intelligence models · Embodied agents · Robots · Drones

P. Chitra (✉)
GITAM University, Bangalore, India
e-mail: bchitra.tr@gmail.com

A. Saleem Raja
University of Technology and Applied Science, Shinas, Oman

P. Raj et al. (eds.), *Building Embodied AI Systems: The Agents, the Architecture Principles, Challenges, and Application Domains*, Information Systems Engineering and Management 14, https://doi.org/10.1007/978-3-031-68256-8_19

1 Introduction

The realm of artificial intelligence has witnessed remarkable progress in recent years, giving rise to intelligent systems that not only comprehend and reason about the world but can also interact with it in meaningful ways. One of the most fascinating and promising frontiers in AI is the development of embodied agents—autonomous entities equipped with physical or virtual bodies that enable them to navigate, interact with, and make decisions in the real or simulated world. This paradigm shift from passive software systems to agents with a sense of embodiment has opened up exciting possibilities in fields ranging from robotics and autonomous vehicles to virtual assistants and digital avatars [1–5]. Crucial to the success of these embodied agents are the underlying AI algorithms and models that empower them. These algorithms serve as the brainpower that drives these agents to perceive their surroundings, make intelligent decisions, and adapt to diverse environments. In this chapter, we delve into the world of "Empowering Embodied Agents: The Role of AI Algorithms and Models," where we explore the fundamental techniques and methodologies that give life to these agents and enable them to perform tasks that were once the stuff of science fiction (Fig. 1).

The marriage of artificial intelligence and embodied agents brings us to the frontier of AI research, where machines not only think but also act, learn, and interact in dynamic and unpredictable environments. This new era has the potential to transform industries, from healthcare and education to entertainment and manufacturing. However, it also presents unique challenges, pushing researchers to continually innovate and develop novel AI approaches.

In the following sections, we will delve into specific AI techniques and models that play a crucial role in shaping embodied agents. We will discuss how reinforcement learning empowers agents to learn through interaction, sim-to-real transfer enables safe training in virtual environments, and imitation learning enables agents

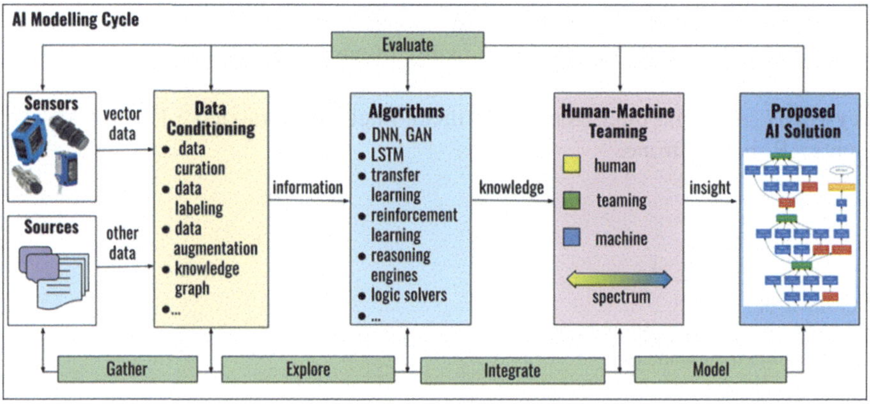

Fig. 1 A structure of AI modelling cycle

to mimic human expertise. We will also explore model-based reinforcement learning, evolutionary algorithms, deep learning in perception and control, 3D simulation environments, hybrid models that combine symbolic reasoning with neural networks, and the potential of transfer learning and embodied cognitive architectures [6–10].

These AI algorithms and models serve as the building blocks that give rise to agents capable of autonomous navigation, problem-solving, and decision-making. Through these agents, we are pushing the boundaries of AI, and we are on the verge of creating systems that can work alongside humans, assist in dangerous tasks, and even provide companionship and support in a variety of contexts.

Join us on this journey through the evolving landscape of embodied agents and discover how AI algorithms and models are shaping the future of intelligent, physically or virtually embodied entities.

2 Reinforcement Learning

Reinforcement Learning (RL) is a fundamental paradigm in artificial intelligence that has found extensive application in training embodied agents. In this chapter, we will explore the role of Reinforcement Learning for Embodied Agents, discussing the principles, algorithms, and applications of RL in enabling autonomous entities to interact with and learn from their environment.

2.1 Introduction to Reinforcement Learning a Subsection Sample

Reinforcement Learning is a machine learning approach where an agent learns to make sequences of decisions through interaction with an environment. The agent aims to maximize a cumulative reward signal, making decisions to achieve its goals and adapt to varying conditions.

Key Components of Reinforcement Learning:

Agent: The learner or decision-maker that interacts with the environment.

Environment: The external system with which the agent interacts and learns from.

State (s): A representation of the current situation or configuration of the environment.

Action (a): Choices made by the agent that influence the environment.

Reward (r): A numerical value indicating the immediate feedback from the environment.

Policy (π): A strategy or mapping from states to actions that the agent follows.

Value Function (V): An estimate of the expected cumulative reward an agent can achieve from a given state or state-action pair.

Q-Value Function (Q): An estimate of the expected cumulative reward an agent can achieve starting from a state, taking a specific action, and following a policy thereafter.

2.2 Reinforcement Learning Algorithms for Embodied Agents

There are various RL algorithms that can be applied to train embodied agents. Some of the most commonly used algorithms include:

Deep Q-Networks (DQN): DQN combines Q-learning with deep neural networks, enabling agents to handle high-dimensional input data. It's widely used in training agents for tasks like game playing and robotic control.

Proximal Policy Optimization (PPO): PPO is a policy optimization method that iteratively updates the agent's policy to maximize rewards. It is known for its stability and suitability for continuous action spaces.

Trust Region Policy Optimization (TRPO): TRPO optimizes policies while ensuring that the changes are not too radical. It is favored for its safety and reliability in training embodied agents.

2.3 Applications of Reinforcement Learning in Embodied Agents

Reinforcement Learning has been successfully employed in various applications involving embodied agents, including:

Robotics: Embodied agents in the form of robots can learn to perform tasks like autonomous navigation, manipulation of objects, and even complex assembly operations through RL.

Autonomous Vehicles: RL plays a crucial role in training autonomous vehicles to make driving decisions, navigate complex road networks, and avoid obstacles.

Video Games: RL agents are used to develop game-playing agents that can achieve human-level performance in a wide range of video games.

Simulated Environments: Many researchers use 3D simulation environments to train embodied agents, facilitating a safe and cost-effective way to experiment with various tasks and challenges.

Humanoid and Virtual Agents: Embodied agents in the form of virtual characters and humanoid robots use RL to learn natural movement, interaction, and problem-solving skills.

Healthcare and Education: Embodied agents can serve as personalized tutors or healthcare companions, adapting their behavior based on the user's needs and progress.

2.4 Challenges and Future Directions

While RL has made significant progress in training embodied agents, challenges remain, including sample inefficiency, safety concerns, and exploration–exploitation trade-offs. Future directions in RL for embodied agents involve addressing these challenges, incorporating advanced techniques like meta-learning, and extending the applicability to a wider range of real-world scenarios [11–15].

Reinforcement learning for embodied agents represents a promising avenue of research that continues to shape the landscape of AI and robotics. As technology advances, we can expect these intelligent agents to become increasingly capable and integrated into our daily lives, transforming industries and opening new horizons for AI-driven decision-making and automation.

3 Sim-To-Real Transfer: Bridging the Gap Between Simulation and Reality

Sim-to-Real Transfer is a critical concept in the development of embodied agents, such as robots and autonomous systems. It addresses the challenge of training agents in simulated environments and successfully deploying them in the real world. In this chapter, we explore the principles, methods, and applications of Sim-to-Real Transfer and its significance in AI and robotics.

3.1 Introduction to Sim-To-Real Transfer

Sim-to-Real Transfer refers to the process of training agents, typically using simulated environments, and then transferring the learned policies or behaviors to the physical world. This approach is essential for various reasons:

1. Cost-Efficiency: Simulated training is often more cost-effective and safer than real-world experimentation. It allows for a high number of trials and rapid iterations.

2. Safety: Testing and refining an agent in a simulated environment reduce the risk of damage or harm to physical systems or the environment.
3. Data Efficiency: Simulation provides a controlled setting, which can generate ample data for training, making it feasible to learn complex behaviors.
4. Rapid Prototyping: Simulations facilitate rapid prototyping of agent behavior, enabling quicker development.

3.2 Challenges in Sim-To-Real Transfer

Sim-to-Real Transfer presents several challenges, including:

1. Reality Gap: Simulated environments often lack the complexity and fidelity of the real world, leading to a reality gap where learned behaviors may not generalize well.
2. Domain Shift: Differences in dynamics, sensor noise, and unmodeled factors between simulation and reality can lead to domain shift issues.
3. Transferability: It is not always clear how well policies learned in a simulated environment will perform in the real world.

3.3 Methods and Approaches

To address these challenges, researchers have developed various techniques for successful Sim-to-Real Transfer, including:

1. Domain Adaptation: Techniques like domain randomization, domain adversarial training, and domain adaptation networks help reduce the domain gap between simulation and reality.
2. Data Augmentation: Injecting real-world data into the simulation can enhance the realism of the training data and improve transferability.
3. Model Calibration: Fine-tuning the learned policies in the real world to align with real-world dynamics is another approach.
4. Hybrid Simulation: Combining real and simulated components can lead to more accurate representations of the real environment.

3.4 Applications of Sim-To-Real Transfer

Sim-to-Real Transfer has applications in various fields, including:

1. Robotics: Simulated training is widely used for robotic arm manipulation, locomotion, and path planning. Robots trained in simulations can adapt to the real-world environment more effectively.

2. Autonomous Vehicles: Self-driving cars and drones leverage sim-to-real transfer to develop and validate their control algorithms in a safe and controlled environment.
3. Healthcare: Simulated training environments are used to train medical robots for surgeries, diagnostics, and rehabilitation.
4. Industrial Automation: Robots used in manufacturing and assembly processes can be trained and optimized in simulation before being deployed in factories.

3.5 Future Directions

The future of Sim-to-Real Transfer lies in advancing the fidelity and realism of simulations, improving domain adaptation methods, and scaling its application to more complex and unstructured real-world scenarios. As AI and robotics continue to evolve, the ability to bridge the gap between simulation and reality will be crucial in bringing intelligent embodied agents into the mainstream.

In conclusion, Sim-to-Real Transfer represents a crucial step in the development of embodied agents, enabling safer, more cost-effective, and efficient training processes. This approach is at the forefront of AI and robotics research, promising to reshape industries and drive innovation in the coming years.

4 Imitation Learning and Learning from Demonstrations: Harnessing Human Expertise for Embodied Agents

Imitation Learning and Learning from Demonstrations are powerful techniques in the field of artificial intelligence, enabling embodied agents to acquire complex skills and behaviors by observing and mimicking human or expert demonstrations. In this chapter, we delve into these approaches, exploring their principles, methods, and applications in the development of intelligent agents [16–18].

4.1 Introduction to Imitation Learning and Learning from Demonstrations

Imitation Learning, also known as learning from Demonstrations, is a subfield of machine learning where an agent learns to perform tasks by observing and emulating the actions of human experts or pre-recorded demonstrations. These approaches offer several advantages:

Efficient Skill Transfer: Imitation learning allows agents to rapidly acquire skills without the need for extensive trial and error.

Safety: By learning from expert demonstrations, agents can adopt behaviors that are safe and reliable, reducing the risk of accidents.

Generalization: Imitation learning can enable agents to generalize from a limited set of demonstrations and adapt to new, similar tasks.

Key Components of Imitation Learning:

Expert Demonstrations: High-quality demonstrations, typically provided by human experts, serve as the training data for the agent.

Policy Network: The agent uses a policy network, often a neural network, to map observations to actions, effectively replicating the behavior seen in the demonstrations.

Loss Function: A loss function is used to minimize the discrepancy between the agent's actions and the demonstrated actions, allowing the agent to fine-tune its policy.

4.2 Imitation Learning Methods

Several methods are employed for imitation learning and learning from demonstrations, including:

Behavioral Cloning: The agent directly imitates the actions demonstrated by experts. It's a simple and intuitive approach but can suffer from compounding errors if the agent encounters states not seen during training.

Inverse Reinforcement Learning (IRL): IRL infers the underlying reward function that guided the expert's behavior and then learns a policy that optimizes this reward. It is more robust to unseen states but requires solving a complex optimization problem.

4.3 Applications of Imitation Learning and Learning from Demonstrations

Imitation learning and learning from demonstrations have a wide range of applications in embodied agents, including:

Robotics: Robots can be trained to perform intricate tasks such as pick-and-place, object manipulation, and assembly by observing and replicating human demonstrations.

Autonomous Vehicles: Self-driving cars learn safe driving behaviors from human drivers, reducing the need for exhaustive testing and manual rule specification.

Humanoid and Virtual Agents: Virtual characters, avatars, and humanoids can acquire lifelike behaviors, making them more interactive and realistic in virtual environments.

Healthcare: Medical robots and devices can be trained to perform delicate procedures or provide assistance to healthcare professionals.

4.4 Challenges and Future Directions

Challenges in imitation learning include handling the distribution mismatch between training data and the real-world environment and addressing issues related to exploration and uncertainty. Future research directions involve combining imitation learning with reinforcement learning to improve adaptability and developing more sample-efficient methods for imitation [19–22].

In conclusion, Imitation Learning and Learning from Demonstrations offer a compelling means of imparting human-like expertise to embodied agents. These techniques enable agents to perform tasks with precision, safety, and efficiency, making them indispensable in a wide range of real-world applications and a key driver of progress in the field of AI and robotics.

5 Model-Based Reinforcement Learning: Shaping Embodied Agents with Environment Models

Model-Based Reinforcement Learning (MBRL) stands as a pivotal paradigm in the development of embodied agents. This chapter explores the core principles, methods, and applications of MBRL, revealing how it equips agents with the ability to build and leverage environment models to make intelligent decisions and adapt to complex, dynamic settings.

5.1 Introduction to Model-Based Reinforcement Learning

Model-Based Reinforcement Learning is an approach where an agent employs a learned model of its environment to make decisions and maximize cumulative rewards. It bridges the gap between Reinforcement Learning (RL) and traditional control theory by introducing a predictive model of the environment. Key characteristics of MBRL include:

Model of the Environment: MBRL agents create a model that represents the dynamics of their environment, which predicts how the environment will evolve in response to their actions.

Decision-Making: The agent employs this model for decision-making, which can involve planning, exploration, and optimization, to maximize rewards over time.

Sample Efficiency: MBRL can be more sample-efficient than pure RL, as it uses its model to simulate various scenarios without requiring real-world trial and error.

Key Components of Model-Based Reinforcement Learning:

Model of the Environment: This model can take the form of a neural network, a physics-based simulator, or other representations, and it predicts the next state and reward given the current state and action.

Planner or Optimization Algorithm: The agent uses a planner or optimization algorithm to decide which actions to take based on the predictions of its environment model.

Value Function or Policy Network: MBRL can also incorporate traditional RL components, such as value functions or policy networks, to further enhance decision-making.

5.2 Methods and Approaches in Model-Based Reinforcement Learning

Several methods and approaches are used in MBRL, including:

Monte Carlo Tree Search (MCTS): MCTS is a tree-based search algorithm that simulates possible trajectories in the environment model and guides the agent toward promising actions.

Model Predictive Control (MPC): MPC involves optimizing a sequence of actions in the model to find an optimal policy for the agent.

Ensemble Models: Ensemble methods, where multiple environment models are used and their predictions are aggregated, can enhance robustness and performance.

5.3 Applications of Model-Based Reinforcement Learning

Model-Based Reinforcement Learning has found application in various domains and embodied agents, including:

Robotics: Robots employ MBRL to perform precise manipulation, navigation in complex environments, and exploration of unknown terrain.

Autonomous Vehicles: Self-driving cars use MBRL to make decisions in dynamic traffic scenarios, enhancing safety and adaptability.

Game Playing: MBRL techniques are applied in playing complex games, achieving superhuman performance in board games, video games, and real-time strategy games.

Healthcare: Medical robots utilize MBRL for minimally invasive surgeries, rehabilitation, and other healthcare applications.

6 Evolutionary Algorithms: Shaping Embodied Agents Through Evolutionary Computation

Evolutionary Algorithms (EAs) stand as a powerful computational approach for the development of embodied agents, offering a unique perspective on learning and optimization inspired by the principles of biological evolution. This chapter explores the fundamental concepts, methods, and applications of Evolutionary Algorithms, revealing how they enable agents to adapt, learn, and optimize in complex and dynamic environments [23–26].

6.1 Introduction to Evolutionary Algorithms

Evolutionary Algorithms are a family of optimization and search techniques that draw inspiration from the process of natural selection and evolution. They operate by evolving a population of candidate solutions over multiple generations, gradually improving the solutions to better fit a given objective function or fitness landscape. Key characteristics of EAs include:

Population-Based Search: EAs maintain a population of individuals, representing potential solutions, and iteratively evolve and select individuals to improve their performance.

Variation Operators: EAs employ genetic operators such as mutation, crossover, and selection to create diversity, explore the solution space, and refine candidate solutions.

Objective Function: EAs optimize solutions based on a specified objective function or fitness criteria, with the goal of finding the best possible solution.

Key Components of Evolutionary Algorithms:

Population: A collection of individuals, each representing a potential solution to the problem at hand.

Fitness Function: An objective function that evaluates how well each individual performs in solving the problem.

Genetic Operators: Variation operators like mutation and crossover that introduce genetic diversity into the population.

Selection Mechanism: A method for choosing individuals for the next generation based on their fitness.

6.2 Methods and Approaches in Evolutionary Algorithms

Various Evolutionary Algorithms exist, tailored to different problem domains and requirements. Some common EA variants include:

Genetic Algorithms (GAs): GAs are a widely used variant of EAs that apply selection, crossover, and mutation operators to evolve solutions to optimization problems. They have been applied in a variety of domains, from function optimization to machine learning.

Evolutionary Strategies (ES): ES focus on optimizing continuous variables and are commonly used in robotics, reinforcement learning, and optimization of neural network parameters.

Genetic Programming (GP): GP extends the idea of EAs to evolve computer programs or trees, often used in symbolic regression, automatic code generation, and evolving decision-making rules.

6.3 Applications of Evolutionary Algorithms

Evolutionary Algorithms find applications in a range of domains, including:

Robotics: EAs can be used to evolve robot controllers, enabling robots to adapt to diverse environments, navigate autonomously, or optimize physical movements.

Optimization: EAs are employed for complex optimization problems in engineering, logistics, finance, and manufacturing.

Game Playing: Evolutionary Algorithms can optimize game strategies in both traditional board games and video games.

Artificial Life: EAs are used to simulate and study the emergence of complex behavior in artificial life and agent-based models.

6.4 Challenges and Future Directions

EAs face challenges related to computational efficiency, scalability, and the need for expert parameter tuning. Future research in Evolutionary Algorithms focuses on developing more efficient variants, incorporating machine learning techniques, and applying them to larger and more complex problem domains [27–30].

In conclusion, Evolutionary Algorithms offer a unique and powerful approach to shaping embodied agents. By drawing inspiration from the principles of evolution, EAs enable agents to adapt, learn, and optimize in dynamic and complex environments, opening up possibilities in a wide range of applications in AI and robotics research.

7 Deep Learning in Perception and Control for Embodied Agents

Deep Learning has emerged as a transformative force in the field of artificial intelligence, enabling embodied agents to perceive and control their environments more effectively. This chapter explores the integral role of Deep Learning in perception and control, shedding light on its fundamental principles, methodologies, and applications in creating intelligent, adaptable agents.

7.1 Introduction to Deep Learning in Perception and Control Challenges and Future Directions

Deep Learning is a subfield of machine learning that leverages artificial neural networks with many layers (deep neural networks) to automatically extract and learn features from large and complex datasets. In the context of embodied agents, Deep Learning plays a crucial role in perception and control, serving as the foundation for understanding the environment and executing actions effectively.

Key Components of Deep Learning:

- Neural Networks: Deep Learning employs artificial neural networks, comprising multiple layers of interconnected neurons, to model complex functions and relationships within data.
- Perception Networks: Neural networks, often convolutional neural networks (CNNs), are used to process sensory data, such as images, speech, and sensor readings, enabling agents to make sense of their surroundings.
- Control Networks: Recurrent neural networks (RNNs) and other architectures are used for decision-making and control, allowing agents to choose actions based on sensory input.

7.2 Deep Learning Models and Techniques

Deep Learning in perception and control encompasses various models and techniques, including:

1. Convolutional Neural Networks (CNNs): CNNs are widely used for image and video analysis, enabling agents to recognize objects, scenes, and patterns in visual data.
2. Recurrent Neural Networks (RNNs): RNNs are employed for sequential tasks and temporal data, enabling agents to process time-series data and perform actions based on dynamic context.
3. Reinforcement Learning with Deep Q-Networks (DQNs): DQNs combine Deep Learning with Reinforcement Learning to enable agents to make decisions and learn policies from sensory input.
4. Generative Adversarial Networks (GANs): GANs can be used to generate synthetic data or augment real data for training, improving agent performance in perception tasks.

7.3 Applications of Deep Learning in Perception and Control

Deep Learning in perception and control has a profound impact on various domains, including:

1. Computer Vision: Deep Learning enables agents to recognize and locate objects, track movements, and understand visual scenes in real-time. It finds applications in autonomous vehicles, surveillance, and robotics.
2. Natural Language Processing (NLP): Deep Learning models, such as recurrent neural networks and transformers, enhance language understanding and generation, empowering virtual assistants and chatbots.
3. Robotics: Agents equipped with Deep Learning can perform object manipulation, navigate through complex environments, and interact with humans more effectively.
4. Autonomous Systems: Self-driving cars and drones rely on Deep Learning to perceive their surroundings and make real-time decisions, ensuring safety and efficiency.
5. Healthcare: Deep Learning plays a role in medical image analysis, disease diagnosis, and drug discovery, improving patient care and drug development.

7.4 Challenges and Future Directions

Deep Learning in perception and control faces challenges such as the need for large datasets, potential biases in data, and the interpretability of deep models. Future research focuses on addressing these challenges, developing more efficient and interpretable models, and applying Deep Learning to more complex, unstructured, and multi-modal data [31–33].

In conclusion, Deep Learning in perception and control is pivotal in shaping the capabilities of embodied agents. By harnessing the power of neural networks, these agents can perceive their environment and execute actions with increasing accuracy and autonomy, paving the way for advances in AI, robotics, and autonomous systems.

8 3D Simulation Environments: Accelerating Progress in Embodied Agent Development

3D Simulation Environments are indispensable tools in the realm of artificial intelligence and robotics, providing a safe and cost-effective means to train, test, and fine-tune embodied agents. In this chapter, we delve into the significance, design principles, applications, and future prospects of 3D simulation environments in the development of intelligent agents.

8.1 Introduction to 3D Simulation Environments

3D Simulation Environments are computer-based platforms that replicate three-dimensional physical spaces and allow embodied agents, such as robots or virtual entities, to interact with, navigate, and learn from these spaces. These environments offer several advantages:

Safe Testing: Simulations provide a risk-free setting for training and testing agents, reducing the potential for damage or harm to physical systems or the environment.

Cost-Efficiency: Running experiments in simulations is often more affordable than conducting real-world trials, especially for expensive hardware like robots.

Unlimited Reproducibility: Simulations allow for the creation of controlled and reproducible experiments, where conditions can be precisely set and repeated.

Accessibility: Access to 3D simulation environments is more convenient, making it easier for researchers, developers, and learners to experiment and innovate.

8.2 Design Principles of 3D Simulation Environments

Effective 3D simulation environments incorporate the following design principles:

Realism: To accurately model the real world, these environments should replicate physics, lighting, and object interactions as closely as possible.

Scalability: The ability to model and simulate a wide range of scenarios and agent interactions, from simple tasks to complex, multi-agent environments.

Customizability: Providing tools to modify, extend, or create new scenarios, objects, and agents to suit specific research needs.

Interactivity: Allowing agents to perceive their environment, make decisions, and execute actions in real-time, closely emulating real-world interactions.

8.3 Applications of 3D Simulation Environments

3D simulation environments find applications in various fields, including:

Robotics: Agents, especially robots, are trained in simulation before transitioning to the real world. This applies to tasks like path planning, object manipulation, and navigation.

Autonomous Vehicles: Self-driving cars and drones are tested in 3D simulations to validate their control algorithms and decision-making processes.

Gaming and Entertainment: Video game developers use these environments for level design, character animation, and testing game mechanics.

Research and Education: Researchers explore AI, machine learning, and robotics concepts, while educators use them as teaching aids to instruct students in various subjects.

8.4 Challenges and Future Directions

While 3D simulation environments offer numerous advantages, they also present challenges such as the difficulty of modeling all real-world complexities accurately. Future research in this area focuses on:

Enhanced Realism: Developing more accurate physics engines, rendering techniques, and sensory models to increase the realism of simulations.

Domain Adaptation: Improving techniques for transferring knowledge and learned policies from simulations to the real world effectively.

Complex Multi-Agent Scenarios: Enabling simulations to handle multi-agent environments and dynamic, unpredictable situations.

Generalization: Ensuring that agents trained in simulations can generalize to diverse and unstructured real-world settings.

In conclusion, 3D Simulation Environments have become indispensable tools for the development of embodied agents, offering a valuable bridge between theory and practical application. As these environments continue to evolve, they will drive innovation in artificial intelligence, robotics, and autonomous systems, making them central to the progress of intelligent agent development.

9 Hybrid Models and Symbolic Reasoning in Embodied Agents: Integrating Logic and Neural Networks

Hybrid Models and Symbolic Reasoning represent a convergence of traditional symbolic AI and modern neural networks, offering a powerful approach for creating intelligent embodied agents. In this chapter, we explore the significance, principles, methods, and applications of hybrid models and symbolic reasoning in advancing the capabilities of these agents.

9.1 Introduction to Hybrid Models and Symbolic Reasoning

Hybrid Models and Symbolic Reasoning involve the integration of symbolic AI, which relies on logic, rules, and formal reasoning, with neural networks, which excel at learning from data. This fusion enables embodied agents to harness the strengths of both paradigms and effectively reason about their environment while learning from sensory input [34].

Key Components of Hybrid Models and Symbolic Reasoning:

Symbolic Component: This component typically involves symbolic knowledge representation, rules, logic, and reasoning engines, allowing agents to perform high-level reasoning and abstraction.

Neural Component: The neural component comprises artificial neural networks, including convolutional neural networks (CNNs) and recurrent neural networks (RNNs), used for perception, feature extraction, and low-level processing.

Integration Layer: The integration layer facilitates seamless communication between the symbolic and neural components, ensuring that high-level reasoning can guide low-level sensory processing.

9.2 Methods and Approaches in Hybrid Models

Hybrid models can take various forms, depending on the problem and agent's needs. Some common approaches include:

1. Neuro-Symbolic Integration: This approach combines neural networks and symbolic reasoning engines to allow neural components to feed information to the symbolic component for high-level decision-making.
2. Cognitive Architectures: Cognitive architectures, such as Soar and ACT-R, provide a framework for integrating symbolic reasoning with neural processing.
3. Explainable AI (XAI): Hybrid models support the development of interpretable AI systems, which are crucial for understanding agent behavior and decisions.

9.3 Applications of Hybrid Models and Symbolic Reasoning

Hybrid Models and Symbolic Reasoning have applications in various domains, including:

1. Robotics: Embodied agents in the form of robots use hybrid models to combine symbolic knowledge with sensory perception to plan and execute complex tasks.
2. Autonomous Systems: Self-driving cars benefit from symbolic reasoning for high-level decision-making, while neural networks handle perception and control.
3. Virtual Assistants: Virtual assistants, like chatbots, leverage symbolic reasoning to understand user queries and provide meaningful responses based on structured knowledge.
4. Healthcare: Medical diagnosis and treatment planning can benefit from hybrid models, as they combine medical expertise with machine learning for precise decision-making.

9.4 Challenges and Future Directions

Hybrid Models and Symbolic Reasoning face challenges related to knowledge representation, system complexity, and the need for tight integration between symbolic and neural components. Future research focuses on:

1. Knowledge Acquisition: Developing methods for acquiring symbolic knowledge from data or human experts.
2. Efficient Integration: Improving the efficiency and seamless communication between symbolic and neural components.
3. Explainability: Advancing techniques for explaining agent behavior and decisions, enhancing transparency and trust.

In conclusion, Hybrid Models and Symbolic Reasoning represent a promising approach for creating intelligent embodied agents that can reason logically and learn from sensory input. The integration of symbolic and neural AI paradigms offers a powerful framework for developing agents that are not only smart but also capable of clear, interpretable decision-making, thereby advancing the capabilities of AI, robotics, and autonomous systems.

10 Transfer Learning and Few-Shot Learning: Adapting Knowledge for Embodied Agents

Transfer learning and Few-Shot Learning are pivotal techniques in the realm of artificial intelligence, enabling embodied agents to leverage existing knowledge and adapt quickly to new tasks and environments. In this chapter, we delve into the principles, methods, and applications of Transfer Learning and Few-Shot Learning in the context of intelligent agents.

10.1 Introduction to Transfer Learning and Few-Shot Learning

Transfer Learning is the process of using knowledge gained from one task or domain to improve learning and performance in a different but related task or domain. Few-Shot Learning, a subset of Transfer Learning, focuses on training agents with very limited labeled data, often as few as one or a few examples per class. These techniques have become crucial in the development of embodied agents, particularly in scenarios where collecting vast amounts of data is impractical or expensive.

Key Concepts in Transfer Learning:

Source and Target Domains: In Transfer Learning, the source domain refers to the domain where pre-existing knowledge is available, and the target domain is the domain in which the agent aims to perform a task.

Fine-Tuning: Agents often fine-tune their neural network models, pretrained on the source domain, to adapt them to the target domain or specific task.

Knowledge Transfer: Knowledge can be transferred at various levels, from low-level features to high-level representations, enabling agents to adapt to different aspects of the target domain.

10.2 Few-Shot Learning Techniques

Few-Shot Learning encompasses several methods, including:

1. Meta-Learning: Meta-learning involves training agents to learn how to learn. Agents are exposed to many tasks in the source domain to learn a good initialization that allows rapid adaptation to new tasks.
2. Siamese Networks: Siamese networks learn to discriminate between pairs of examples, making them suitable for tasks like face recognition and one-shot image classification.
3. Prototypical Networks: These networks compute a prototype for each class, enabling agents to classify new examples based on their similarity to the prototypes.

10.3 Applications of Transfer Learning and Few-Shot Learning

Transfer Learning and Few-Shot Learning find applications in various domains, including:

1. Computer Vision: Agents can adapt pretrained models to perform tasks like object recognition, image segmentation, and even facial recognition.
2. Natural Language Processing: Transfer Learning is widely used for tasks like sentiment analysis, language translation, and chatbot development.
3. Robotics: Embodied agents, like robots, can leverage Transfer Learning to adapt to new environments, tasks, and objects quickly.
4. Healthcare: Few-Shot Learning is beneficial for medical image analysis and diagnosis, where labeled data is often scarce.

10.4 Challenges and Future Directions

Challenges in Transfer Learning and Few-Shot Learning include knowledge transfer, domain adaptation, and achieving high performance with limited data. Future research focuses on developing more efficient and effective methods for knowledge transfer, robust domain adaptation, and addressing the challenges of few-shot learning, pushing the boundaries of what agents can achieve with limited data [35].

In conclusion, Transfer Learning and Few-Shot Learning are critical techniques for empowering embodied agents with the ability to adapt rapidly and effectively to new tasks and domains. As these methods continue to advance, they hold the potential to revolutionize AI and robotics by making it easier and more practical to deploy intelligent agents in diverse real-world scenarios.

11 Embodied Cognitive Architectures: The Blueprint for Intelligent Agents

Embodied Cognitive Architectures represent a foundational framework in the development of intelligent embodied agents. In this chapter, we explore the principles, components, and applications of embodied cognitive architectures, which serve as the blueprint for agents that can perceive, reason, and act in their environment.

11.1 Introduction to Embodied Cognitive Architectures

Embodied Cognitive Architectures are frameworks that integrate cognitive processing and physical embodiment. These architectures provide agents with the ability to perceive, interact, and reason about their environment, enabling them to adapt to complex, dynamic scenarios. Key features of embodied cognitive architectures include:

1. Sensory Perception: The architecture incorporates sensory input from the environment, such as vision, audition, and touch, allowing the agent to perceive the world.
2. Internal Representations: Agents maintain internal representations of their environment, themselves, and their goals. These representations are critical for cognitive processing and decision-making.
3. Action Generation: The architecture generates actions that allow the agent to interact with the environment, including planning, motor control, and coordination.

Key Components of Embodied Cognitive Architectures:

Embodied Cognitive Architectures are composed of various components that work in concert to facilitate agent-environment interactions:

1. Perception Module: This module processes sensory data and extracts relevant information about the environment, enabling the agent to form a coherent perception of its surroundings.
2. Cognition Module: The cognition module encompasses memory, reasoning, planning, and decision-making processes, allowing the agent to think, learn, and make decisions.
3. Action Generation Module: Agents rely on this module to generate actions based on their goals, perceptions, and cognitive processes. It includes motor control and coordination.
4. Knowledge Base: The knowledge base stores information and models of the agent's world, which can be used for reasoning and problem-solving.
5. Communication Module: This component enables agents to communicate with other agents or entities, facilitating collaboration and cooperation.

11.2 Applications of Embodied Cognitive Architectures

Embodied Cognitive Architectures have a wide range of applications, including:

1. Robotics: These architectures are fundamental in developing intelligent robots capable of autonomous navigation, manipulation of objects, and human–robot interaction.
2. Autonomous Systems: Embodied cognitive architectures enable autonomous vehicles to perceive their surroundings, plan routes, and make driving decisions.
3. Virtual Agents: In virtual environments, embodied cognitive architectures enhance the realism and interactivity of virtual characters and avatars, making them more human-like.
4. Education and Healthcare: Embodied agents with cognitive architectures can serve as personalized tutors, healthcare companions, and assistive devices, enhancing the learning and well-being of users.

11.3 Challenges and Future Directions

Challenges in embodied cognitive architectures include the need for efficient memory management, robust perception in dynamic environments, and enhancing the adaptability of agents to diverse scenarios. Future research in this field focuses on improving architectures to support more complex, realistic environments, enhancing their reasoning and learning capabilities, and expanding their application in various domains.

In conclusion, Embodied Cognitive Architectures serve as the foundation for the development of intelligent embodied agents. These architectures enable agents to perceive their environment, reason about it, and take actions to achieve goals, opening up a wide range of applications and propelling the field of AI and robotics forward.

12 Conclusion

In conclusion, the development of embodied agents represents a fascinating and dynamic field within artificial intelligence and robotics. Embodied agents are designed to interact with and adapt to their physical or virtual environments, making them capable of performing a wide range of tasks, from autonomous navigation to complex decision-making. To equip these agents with the necessary skills, various techniques and models are employed. Here, we've explored key concepts, methods, and applications of these techniques, including Reinforcement Learning, Imitation Learning, Model-Based Reinforcement Learning, Evolutionary Algorithms, Deep Learning, 3D Simulation Environments, Hybrid Models and Symbolic Reasoning,

Transfer Learning, Few-Shot Learning, and Embodied Cognitive Architectures. Each of these techniques plays a vital role in enhancing the capabilities of embodied agents: Reinforcement Learning enables agents to learn from interaction with their environment and make sequential decisions to maximize rewards. Imitation Learning empowers agents to mimic expert behavior and rapidly acquire skills through observation. Model-Based Reinforcement Learning allows agents to build predictive models of their environment to make informed decisions. Evolutionary Algorithms harness the power of evolution for optimizing agent behavior. Deep Learning in perception and control enhances an agent's ability to interpret and interact with its surroundings. 3D Simulation Environments provide safe and cost-effective training and testing grounds for these agents. Hybrid Models and Symbolic Reasoning merge symbolic logic with neural networks to enable interpretable and logical decision-making. Transfer Learning and Few-Shot Learning enable agents to adapt to new tasks and domains quickly, even with limited data. Embodied Cognitive Architectures form the framework that allows agents to perceive, reason, and act in their environment, driving advancements in robotics, autonomous systems, virtual agents, education, and healthcare.

As technology and research continue to advance, the capabilities of embodied agents will expand, making them increasingly valuable in a wide range of real-world applications. These techniques and models are the building blocks of the next generation of intelligent, adaptable, and capable agents, and they hold the promise of revolutionizing industries and improving the quality of our lives. The future of embodied agents is bright, as researchers and developers continue to push the boundaries of what these agents can achieve.

References

1. Franklin, S.: Autonomous agents as embodied AI. Cybern. Syst. **28**(6), 499–520 (1997)
2. Weihs, L., Salvador, J., Kotar, K., Jain, U., Zeng, K.H., Mottaghi, R., Kembhavi, A.: Allenact: a framework for embodied AI research (2020). arXiv:2008.12760
3. Duan, J., Yu, S., Tan, H.L., Zhu, H., Tan, C.: A survey of embodied AI: from simulators to research tasks. IEEE Trans. Emerg. Top. Comput. Intell. **6**(2), 230–244 (2022)
4. Savva, M., Kadian, A., Maksymets, O., Zhao, Y., Wijmans, E., Jain, B., Straub, J., Liu, J., Koltun, V., Malik, J., Parikh, D., Batra, D.. Habitat: A platform for embodied AI research. In Proceedings of the IEEE/CVF International Conference on Computer Vision, pp. 9339–9347 (2019)
5. Savva, M., Kadian, A., Maksymets, O., Zhao, Y., Wijmans, E., Jain, B., Straub, J., Liu, J., Koltun, V., Malik, J., Parikh, D., Batra, D.: Habitat: a platform for embodied AI research. In: Proceedings of the IEEE/CVF International Conference on Computer Vision, pp. 9339–9347 (2019)
6. Deitke, M., Han, W., Herrasti, A., Kembhavi, A., Kolve, E., Mottaghi, R., Salvador, J., Schwenk, D., VanderBilt, E., Wallingford, M., Weihs, L., Yatskar, M., & Farhadi, A.: Robothor: an open simulation-to-real embodied AI platform. In: Proceedings of the IEEE/CVF conference on computer vision and pattern recognition, pp. 3164–3174 (2020)
7. Chrisley, R.: Embodied artificial intelligence. Artif. Intell. **149**(1), 131–150 (2003)

8. Pfeifer, R., Iida, F.: Embodied artificial intelligence: trends and challenges. Lecture Notes in Computer Science, pp. 1–26 (2004)
9. Jain, U., Liu, I.J., Lazebnik, S., Kembhavi, A., Weihs, L., & Schwing, A.G.: Gridtopix: training embodied agents with minimal supervision. In: Proceedings of the IEEE/CVF International Conference on Computer Vision, pp. 15141–15151 (2021)
10. Batra, D., Chang, A.X., Chernova, S., Davison, A.J., Deng, J., Koltun, V., Levine, S., Malik, J., Mordatch, I., Mottaghi, R., Su, H.: Rearrangement: a challenge for embodied AI (2020). arXiv:2011.01975
11. Divekar, R.R., Mou, X., Chen, L., De Bayser, M. G., Guerra, M.A., Su, H.: Embodied Conversational AI Agents in a Multi-modal Multi-agent Competitive Dialogue. In: IJCAI, pp. 6512–6514, August 2019 (2019)
12. Petrenko, A., Wijmans, E., Shacklett, B., Koltun, V.: Megaverse: simulating embodied agents at one million experiences per second. In: International Conference on Machine Learning, pp. 8556–8566, July 2021. PMLR (2021)
13. Fan, L., Wang, G., Jiang, Y., Mandlekar, A., Yang, Y., Zhu, H., Tang, A., Huang, D.A., Zhu, Y., Anandkumar, A.: Minedojo: Building open-ended embodied agents with internet-scale knowledge. Adv. Neural. Inf. Process. Syst. **35**, 18343–18362 (2022)
14. Poole, D.L., Mackworth, A.K.: Artificial Intelligence: Foundations of Computational Agents. Cambridge University Press (2010)
15. Roy, N., Posner, I., Barfoot, T., Beaudoin, P., Bengio, Y., Bohg, J., Brock, O., Depatie, I., Fox, D., Koditschek, D., Lozano-Perez, T., Mansinghka, V., Pal, C., Richards, B., Sadigh, D., Schaal, S., Sukhatme, G., Therien, D., Toussaint, M., Van de Panne, M.: From machine learning to robotics: challenges and opportunities for embodied intelligence (2021). arXiv:2110.15245
16. Cassell, J.: Embodied conversational agents: representation and intelligence in user interfaces. AI Mag. **22**(4), 67–67 (2001)
17. Wallkötter, S., Tulli, S., Castellano, G., Paiva, A., Chetouani, M.: Explainable embodied agents through social cues: a review. ACM Trans. Hum.-Robot Interact. (THRI) **10**(3), 1–24 (2021)
18. Manzotti, R.: Embodied AI beyond embodied cognition and enactivism. Philosophies **4**(3), 39 (2019)
19. Gan, C., Zhou, S., Schwartz, J., Alter, S., Bhandwaldar, A., Gutfreund, D., Yamins, D.L.K., DiCarlo, J.J., McDermott, J., Torralba, A., Tenenbaum, J.B.: The ThreeDWorld transport challenge: a visually guided task-and-motion planning benchmark towards physically realistic embodied AI. In: 2022 International Conference on Robotics and Automation (ICRA), pp. 8847–8854, May 2022. IEEE (2022)
20. Dautenhahn, K.: Embodiment and interaction in socially intelligent life-like agents. In: International Workshop on Computation for Metaphors, Analogy, and Agents, pp. 102–141, April 1998. Springer, Berlin, Heidelberg (1998)
21. Geiselmann, R., Tsourgianni, A., Deroy, O., Harris, L.T.: Interacting with agents without a mind: the case for artificial agents. Curr. Opin. Behav. Sci. **51**, 101282 (2023)
22. Singh, K.P., Weihs, L., Herrasti, A., Choi, J., Kembhavi, A., Mottaghi, R.: Ask4help: learning to leverage an expert for embodied tasks. Adv. Neural. Inf. Process. Syst. **35**, 16221–16232 (2022)
23. Xie, X., Liu, H., Zhang, Z., Qiu, Y., Gao, F., Qi, S., Zhu, Y., Zhu, S.C.: VRGym: a virtual testbed for physical and interactive AI. In: Proceedings of the ACM Turing Celebration Conference-China, pp. 1–6, May 2019 (2019)
24. Shenavarmasouleh, F., Mohammadi, F.G., Amini, M.H., Reza Arabnia, H.: Embodied AI-driven operation of smart cities: a concise review. In: Cyberphysical Smart Cities Infrastructures: Optimal Operation and Intelligent Decision Making, pp. 29–45 (2022)
25. Paparo, G.D., Dunjko, V., Makmal, A., Martin-Delgado, M.A., Briegel, H.J.: Quantum speedup for active learning agents. Phys. Rev. X **4**(3), 031002 (2014)
26. Zheng, K., Zhou, K., Gu, J., Fan, Y., Wang, J., Di, Z., He, X., Wang, X.E.: Jarvis: a neuro-symbolic common sense reasoning framework for conversational embodied agents (2022). arXiv:2208.13266

27. Kovalev, A.K., Panov, A.I.: Application of pretrained large language models in embodied artificial intelligence. In: Doklady Mathematics, vol. 106, No. Suppl 1, pp. S85–S90. Pleiades Publishing, Moscow (2022)
28. Beer, R.D.: The dynamics of active categorical perception in an evolved model agent. Adapt. Behav. **11**(4), 209–243 (2003)
29. Rajan, K., Saffiotti, A.: Towards a science of integrated AI and robotics. Artif. Intell. **247**, 1–9 (2017)
30. Xia, F., Zamir, A. R., He, Z., Sax, A., Malik, J., Savarese, S.: Gibson Env: real-world perception for embodied agents. In: Proceedings of the IEEE conference on computer vision and pattern recognition, pp. 9068–9079 (2018)
31. Song, C.H., Wu, J., Washington, C., Sadler, B.M., Chao, W.L., Su, Y.: LLM-planner: few-shot grounded planning for embodied agents with large language models. In: Proceedings of the IEEE/CVF International Conference on Computer Vision, pp. 2998–3009 (2023)
32. De Rosis, F., Pelachaud, C., Poggi, I., Carofiglio, V., De Carolis, B.: From Greta's mind to her face: modelling the dynamics of affective states in a conversational embodied agent. Int. J. Hum. Comput. Stud. **59**(1–2), 81–118 (2003)
33. Peng, Y., Han, J., Zhang, Z., Fan, L., Liu, T., Qi, S., Feng, X., Ma, Y., Wang, Y., Zhu, S.C.: The tong test: evaluating artificial general intelligence through dynamic embodied physical and social interactions. Engineering (2023)
34. Briegel, H.J., De las Cuevas, G.: Projective simulation for artificial intelligence. Sci. Rep. **2**(1), 400 (2012)
35. Gan, C., Zhou, S., Schwartz, J., Alter, S., Bhandwaldar, A., Gutfreund, D., Yamins, D.L.K., DiCarlo, J.J., McDermott, J., Torralba, A., Tenenbaum, J.B.: The ThreeDWorld transport challenge: a visually guided task-and-motion planning benchmark for physically realistic embodied AI (2021). arXiv:2103.14025

P. Chitra is an Assistant Professor in Computer Science at GITAM, University, Bangalore, India. Holding a Ph.D. in Computer Science from The Gandhigram Rural University, India, Dr. Chitra specializes in Image compression and recent advances in computer sciences. Their research interests encompass Image processing, Machine learning and Deep Learning, and they have contributed significantly to the field through publications in esteemed journals and conferences. Dr. P. Chitra is committed to nurturing a vibrant academic atmosphere, fostering curiosity and innovation among students. With a passion for teaching and mentorship, they strive to empower aspiring computer scientists with the knowledge and skills essential for success in the rapidly evolving technology landscape. Driven by a dedication to excellence in both research and education, Dr. P. Chitra is a valuable asset to the academic community, shaping the future of computer science through their expertise and leadership.

A. Saleem Raja, Faculty and researcher with 17+ years of experience teaching courses in both undergraduate and postgraduate levels. Published over 30+ articles in peer-reviewed journals and conferences. Trained project students in bachelor and master level in field of web application development, mobile application development, IoT application development, cisco networks and python data science. Organized practical workshops for students and teachers in reputed colleges. Supervised more than 5 applied research projects in Oman.

Entertainment Recommendation and Rating System Based on Emotions

Geetraj Kumar Shaw, Romit Chattopadhyay, Anubhab Paul, Kumar Rounik, Sawan Kumar, Biswapa Saha, and Pabak Indu

Abstract In today's digital landscape, the abundance of entertainment content on streaming platforms, music services, and e-book libraries presents challenges and opportunities. Traditional recommendation systems often fail to capture users' dynamic preferences, including their emotional states. Our project addresses this gap by introducing an Emotion-based rating and recommendation system. By integrating real-time facial expression detection, we aim to understand users' emotional reactions to content. This system responds to the demand for personalized experiences in digital entertainment. Leveraging computer vision and machine learning, our research provides personalized suggestions based on users' emotional responses, surpassing surface-level preferences. This initiative transforms entertainment interaction by acknowledging emotions' role in choices and experiences. It revolutionizes content discovery in the digital age, offering holistic entertainment experiences. Experiment results demonstrate the accuracy of emotion recognition and the effectiveness of personalized recommendations. Positive user feedback indicates high satisfaction and engagement, guiding future enhancements.

Keywords Digital entertainment · Recommendation systems · Emotion-based rating · Facial expression detection · Personalized experiences · Computer vision · Machine learning · User engagement · Content discovery · Entertainment interaction · Emotion recognition · User feedback

---

G. K. Shaw · R. Chattopadhyay · A. Paul · K. Rounik · S. Kumar · B. Saha · P. Indu (✉)
Department of CSE(AIML) and CSBS, Institute of Engineering and Management, University of Engineering and Management, Kolkata, India
e-mail: pabakindu@yahoo.co.in

P. Indu
IEM Centre of Excellence for Cloud Computing and IoT, Department of CSE (AIML), Institute of Engineering and Management, University of Engineering & Management, Kolkata 700091, India

IEM Centre of Excellence for Data Science, Department of CSE (AIML), Institute of Engineering and Management, University of Engineering & Management, Kolkata 700091, India

P. Raj et al. (eds.), *Building Embodied AI Systems: The Agents, the Architecture Principles, Challenges, and Application Domains*, Information Systems Engineering and Management 14, https://doi.org/10.1007/978-3-031-68256-8_20

1 Introduction

1.1 Background

In today's digital landscape, the abundance of entertainment content presents both opportunities and challenges. Traditional recommendation systems relying on historical data often struggle to capture users' evolving preferences and emotions. To address this, we propose an emotion-based entertainment recommendation system integrating real-time facial expression detection technology. By understanding users' emotional reactions, we aim to deliver personalized experiences, aligning content with their moods. Leveraging advanced technologies like computer vision and machine learning, our initiative offers tailored suggestions that consider users' emotional responses. Ultimately, our goal is to revolutionize content discovery by recognizing the significance of emotions in shaping user choices and experiences.

1.2 Problem Statement

In the realm of digital entertainment, users often face challenges finding content that resonates with their current emotional state amid vast content libraries. Traditional recommendation systems, reliant on past viewing habits, frequently miss the mark in capturing users' present moods and preferences, resulting in disjointed suggestions. An inherent limitation lies in these systems' inability to discern the dynamic nature of human emotions, overlooking their profound impact on content engagement. Integrating facial expression analysis, a potent indicator of behavior and emotional well-being holds promise in enhancing recommendation accuracy by aligning suggestions with users' current emotional states. By bypassing reliance solely on historical data, our project seeks to revolutionize the entertainment experience, offering personalized recommendations tailored to users' real-time emotional expressions.

1.3 Aims and Objectives

The Entertainment Recommendation and Rating System Based on Emotions aims to revolutionize content discovery by integrating emotional intelligence into recommendations. Leveraging facial expression recognition, the project tailors personalized suggestions for movies, music, and books based on users' emotional states and preferences, fostering satisfaction and engagement. Objectives include enhancing user engagement by offering captivating recommendations, achieving accurate real-time emotion detection, integrating user feedback for algorithm refinement, seamlessly integrating with external platforms, and contributing to mental health research. By providing personalized recommendations, ensuring accurate emotion detection, and integrating user feedback, the system offers a holistic entertainment discovery experience, fostering satisfaction, engagement, and well-being.

2 Literature Review

2.1 Facial Emotion Recognition

Facial emotion recognition (FER) plays a crucial role in mental health assessment and human–computer interfaces, enhancing communication and user experiences. Researchers have developed advanced systems and algorithms to handle real-time processing and diverse facial expressions, ensuring accurate FER. Happy et al. [1] utilize precise facial landmark detection for consistent performance, while Kim et al. [2] propose a hierarchical deep learning-based FER system for significant accuracy gains. Pranav et al. [3] contribute a Deep Convolutional Neural Network (DCNN) model trained on curated image data for emotion recognition. Kartali et al. [4] analyze real-time emotion recognition methods, favoring CNN-based models like Affdex CNN and AlexNet for classifying basic emotions. Deshmukh et al. [5] aim to integrate music therapy into an automated system for recognizing facial emotions in stressed individuals, addressing gender, age, and facial expression challenges. Gill et al. [6] explore deep learning techniques for FER, demonstrating high accuracy with CNNs applied to six discrete emotions, showcasing the potential of deep learning in understanding human emotions from facial expressions [7].

2.2 Entertainment Recommendation

In our interconnected world, music, movies, and books shape our cultural experiences and identities significantly. Recommendation systems (RSs) are crucial tools for navigating this vast content landscape. Kumar et al. [8] propose a hybrid RS for movies integrating sentiment analysis of tweets with content-based and collaborative filtering. Balfaqih et al. [9] introduce a personalized movie recommendation system based on demographics and emotions. Fessahaye et al. [10] present T-RECSYS, a precise music recommendation system merging content-based and collaborative filtering approaches. Chang et al. [11] developed a personalized music recommendation system using collaborative filtering and neural networks. Hasib et al. [12] tackle Bengali music classification with BMNet-5, a neural network model. Balakrishnan et al. [13] combine Matrix Factorization and Singular Value Decomposition for accurate and diverse book recommendations. Jayapal et al. [14] propose a hybrid recommendation system tailored for students, while Nawar et al. [15] introduce a cross-content recommendation system leveraging natural language processing.

2.3 Mood Based Entertainment Recommendation

Mood-based recommendation systems leveraging facial expression recognition offer personalized content suggestions, enhancing user engagement. Pioneering studies by Elias et al. [16] demonstrate the effectiveness of emotion-driven recommendation systems, achieving superior accuracy in inferring movie preferences from facial expressions. Chauhan et al. [17] propose hybrid recommendation systems for personalized movie suggestions, while Iyer et al. [18] introduce 'EmoPlayer,' an Android app curating real-time playlists aligned with users' moods. Priyanka et al. [19] emphasize neural networks' significance in recommending songs based on facial expressions, advancing automation in music curation. Krupa et al. [20] integrate chatbot interactions with emotion recognition for personalized song suggestions, showcasing potential applications beyond leisure. Negandhi et al. [21] employ machine learning to recommend books tailored to users' preferences and moods, aiming to enhance reading experiences. Fujimoto et al. [22] analyze social media data to recommend books based on individual readers' preferences and emotional states, with ongoing efforts to improve usability. Gupta et al. [23] introduce Mind Frame, achieving remarkable accuracy in recommending movies and music based on users' current moods. Kumar et al. [24] present Emotion Booster, highlighting the emotional context's importance in personalized recommendations. Mariappan et al. [25] demonstrate Face Fetch, a multimedia content suggestion system driven by real-time facial expression recognition, paving the way for further improvements. These studies illustrate the transformative potential of aligning content recommendations with users' emotional states, reshaping how we interact with entertainment and literature.

2.4 Reviews Based on Emotions

Product reviews are vital for understanding customer preferences and satisfaction and improving recommendation systems and user experience. Advanced techniques showcased in the following studies enhance recommendation system accuracy by leveraging sentiment analysis, user reviews, and facial expression analysis. Chaugule et al. [26] propose a method to evaluate product acceptance in retail settings by analyzing facial expressions, suggesting further enhancements by incorporating factors like viewing time and online reviews. Chaurasiya et al. [27] emphasize the importance of user reviews in e-commerce recommendation systems, introducing a sentiment intensity-based approach for better accuracy. Sharma et al. [28] address fake reviews using a Visitor Rating System with facial emotion recognition, ensuring genuine ratings and fostering trust between businesses and consumers. These studies demonstrate innovative strategies to improve recommendation systems and combat fraudulent ratings, ultimately benefiting businesses and consumers alike.

3 Materials

3.1 Proposed Methodology

Facial Emotion Recognition. Facial emotion recognition (FER) is a crucial component of the Entertainment Recommendation and Rating System. FER utilizes facial expressions to automatically recognize and understand human emotions, enabling the system to make appropriate content recommendations based on users' emotional states. Real-time emotion identification and analysis are facilitated by deep learning models trained on annotated face image datasets, categorizing emotions into different groups.

1. *Model Training:* The initial stage of FER involves training a model to record, analyze, detect, and identify emotions. This begins with obtaining and preprocessing a sizable database, such as the FER2013 picture dataset, consisting of grayscale face images representing seven emotional states. The dataset is split into training and testing sets, with real-time data augmentation used to generate batches of tensor image data (Fig. 1).

The model architecture is defined, typically comprising convolutional blocks, pooling layers, batch normalization, dropout layers, and fully connected layers. Training involves iteratively feeding batches of labeled face pictures to the model, adjusting weights using optimization techniques like stochastic gradient descent and Adam optimizer. Performance is evaluated using metrics such as recall, accuracy, precision, and F1 score. Techniques like dropout and batch normalization are employed to improve generalization and avoid overfitting. Training continues for multiple epochs until satisfactory performance is achieved on the validation dataset.

Once training is complete, the trained model is saved for inference during the model usage phase. This model forms the core of the Facial Emotion Recognition module, enabling real-time detection and interpretation of users' emotional states from facial images.

In summary, the model training phase involves acquiring and preprocessing datasets, defining and initializing the model architecture, iterative training using

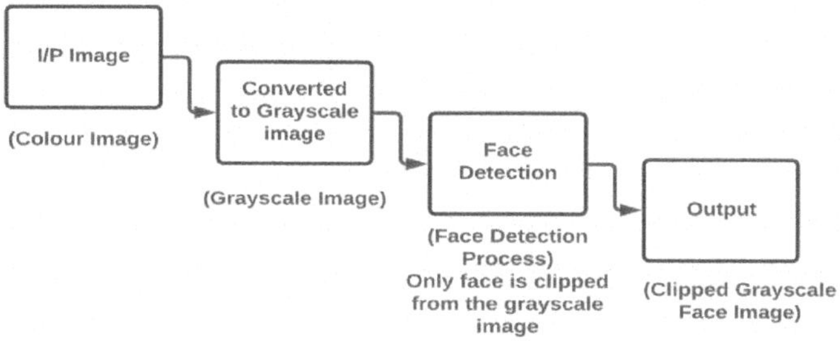

Fig. 1 Face recognition process

optimization techniques, and evaluating performance metrics. This phase prepares the trained model for deployment and utilization within the Entertainment Recommendation and Rating System.

2. *Model Usage (Detection).* In the model usage phase, the trained Facial Emotion Recognition (FER) model is applied to detect and interpret users' emotional expressions in real-time. Firstly, the pre-trained ONNX model is loaded for its benefits in interoperability, optimization, and deployment flexibility. Using the OpenCV Python library, live video of the user's face is captured through the camera, with each frame analyzed and stored as an image file.

The captured frame is converted to grayscale for faster processing, and faces are identified using the Haar cascade frontal face classifier from OpenCV. A region of interest is then shrunk to 48 $\times$ 48 pixels for consistent facial recognition. The pixel values are normalized by dividing them by 255, and the normalized pixels are fed into the ONNX model for emotion prediction [29, 30].

The FER model extracts features related to various emotional expressions from the input images through pooling layers, activation functions, and convolutional layers. By utilizing acquired weights from training, it categorizes photos into one of seven predetermined emotional groups.

This process iterates in a loop for a maximum of 30 frames, with each emotion prediction stored in an array. The final predicted emotion is determined by finding the index of the maximum value in the predictions array, representing the dominant emotion captured during the run. Probabilistic predictions for each emotional category are generated, indicating the likelihood of the user exhibiting a specific emotion. These predictions inform content recommendations tailored to the user's mood.

Evaluation criteria that evaluate the real-time performance of the FER model include precision, accuracy, and recall. By deploying this model, the Entertainment Recommendation and Rating System can offer personalized content suggestions aligned with users' emotional states, enhancing engagement and satisfaction (Fig. 2).

Recommendation System. The Entertainment Recommendation and Rating System employs pivotal recommendation algorithms to offer personalized content suggestions to users. It utilizes mood-based and popularity-based algorithms to ensure diverse and relevant recommendations for books, music, and movies.

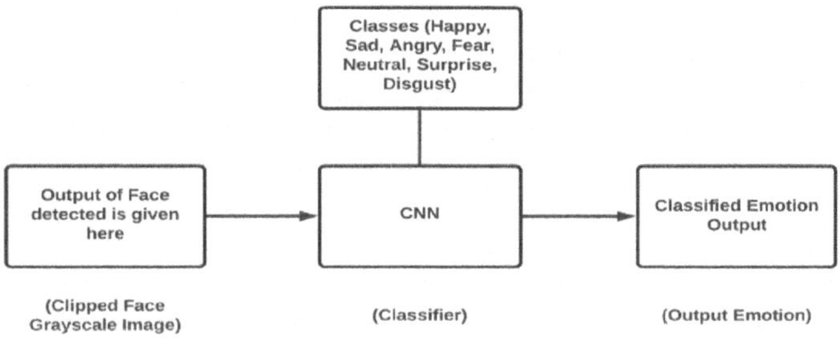

Fig. 2 Emotion recognition process

1. *Mood-Based Recommendation Algorithm:* This algorithm aligns content with users' emotional states by categorizing items into mood-specific categories based on topics and genres. It analyzes facial expressions from the Facial Emotion Recognition module to customize recommendations, enhancing user experience.
2. *Popularity-Based Recommendation Algorithm:* This algorithm identifies popular content based on metrics like views and ratings, ensuring widely recognized items appear prominently. It provides users with appealing content options based on popularity.

By combining these algorithms, the system offers a comprehensive and tailored content discovery experience, considering users' emotional states and preferences while also exposing them to popular content. It utilizes a curated dataset to procure top-rated content, enhancing the recommendation process. The features that allow for individualized recommendations for each genre of entertainment are listed below:

1. Personalized Movie Recommendations: The system delivers tailored movie suggestions to users based on their emotional states, detected through facial expression recognition. Users receive dynamic recommendations on the frontend interface, enhancing their viewing experience. They can explore recommended movies and access IMDb pages for more details, facilitating informed decision-making.
2. Personalized Song Recommendations: Similarly, the system offers customized music suggestions by analyzing users' emotional profiles and music libraries. Users discover new songs and playlists aligned with their mood preferences, presented seamlessly on the frontend interface. Integration with platforms like Spotify allows direct access to recommended songs, enhancing accessibility (Fig. 3).
3. *Personalized Book Recommendations:* Expanding its scope to literature, the system provides personalized book suggestions by analyzing users' emotional profiles and digital libraries. Users explore recommended books and access details on platforms like Goodreads, streamlining discovery and engagement.

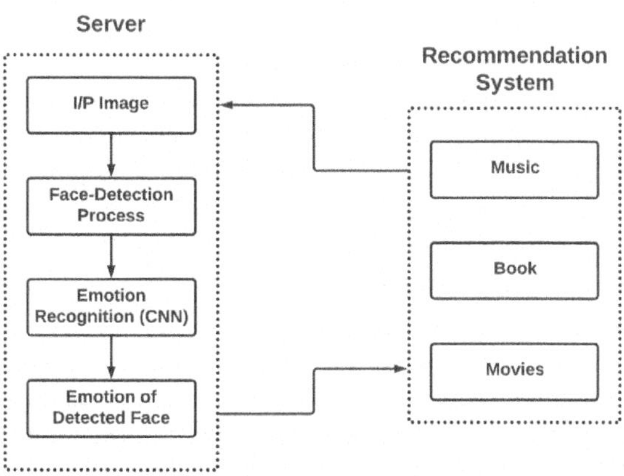

Fig. 3 Backend of the proposed system

By integrating these personalized features across movies, songs, and books, the Entertainment Recommendation and Rating System enhances user experiences and fosters deeper engagement with diverse media forms. Custom-made databases for each category ensure comprehensive content coverage and seamless integration with platforms like IMDb, Spotify, and Goodreads, enriching users' interactions with recommended content [31].

Rating System. In the Entertainment Recommendation and Rating System, user ratings significantly enhance the content discovery experience by providing valuable feedback on recommended movies, songs, and books. Here's how it works:

1. *Rating Collection Process:* Users are prompted to rate recommended content based on their emotional responses, captured through facial expression detection. They express satisfaction or dissatisfaction, choosing from options corresponding to positive, negative, or neutral emotional states. This intuitive process, accessible via a 'Rate Now' button, updates the content's database with user feedback.
2. *Utilization of Ratings:* Collected ratings refine recommendation algorithms, ensuring future suggestions better align with users' emotional states and preferences. They also guide other users by indicating content quality and relevance. The system analyzes ratings to discern patterns and tailor recommendations accordingly.
3. *Enhancing User Experience:* The rating system empowers users to shape their entertainment preferences and actively participate in the content discovery process. By providing feedback, users gain a sense of ownership over their choices and can make informed decisions based on their tastes and emotional states.

Overall, user feedback gathered through ratings enhances the relevancy of content recommendations and fosters user engagement, contributing to an enriched entertainment experience within the Entertainment Recommendation and Rating System.

3.2 System Architecture

The Entertainment Recommendation and Rating System's system architecture describes how data moves through and interacts with different parts to offer users individualized content recommendations depending on their emotional states. This design seeks to optimize the processing of deep learning algorithms for facial expression identification, speed up user replies, and streamline the content discovery process. The system consists of a straightforward frontend interface where users may take pictures and a backend server that processes photographs and uses convolutional neural networks (CNNs) to detect emotions. Additionally, a filter manager component processes the detected emotions and provides tailored recommendations for music, movies, and books. Through the coordination of data movement across

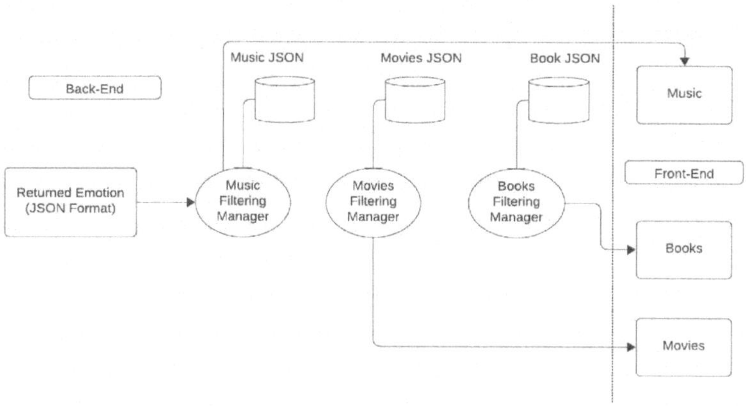

Fig. 4 System architecture

these elements, the system guarantees precise and effective content recommendations that correspond with users' emotional states, thereby augmenting their platform experience as a whole (Fig. 4).

Datasets Used. The Entertainment Recommendation and Rating System relies on diverse datasets to train its machine-learning models for content recommendation and facial expression identification [32].

For facial emotion identification, the system utilizes the FER-2013 dataset, which includes grayscale facial photos depicting seven fundamental emotions: anger, contempt, fear, happiness, sorrow, surprise, and neutrality. This dataset is divided into training, public test, and private test subsets, comprising 28,709, 3,589, and 3,589 photos, respectively, to train and validate the CNN models effectively (Fig. 5).

For content recommendation, the system carefully selects datasets to ensure diversity and applicability across books, music, and movies. It utilizes datasets such as MovieLens and IMDB Movies for movie recommendations, Spotify Million Songs for song recommendations, and Goodbooks-10 k for book recommendations. These datasets are analyzed, categorized, and incorporated into the system's custom-made datasets, ensuring personalized recommendations aligned with users' preferences and emotional states. Each dataset undergoes thorough examination and categorization to ensure suitability for inclusion in the recommendation system (Fig. 6).

Fig. 5 Sample of FER2013 dataset

Emotions	Genre
Happy	Musical, Comedy, Family, Sports, Fantasy, Action, Animation, Travel
Sad	Drama, Romance, Tragedy, Crime, Documentary, Mystery, Dark Comedy, War
Surprised	Adventure, Thriller, Science Fiction, Mystery, Fantasy, Animation, Experimental
Angry	Action, Crime, Mystery, Thriller, Drama, Horror, Documentary, Romance
Fear	Horror, Science Fiction, Thriller, Mystery, Crime, Experimental, War
Disgust	History, Tragedy, Drama, Crime, Dark Comedy
Neutral	Musical, Comedy, Romance, Mystery, Science Fiction, History, Thriller

Fig. 6 Emotion-based list of genres

Model Architecture. The model architecture of the Entertainment Recommendation and Rating System is that of a Convolution Neural Network (CNN) [7]. It comprised multiple layers of convolutional and pooling operations, followed by fully connected layers for classification. The architecture is designed to effectively extract features from facial images and classify them into distinct emotional categories. The following is an overview of the key components of the model architecture:

1. *Face Emotion Recognition Process.* The face emotion recognition process within the Entertainment Recommendation and Rating System involves several key steps aimed at accurately detecting users' emotional states from facial expressions. The following outlines the architecture of the facial emotion recognition model in detail:

   - İnput Layer.

     The input layer (visible) accepts grayscale images of size 48 × 48 pixels.

   - 1st Block.

     This block consists of two convolutional layers (conv1_1, conv1_2) with 64 filters each, followed by batch normalization (BatchNormalization()) and ReLU activation (activation = 'relu').
     Max pooling (MaxPooling2D()) with a pool size of 2 × 2 is applied to down sample the feature maps obtained after convolution and reduce spatial dimensions.
     Dropout (Dropout()) with a rate of 0.3 is used for regularization to prevent overfitting.

   - 2nd Block.

This block includes three convolutional layers (conv2_1, conv2_2, conv2_3) with 128 filters each, followed by batch normalization and ReLU activation.
Max pooling and dropout are applied similarly to the 1st block.

- 3rd Block.

  This block comprises four convolutional layers (conv3_1, conv3_2, conv3_3, conv3_4) with 256 filters each, followed by batch normalization and ReLU activation.
  Max pooling and dropout are applied similarly to the previous blocks.

- 4th Block.

  Similar to the previous blocks, this block consists of four convolutional layers (conv4_1, conv4_2, conv4_3, conv4_4) with 256 filters each, followed by batch normalization and ReLU activation.
  Max pooling and dropout are applied similarly to the previous blocks.

- 5th Block.

  The final block consists of four convolutional layers (conv5_1, conv5_2, conv5_3, conv5_4) with 512 filters each, followed by batch normalization and ReLU activation.
  Max pooling and dropout are applied similarly to the previous blocks.

- Output Layer.

  The last convolutional block's output is flattened (Flatten()), then probabilities for each of the seven facial expressions (the number of classes) are output using a fully connected layer (Dense()) using softmax activation.
  The number of classes, which in this case is seven, determines the number of output-classes.

The number of output classes is determined by the number of classes, which in this example is seven. This architecture is made up of several convolutional blocks, each of which has a number of convolutional layers with regularization features such as dropout layers, max pooling, ReLU activation, and batch normalization. The probabilities for various facial expressions are predicted by the final output layer.

2. *Frontend.* The front end of the Entertainment Recommendation and Rating System is a user-facing interface facilitating image capture, material suggestions, and comments. Constructed with Flask, a Python-based lightweight web application framework, it serves dynamic web pages, simplifying interactive web app development with features like routing and template rendering. HyperText Markup Language (HTML) organizes content and structure, defining elements such as headings and forms, while Cascading Style Sheets (CSS) enhance appearance and usability. Users can capture facial images via OpenCV, ensuring compatibility across devices and browsers, while content filtering and display provide

personalized recommendations based on detected emotions. This frontend offers a responsive and user-friendly interface for entertainment exploration.

3. Backend. The backend of the Entertainment Recommendation and Rating System acts as the computational engine for data processing, algorithm execution, and personalized recommendation generation based on users' emotional states. Leveraging technologies like OpenCV for face and landmark detection, TensorFlow & Keras for neural network model development, and NumPy for numerical computing, the system effectively processes data and manipulates information. Matplotlib enables informative data visualization, while ONNX Runtime facilitates efficient model inference across platforms. By utilizing these frameworks, the backend performs complex tasks efficiently, delivering personalized recommendations that enhance users' entertainment experience by aligning with their emotional states.
4. *Recommendation & Rating System.* The Recommendation System within the Entertainment Recommendation and Rating System is designed to offer personalized content recommendations based on users' facial emotions. Here's how the architecture works:

   - *Facial Emotion Detection Module:* This module captures the user's facial emotions using facial detection technology and analyzes facial expressions to determine their current mood or emotional state.
   - *Content Recommendation Engine:* Based on the detected facial emotions, the system generates personalized recommendations for movies, books, and songs that match the user's mood. Algorithms match the emotional state to relevant content categories stored in the system's database.
   - *User Interface:* The interface presents recommendations intuitively, allowing users to choose from movies, books, or songs based on their current mood through clickable buttons.
   - *Content Selection and External Integration:* After selecting a content category, users receive 12 recommended options with corresponding ratings, and clicking on a recommendation redirects them to an external website like IMDb, Spotify, or Amazon.
   - *Rating Collection System:* Users can rate their experience, with the rating process facilitated by capturing their facial emotions again. Positive emotions indicate a higher rating, while negative emotions correspond to a lower rating.
   - *Feedback Loop Implementation:* Collected ratings continuously refine and improve the system's algorithms. By analyzing user feedback and preferences, the system adapts recommendations to better align with users' evolving emotional states and content preferences (Fig. 7).

Overall, the architecture of the Recommendation System seamlessly integrates facial emotion detection, content recommendation algorithms, user interface design, external content integration, and feedback mechanisms to provide users with personalized and emotionally resonant content recommendations.

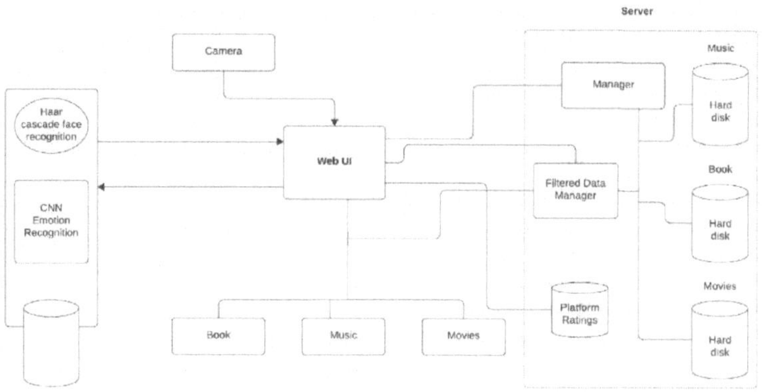

Fig. 7 Architecture of recommendation system

4 Results and Analysis

4.1 Trained Model Results

To efficiently propose entertainment items based on emotion, we developed a deep Convolutional Neural Network (CNN) model for facial emotion recognition. Using the FER2013 dataset with images sized at 48×48 pixels, we trained our model and monitored both validation and training loss at each epoch. Figure 8 illustrates the comparison of these losses, while Fig. 9 depicts validation and training accuracy across epochs. Our assessment of the learning curves shows positive results, with close alignment between validation and training loss curves across most epochs, indicating a favorable balance between underfitting and overfitting. Around the 40th epoch, we observed an upward trend and random fluctuations in the validation curve, coinciding with the plateauing of the training curve, suggesting that the training process reached a stable point. In other words, the model efficiently learns patterns in the training data and performs well on unseen validation data, achieving around 70% accuracy.

4.2 Proposed Approach Results

Upon landing on the Home page of our project as shown in Fig. 10, the user is prompted to scan their face for the system to capture, detect and predict the emotional status of the user. The detected emotion is then shown to the user and they are asked to choose their preferred medium of source of entertainment like movies, songs and books. The user then proceeds to select an option of his liking after which they receive a curated list of top-rated recommendations in line with their

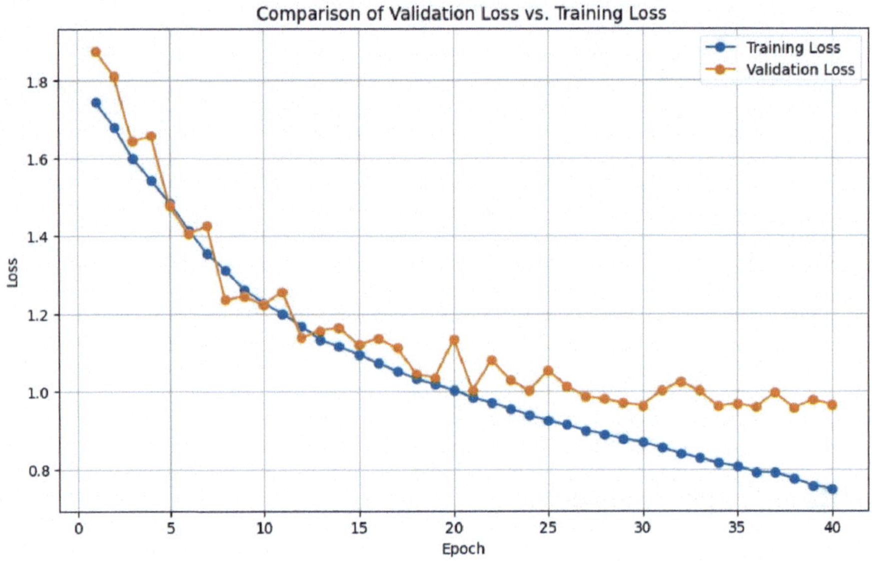

Fig. 8 Comparison of validation loss versus training loss

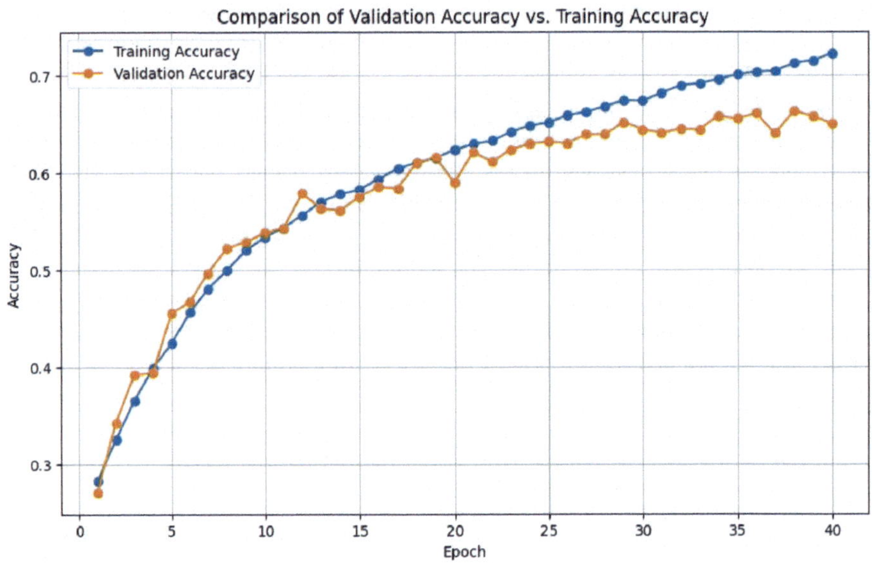

Fig. 9 Comparison of validation accuracy versus training accuracy

Fig. 10 Home page

Fig. 11 Predicted emotion (detected as 'happy')

emotional state. Figure 11 shows an example of the working of the system. The model detects an emotion of 'happy' which was found to be correct for the testing user. Figures 12, 13 and 14 shows the top-rated recommendations for movies, music and books respectively in line with their predicted-emotion.

5 Discussion and Conclusion

The Entertainment Recommendation and Rating System's effectiveness hinges on the accuracy of its facial emotion recognition model, ensuring reliable detection of users' emotional states for pertinent content suggestions. Analyzing recommendation algorithms provides insights into their ability to align users with emotionally relevant content, enhancing satisfaction through tailored preferences. Leveraging facial expression detection, the system accurately identifies users' emotional states, offering relevant content in real-time, and enhancing adaptability through various frameworks.

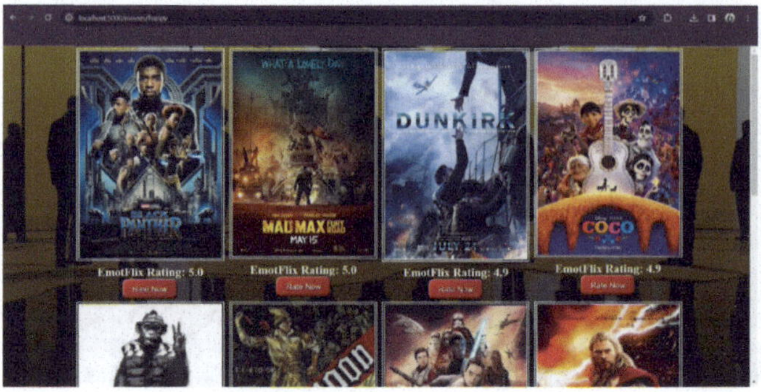

Fig. 12 Movie recommendation for emotion happy

Fig. 13 Song recommendation for emotion happy

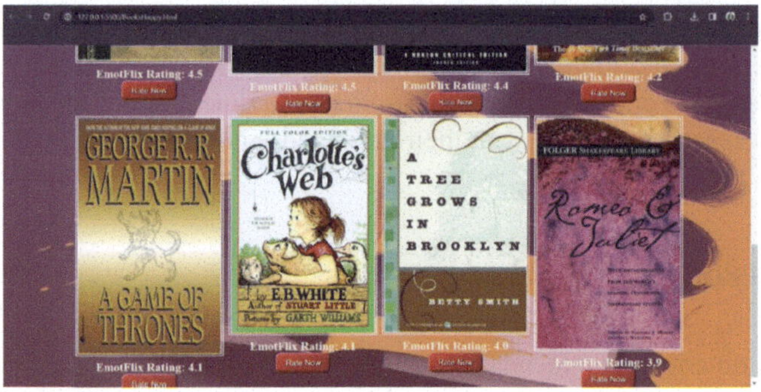

Fig. 14 Book recommendation for emotion happy

However, the system faces limitations such as reliance on facial expression recognition, potentially leading to errors due to cultural variations. Addressing data security and user privacy is crucial to maintaining confidence, necessitating robust measures to mitigate risks and increase trust. The Entertainment Recommendation and Rating

System pioneers personalized entertainment discovery, leveraging facial expression recognition and advanced algorithms to enhance user satisfaction and engagement. Continuous improvement is vital to overcome challenges and ensure a superior user experience.

The experiment's main conclusions emphasize the effectiveness of the emotion recognition model in identifying users' emotional states and the impact of personalized recommendations on engagement and satisfaction. Evaluation of the emotion recognition model shows promising results in accuracy, precision, and recall rates, indicating reliability in capturing users' emotional responses for tailored recommendations. User feedback provides insights into performance, usability, and impact, guiding future enhancements for increased adoption and retention.

6 Future Work

Exploring potential improvements and future directions for the Entertainment Recommendation and Rating System presents opportunities to enhance its performance, usability, and impact on users' entertainment discovery experience. Key areas for improvement include refining the emotion recognition model to enhance accuracy and robustness across diverse demographic groups and environmental conditions. Additionally, improving the recommendation algorithms to consider user preferences and contextual information can increase the relevancy and variety of recommendations.

Expanding the system's scope to encompass additional types of entertainment content and exploring personalized and customized recommendation approaches through user profiling and adaptive learning techniques are also crucial for future enhancements. By continuously innovating and iterating based on user feedback and emerging technologies, the system can maintain its position at the forefront of entertainment discovery, providing personalized, relevant, and engaging recommendations that delight and inspire users.

In conclusion, continuous innovation and iteration based on user feedback and emerging technologies are vital for the Entertainment Recommendation and Rating System to evolve into a more sophisticated, adaptive, and user-centric platform. By delivering personalized, relevant, and engaging recommendations, the system can enrich users' lives and enhance their overall satisfaction and enjoyment of entertainment content.

References

1. Happy, S.L., Routray, A.: Automatic facial expression recognition using features of salient facial patches. IEEE Trans. Affect. Comput. **6**(1), 1–12 (2015). https://doi.org/10.1109/TAFFC.2014.2386334

2. Kim, J.-H., Kim, B.-G., Roy, P.P., Jeong, D.-M.: Efficient facial expression recognition algorithm based on hierarchical deep neural network structure. IEEE Access **7**, 41273–41285 (2019). https://doi.org/10.1109/ACCESS.2019.2907327
3. Pranav, E., Kamal, S., Satheesh Chandran, C., Supriya, M.H.: Facial emotion recognition using deep convolutional neural network. In: 2020 6th International Conference on Advanced Computing and Communication Systems (ICACCS), Coimbatore, India, pp. 317–320 (2020). https://doi.org/10.1109/ICACCS48705.2020.9074302
4. Kartali, A., Roglić, M., Barjaktarović, M., Đurić-Jovičić, M., Janković, M.M.: Real-time algorithms for facial emotion recognition: a comparison of different approaches. In: 2018 14th Symposium on Neural Networks and Applications (NEUREL), Belgrade, Serbia, pp. 1–4 (2018). https://doi.org/10.1109/NEUREL.2018.8587011
5. Deshmukh, R.S., Jagtap, V., Paygude, S.: Facial emotion recognition system through machine learning approach. In: 2017 International Conference on Intelligent Computing and Control Systems (ICICCS), Madurai, India, pp. 272–277 (2017). https://doi.org/10.1109/ICCONS.2017.8250725
6. Gill, R., Singh, J.: A deep learning approach for real time facial emotion recognition. In: 2021 10th International Conference on System Modeling & Advancement in Research Trends (SMART), MORADABAD, India, pp. 497–501 (2021). https://doi.org/10.1109/SMART52563.2021.9676202
7. Shi, W., Wang, L., Qin, J.: Extracting user influence from ratings and trust for rating prediction in recommendations. Sci. Rep. **10**, 13592 (2020). https://doi.org/10.1038/s41598-020-70350-1
8. Kumar, S., De, K., Roy, P.P.: Movie recommendation system using sentiment analysis from microblogging data. IEEE Trans. Comput. Soc. Syst. **7**(4), 915–923 (2020). https://doi.org/10.1109/TCSS.2020.2993585
9. Balfaqih, M., Altwaim, A., Almohammedi, A.A., Yusof, M.H.M.: An ıntelligent movies recommendation system based facial attributes using machine learning. In: 2023 3rd International Conference on Emerging Smart Technologies and Applications (eSmarTA), Taiz, Yemen, pp. 1–6 (2023). https://doi.org/10.1109/eSmarTA59349.2023.10293398
10. Fessahaye, F., et al.: T-RECSYS: a novel music recommendation system using deep learning. In: 2019 IEEE International Conference on Consumer Electronics (ICCE), Las Vegas, NV, USA, pp. 1–6 (2019). https://doi.org/10.1109/ICCE.2019.8662028
11. Chang, S.-H., Abdul, A., Chen, J., Liao, H.-Y.: A personalized music recommendation system using convolutional neural networks approach. In: 2018 IEEE International Conference on Applied System Invention (ICASI), Chiba, Japan, pp. 47–49 (2018). https://doi.org/10.1109/ICASI.2018.8394293
12. Hasib, K.M., Tanzim, A., Shin, J., Faruk, K.O., Mahmud, J.A., Mridha, M.F.: BMNet-5: a novel approach of neural network to classify the genre of Bengali music based on audio features. IEEE Access **10**, 108545–108563 (2022). https://doi.org/10.1109/ACCESS.2022.3213818
13. Balakrishnan, S., Naulegari, J., Dharanyadevi, P., Manikumar, B.: Book recommendation system using matrix factorization with SVD. In: 2023 Second International Conference on Advances in Computational Intelligence and Communication (ICACIC), Puducherry, India, pp. 1–5 (2023). https://doi.org/10.1109/ICACIC59454.2023.10435212
14. Jayapal, C., Gokul, S., Harshavardhan S.V.: Book recommendation system using hybrid filtering. In: 2023 2nd International Conference on Advancements in Electrical, Electronics, Communication, Computing and Automation (ICAECA), Coimbatore, India, pp. 1–5 (2023) https://doi.org/10.1109/ICAECA56562.2023.10199590.
15. Nawar, A., Toma, N.T., Al Mamun, S., Kaiser, M.S., Mahmud, M., Rahman, M.A.: Cross-content recommendation between movie and book using machine learning. In: 2021 IEEE 15th International Conference on Application of Information and Communication Technologies (AICT), Baku, Azerbaijan, pp. 1–6 (2021). https://doi.org/10.1109/AICT52784.2021.9620432
16. Elias, T., Rahman, U.S., Ahamed, K.A.: Movie recommendation based on mood detection using deep learning approach. In: 2022 Second International Conference on Advances in Electrical, Computing, Communication and Sustainable Technologies (ICAECT), Bhilai, India, pp. 1–6 (2022). https://doi.org/10.1109/ICAECT54875.2022.9807654

17. Chauhan, S., Mangrola, R., Viji, D.: Analysis of Intelligent movie recommender system from facial expression. In: 2021 5th International Conference on Computing Methodologies and Communication (ICCMC), Erode, India, pp. 1454–1461 (2021). https://doi.org/10.1109/ICCMC51019.2021.9418421
18. Iyer, A.V., Pasad, V., Sankhe, S.R., Prajapati, K.: Emotion based mood enhancing music recommendation. In: 2017 2nd IEEE International Conference on Recent Trends in Electronics, Information & Communication Technology (RTEICT), Bangalore, India, pp. 1573–1577 (2017). https://doi.org/10.1109/RTEICT.2017.8256863
19. Tejaswini Priyanka, V., Reshma Reddy, Y., Vajja, D., Ramesh, G., Gomathy, S.: A novel emotion based music recommendation system using CNN. In: 2023 7th International Conference on Intelligent Computing and Control Systems (ICICCS), Madurai, India, pp. 592–596 (2023). https://doi.org/10.1109/ICICCS56967.2023.10142330
20. Krupa, K.S., Ambara, G., Rai, K., Choudhury, S.: Emotion aware smart music recommender system using two level CNN. In: 2020 Third International Conference on Smart Systems and Inventive Technology (ICSSIT), Tirunelveli, India, pp. 1322–1327 (2020). https://doi.org/10.1109/ICSSIT48917.2020.9214164
21. Negandhi, T., Dhumale, P., Vachhani, M., Gawande, K.: Bookbuddy: a mood based book recommendation system. In: 2023 7th International Conference On Computing, Communication, Control and Automation (ICCUBEA), Pune, India, pp. 1–7 (2023). https://doi.org/10.1109/ICCUBEA58933.2023.10392250
22. Fujimoto, T., Murakami, H.: A book recommendation system considering contents and emotions of user interests. In: 2022 12th International Congress on Advanced Applied Informatics (IIAI-AAI), Kanazawa, Japan, pp. 154–157 (2022). https://doi.org/10.1109/IIAIAAI55812.2022.00039
23. Gupta, M., Amirishetty, S.V., Nithin, B., Kurakula, H.: Mind frame music and movie recommendations to uplift the current mood using deep learning. In: 2021 International Conference on System, Computation, Automation and Networking (ICSCAN), Puducherry, India, pp. 1–6 (2021). https://doi.org/10.1109/ICSCAN53069.2021.9526435
24. Kumar, B.S.A., Kumar, P.M., Teja, G.V., Pavan, G., Shashank, G.: Deep Learning based content recommendation using facial emotions. In: 2023 7th International Conference on Trends in Electronics and Informatics (ICOEI), Tirunelveli, India, pp. 1412–1416 (2023). https://doi.org/10.1109/ICOEI56765.2023.10125608
25. Mariappan, M.B., Suk, M., Prabhakaran, B.: FaceFetch: a user emotion driven multimedia content recommendation system based on facial expression recognition. In: 2012 IEEE International Symposium on Multimedia, Irvine, CA, USA, pp. 84–87 (2012). https://doi.org/10.1109/ISM.2012.24
26. Chaugule, V., Abhishek, D., Vijayakumar, A., Ramteke, P.B., Koolagudi, S.G.: Product review based on optimized facial expression detection. In: 2016 Ninth International Conference on Contemporary Computing (IC3), Noida, India, pp. 1–6 (2016). https://doi.org/10.1109/IC3.2016.7880213
27. Chaurasiya, R.K., Sahu, U.: Improving performance of product recommendations using user reviews. In: 2018 3rd International Conference and Workshops on Recent Advances and Innovations in Engineering (ICRAIE), Jaipur, India (2018), pp. 1–4. https://doi.org/10.1109/ICRAIE.2018.8710414
28. Sharma, H., Sharma, D., Bhatt, K., Shah, B.: Facial emotion based review accumulation system. In: 2020 IEEE 17th India Council International Conference (INDICON), New Delhi, India, pp. 1–6 (2020). https://doi.org/10.1109/INDICON49873.2020.9342467
29. Zhu, Q., Zhuang, H., Zhao, M., et al.: A study on expression recognition based on improved mobilenetV2 network. Sci. Rep. **14**, 8121 (2024). https://doi.org/10.1038/s41598-024-58736-x
30. Huang, Z.Y., Chiang, C.C., Chen, J.H., et al.: A study on computer vision for facial emotion recognition. Sci. Rep. **13**, 8425 (2023). https://doi.org/10.1038/s41598-023-35446-4
31. Wayesa, F., Leranso, M., Asefa, G., et al.: Pattern-based hybrid book recommendation system using semantic relationships. Sci. Rep. **13**, 3693 (2023). https://doi.org/10.1038/s41598-023-30987-0

32. Keating, C.T., Cook, J.L.: The inside out model of emotion recognition: how the shape of one's internal emotional landscape influences the recognition of others' emotions. Sci. Rep. **13**, 21490 (2023). https://doi.org/10.1038/s41598-023-48469-8

Securing Embodied AI: Addressing Cybersecurity Challenges in Physical Systems

Fatema khalifa said ALSaidi, Sabah Ali AL'Abd AL-Busaidi, and B. Sundaravadivazhagan

Abstract In the digital era, artificial intelligence (AI) is a revolutionary force that holds unimaginable potential when it comes to cybersecurity. This chapter examines the interaction among artificial intelligence (AI) and cybersecurity, highlighting the critical role that AI techniques play in supporting security measures when more conventional systems fall short. The application of AI stands out as a game-changer in sectors where traditional security procedures are found to be ineffective, offering improved overall security performance against a variety of sophisticated cyber-attacks. Security considerations are of utmost importance, especially when it comes to embedded AI systems, where resilience and dependability are critical requirements. This chapter explores the complex world of security issues that are inherent to embodied AI systems, which are actual objects or robots that are deeply ingrained in our surroundings. These systems present distinct weaknesses that require specialized care, even though they hold great promise for breakthroughs in a variety of applications. The investigation includes a thorough examination of all possible dangers, ranging from worries about data integrity and privacy to the more concrete dangers of physical assaults. As we assign more complicated jobs to autonomous machines, it is critical to comprehend and reduce these dangers. The need for an all-encompassing approach to AI security is highlighted by the discussion's deeper exploration of the wider ethical implications of protecting autonomous humans. Along with practical remedies, the chapter offers a road map for defending embodied AI systems against hostile activity. This chapter is a useful resource for researchers, practitioners, and policymakers navigating the changing field of AI cybersecurity by breaking down security issues and offering workable solutions.

Keywords Artificial intelligence · Cyber security · Cyber-attacks · Embodied AI systems

---

F. khalifa said ALSaidi (✉) · S. Ali AL'Abd AL-Busaidi · B. Sundaravadivazhagan
Department of Information Technology, University of Technology and Applied Sciences-Al Mussanah, Muladdah, Oman
e-mail: fatema.alsaidi@utas.edu.om

P. Raj et al. (eds.), *Building Embodied AI Systems: The Agents, the Architecture Principles, Challenges, and Application Domains*, Information Systems Engineering and Management 14, https://doi.org/10.1007/978-3-031-68256-8_21

1 Introduction

The age where digitization never stops and AI is the leader we find ourselves standing at that point which was intersection between technology and physical world. This chapter begins a tour of the design challenges facing embodied AI systems, those tech smart assistants allowing us to smoothly live our everyday existence. In contrast with virtual services, these entities subject their users to a new set of security barriers that require further analysis into intricacies. In this story, we unfold the elaborate tango between AI and cybersecurity, highlighting these real problems when one incorporates intelligent agents into our built environment. This chapter does not merely analyze technical complexity but acts as a guide for securing AI's future in the physical world. In our efforts to ensure that systems are protected, it's also about protecting the autonomous individuals with whom these said systems interact and therefore humanizing AI.

1.1 Defining Embodied AI Systems

Embodied AI systems are not mere lines of code; they are the avant-garde pioneers of a tech revolution stepping boldly into the physical world. As we greet them, it is important to regard them as something more than technical marvels they are our neighbors, members of the community. [11] are our veteran narrators as they lead us through this developing story. As we confront this new reality where the digital and physical worlds converge, it becomes vital to understand their strengths, quirks, and inadequacies [5].

1.2 Significance of Security in Embodied AI

Security is the core of embedded AI systems [1, 2] which allows technology and human life to coincide. Values such as resilience and reliability are important which require a broad strategy that can be used to address any vulnerability in the system while preparing for potential threats. This AI to real-life partnership is an amalgamation of creativity and the safety measure that combines with each other for a peaceful coexistence.

In above diagram Embodied AI systems such as robots, autonomous vehicles, as well as smart appliances rely on a compound integration of physical in addition to digital components that make them vulnerable to cybersecurity threats. These systems use sensors to take in data about the surrounding environment. This sensor data gets processed by the core AI system that controls the system's behavior and movement through actuators and control logic. Most embodied systems also have network connectivity through Wi-Fi, Bluetooth, or cellular networks for communication and software updates. However, this connectivity exposes more avenues for potential cyber-attacks. Securing embodied AI systems is challenging because threats can manifest both digitally and physically. It requires building resilient control logic, hardening data pipelines and networks, continuous authentication of all components, and overall system designs that mitigate risk even when parts of the system are compromised. If vulnerability is found in any component, it threatens the safe and reliable operation of the AI system as a whole as well as the safety of human users.

1.3 Overview of Applications for Embedded AI

Embedded AI systems operate in various sectors, such as Healthcare and manufacturing industries requiring customized safety protocols [9]. It is the breadth of applications that become critical and guide in formulating security measures designed to suit each case, thereby minimization risks. Understanding the different contexts in which these systems function is a key to ensure that effective and relevant security measures are implemented.

1.4 Cybersecurity and AI's Convergence

The symbiosis of cybersecurity and AI begins, whereby the latter takes on twofold duties: defender and target [3]. This convergence mandates a deep understanding of unlocking strategies that increase security through AI's common features. It's a dance where AI under the guise of enhancing security is actually actively dealing with and mitigating cyber threats aimed at altering robot target systems.

1.5 Vulnerabilities in Systems with Embodied AI

While embodied AI systems provide unparalleled benefits, they present clear weaknesses. This chapter carefully analyses possibilities for threats, ranging from privacy and data integrity concerns to concrete risks of assaulting such systems physically [12]. This knowledge enables researchers, practitioners, and policymakers to navigate through the complex terrain of cybersecurity like a compass.

1.6 Key Challenges in Designing Embodied AI Systems

Before proceeding to the detailed consideration of design peculiarities of embodied AI systems, defining application cases and essential tasks that pave our way has primary importance. For instance, picture an embodied agent transiting smoothly through a challenging non-structured environment and reacting to dynamic changes in its surroundings while engaging with it. Imagine an agent navigating through difficult and potentially hazardous unlevel or uneven terrain, reacting to sensory inputs that reflect the reality of physical nature [5].

1.6.1 Sensory Realism

The adversary that our protagonist seems to face is not powerful than the limitations of current sensor technologies dealing with problems related by resolution, accuracy and resilience. The battlefield extends as the challenge of aggregating numerous sensor modalities arises with a careful balance between capabilities and processing power limits. In this technological standoff, the hero has to strike a delicate balance between extracting sensor data's wealth and meeting the practical constraints of processing resources. The story tells of the fight for balance and ultimately victory in embodied AI systems through sensor efficiency having overcome computational limitations.

1.6.2 Real-World Navigation

In the elaborate dance of embodied AI systems, passing through complex environments that have no structure demands a highly coordinated group of algorithms. The hero has to face challenges of responding in real-time, turning each moment into a subtly different step within the larger principle in algorithmic choreography. Challenge layers are woven in by the challenges of explorations and irregular terrains as each decision is a step on this platform of indeterminacy. The story is sketched out with the agent, like a masterful dancer but here translating algorithmic accuracy into almost flawless performance on an always changing stage of reality.

1.6.3 Safety and Ethics

Building ethical embodied AI systems requires a strong basis, which includes fail-safe designs, thorough validation testing and dealing with security loopholes in interrelated systems. This ethical spine guarantees the reliability of an agent and provides protection from threats. An elaborate ethical landscape is developed by such directives for proper data collection, respect towards privacy and obtaining consent which dictate the entire performance of this ballet. Therefore, as the protagonist negotiates through interconnected landscape ethic norms constitute a guiding force stressing designer and developers in creating smart agents that respect user privacy, security is multi-stakeholder in genesis of self-choreographed technology.

1.6.4 Complexity Versus Scalability

In the embodiment narrative of AI systems, complexity management and intricate balances between customization versus generalization may appear to be challenging. The character's trip is made somewhat more complicated by scaling down to cost-effective forms. The turning point of this story is the subtle interplay between reach depth and practical availability by an agent. Achieving this balance serves as the protagonist's breakthrough, where technological advancement meets real-life limitations. It is a story of crossing the tightrope between elegance and utility, crafting a narrative that finds balance in designing for intelligent agents on an inclusive field of technologies [13, 14].

1.6.5 Hardware Design

The design saga comes to its final chapter and the focus shifts toward an embodiment of our concrete hardware. Constraints lead here in such features as size, weight, the number of degrees of freedom and power play important roles that define what a subject is. In the end, there is a true finesse as production cost, search for special materials along with rigorous requirements of quality are truly highlighted. This is an embodiment that rises from being a mere abstract idea to become the essence of material tech industry into which we design our narrative at its peak.

2 Security Threats and Vulnerabilities

The introduction of embodied AI systems into our physical surroundings within the dynamic world of the artificial intelligence (AI) raises a slew of security concerns. This exploration delves into the battlefield of threats and vulnerabilities faced by these systems, each demanding a unique defense strategy. Leveraging insights from recent research, this chapter strives to offer a thorough understanding of these challenges and the corresponding strategies for robust fortification.

2.1 Physical Attacks and Tampering

Picture this: embodied AI systems, like futuristic sentinels, standing guard in our physical spaces. Yet, this very tangible presence exposes them to a unique vulnerability physical attacks and tampering. [11] act as our seasoned generals, conducting a thorough reconnaissance of this battlefield. Their recent research offers a comprehensive review of physical threats, acting as a strategic guide for formulating preventive measures. Physical attacks on embodied AI aren't theoretical, they're tangible risks demanding a robust defense, akin to securing a fortress against potential infiltrators. Smith and Johnson's insights illuminate potential breach points, helping us reinforce the defense lines. Their research acts as a shield, protecting these AI sentinels from the tangible dangers lurking in the physical world.

2.2 Privacy and Data Security

In the interconnected landscape of embodied AI, data is the lifeblood. [9] emerge as the guardians of this lifeblood, contributing invaluable insights into privacy-preserving techniques. Consider them to be diligent gatekeepers, standing at the crossroads of data integrity and privacy issues. Addressing concerns about unauthorized access and data breaches is about more than simply securing information,it's about protecting the fundamental core of trust. The study of Patel et al. digs into the delicate tango between the requirement for data accessibility as well as the need of privacy. It's like crafting a delicate balance between transparency and protection. Their insights inform strategies, not just for safeguarding sensitive information, but for nurturing an environment where users can trust that their data is in capable hands.

2.3 Communication Interception

Now, imagine embodied AI systems communicating with each other, orchestrating a symphony of information. However, this communication, if intercepted, becomes vulnerability. [8] take on the role of architects, designing fortifications for these communication pathways. Their research on securing communication channels in swarm robotics is like a blueprint for building secure communication highways in the realm of embodied AI. Think of it as laying down secure pipelines where information flows freely but remains shielded from prying eyes. Park et al.'s insights guide the construction of these secure communication pathways, ensuring that the vital information exchanged between embodied AI systems remains confidential and un-tampered.

2.4 *Adversarial Machine Learning*

In the world of cyber threats, adversarial machine learning emerges as a sophisticated adversary, targeting the very sensors that form the sensory backbone of embodied AI. Chen et al. [1, 2] take on the roles of cyber tacticians, exploring the intricacies of these attacks. Their research lays bare the evolving landscape of adversarial threats on physical sensors in robotic systems.

Imagine defending against adversaries who constantly adapt and evolve. Chen et al.'s exploration is like studying the enemy's playbook, understanding their tactics, and formulating a counter-strategy. But defense against adversarial attacks is not a one-size-fits-all approach. Wang et al. [13, 14] step in as the AI strategists, demonstrating that a deep learning approach is essential. It's akin to deploying an advanced AI defense system that evolves alongside the adversaries, adapting to new threats in real-time.

As we navigate this intricate battlefield of threats and vulnerabilities, it becomes evident that a unified defense is paramount. Just as a medieval fortress required strong walls, vigilant guards, and an evolving strategy, embodied AI systems demand a holistic approach. Imagine this defense as a tapestry woven with threads of preventive measures, privacy-preserving techniques, secure communication channels, and adaptive AI strategies. Smith and Johnson, Patel et al., Park et al., Chen et al., and Wang et al. collectively act as the architects and generals, contributing to the construction of this fortified defense. The insights from each research piece are not isolated strategies; they're interconnected components of a resilient defense mechanism. It's about creating an ecosystem where embodied AI systems can thrive, free from the shackles of threats, and where users can trust in the security of the AI-infused world.

3 Risk Assessment in Embodied AI

Embodied AI systems evolve as transformational creatures, smoothly integrating into our physical environment, in the delicate dance between artificial intelligence (AI) and the real world. With this integration, however, comes an urgent requirement for a detailed awareness of security concerns. This chapter dives into risk assessment in embodied AI, utilizing frameworks proposed by Garcia and Rodriguez [3] and insights from [12] to navigate the complexities and prioritize mitigation strategies.

In the Fig. 1 shows Embodied AI systems such as robots and autonomous vehicles interact closely with the physical world, which necessitates comprehensive risk assessment across every component. Evaluating risks from potential sensor malfunctions, actuator errors, software flaws in perception/reasoning, and overall system behaviors is crucial for safety. Continuous simulation of risks the AI system may pose to humans in its environment is also critical. Safety validation accounting for changing environments needs to occur before and during live system operation. As

Fig. 1 Embodied AI systems

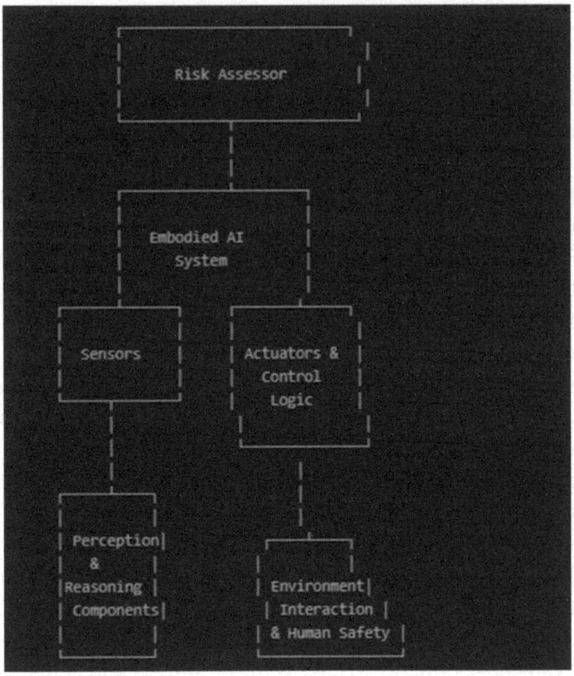

embodied systems fuse the digital and physical in complex ways, holistic risk analyses spanning technological capabilities, real-world interactions, and human factors are essential to mitigate hazards.

3.1 Frameworks for Evaluating Security Risks

Security risks in embodied AI are not to be taken lightly, and Garcia and Rodriguez [3] provide a crucial framework for structured risk assessment. Consider this framework to be a road map that will lead us through the terrain of possible hazards. Organizations and researchers may use Garcia and Rodriguez's methodology to discover, assess, and prioritize security concerns in embodied AI systems, much as a tourist depends on a thorough map to traverse unknown territory.

A systematic risk assessment is similar to putting a light on possible flaws. Organizations receive a comprehensive understanding of the risks they face by using Garcia and Rodriguez's approach. It is not only about dealing with recognized hazards, but also about predicting those that may lie in the shadows. The organized method provides for a thorough knowledge, ensuring that no possible danger is overlooked. In a world where technology is continuously evolving, having a dynamic risk assessment methodology is critical. Garcia and Rodriguez's work shines a light on the way to a resilient security posture. It's about being proactive in preparing for prospective problems, similar to how a strong chess player anticipates the opponent's movements.

3.2 *Identifying Critical Assets and Functions*

[12] emphasize the importance of understanding critical assets and functions in the realm of embodied AI. Imagine this as identifying the key players and vital components in a complex orchestra. Knowing which instruments are fundamental allows for targeted strategies to ensure the harmony of the entire composition. In the context of embodied AI, critical assets and functions are the pillars supporting the system's functionality. Tan et al.'s cyber-physical systems viewpoint gives a comprehensive picture, acknowledging the interdependence of digital and physical factors. It's not just about conserving lines of code, it's about preserving the core of functionality that makes embodied AI systems worthwhile. The cyber-physical systems concept shifts risk identification from a general procedure to a customized strategy. Like a doctor diagnosing a patient, understanding critical assets allows for precise prescriptions for security measures. Tan et al.'s insights empower organizations to allocate resources efficiently, ensuring that the most critical aspects receive heightened protection.

As we navigate the landscape of risk assessment in embodied AI, the collaboration of frameworks by Garcia and Rodriguez [3] and insights from [12] becomes paramount. Consider it the creation of a complete defense plan, one that not only detects possible dangers but also surgically prioritizes preventative actions. This comprehensive strategy is analogous to constructing a castle with numerous levels of protection in the ever-changing terrain of cybersecurity. The outer walls are formed by Garcia and Rodriguez's framework, which provides a strong basis for risk assessment. Tan et al.'s insights serve as the inner sanctum, indicating essential assets and functions that must be strengthened. Organizations and researchers establish a synergy that transcends individual contributions by incorporating these concepts. It's not just about identifying dangers in isolation, it's about composing a symphony of security solutions that work in tandem with the distinct characteristics of embodied AI systems.

4 Countermeasures and Security Measures

The incorporation of embodied AI systems into our physical environment offers a dynamic area of security problems in the ever-changing paradigm of artificial intelligence (AI). As we dive into the complexities of safeguarding these organizations, a multi-pronged strategy is essential. This chapter delves into a toolbox of countermeasures including security measures based on current research to strengthen the resistance of embodied AI systems against cyber-attacks.

4.1 Encryption and Secure Communication Protocol

[15] emphasize the critical function of encryption and secure communication methods in guarding against cyber-attacks. Focusing on wireless communication in robotic systems, their insights illuminate the significance of these measures. Much like the protective shield of encryption around sensitive information, securing communication channels becomes an impervious barrier, preserving the integrity of data within embodied AI systems.

Implementing these measures is akin to fortifying the communication highways that connect AI entities. Zhu and Li's recommendations serve as a guide, enabling organizations to erect virtual impenetrable walls against potential cyber adversaries. Just as a fortified fortress deters intruders, the integration of encryption and secure communication protocols ensures that the vulnerabilities in wireless communication become formidable barriers. This proactive defense not only safeguards the information exchanged by embodied AI systems but also reinforces the resilience of the entire network, fostering a secure environment in the face of evolving cyber threats.

4.2 Intrusion Detection and Prevention Systems

In the defense against potential threats, [10] play a crucial role by presenting a comprehensive survey of intrusion detection systems customized for embodied AI. Envision these systems as vigilant guardians, tirelessly scanning the environment for any signs of unauthorized access. The insights provided by Rodriguez and team offer a thorough understanding of effective mechanisms, forming a critical part of the defense playbook.

These intrusion detection and prevention systems act as sentinels, standing guard to ensure that only authorized entities gain access to the sensitive domains within embodied AI systems. Much like watchful protectors, their role extends beyond detection; they actively prevent potential intrusions, fortifying the security posture of the entire system. Rodriguez et al.'s survey serves as a guiding light for organizations, equipping them with the necessary knowledge to deploy these guardians strategically. Through this knowledge, organizations can erect an effective barrier against unauthorized access, fostering a secure environment for embodied AI systems to operate with confidence.

4.3 Secure Firmware and Software Development

The security of an embodied AI system is intricately woven into the fabric of its firmware and software, as emphasized by [5]. Their insights stress the significance of secure development practices, advocating for the integration of security measures throughout the entire development lifecycle. Picture this as imbuing the very code that dictates the actions of autonomous entities with a security consciousness.

Secure firmware and software development serve as the protective armor for these AI systems. Kim et al.'s insights act as a blueprint, guiding the creation of this armor to not only withstand known threats but also evolve to counter new and unforeseen challenges. Much like a living, adaptive shield, the proactive approach to development becomes a cornerstone in crafting a resilient defense. In the dynamic realm of cyber security, where threats constantly evolve, the incorporation of security measures at the developmental core ensures that these AI entities are equipped to navigate the ever-changing landscape with robust and adaptive resilience.

4.4 Ethical Considerations in AI Security

In the quest to fortify the defenses of embodied AI, [4] illuminate the indispensable role of a multidisciplinary perspective, particularly in ethical considerations. Their insights guide us into a realm where security measures are entwined with ethical principles, emphasizing the imperative for responsible AI deployment. Picture this multidisciplinary perspective as a compass, steering the trajectory of security measures to ensure alignment with societal values and norms. The balance between security and ethics is not a mere consideration but a fundamental imperative for the responsible integration of AI into our lives. Johnson et al.'s exploration serves as an invitation to organizations and researchers to navigate this delicate equilibrium. It asks for a concerted effort to ensure that security measures do more than just safeguard systems and data, they have to actively contribute to and improve AI's ethical fabric [6].

This interdisciplinary approach invites stakeholders to think about more than just the technical issues of security. It refutes the notion that security and ethics are mutually contradictory, arguing that a harmonic integration of the two is required to establish a trustworthy and responsible AI environment. Organizations may manage the challenges of AI deployment by following this ethical compass, enabling innovation while preserving the ethical ideals that underlie a technologically sophisticated society.

4.5 Strategies for Mitigation and Countermeasures

In the realm of cyber threats, where change is constant, Chen et al. [1, 2] inject a crucial perspective into the discourse that of resilient embodied AI from a cyber-physical systems standpoint. Envision this perspective as a strategic outlook, a panoramic view where mitigation strategies synchronize with the dynamic nature of evolving cyber threats. Mitigation strategies are not static safeguards but dynamic responses, akin to a choreographed dance, adapting to the ever-changing threat landscape. Chen et al.'s contributions act as a compass, providing organizations with a directional guide to formulate strategies that extend beyond merely reacting to current threats. Instead, these strategies anticipate and prepare for the unseen challenges on the horizon.

It's a strategic dance where organizations take the lead, adjusting their steps in harmony with the rhythm of evolving cyber threats. Chen et al.'s insights catalyze a proactive approach, urging organizations to be anticipatory rather than reactionary. In this dynamic dance of cybersecurity, the ability to foresee and prepare for future threats becomes the hallmark of a resilient defense. By adopting this strategic perspective, organizations can navigate the intricate patterns of cyber threats, ensuring that their mitigation strategies remain not only effective but also ahead of the evolving threat curve.

4.6 Techniques for Encryption and Privacy Preservation

In the pursuit of fortifying embodied AI, Wang et al.'s [13, 14] groundbreaking exploration of machine learning for anomaly detection establishes a foundation for techniques that ensure robust encryption and privacy preservation. Picture these techniques as dynamic tools, continuously refining and adapting encryption mechanisms to safeguard privacy while countering emerging threats. Wang et al.'s work unfolds like an evolving textbook, offering a repertoire of techniques that elevate the encryption and privacy preservation capabilities of embodied AI systems. The continuous evolution of these techniques aligns with the ever-changing landscape of cyber security, ensuring that the defense mechanisms remain not only effective but also ahead of potential threats.

Imagine machine learning algorithms as vigilant guardians of privacy. As they detect anomalies in the vast data streams processed by embodied AI systems, they actively intervene to preserve the confidentiality of sensitive information. This proactive stance position of machine learning algorithms as dynamic defenders is capable of upholding privacy even in the face of sophisticated and evolving cyber threats. Wang et al.'s contribution, akin to a guidebook, equips organizations with the knowledge to employ machine learning for anomaly detection effectively. By adopting these techniques, organizations can ensure that their embodied AI systems not only thrive in innovation but also stand resilient against potential breaches, embodying a future where privacy is a non-negotiable aspect of AI integration.

4.7 Physical Threat Detection Through Intrusion

In the landscape of embodied AI, [6] underscore the critical necessity for physical threat detection. Their insights emphasize the pivotal role of intrusion detection systems as the frontline defense against physical threats, envisioning these systems as vigilant guards armed with sensors to identify and thwart any potential physical intrusion. The concept of physical threat detection transcends the virtual space, reaching into the very essence of these AI systems' physical existence. Kumar et al.'s insights serve as a valuable guide for organizations, offering the knowledge required to implement intrusion detection systems that extend beyond the digital realm. These systems become active protectors, ensuring the tangible presence of embodied AI entities is safeguarded against physical threats.

Imagine these intrusion detection systems as sentinels standing guard over the physical domain of embodied AI, equipped not only to detect but also to actively prevent unauthorized access. Kumar et al.'s contribution to this realm acts as a beacon, illuminating the path for organizations to fortify the physical security of their AI systems. By implementing these insights, organizations can ensure a holistic defense, where the protection of both the virtual and physical aspects of embodied AI becomes integral to a resilient security strategy.

5 Case Studies and Real-World Examples

5.1 Security Incidents in Embodied AI Systems

The growing integration of artificial intelligence (AI) into our physical environments needs a close consideration of security issues in embodied AI systems. Various academics' case studies give significant insights into real-world circumstances, giving light on the ramifications of security flaws within these complex systems. Smith and Johnson [11] highlight one significant case study that goes into a security breach using a physically installed robot that is vulnerable to a cyber-attack. This incident underscores the tangible risks associated with embodied AI, emphasizing the critical need for a robust defense mechanism in the physical world.

Chen et al. [1, 2] contribute another case study focusing on adversarial attacks on physical sensors in robotic systems. This study illuminates the evolving landscape of cyber threats, emphasizing the vulnerabilities inherent in the physical components of embodied AI. The instances mentioned by Chen et al. highlight the necessity for adaptive security solutions to thwart hostile assaults on these systems' physical infrastructure. These real-world cases serve as cautionary stories, emphasizing the varied nature of embodied AI security issues. These occurrences, which range from physical weaknesses to sophisticated cyber-attacks, highlight the dynamic and complicated nature of the risks that these systems confront.

5.2 Lessons Learned and Best Practices

In analyzing security incidents within embodied AI systems, researchers and practitioners derive valuable insights to formulate best practices. Integrating these practices into security protocols becomes imperative, offering a proactive approach to enhance the overall resilience of embodied AI systems. Patel et al. [9] provide insights into privacy-preserving techniques for embodied AI applications. Their case study reveals instances where lapses in data security and privacy led to tangible consequences. A key lesson emerges—the critical importance of incorporating encryption and secures communication protocols. Best practices derived from this lesson include regular assessment and upgrading of encryption methods to stay ahead of emerging threats.

Wang et al. [13, 14] provide crucial insights into defending embodied AI systems against adversarial attacks through a deep learning approach. Their case study underscores the necessity for continuous monitoring and adaptation of defense mechanisms in the face of evolving cyber threats. In the pursuit of best practices, Wang et al.'s findings advocate for the integration of machine learning-based anomaly detection systems. These systems act as dynamic guardians, evolving in tandem with the ever-changing threat landscape. By employing continuous monitoring and adaptive defense measures, organizations can proactively identify and counter adversarial attempts on embodied AI systems. Wang et al.'s work serves as a beacon for establishing robust security practices, emphasizing the importance of staying ahead in the cybersecurity arms race. The integration of machine learning-based anomaly detection not only enhances the immediate defense capabilities of embodied AI systems but also positions organizations to anticipate and respond effectively to emerging adversarial tactics.

Garcia and Rodriguez [5] provide a framework for assessing security risks in embodied AI systems, emphasizing the identification of critical assets and functions. Their case study elucidates instances where a lack of comprehensive risk assessment led to vulnerabilities. Best practices derived from this lesson include the implementation of thorough risk assessment frameworks that consider both virtual and physical components of embodied AI. Incorporating these lessons into best practices forms a proactive defense strategy. Regular risk assessments, continuous monitoring, adaptive defense mechanisms, and privacy-preserving techniques collectively contribute to a security posture that not only reacts to incidents but actively prevents and mitigates potential threats.

In general, within the exploration of security incidents and the subsequent derivation of lessons learned and best practices in embodied AI systems offer a holistic understanding of the challenges faced by these systems. By addressing vulnerabilities, understanding the dynamic threat landscape, and implementing proactive security measures, organizations can fortify the resilience of embodied AI, ensuring its safe integration into various applications. The multifaceted nature of these lessons and best practices underscores the need for a comprehensive and adaptable approach to security in the ever-evolving field of embodied AI.

6 Future Trends and Emerging Challenges

6.1 Evolution of Security Threats

Anticipating the evolution of security threats is critical in the dynamic environment of cybersecurity for the creation of proactive defensive systems. Ongoing research, such as that done by [5], is critical in offering insights into new threats and guiding adaptive security methods. The study of Kim et al. offers light on the ever-changing nature of cyber threats, underlining the need of constant innovation in protection methods. Organizations may strengthen their security posture by studying the strategies and procedures used by attackers. This study not only detects existing hazards, but also forecasts the path of future issues, allowing for the proactive creation of responses. Security concerns in the context of embodied AI systems are evolving beyond traditional cyber-attacks. The incorporation of AI into physical things brings new vulnerabilities, necessitating a more in-depth awareness of possible dangers. The observations of Kim et al. serve as a road map, leading businesses through the complex environment of developing security concerns in the field of embodied AI.

6.2 Advances in AI Security Research

Keeping up with advancements in AI security research is critical for firms seeking to remain ahead of cyber threats. The thorough assessment of security with autonomous robots conducted by [7] adds greatly to a holistic knowledge of the emerging security environment. Li and Kim look into the complexities of protecting autonomous robots, an area where AI functions in a dynamic and sometimes unexpected environment. Their poll not only reveals existing security vulnerabilities, but also new solutions to these problems. Organizations may get insights into cutting-edge technologies and approaches that improve the robustness of their security procedures by investigating breakthroughs in AI security research. The poll underlines the multidisciplinary character of AI security, acknowledging the need of cooperation among AI professionals, cybersecurity experts, especially roboticists. This collaborative approach enables a thorough grasp of the security implications of autonomous robots, making it easier to create effective and adaptable security solutions.

6.3 Regulatory and Ethical Implications

The regulatory and ethical implications of AI security are crucial considerations that must not be disregarded. The examination of these factors by [4] emphasizes the significance of a balanced approach. Navigating legal frameworks and ethical concerns is critical for assuring responsible AI implementation in security scenarios.

Because AI technologies are rapidly evolving, regulatory frameworks must be developed in tandem to protect against possible exploitation. Johnson et al.'s findings give a thorough evaluation of current regulatory systems and ethical issues, as well as suggestions on how firms may align their AI security procedures with legal and ethical norms. The ethical component is especially essential in the context of privacy, where the use of AI has far-reaching ramifications for privacy, autonomy, including human rights. Johnson et al.'s investigation encourages firms to take a multidisciplinary approach to AI security, including not just the technological issues but also the social effect and ethical implications.

7 Conclusion

To sum it up, protecting embodied AI systems requires a multidimensional strategy that combines technological answers, ethical concerns, and regulatory compliance. The insights gained from this chapter's extensive study and research serve as a road map for academics, practitioners, especially policymakers navigating the changing area of AI cybersecurity. By humanizing the conversation around safeguarding embodied AI, we highlight the need to secure not just systems but additionally the autonomous individuals who engage with them. This chapter is a wonderful resource, giving a comprehensive grasp of security concerns as well as practical ways to secure AI's future in the physical environment.

References

1. Chen, X., et al.: Adversarial attacks on physical sensors in robotic systems (2023)
2. Chen, Y., et al.: Towards resilient embodied AI: a cyber-physical systems perspective (2023)
3. Garcia, M., Rodriguez, A.: A framework for assessing security risks in embodied AI systems (2023)
4. Johnson, M., et al.: Ethical considerations in securing embodied AI: a multidisciplinary perspective (2023)
5. Kim, J., et al.: Secure firmware development for autonomous robots (2023)
6. Kumar, S., et al.: Security challenges in human-robot interaction: a case study (2023)
7. Li, W., Kim, Y.: Towards trustworthy AI: a survey of security in autonomous robots (2023)
8. Park, S., et al. (2023). "Securing Communication Channels in Swarm Robotics."
9. Patel, R., et al.: Privacy-preserving techniques for embodied AI applications (2023)
10. Rodriguez, P., et al.: Privacy challenges and solutions in human-robot interaction (2023)
11. Smith, J., Johnson, A.: Securing embodied AI: a comprehensive review (2023)
12. Tan, Z., et al.: Physical layer security for wireless communication in robotic systems (2023)
13. Wang, L., et al.: Defending against adversarial attacks in embodied AI: a deep learning approach (2023)
14. Wang, Y., et al.: Machine learning for anomaly detection in embodied AI systems (2023)
15. Zhu, H., Li, Q.: A survey on intrusion detection systems for embodied AI (2023)

Fatema khalifa said ALSaidi holds a Master's degree in Computer Science from the University of Bedfordshire, UK. She currently serves as faculty in the Department of Information Technology at the University of Technology and Applied Science - Al Mussanah in Oman. With over 8 years of academic and research experience across multiple institutions, she is actively engaged in research and has contributed to two book chapters, one of which is published in Springer. Her research interests encompass Cyber Security, Artificial Intelligence, Machine Learning, Deep Learning, and Cloud Computing. Academic URL: https://spd.act.edu.om/index.aspx?XY=3797, LinkedIn: https://www.linkedin.com/in/fatma-alsaidi-5688a1117.

Sabah Ali AL'Abd AL-Busaidi holds a Master's degree in Computer Science from the Sultan Qaboos University, Oman. She currently serves as Head of Innovation and Technology Transfer Department at the University of Technology and Applied Science - Al Mussanah in Oman. With over 8 years of academic and research experience across multiple institutions, she is actively engaged in research and has contributed to two book chapters, one of which is published in Springer. Her research interests encompass Cyber Security, Artificial Intelligence, Data Sciences, IoT, and Cloud Computing. Academic URLETIMAD (act.edu.om), LinkedIn: https://www.linkedin.com/in/sabah-al-busaidi-30b418b9ORCID ID: http://orcid.org/0009-0002-4603-0128.

B. Sundaravadivazhagan has obtained Ph.D. in computer science from Anna University, Chennai India. Currently, he is faculty in the Department of Information Technology at the University of Technology, and Applied Science-AL Mussanah in Oman He has professional memberships in ISACA and ISTE and is a senior member of IEEE. His academic and research background spans more than 23 years at many institutions. Working on two financed research projects for the Ministry of Higher Education Research Innovation in Oman via TRC. His research interests include the Cyber Security, Artificial Intelligence and machine learning, deep learning, and cloud computing. He has published more than 55 technical articles in journals and conferences throughout the world. Orcid Id: https://orcid.org/my-orcid?orcid=0000-0002-5515-5769, Academic URL: https://www.act.edu.om/staffdirectory/index.aspx?XY=496LinkedIn: https://www.linkedin.com/in/dr-b-sundaravadivazhagan-a3562b136/Google Scholar: https://scholar.google.com/citations?user=gWbzTjMAAAAJ&hl=en.

The 5G Era: Transforming Connectivity and Enabling New Use Cases Across Industries

Maad M. Mijwil, Mostafa Abotaleb, and Pushan Kumar Dutta

Abstract The advent of 5G technology marks a significant leap in connectivity, promising to revolutionize both consumer and industrial applications. This paper provides an overview of the 5G landscape, focusing on its key applications and potential impact across various sectors. According to the 3rd Generation Partnership Project (3GPP), 5G primarily enhances three critical areas: Enhanced Mobile Broadband (EMBB), Ultra-Reliable Low-Latency Communication (URLLC), and Massive Machine-Type Communication (MMTC). These applications are expected to be adopted in waves, with EMBB leading the way, followed by URLLC and MMTC. While 5G presents significant opportunities, it's crucial to distinguish between myth and reality. Contrary to some beliefs, 5G won't immediately close 4G coverage gaps, isn't always necessary in the short term, and is not currently required for autonomous driving. However, particularly in industrial settings, 5G represents more than just an incremental upgrade from 4G. The analysis identified over 60 distinctive B2B use cases across various industries where 5G will be the preferred communication technology. These span sectors such as Industry 4.0, smart cities, smart energy, connected offices, smart security, and connected health. Each of these areas stands to benefit from 5G's enhanced capabilities, enabling new applications and improving existing ones. As 5G technology continues to evolve, it's expected to enable innovative use cases and transform connectivity across these sectors. This presents significant opportunities for businesses to innovate and improve their operations. However, the full realization of 5G's potential will depend on factors such as chipset availability, network deployment, and regulatory evolution.

M. M. Mijwil (✉)
Computer Techniques Engineering Department, Baghdad College of Economic Sciences University, Baghdad, Iraq
e-mail: mr.maad.alnaimiy@baghdadcollege.edu.iq

M. Abotaleb
Department of System Programming, South Ural State University, Chelyabinsk, Russia
e-mail: abotalebmostafa@bk.ru

P. K. Dutta
School of Engineering and Technology, Amity University Kolkata, Kolkata, India

P. Raj et al. (eds.), *Building Embodied AI Systems: The Agents, the Architecture Principles, Challenges, and Application Domains*, Information Systems Engineering and Management 14, https://doi.org/10.1007/978-3-031-68256-8_22

Keywords Artificial intelligence · 5G mobile · Communications · Antenna · Virtual reality

1 Introduction

The dawn of the 5G era heralds a new chapter in the evolution of mobile communications, promising unprecedented connectivity, speed, and transformative potential across industries. As we stand at this technological crossroads, the integration of Artificial Intelligence (AI) with 5G emerges as a crucial nexus, poised to redefine the boundaries of what's possible in our increasingly connected world. This chapter explores the symbiotic relationship between AI and 5G, delving into how this powerful convergence is set to revolutionize not only our digital interactions but also the very fabric of our physical reality through the lens of embodied AI.

Embodied AI, a field that seeks to ground artificial intelligence in physical or simulated bodies, stands to benefit enormously from the capabilities of 5G networks. The ultra-low latency, high bandwidth, and massive connectivity offered by 5G create an ideal environment for embodied AI systems to flourish. From autonomous robots navigating complex urban environments to AI-powered augmented reality experiences that seamlessly blend the digital and physical worlds, the potential applications are as vast as they are transformative. The integration of AI into 5G systems goes beyond merely enhancing network performance; it promises to create a new paradigm of intelligent connectivity. AI algorithms can optimize network resource allocation in real-time, predict and mitigate potential disruptions, and enable dynamic spectrum sharing, all of which are critical for supporting the diverse and demanding applications of embodied AI. Moreover, edge computing, a key component of 5G architecture, allows for the distribution of AI processing closer to the end-users, enabling faster response times and more efficient use of network resources. The convergence of Artificial Intelligence (AI) and 5G technology marks a pivotal moment in the evolution of mobile communications, offering unprecedented potential to redefine connectivity and unlock a myriad of possibilities for users and industries alike [1–3]. As we stand on the cusp of this transformative era, it becomes imperative to explore the synergistic relationship between AI and 5G, understanding how the fusion of these cutting-edge technologies is poised to revolutionize the way we interact, communicate, and innovate in the digital age. 5G, the fifth generation of cellular networks, promises lightning-fast speeds, ultra-low latency, and massive connectivity, laying the foundation for a hyper-connected world where devices, machines, and sensors seamlessly communicate in real-time. Meanwhile, AI, with its ability to analyze vast amounts of data, learn from patterns, and make intelligent decisions, emerges as the catalyst for unlocking the full potential of 5G networks [4, 4]. Together, AI and 5G form a symbiotic relationship, where AI-powered algorithms optimize network performance, enhance reliability, and enable innovative applications that were once deemed unattainable [6–8]. From autonomous vehicles and smart cities to immersive virtual experiences and remote healthcare, the integration of AI into 5G systems promises to transform industries, improve quality of life, and drive economic growth on a global scale. However, realizing the full potential of AI in 5G applications requires a nuanced understanding of the technical challenges, ethical considerations, and regulatory frameworks that govern its implementation [9, 9]. As we embark on this journey towards next-level connectivity, it is essential to navigate these complexities with foresight, collaboration, and a commitment to harnessing the power of AI and 5G for the greater good.

The burgeoning integration of Artificial Intelligence (AI) technologies into the development of 5G mobile communications technology is emblematic of a seismic shift in the telecommunications landscape. This convergence underscores the recognition that to fully realize the transformative potential of 5G, leveraging AI is not just beneficial but imperative. As the demand for faster speeds, lower latency, and enhanced connectivity escalates, AI emerges as a critical enabler, augmenting the capabilities of 5G networks and unlocking new avenues of innovation. One compelling reason for integrating AI into 5G development lies in its ability to optimize network performance and efficiency. Through AI-driven algorithms, networks can dynamically adapt to fluctuating demands, allocating resources intelligently to ensure optimal throughput and seamless user experiences [11, 12]. This proactive approach to network management minimizes congestion, reduces latency, and enhances overall reliability, thereby fulfilling the promise of 5G's unparalleled speed and responsiveness. Moreover, AI facilitates predictive maintenance and proactive troubleshooting, mitigating potential disruptions before they impact users. By analyzing vast amounts of data in real-time, AI algorithms can detect anomalies, identify potential failures, and recommend preemptive actions, thereby enhancing network resilience and uptime. This predictive capability not only improves network reliability but also reduces operational costs and enhances customer satisfaction. Furthermore, the integration of AI into 5G technology holds immense promise for unlocking new use cases and driving innovation across various industries [13–15]. From smart cities and autonomous vehicles to remote healthcare and industrial automation, the combination of AI and 5G enables transformative applications that were previously inconceivable. By harnessing the power of AI-driven insights and analytics, organizations can capitalize on the wealth of data generated by 5G networks to deliver personalized services, optimize processes, and create new revenue streams. In essence, the increasing importance of integrating artificial intelligence technologies in the development of 5G mobile communications technology is a testament to the symbiotic relationship between these two transformative forces [16–18]. As we journey towards a hyper-connected future, AI will continue to play a pivotal role in realizing the full potential of 5G, driving innovation, empowering societies worldwide, and reshaping the digital landscape in profound ways.

2 5G Mobile Communications Technology

Given the frequent use of mobile phones among individuals, mobile technologies are advancing rapidly. While 4G had recently been introduced, the emergence of 5G has quickly become a prominent topic of discussion, underscoring the swift pace of mobile technology development. The advent of 5G enhances IoT networks and enables the adoption of new bandwidth-intensive applications [19–22]. This article aims to provide insights into 5G technology and essential information regarding its functionalities. 5G encompasses a suite of standardized applications and significantly accelerated wireless internet technologies. Users utilizing 5G experience notably faster internet speeds and reduced latency. Additionally, as the fifth generation, 5G signifies a transformation in infrastructure for the fifth time. With each generational shift in technology, such as with 5G, alterations or additions to the infrastructure become necessary. Furthermore, 5G endeavors to enhance the utility of present-day technologies like VR and autonomous vehicles by reinforcing IoT networks [23–25]. One of the most remarkable aspects of transitioning to a new wireless technology generation is the enhanced internet speed. 5G networks boast the potential to achieve speeds ranging from 10 to 20 Gbps, surpassing not only the speeds of 4G networks but also those of many wired internet connections reaching households. Consequently, 5G provides network speeds comparable to those of fiber optic connections. While speed stands out as a primary feature of 5G, another significant aspect is the reduction in network latency, leading to minimized ping times. 5G substantially diminishes the duration required to transmit data between mobile devices and remote servers, facilitating swifter data access. Although network latency varies based on numerous factors, 5G networks offer latency rates 60–120 times lower than 4G networks under optimal conditions. Such reduced latency holds the promise of enabling various applications, such as virtual reality, to operate efficiently within contemporary environments.

One of the primary drivers behind the increasing importance of 5G is its unparalleled capacity to deliver faster speeds and lower latency. With data traffic skyrocketing due to the widespread adoption of bandwidth-intensive applications such as video streaming, online gaming, and augmented reality, traditional networks are struggling to keep pace with user demands. 5G addresses this challenge by offering significantly higher data rates and ultra-low latency, enabling seamless real-time communication and immersive multimedia experiences. This enhanced connectivity is not only crucial for meeting consumer expectations but also for unlocking new use cases in areas such as autonomous vehicles, remote healthcare, and smart manufacturing. Moreover, 5G plays a pivotal role in driving economic growth and innovation by fostering the development of new technologies and business models [26, 27]. The deployment of 5G networks creates a fertile ground for experimentation and entrepreneurship, empowering startups and enterprises to explore novel applications and services that leverage the capabilities of high-speed, low-latency connectivity. From edge computing and Internet of Things (IoT) deployments to advanced analytics and artificial intelligence, the possibilities enabled by 5G are virtually limitless, fuelling a wave of innovation across industries and driving digital transformation at an unprecedented pace. Furthermore, the increasing importance of 5G extends beyond consumer and enterprise applications to encompass broader societal benefits [28–30]. 5G has the potential to bridge the digital divide by providing high-speed connectivity to underserved communities and rural areas, enabling greater access to educational resources, healthcare services, and economic opportunities. Additionally, 5G networks play a critical role in enabling smart city initiatives, facilitating the deployment of intelligent infrastructure and services that enhance urban livability, sustainability, and resilience. The importance of 5G mobile communications technology continues to grow exponentially, driven by its capacity to deliver faster speeds, lower latency, and enable transformative applications and services across various industries. As we embrace the era of hyper-connectivity and digital transformation, 5G serves as a catalyst for innovation, economic growth, and societal progress, shaping the way we live, work, and interact in the digital age.

A millimeter wave patch antenna is a type of antenna designed to operate within the millimeter wave frequency range, typically spanning from 30 to 300 GHz. These antennas are commonly used in wireless communication systems, radar systems, and other applications where high-frequency, high-bandwidth transmission is required. The design of a millimeter wave patch antenna typically consists of a metallic patch, often rectangular or circular, mounted on a dielectric substrate. The patch is usually fed by a coaxial cable or a microstrip transmission line, which connects the patch to the radio frequency (RF) circuitry. One of the key advantages of millimeter wave patch antennas is their compact size relative to the wavelength of the operating frequency. This allows for the development of small, lightweight antennas suitable for integration into portable devices or arrays for beamforming applications. Millimeter wave patch antennas are also capable of achieving high directivity and gain, making them suitable for point-to-point communication links and radar systems. However, they are also highly sensitive to alignment and environmental factors due to the short wavelength of millimeter wave signals. To mitigate these challenges, precise design and manufacturing techniques are employed to optimize the performance of millimeter wave patch antennas. This includes careful selection of substrate materials, antenna dimensions, and feed mechanisms to achieve desired radiation patterns, impedance matching, and efficiency.

3 Challenges and Requirements

The integration of AI into 5G systems for advanced applications like embodied AI presents a complex set of challenges and requirements. Key among these is the need for ultra-low latency to support near-instantaneous response times, massive connectivity to handle the influx of data from numerous devices, and edge computing integration for real-time processing. Security and privacy concerns are paramount, necessitating AI-powered security measures that can protect sensitive data while ensuring compliance. Energy efficiency, spectrum management, and network resilience pose additional challenges, requiring sophisticated AI algorithms for optimization and predictive maintenance. Furthermore, interoperability between diverse systems, quality of service management, and scalability are crucial factors that demand flexible AI architectures and standardization efforts. To begin, the comparison between 5G technology and the Marvel character Flash has recently surfaced. This analogy arises from the capability of 5G to download HD movies within mere minutes, and even seconds for certain films (see Fig. 1). Typically, the process of downloading a high-definition movie entails a significant time investment, regardless of internet speed. However, with 5G technology, the download duration for substantial files is markedly reduced. Furthermore, the duration required to download an HD movie using standard internet aligns with the time it takes to download an entire season or two of a television series in HD quality when utilizing 5G. This example vividly illustrates the revolutionary speed enhancements facilitated by 5G. Beyond these advancements, the advent of 5G promises to eradicate the latency issues encountered in website browsing and online gaming experiences. Gamers commonly express frustration over ping and lag problems, which 5G technology aims to address and resolve, marking a significant improvement in online gaming performance. As commonly acknowledged, contemporary smartphones boast sleek and delicate designs. The evolution of virtual reality (VR) technologies is poised for significant advancement alongside the introduction of new-generation smartphones and 5G networks. With the advent of 5G, VR applications have the potential to seamlessly integrate into our daily routines. For instance, the incorporation of 5G into our lives opens avenues for immersive experiences such as virtually attending a basketball match, providing an experience akin to being present on the court. Additionally, real-time access to player information and match statistics enhances the viewing experience without any perceptible delay. Furthermore, as devices become increasingly compact, even referees and players could leverage this technology during matches. For instance, Video Assistant Referee (VAR) technology in football could be presented to sports enthusiasts in a significantly enhanced manner [31–34]. Beyond these developments, 5G technology stands poised to revolutionize various domains. In essence, the advent of 5G heralds a new era across diverse sectors including business, education, entertainment, advertising, and medicine, potentially rewriting existing norms and practices. Recommendation for Methodology Section:

In light of the challenges and opportunities presented by the integration of AI, 5G, and IoT technologies, we recommend a multi-faceted methodological approach for future research and development in this field: In light of the challenges and opportunities presented by the integration of AI, 5G, and IoT technologies, we recommend a multi-faceted methodological approach for future research and development in this field:

Cross-disciplinary collaboration: Promote collaboration between developers of AI algorithms, network engineers and specialists working in fields that AI is applied to address the problems from all angles.

Iterative prototyping: Incorporate rapid prototyping and testing frameworks to determine the problems in AI applied to 5G and IoT as soon as possible.

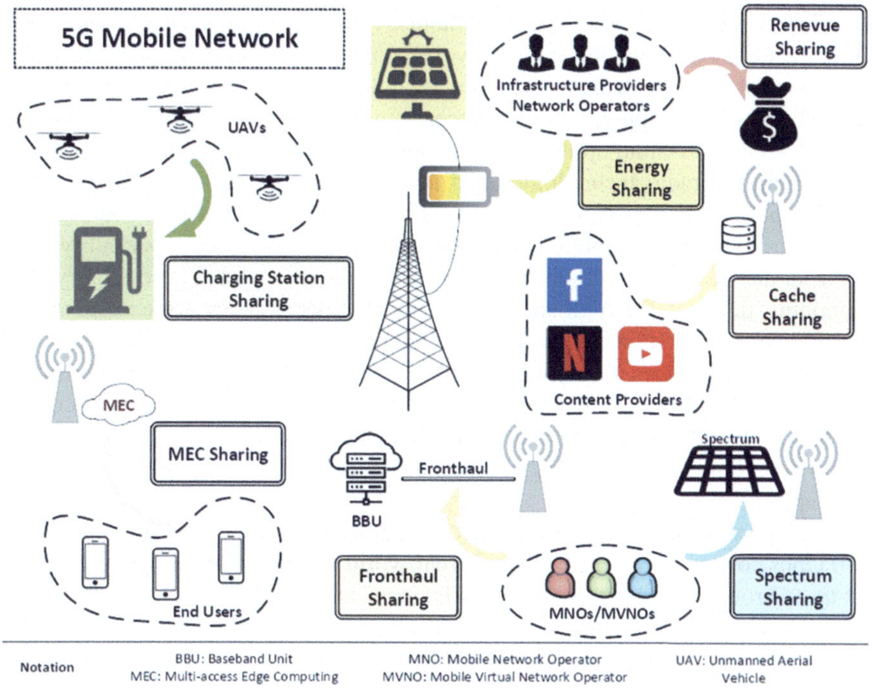

Fig. 1 5G network sharing use cases

Edge-centric design: Adopt the edge computing architectural approach more often in the methodology, as it will reduce latency while deploying the embodied AI.
Adaptive AI algorithms: Congenial and calibrate algorithmic models that can flexibly adapt to the network environments and the applications.
Standardized benchmarking: Set elaborate reference points in terms of performance, security and energy consumption for the AI-5G-IoT architectures.
Ethical framework integration: It is necessary to apply ethical factors and privacy measures throughout the process of developing the application.
Scalability testing: Prototyping methods that subject the systems to scaled-up or scaled-down instances of usage to gain realism.
Interdisciplinary validation: Check findings of proposed solutions in various scenarios and industries to guarantee widespread applicability of the work.

With the help of these methodological recommendations, the researchers and developers will be able to avoid certain pitfalls and ensure the responsible integration of AI into 5G and IoT systems, which share the similar aims and objectives in the context of next-generation connected society and embodied AI applications.

Deployment challenges in 5G mobile technologies encompass several key areas, including network constraints, device physicality, high mobility considerations, flexibility requirements, data speed demands, compatibility concerns, and affordability factors. Network challenges involve coping with limited spectrum availability amidst soaring demand, ensuring seamless operation across diverse environments, and addressing reliability concerns in the face of channel effects. Device physical challenges entail optimizing size, power consumption, and cost, while integrating multiple radio technologies and accommodating various standards. High mobility poses complexities in transmission, affecting addressing, routing, and overall throughput for mobile users. Flexibility mandates interoperability with terrestrial and satellite communication infrastructures for global accessibility. The burgeoning demand for high-speed data transfer necessitates wide bandwidths, GHz-frequency operation, and advanced antenna configurations, all within compact, energy-efficient devices. Compatibility with existing networks and affordability for widespread adoption are additional imperatives for the commercial success of 5G systems. Thus, addressing these multifaceted challenges is essential for realizing the full potential of 5G technology and ensuring its viability in the evolving mobile landscape.

The requirements for 5G connectivity encompass eight key aspects: achieving 1–10 Gbps connections to end points in real-world scenarios, ensuring a mere 1 ms end-to-end round trip delay for minimal latency, delivering a remarkable 1000-fold increase in bandwidth per unit area, accommodating an exponential surge of 10–100 times more connected devices, striving for an impeccable (perceived) availability rate of 99.999%, ensuring ubiquitous (perceived) coverage extending to every corner, aiming for a substantial 90% reduction in network energy consumption, and facilitating a battery life of up to ten years for low-power, machine-type devices. Given that these requirements are delineated from diverse perspectives, they form a multifaceted set of objectives that may not seamlessly align. The challenge lies in envisioning a singular technology capable of meeting all these conditions concurrently, underscoring the complexity inherent in the pursuit of 5G innovation.

- Dynamic Resource Allocation: Artificial Intelligence (AI) enables dynamic allocation of network resources in 5G applications. Through AI algorithms, network resources such as bandwidth, spectrum, and computing power can be intelligently distributed based on real-time demand and usage patterns. This optimization ensures efficient resource utilization, minimizes congestion, and enhances overall network performance, leading to improved user experiences.
- Predictive Maintenance and Proactive Optimization: AI-driven predictive maintenance techniques play a crucial role in optimizing 5G applications. By analyzing vast amounts of data from network components and devices, AI algorithms can identify potential failures or degradation in performance before they occur. This proactive approach allows network operators to preemptively address issues, minimize downtime, and optimize the reliability and availability of 5G services.
- Intelligent Traffic Management: AI enables intelligent traffic management in 5G networks, optimizing the routing and prioritization of data packets based on factors such as network conditions, application requirements, and user preferences. By

dynamically adjusting traffic flows in real time, AI algorithms can mitigate congestion, reduce latency, and ensure consistent service quality for different applications and users.
- Network Slicing and Service Orchestration: AI facilitates network slicing and service orchestration in 5G applications, allowing for the creation of virtualized network instances tailored to specific use cases or service requirements. AI algorithms can automate the provisioning, management, and optimization of network slices, enabling efficient resource allocation and rapid deployment of new services. This flexibility and agility are essential for supporting diverse applications and use cases in 5G networks.
- Enhanced Security and Threat Detection: AI-powered security solutions play a vital role in optimizing security in 5G applications. By leveraging machine learning algorithms, AI can detect and mitigate various security threats, including malware, intrusion attempts, and network vulnerabilities. AI-driven security analytics enable real-time threat detection, rapid response, and continuous adaptation to evolving cyber threats, ensuring the integrity and confidentiality of 5G networks and services.
- Personalized User Experiences: AI enables personalized user experiences in 5G applications by leveraging data analytics and machine learning techniques to understand user behavior, preferences, and context. By analyzing user data in real time, AI algorithms can deliver customized content, services, and recommendations tailored to individual users, enhancing engagement, satisfaction, and retention in 5G-enabled applications and services.

4 Conclusions

Optimizing 5G applications through Artificial Intelligence (AI) involves leveraging AI-driven techniques across various aspects of network management and service delivery. Dynamic resource allocation ensures efficient utilization of network resources, minimizing congestion and enhancing performance. Predictive maintenance and proactive optimization preemptively address potential issues, improving reliability and availability. Intelligent traffic management prioritizes and routes data packets in real time, reducing latency and ensuring consistent service quality. Network slicing and service orchestration enable the creation of virtualized network instances tailored to specific use cases, enhancing flexibility and agility. AI-powered security solutions detect and mitigate security threats, safeguarding network integrity and confidentiality. Personalized user experiences leverage data analytics to deliver customized content and services, enhancing user engagement and satisfaction. Through these AI-driven optimizations, 5G networks can fulfill the diverse requirements of applications and users, realizing their full potential in the digital landscape.

References

1. Attaran, M.: The impact of 5G on the evolution of intelligent automation and industry digitization. J. Ambient. Intell. Humaniz. Comput. **14**, 5977–5993 (2021). https://doi.org/10.1007/s12652-020-02521-x
2. You, X., Zhang, C., Tan, X., Jin, S., Wu, H.: AI for 5G: research directions and paradigms. Sci. China Inf. Sci. **62**(21301), 1–13 (2018). https://doi.org/10.1007/s11432-018-9596-5
3. Liu, B., Han, C., Liu, X., Li, W.: Vehicle artificial intelligence system based on intelligent image analysis and 5G network. Int. J. Wirel. Inf. Netw. **30**, 86–102 (2021). https://doi.org/10.1007/s10776-021-00535-6
4. Yazici, İ, Shayea, I., Din, J.: A survey of applications of artificial intelligence and machine learning in future mobile networks-enabled systems. Eng. Sci. Technol. Int. J. **44**, 101455 (2023). https://doi.org/10.1016/j.jestch.2023.101455
5. Mijwil, M.M., Gök, M., Doshi, R., Hiran, K.K., Kösesoy, I.: Utilizing artificial intelligence techniques to improve the performance of wireless nodes. In: Applications of Artificial Intelligence in Wireless Communication Systems, pp.150–162 (2023). https://doi.org/10.4018/978-1-6684-7348-1.ch010
6. Pupo, E.F., González, C.C., Montalban, J., Angueira, P., Murroni, M., Iradier, E.: Artificial intelligence aided low complexity RRM algorithms for 5G-MBS. IEEE Trans. Broadcast. 1 (2023). https://doi.org/10.1109/TBC.2023.3311337
7. Kumar, A., Nanthaamornphong, A., Selvi, R., Venkatesh, J., Alsharif, M.H., Uthansakul, P., Uthansakul, M.: Evaluation of 5G techniques affecting the deployment of smart hospital infrastructure: Understanding 5G, AI and IoT role in smart hospital. Alex. Eng. J. **83**, 335–354 (2023). https://doi.org/10.1016/j.aej.2023.10.065
8. Xin, Y., Yang, K., Shamsunder, C.I.S., Lin, X., et al. (2023). Guest editorial: AI-powered telco network automation: 5G evolution and 6G. IEEE Wirel. Commun. **30**(1), 68–69 (2023). https://doi.org/10.1109/MWC.2023.10077118
9. Kinder, T., Stenvall, J., Koskimies, E., Webb, H., Janenova, S.: Local public services and the ethical deployment of artificial intelligence. Gov. Inf. Q. **40**(4), 101865 (2023). https://doi.org/10.1016/j.giq.2023.101865
10. Kulkov, I., Kulkova, J., Rohrbeck, R., Menvielle, L., Kaartemo, V., Makkonen, H.: Artificial intelligence - driven sustainable development: examining organizational, technical, and processing approaches to achieving global goals. Sustain. Dev. **32**, 2253–2267 (2023). https://doi.org/10.1002/sd.2773
11. Chahed, H., Usman, M., Chatterjee, A., Bayram, F., Chaudhary, R., et al.: AIDA—A holistic AI-driven networking and processing framework for industrial IoT applications. Internet of Things **22**, 100805 (2023). https://doi.org/10.1016/j.iot.2023.100805
12. Doshi, R., Hiran, K.K., Mijwil, M.M., Anand, D.: To that of artificial intelligence, passing through business intelligence. In: Handbook of Research on AI and Knowledge Engineering for Real-Time Business Intelligence, vol. 01, pp. 1–16. https://doi.org/10.4018/978-1-6684-6519-6.ch001
13. Ghildiyal, Y., Singh, R., Alkhayyat, A., Gehlot, A., Malik, P., et al.: An imperative role of 6G communication with perspective of industry 4.0: challenges and research directions. Sustainable Energy Technol. Assess. **56**, 103047 (2023). https://doi.org/10.1016/j.seta.2023.103047
14. Unogwu, O., Doshi, R., Hiran, K.K., Mijwil, M.M.: Introduction to quantum-resistant blockchain. In: Advancements in Quantum Blockchain with Real-Time Applications, vol. 01, pp. 36–55 (2022). https://doi.org/10.4018/978-1-6684-5072-7.ch002
15. Jan, Z., Ahamed, F., Mayer, W., Patel, N., Grossmann, G., et al.: Artificial intelligence for industry 4.0: systematic review of applications, challenges, and opportunities. Expert Syst. Appl. **216**, 119456 (2023). https://doi.org/10.1016/j.eswa.2022.119456
16. Al-Shahwani, H.I.W., Mijwil, Doshi, M.M.R., Hiran, K.K., Bala, I.: Evaluating antivirus effectiveness against malware in ascending order for increase blockchain endpoint protection. In:

Bio-Inspired Optimization Techniques in Blockchain Systems, pp.150–156 (2024). https://doi.org/10.4018/979-8-3693-1131-8.ch008
17. Eswaran, S., Honnavalli, P.: Private 5G networks: a survey on enabling technologies, deployment models, use cases and research directions. Telecommun. Syst. **82**, 3–26 (2022). https://doi.org/10.1007/s11235-022-00978-z
18. Sallam, M., Mousa, D.: Evaluating ChatGPT performance in Arabic dialects: a comparative study showing defects in responding to Jordanian and Tunisian general health prompts. Mesopotamian J. Artif. Intell. Healthc. 1–7 (2024). https://doi.org/10.58496/MJAIH/2024/001%20
19. Pons, M., Valenzuela, E., Rodríguez, B., Nolazco-Flores, J.A., Del-Valle-Soto, C.: Utilization of 5G technologies in IoT applications: current limitations by interference and network optimization difficulties—a review. Sensors **23**(8), 3876 (2023). https://doi.org/10.3390/s23083876
20. Saravanan, V., Sreelatha, P., Atyam, N.R., Madiajagan, M., Saravanan, D., et al.: Design of deep learning model for radio resource allocation in 5G for massive IoT device. Sustain. Energy Technol. Assess. **56**, 103054 (2023). https://doi.org/10.1016/j.seta.2023.103054
21. Arora, K., Bala, I., Mijwil, M.M.: Energy optimized LDPC codes for next-generation MIMO OFDM systems. Iraqi J. Comput. Sci. Math. **4**(4), 1–12 (2023). https://doi.org/10.52866/ijcsm.2023.04.04.001
22. Benlloch-Caballero, P., Wang, Q., Alcaraz-Calero, J.M.: Distributed dual-layer autonomous closed loops for self-protection of 5G/6G IoT networks from distributed denial of service attacks. Comput. Netw. **222**, 109526 (2023). https://doi.org/10.1016/j.comnet.2022.109526
23. Salor, L.C., Baeza, V.M.: Harnessing the potential of emerging technologies to break down barriers in tactical communications. Telecom **4**(4), 709–731 (2023). https://doi.org/10.3390/telecom4040032
24. Fakhouri, H.N., Alawadi, S., Awaysheh, F.M., Hani, I.B., Alkhalaileh, M., Hamad, F.: A comprehensive study on the role of machine learning in 5G security: challenges, technologies, and solutions. Electronics **12**(22), 4604 (2023). https://doi.org/10.3390/electronics12224604
25. Desai, A., Desai, M.: A review of the state of cybersecurity in the healthcare industry and propose security controls. Mesopotamian J. Artif. Intell. Healthc. 82–84 (2023). https://doi.org/10.58496/MJAIH/2023/016
26. Beltozar-Clemente, S., Iparraguirre-Villanueva, O., Pucuhuayla-Revatta, F., Sierra-Liñan, F., Zapata-Paulini, J., Cabanillas-Carbonell, M.: Contributions of the 5G network with respect to decent work and economic growth (sustainable development goal 8): a systematic review of the literature. Sustainability **15**(22), 15776 (2023). https://doi.org/10.3390/su152215776
27. Parcu, P.L., Pisarkiewicz, A.R., Carrozza, C., Innocenti, N.: The future of 5G and beyond: leadership, deployment and European policies. Telecommun. Policy **47**(9), 102622 (2023). https://doi.org/10.1016/j.telpol.2023.102622
28. Palattella, M.R., Dohler, M., Grieco, A., Rizzo, G., Torsner, J., et al.: Internet of things in the 5G era: enablers, architecture, and business models. IEEE J. Sel. Areas Commun. **34**(3), 510–527 (2016). https://doi.org/10.1109/JSAC.2016.2525418
29. Varga, P., Peto, J., Franko, A., Balla, D., Haja, D., et al.: 5G support for industrial IoT applications—challenges, solutions, and research gaps. Sensors **20**(3), 828 (2020). https://doi.org/10.3390/s20030828
30. Hui, H., Ding, Y., Shi, Q., Li, F., Song, Y., Yan, J.: 5G network-based internet of things for demand response in smart grid: a survey on application potential. Appl. Energy **257**, 113972 (2020). https://doi.org/10.1016/j.apenergy.2019.113972
31. Guo, Y.: Pattern recognition and neural network-driven roller track analysis via 5G network. Mob. Inf. Syst. **2020**(6675140), 1–8 (2020). https://doi.org/10.1155/2020/6675140
32. Aljanabi, M.: Safeguarding connected health: leveraging trustworthy AI techniques to harden intrusion detection systems against data poisoning threats in IoMT environments. Babylon. J. Netw. 31–37 (2023). https://doi.org/10.58496/BJIoT/2023/005

33. Qasim, S. S., Nsalif, S. M.: Advancements in time series-based detection systems for Distributed Denial-of-Service (DDoS) attacks: a comprehensive review. Babylon. J. Netw. 9–17 (2024). https://doi.org/10.58496/BJN/2024/002
34. Zhong, H., Wang, L., Cui, J., Zhang, J., Bolodurina, I.: Secure edge computing-assisted video reporting service in 5G-enabled vehicular networks. IEEE Trans. Inf. Forensics Secur. **18**, 3774–3786 (2023). https://doi.org/10.1109/TIFS.2023.3287731

The manufacturer's authorised representative in the EU is Springer Nature Customer Service Centre GmbH, Europaplatz 3, 69115 Heidelberg, Germany. If you have any concerns regarding our products, please contact ProductSafety@springernature.com

Printed and bound by CPI Group (UK) Ltd, Croydon, CR0 4YY

30/09/2026

02225562-0015